MESOSCALE-CONVECTIVE PROCESSES IN THE ATMOSPHERE

This modern textbook is devoted to a deep understanding of mesoscale-convective processes in the atmosphere. Mesoscale-convective processes are commonly realized in the form of thunderstorms, which are dynamic, fast evolving, and assume a broad range of sizes and severity. Indeed, convective storms have the potential to spawn tornadoes and generate damaging "straight-line" winds, and are additionally responsible for the rainfall that can be detrimental but also immensely beneficial to society.

To facilitate this understanding, descriptions of the formation, dynamics, and qualitative characteristics of specific convective phenomena such as supercell thunderstorms and mesoscale-convective systems are provided. Although the descriptions pertain largely to the extratropical atmosphere, examples of related tropical phenomena are given for comparison and contrast. To provide a further holistic perspective, separate chapters are included on mesoscale observations and data analysis, numerical modeling, and the theoretical predictability and actual numerical prediction of mesoscale weather. An additional chapter on interactions and feedbacks addresses ways in which convective storms affect and are affected by external processes, particularly on the synoptic and planetary scales.

This textbook provides advanced students, researchers, and weather professionals with a modern, accessible treatment of the convective processes that lie within the range of the atmospheric mesoscale.

ROBERT J. TRAPP received his PhD in Meteorology from The University of Oklahoma in 1994. He was a National Research Council Postdoctoral Fellow, and then was appointed as a research scientist at the Cooperative Institute for Mesoscale Meteorological Studies and the National Severe Storms Laboratory; part of this appointment was served at the National Center for Atmospheric Research. Trapp joined the faculty at Purdue University in 2003, where he is currently Professor and Associate Head of the Department of Earth, Atmospheric, and Planetary Sciences. He is a University Faculty Scholar, which recognizes outstanding faculty members who are on an accelerated path for academic distinction. Professor Trapp has also been recognized as an Outstanding Teacher in the College of Science and has appeared on his departmental teaching honor roll in every semester of his tenure at Purdue. He is an expert on convective storms, their attendant hazards, and their two-way interaction with the larger-scale atmosphere.

MESOSCALE-CONVECTIVE PROCESSES IN THE ATMOSPHERE

ROBERT J. TRAPP
Purdue University, West Lafayette, Indiana

Shaftesbury Road, Cambridge CB2 8EA, United Kingdom

One Liberty Plaza, 20th Floor, New York, NY 10006, USA

477 Williamstown Road, Port Melbourne, VIC 3207, Australia

314–321, 3rd Floor, Plot 3, Splendor Forum, Jasola District Centre, New Delhi – 110025, India

103 Penang Road, #05–06/07, Visioncrest Commercial, Singapore 238467

Cambridge University Press is part of Cambridge University Press & Assessment, a department of the University of Cambridge.

We share the University's mission to contribute to society through the pursuit of education, learning and research at the highest international levels of excellence.

www.cambridge.org
Information on this title: www.cambridge.org/9780521889421

First published 2013

A catalogue record for this publication is available from the British Library

Library of Congress Cataloging-in-Publication data
Trapp, Robert J. (Robert Jeffrey), 1963–
Mesoscale-convective processes in the atmosphere / Robert J. Trapp.
p. cm.
Includes bibliographical references and index.
ISBN 978-0-521-88942-1 (hardback)
1. Atmospheric thermodynamics. I. Title.
QC880.4.T5T75 2013
551.55–dc23 2012027323

ISBN 978-0-521-88942-1 Hardback

Additional resources for this publication at www.cambridge.org/Trapp

Contents

Preface

The primary resource for students enrolled in my early mesoscale meteorology courses at Purdue University was *Mesoscale Meteorology and Forecasting*, the edited collection of review articles published by the American Meteorological Society in 1986. Though still valuable, it was conspicuously missing a number of important developments that had taken place since its publication, including: (1) major field programs such as IHOP (International H2O Project), BAMEX (Bow Echo and Mesoscale Convective Vortex Experiment), and VORTEX (Verification of the Origins of Rotation in Tornadoes Experiment) and its successor VORTEX2; (2) the maturation and implementation of operational Doppler weather radar, and an equivalent advancement of airborne and ground-based mobile radar systems; and (3) the relative proliferation of open-source community models and a concurrent ability to run such models using accessible computing resources, including desktop systems.

In short, these and other developments have led to significant evolution in the understanding of the atmospheric mesoscale since 1986, and motivated my effort to produce an updated resource. The realization of this effort is *Mesoscale-Convective Processes in the Atmosphere*.

As a perusal of the book shows, a major difference between *Mesoscale-Convective Processes in the Atmosphere* and other newly available mesoscale books is its focus on deep moist convection. This limited focus was driven partly by my perception of student interest, and partly by a philosophical choice to provide a concentrated treatment of a few topics, rather than a diluted treatment of all things mesoscale. Of course, it also follows my own interests, which most certainly biased the directions of some explanations (as in my considerable use of numerical modeling results, for example), although I did strive for balance as much as possible.

The discourse is aimed at upper-division undergraduate/beginning graduate students and assumes a basic knowledge of atmospheric dynamics (and a requisite knowledge of vector differential calculus, differential equations, etc.). When I teach a semester-long course based on this book, I typically begin with Chapter 1, and then endeavor to cover the material in Chapters 5 through 8. Specific sections within Chapters 2 through 4 are referenced as needed. I use the material in Chapters 3, 4, 9, and 10 as the basis for special topics courses.

Supplements to each chapter are provided on the companion website. The supplements include problem sets, exercises, and discussion questions that were purposely omitted from the book itself: my desire is to keep these materials fresh and topical, and also to incorporate book-user contributions (which are encouraged). For each relevant chapter, a list of cases is provided so that the reader can apply the theory to real events. Suggested numerical modeling experiments are also given, as are links to appropriate software, datasets, etc.

Mesoscale-Convective Processes in the Atmosphere represents the fruit of my course-developmental labors at Purdue University, but was influenced by multiple external sources. One, of course, is *Mesoscale Meteorology and Forecasting*, which also served as a blueprint for the organization of my topics. Lectures and notes from courses given by Howie Bluestein, Fred Carr, Chuck Doswell, Kelvin Droegemeier, Brian Fiedler, and Tzvi Gal-Chen shaped my thinking while I was a graduate student at the University of Oklahoma, and are reflected either directly or indirectly in the text. During my early career, discussions and debates with Harold Brooks, Bob Davies-Jones, Joe Klemp, Rich Rotunno, and Morris Weisman further shaped my understanding of deep moist convection and attendant phenomena. Additional understanding has come from other people and sources too numerous to list here, but I feel compelled to acknowledge the particular influence of books written by Robert Houze, Kerry Emanuel, and James Holton, which I cite heavily.

I benefited immensely from reviews by George Bryan (Chapter 2), Tammy Weckwerth (Chapter 3), Lou Wicker (Chapter 4), Conrad Ziegler (Chapter 5), Sonia Lasher-Trapp (Chapter 6), Morris Weisman (Chapter 7), Matt Parker (Chapter 8), Dave Stensrud (Chapter 9), and Mike Baldwin (Chapter 10). Dave Schultz, Phil Smith, and John Marsham provided helpful comments along the way.

Several of the chapters were written while I was on sabbatical leave at the University of Leeds, U.K. I am deeply appreciative for the discussions with – and support provided by – Doug Parker and Alan Blyth while I was in the U.K. I am also indebted to the staff of the National Centre for Atmospheric Science, and am forever grateful for the hospitality and friendship of Adrian Kybett and his family.

This project would not have been possible without the patience and support of Matt Lloyd and Cambridge University Press, would not have been undertaken without the opportunities for student engagement provided to me at Purdue University, and would not have been tolerable without the constant source of love and understanding from my wife Sonia and children Noah and Nadine.

Robert J. Trapp

1

The Atmospheric Mesoscale

1.1 Introduction

It is fitting to begin this exploration of mesoscale-convective processes with a definition of the *atmospheric mesoscale*. Likely, the reader has at least a vague idea of atmospheric phenomena that are normally categorized as mesoscale. Thunderstorms and the dryline are common examples. What the reader might not yet appreciate, however, is that devising an objective and quantitative basis for such categorization is nontrivial. Indeed, even the more basic practice of separating the atmosphere into discrete intervals can be difficult to rationalize universally, because the atmosphere is, in fact, continuous in time and space in its properties.[1]

Consider the atmospheric measurements represented in Figure 1.1. These have been analyzed to reveal a frequency spectrum of zonal atmospheric kinetic energy.[2] Although the spectrum is continuous, it does exhibit a number of distinct peaks. Conceivably (and arguably), the intervals centered about the peaks represent atmospheric scales. The relatively narrow peak at a frequency of 10^0 (/day) is compelling here, because it indicates the existence of energetic eddies with a diurnal cycle. Dry and moist convective motions that grow and decay with the daily cycle of solar insolation are the presumed manifestations of such eddies, and would fall generally within the atmospheric mesoscale.

The spectral analysis technique used to generate Figure 1.1 involved a transformation of the *temporal distribution* of zonal kinetic energy to a distribution in terms of *frequency* space. A similar technique can be used to transform the *spatial distribution* of some variable to its distribution in *wavenumber* space. The wavenumber spectra in Figure 1.2 are from independent measurements, but similarly reveal continuous, yet separate, regions.[3] Here, the region distinction is given by the slope of statistical fits to the data in wavenumber space. At wavelengths greater than a few hundred kilometers, which includes the *synoptic scale*, the spectral curves have slopes near –3. At wavelengths of a few kilometers to a few hundred kilometers, which spatially

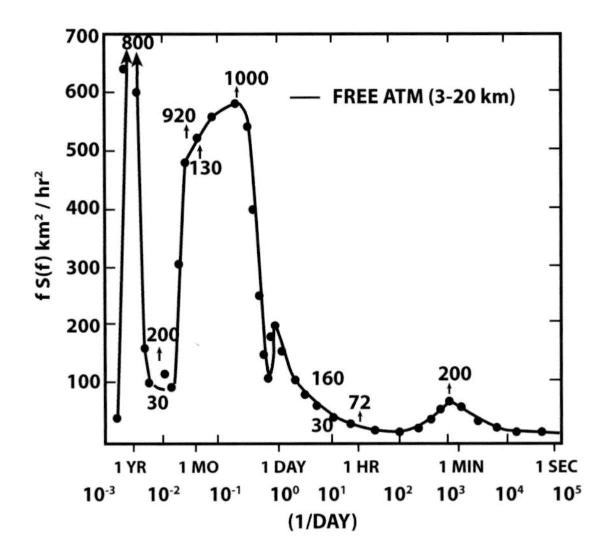

Figure 1.1. Average kinetic energy spectra (spectral density) of the free-atmospheric zonal wind as a function of frequency. Numbers show maximum kinetic energy at particular periods. After Vinnichenko (1970).

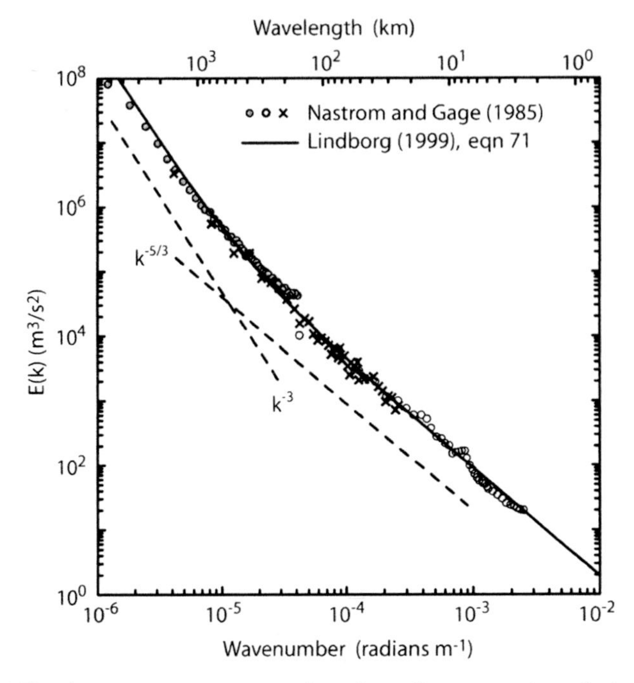

Figure 1.2. Kinetic energy spectra as a function of wavenumber, derived from the measurements of Nastrom and Gage (1985). The solid curves are from Lindborg (1999), which are fits to a separate set of measurements. The dashed lines indicate the slopes discussed in the text. From Skamarock (2004).

will be shown to correspond to mesoscale structures, the slope of the spectral curves is near –5/3. Thus, the spectral slopes help to distinguish between these two scales; we will learn in Chapter 10 that these slopes also have implications on the limits of atmospheric predictability.

Though valuable, the information in Figures 1.1 and 1.2 is still a bit incomplete for the purposes of this chapter, because it discloses nothing explicit about other structural characteristics, such as the variations in temperature, pressure, humidity, and wind intrinsic to the various phenomena. As will be demonstrated in Chapters 2, 3, and 4, respectively, knowledge of these characteristics is paramount to choosing appropriate forms of the dynamical equations, planning observation strategies, and designing numerical simulation and prediction models.

History, as it turns out, was critically influential in how these remaining characteristics were determined and assigned. Technological advances played an obvious role, but so also did the geo politics and events of the time. Let us digress briefly to examine some of this history.

1.2 Historical Perspectives

We begin with the historical origin of the synoptic scale, because in most classification schemes the synoptic scale places an upper limit to the mesoscale. The modern definition of the synoptic scale is usually given in terms of the size of migratory high- and low-pressure systems (midlatitude anticyclones and cyclones), which range from several hundred to several thousand kilometers.[4] Interestingly, the quantitative values attached to the synoptic scale actually arose out of the size of the observing networks in the late 1800s and early 1900s. This is the period when synoptic weather maps were first constructed routinely and the study and prediction of air masses began.[5] Synoptic weather observations were – and still are – made nearly simultaneously, to give a snapshot of the state of an otherwise evolving atmosphere.[6] Hence, the synoptic length scales we accept today effectively came from the practical area of such simultaneous observations as of the late 1800s and early 1900s, as limited by communication technology, and perhaps complicated by geopolitical boundaries.

Meteorological features smaller than these length scales, and therefore not well represented by the synoptic observations, were deemed "noise" or subsynoptic disturbances.[7] As demonstrated in Figure 1.3, at least some of this noise is attributable to the surface outflow of thunderstorms. Nonetheless, thunderstorms effectively kept their status as noise until the hazards they were imposing on the rapidly expanding air transportation industry around1930 (and presumably on military interests, around 1940) were recognized. Indeed, efforts to observe subsynoptic-scale motions – and to describe the associated phenomena – were motivated in part by the rise in weather-related accidents.

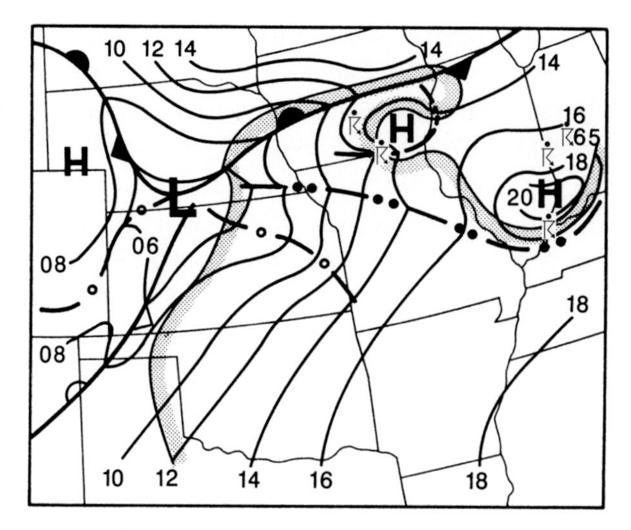

Figure 1.3. Analysis of sea-level pressure (millibars – 1000) and surface fronts at 0000 UTC 24 June 1985. The bold "H"s in southeast Iowa and southeast Illinois show convectively induced mesohigh pressure. Convective-storm outflow boundaries are indicated by lines with a dash-dot-dot pattern, and wind shifts by lines with a dash-dot pattern. The shaded line outlines the region where dewpoint temperatures exceeded 65°F. From Stensrud and Maddox (1988).

One result of such an effort was the subsynoptic observational network deployed during the *Thunderstorm Project*[8] in 1946. Compared with other experimental networks at about this time (see Figure 1.4), the Thunderstorm Project network had surface-station spacings as fine as 2 km, over a domain spanning several tens of kilometers.[9] The surface observations were supplemented with upper-air data collected using radiosondes, multiple coordinated aircraft, and even gliders.[10] When combined, these novel data helped to form a conceptual model of convective-storm evolution that still holds today (see Chapter 6).

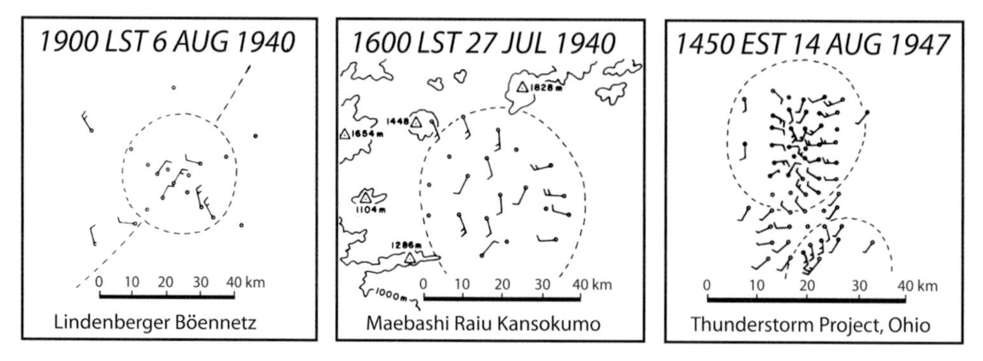

Figure 1.4. Examples of subsynoptic observing networks in the 1940s. The dashed lines indicate locations of thunderstorm outflow boundaries. From Fujita (1986).

In terms of interests specific to the current chapter, the Thunderstorm Project data have been used to quantify the characteristics of deep moist convection in the continental extratropics. The typical updraft length, depth, and vertical speed of ~10 km, ~10 km, and ~10 m s^{-1}, respectively, form the basis of the (sub)classification schemes discussed later; they also allow for interesting comparisons (in Chapter 5) with tropical oceanic convection, as quantified in part during the Global Atmospheric Research Program's Atlantic Tropical Experiment (GATE) almost thirty years after the Thunderstorm Project.[11]

Weather radar was another observing tool used during the Thunderstorm Project. Developed out of a military application in World War II, weather radar was still in its relative infancy during this time. Yet, sufficient data had been collected by 1950 to allow M. G. H. (Herbert) Ligda to posit the mesoscale as an intermediate between the synoptic scale and the microscale:[12]

> It is anticipated that radar will provide useful information concerning the structure and behavior of that portion of the atmosphere which is not covered by either micro- or synoptic-meteorological studies. We have already observed with radar that precipitation formulations which are undoubtedly of significance occur on a scale too gross to be observed from a single station, yet too small to appear even on sectional synoptic charts. Phenomena of this size might well be designated as mesometeorological.[13]

This recognition of the importance of weather radar and precipitation structure helped father the term *mesoscale*, but did not really help quantify it.[14] However, there was another recognition at this time (c. 1950–1960) that did: the importance of surface pressure measurements. This apparently was motivated by observations of abrupt changes in pressure at and just after the onset of thunderstorms (and thunderstorm systems). We will learn in subsequent chapters that pressure jumps – and, in particular, the spatially continuous pressure-jump lines – represent the leading edges of *mesohighs* generated beneath downdrafts, and are now commonly known as *gust fronts* (see also Figure 1.3). Pressure-jump lines were thought to play a vital role in tornado formation, especially when two lines intersected.[15] Although other mechanisms of tornadogenesis have since been established (see Chapter 7), this astute deduction (by M. Tepper) is viewed now as early evidence of tornadogenesis via storm-boundary interaction (see Chapter 9).

To better observe pressure-jump lines and related characteristics in pressure, a network of microbarograph stations was established in the United States (Figure 1.5).[16] This network, which subsequently was operated by the National Severe Storms Project (NSSP), the precursor to the National Severe Storms Laboratory (NSSL), is an early example of the more fully equipped surface mesonetworks now in place in several U.S. regions (see Chapter 3). It consisted of 210 stations in Kansas, Oklahoma, and Texas, and had a station spacing of ~60 km. Because the NSSP network was much coarser than that used during the Thunderstorm Project, yet

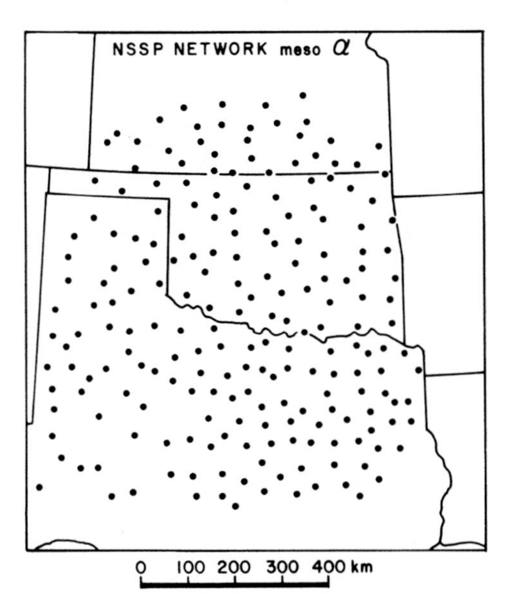

Figure 1.5. The National Severe Storms Project (NSSP) microbarograph network in the early 1960s. From Fujita (1986).

much larger in spatial extent, it served a useful purpose in describing the mesoscale "environment" as well as relatively larger mesoscale phenomena (see Chapter 8). In particular, such mesonetworks led to quantification of the characteristics of mesoscale-convective systems (MCSs), with typical lengths of ~100 km and durations of ~3 h.

Numerical weather prediction (NWP) was also advancing during this period, coincident with the technological advances of digital computing. Although the initial goal of NWP was prediction of planetary- and synoptic-scale waves, it was soon realized that the mesoscale could not be ignored if accurate predictions were to be made. This is especially true in the tropical latitudes, where deep moist convection is essential in redistributing energy and contributing to the global energy balance (Chapter 9). A representation of the effects of convection and other mesoscale processes in terms of relatively larger-scale variables is the premise of *parameterizations* (Chapter 4). Thus, the need to describe and understand the mesoscale goes beyond the hazards of convective weather.

1.3 Atmospheric-Scale Classification Schemes

Ligda's statement was followed by a number of proposals for atmospheric-scale classification schemes that accounted for the mesoscale. An example of a relatively simple three-class scheme of F. Fiedler and H. Panofsky is as follows:[17]

	Synoptic-	Meso-	Micro-
Period	>48 hr	1–48 hr	<1 hr
Wavelength	>500 km	20–500 km	<20 km

Here, wavelength is an average distance, such as that between updrafts or pressure minima, and similarly, period is an average time between wind gust or temperature maxima. The rationalization is as follows: (1) "synoptic scale includes all scales of motion which can be analyzed on the basis of a weather map" (i.e., no discrimination between planetary scale and synoptic scale); (2) "all systems in which the vertical and horizontal velocities are of the same order of magnitude are 'microscale' systems"; and (3) the "mesoscale occurs between the microscale and synoptic scale."[18]

The scheme of I. Orlanski accounts for the characteristic times and lengths, includes examples of phenomena, and also has scale subdivisions, as designated by the Greek letters alpha (α), beta (β), and gamma (γ) (Figure 1.6).[19] T. Fujita's alternative to the Orlanski scheme has five divisions (using the vowels a-e-i-o-u), and each of these has an α and β subdivision (Figure 1.7).[20]

A survey of the literature suggests that Orlanski's scheme is employed fairly widely, particularly when a distinction within the mesoscale (i.e., meso-α, meso-β, or meso-γ) is deemed necessary. Fujita's scheme appears to have been adopted mostly for the characterization of a specific class of vortex, a *miscocyclone* (see Chapter 5), and for the convectively induced, near-surface outflows known as *microbursts* and *macrobursts* (Chapter 6). Herein, we will strive for consistency with the common uses of both schemes, but will have a preference for the Orlanski scheme when reference is made to scale ranges and subdivisions.

1.4 Mesoscale-Convective Processes

Our treatment of the atmospheric mesoscale, as guided by the preceding sections, will include processes with horizontal lengths of ~10 to ~100 km, times of ~1 hr to ~1 day, and speeds of ~1 to ~10 m s^{-1}.

The range of phenomena that possess these characteristics is quite broad. Although it includes mechanically forced flows, such as mountain waves, the discussions herein will be limited largely to flows that arise from buoyancy-driven, or convective, motions. Accordingly, "convective" will be used as a qualifier, to distinguish such motions from those that are mechanically forced.

The convective clouds, precipitation, and associated phenomena have characteristic lengths, times, and speeds that can extend below the lower end of the mesoscale, but can be initiated and forced by mechanisms at or above the upper end of the mesoscale. Subsequent chapters will offer a treatment of the processes that span this spectrum,

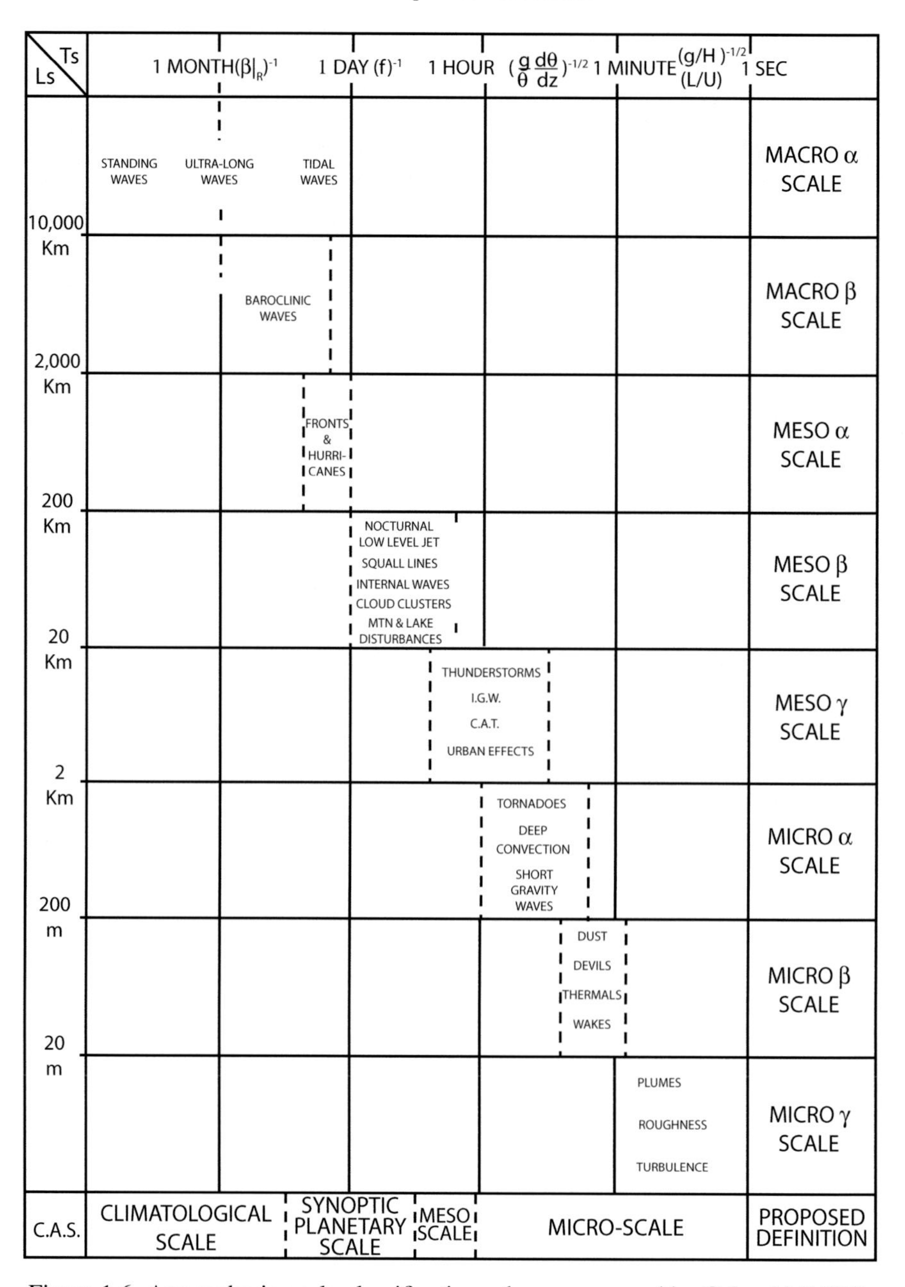

Figure 1.6. Atmospheric-scale classification scheme proposed by Orlanski (1975).

including convection initiation, the subsequent organization and morphology of the clouds and precipitation, and the interaction of these processes with those at the synoptic and larger scales. Other chapters will provide the mathematical theory behind the mesoscale-convective processes, describe how these processes are observed and

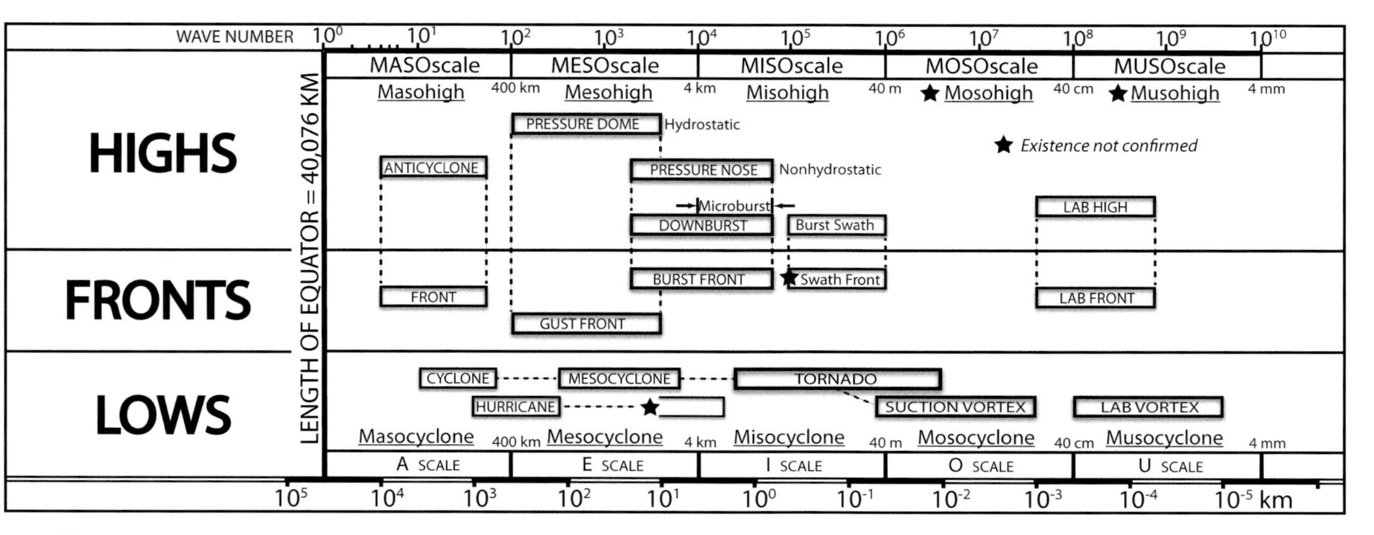

Figure 1.7. Atmospheric-scale classification scheme proposed by Fujita (1981).

numerically modeled, and supply information about the observing and modeling techniques themselves.

Notes

1 The discussion in this section draws from Emanuel (1986).
2 Vinnichenko (1970).
3 Nastrom and Gage (1985); Gage and Nastrom (1986).
4 *Glossary of Meteorology* (Glickman 2000).
5 The discussion in this section is based on Fujita (1986).
6 This practice is consistent with the more general meaning of the word *synoptic* – presenting a general view of the whole – as derived from the Greek *synoptikós*. See *Webster's New World Dictionary of the American Language* (Guralnik 1984).
7 Fujita (1986).
8 Byers and Braham (1949).
9 As compiled by Fujita (1986).
10 This is a configuration that has been copied, and envied, by subsequent field-program designers!
11 A historical perspective of GATE can be found at www.ametsoc.org/sloan/gate/.
12 The atmospheric microscale is typically inclusive of the atmospheric motion referred to as turbulence. Note that this usage of *mesoscale* is based on the Greek *mesos*, which means middle. See *Webster's New World Dictionary of the American Language* (Guralnik 1984).
13 Ligda (1951).
14 Quantitative uses of weather radar will be demonstrated in subsequent chapters, and indeed are now critical in refining the mesoscale.
15 Tepper (1950).
16 See Galway (1992) for more details about the timeline and historical context.
17 Fiedler and Panofsky (1970).
18 Ibid.
19 Orlanski (1975).
20 Fujita (1981).

2

Theoretical Foundations

Synopsis: The basis for theoretical study of mesoscale-convective processes is provided in this chapter. A set of governing equations is developed, followed by a discussion of individual terms and their typical magnitudes based on scale analyses. Approximate forms of the equations are then shown to permit solutions that relate to phenomena described in later chapters.

2.1 Fundamental Set of Equations

2.1.1 Equations

The equations that govern mesoscale-convective processes originate from the laws of conservation of momentum, thermodynamic energy, and mass, both of dry air and of water substance. Let us consider initially a basic set of equations[1] appropriate for a dry atmosphere, beginning with the vector equation of motion,

$$\frac{D\vec{V}}{Dt} = -\frac{1}{\rho}\nabla p + \vec{g} - 2\vec{\Omega} \times \vec{V} + \vec{F}, \tag{2.1}$$

the thermodynamic energy equation,

$$c_v \frac{DT}{Dt} = -\frac{p}{\rho}\nabla \cdot \vec{V} + \dot{Q}, \tag{2.2}$$

and the mass continuity equation,

$$\frac{D\rho}{Dt} = -\rho\nabla \cdot \vec{V}. \tag{2.3}$$

These apply to motion in a reference frame that rotates with the Earth; they assume a 3D orthogonal coordinate system in which the position vector is defined as $\vec{r} = x\hat{i} + y\hat{j} + z\hat{k}$, and thus in which velocity is

$$d\vec{r}/dt \equiv \vec{V} = u\hat{i} + v\hat{j} + w\hat{k}. \tag{2.4}$$

The left-hand side of each equation is the total, or Lagrangian, or material derivative,

$$DA/Dt = \partial A/\partial t + \vec{V} \cdot \nabla A, \tag{2.5}$$

which, for symbolic variable A, represents its change following the motion. We follow typical meteorological convention with the right-hand side terms and variables:

- $-1/\rho \nabla p$ is the pressure gradient force (PGF) per unit mass;
- $\vec{g} = -g\hat{k}$ is the gravity force per unit mass ($g = 9.8$ m s^{-2}), also expressed as $\vec{g} = -\nabla(gz)$;
- $-2\vec{\Omega} \times \vec{V}$ is the Coriolis force (CF) per unit mass, where

$$\vec{\Omega} = \Omega \cos\phi \hat{j} + \Omega \sin\phi \hat{k} \tag{2.6}$$

  is the angular velocity of the rotating Earth ($\Omega = 7.2921 \times 10^{-5}$ s^{-1}), ϕ is latitude, and $f = 2\Omega \sin\phi$ is the Coriolis parameter; and
- $\vec{F}$ is the force due to internal friction per unit mass, and has, for the illustrative case of incompressible flow,[2] the general form

$$\vec{F} = \nabla \cdot \nabla(\mu \vec{V}) \tag{2.7}$$

  where μ is the temperature-dependent, kinematic viscosity coefficient of air, usually assumed constant ($\sim 10^{-5}$ m^2 s^{-1}).[3]

Additional right-hand side terms in (2.1), referred to as *curvature terms*, arise from a transformation to a spherical coordinate system, and thus from the fact that the unit basis vectors are functions of position on the spherical Earth:[4]

$$\left(\frac{uv\tan\phi}{r_E} - \frac{uw}{r_E}\right)\hat{i} + \left(-\frac{u^2\tan\phi}{r_E} - \frac{vw}{r_E}\right)\hat{j} + \left(\frac{u^2+v^2}{r_E}\right)\hat{k}, \tag{2.8}$$

where r_E (= 6,370 km) is the mean radius of Earth at sea level and approximates well the total radial distance $r_E + z$. State variables of pressure (p), density (ρ), and temperature (T) are interrelated through the equation of state for an ideal gas,

$$p = \rho R_d T, \tag{2.9}$$

which is assumed to apply, for now, to dry air; R_d is the gas constant of dry air (287 J kg^{-1} K^{-1}). The thermodynamic energy equation originates from the first law of thermodynamics,

$$c_v dT + p d\alpha = \dot{Q}, \tag{2.10}$$

which also is assumed to apply to dry air, and thus $c_v = 717$ J kg^{-1} K^{-1} is the specific heat of dry air at constant volume, and $\alpha = 1/\rho$ is specific volume. In applications of (2.2) to problems of moist convection, the diabatic heating rate $\dot{Q}$ represents the latent heating due to water phase changes, but also can include radiative heating and conduction as relevant.

2.1.2 Scale Analysis

It is useful to evaluate terms in the various equations and assess their relative importance at the mesoscale. We begin with the x-component of the equation of motion,

$$\frac{\partial u}{\partial t}+u\frac{\partial u}{\partial x}+v\frac{\partial u}{\partial y}+w\frac{\partial u}{\partial z}=-\frac{1}{\rho}\frac{\partial p}{\partial x}+v2\Omega\sin\phi-w2\Omega\cos\phi$$
$$+\frac{uv\tan\phi}{r_E}-\frac{uw}{r_E}+\mu\nabla^2 u. \tag{2.11}$$

For assessment of this and the y-component equation, we introduce length and velocity scales for horizontal (L, U) as well as vertical (H, W) motions, and then use these to normalize the following variables, rendering them dimensionless (as indicated by a tilde):

$$\begin{array}{llll}\tilde{u}=u/U & \tilde{v}=v/U & \tilde{w}=w/W & \delta\tilde{p}/\tilde{\rho}=(\delta p/\rho)/U^2 \\ \tilde{x}=x/L & \tilde{y}=y/L & \tilde{z}=z/H & \tilde{t}=t/(L/U)\end{array}. \tag{2.12}$$

We do not use a separate scale for pressure (and density), but pose instead that pressure changes relate to the velocity scale. After substituting (2.12) into (2.11), dividing through by U^2/L, and recognizing that the vertical velocity scale can be expressed as $W=UH/L$, we obtain the following equation with dimensionless terms:

$$\frac{\partial\tilde{u}}{\partial\tilde{t}}+\tilde{u}\frac{\partial\tilde{u}}{\partial\tilde{x}}+\tilde{v}\frac{\partial\tilde{u}}{\partial\tilde{y}}+\tilde{w}\frac{\partial\tilde{u}}{\partial\tilde{z}}=-\frac{1}{\tilde{\rho}}\frac{\partial\tilde{p}}{\partial\tilde{x}}+\tilde{v}\left(\frac{1}{\mathrm{Ro}}\right)-\tilde{w}\left(\frac{1}{\mathrm{Ro}}\right)\left(\frac{H}{L}\right)\cot\phi$$
$$+\tilde{u}\tilde{v}\left(\frac{L}{r_E}\right)\tan\phi-\tilde{u}\tilde{w}\left(\frac{H}{r_E}\right)+\left(\frac{1}{\mathrm{Re}}\right)\nabla^2\tilde{u}. \tag{2.13}$$

The quantities in parentheses are nondimensional parameters; in particular,

$$\mathrm{Re}=\frac{UL}{\mu}=\text{Reynolds number} \tag{2.14}$$

$$\text{and }\mathrm{Ro}=\frac{U}{fL}=\text{Rossby number}, \tag{2.15}$$

where Re is the ratio between the forces due to inertia and friction, and Ro is the ratio between the forces due to inertia and Coriolis force. Individual terms have unit value except when multiplied by the nondimensional parameters. For much of the mesoscale, $U\sim10^1$ m s^{-1}, $W\sim10^0$ m s^{-1}, $L\sim10^5$ m, and $H\sim10^4$ m, implying that $H/L\sim10^{-1}$, $H/r_E\sim10^{-3}$, $L/r_E\sim10^{-2}$, Ro $\sim10^0$ (at midlatitudes), and Re $\sim10^{11}$. It should be evident that the last four terms on the right-hand side of (2.13) are at least an order of magnitude smaller than all other terms in (2.13) and thus can be neglected if a simplified form of the equation is desired. Such an equation governing the change in x-component motion is

$$\frac{Du}{Dt}=-\frac{1}{\rho}\frac{\partial p}{\partial x}+fv. \tag{2.16}$$

By extension, the simplified y-component equation is

$$\frac{Dv}{Dt} = -\frac{1}{\rho}\frac{\partial p}{\partial y} - fu. \tag{2.17}$$

These equations are further reduced given a smaller length scale, such as $L \sim 10^4$ m, which essentially eliminates the Coriolis force term by virtue of the fact that $\mathrm{Ro} = 10$ and aspect ratio $H/L \sim 1$. A note of caution is appropriate here: elimination by scale analysis does not necessarily mean that the term is unimportant, but rather that its contribution to the rate of change of the variable is relatively small. In fact, as will be demonstrated in Chapter 8, it is possible for the time-integrated contribution from an otherwise neglected term to be comparably large and critically important.

Next we consider the z-component of the equation of motion,

$$\frac{\partial w}{\partial t} + u\frac{\partial w}{\partial x} + v\frac{\partial w}{\partial y} + w\frac{\partial w}{\partial z} = -\frac{1}{\rho}\frac{\partial p}{\partial z} - g + u2\Omega\cos\phi + \frac{u^2 + v^2}{r_E} + \mu\nabla^2 w. \tag{2.18}$$

Following the same procedure as before, except now dividing through by U^2/H, we obtain:

$$\left(\frac{H^2}{L^2}\right)\frac{D\tilde{w}}{Dt} = -\frac{1}{\tilde{\rho}}\frac{\partial \tilde{p}}{\partial \tilde{z}} - \frac{1}{\mathrm{Fr}} + \tilde{u}\left(\frac{1}{\mathrm{Ro}}\right)\left(\frac{H}{L}\right)\cot\phi + \left(\tilde{u}^2 + \tilde{v}^2\right)\left(\frac{H}{r_E}\right) + \left(\frac{1}{\mathrm{Re}}\right)\left(\frac{H^2}{L^2}\right)\nabla^2\tilde{w}. \tag{2.19}$$

Another nondimensional parameter has been introduced:

$$\mathrm{Fr} = \frac{U^2}{gH} = \text{Froude number}, \tag{2.20}$$

where Fr is the ratio between the forces due to inertia and gravity. For the mesoscale, the curvature and friction terms are again comparatively small, but the remaining terms require careful thought. Elimination of terms with aspect-ratio factors is implied by the horizontal and vertical length scales; essentially, this reduces (2.19) to a statement of hydrostatic balance. This is a reasonable approximation for the motions on the upper end of the mesoscale spectrum (see Chapter 1). On the lower end of the spectrum, which includes convective clouds, length scales are of the order of 10 km, and hence the aspect ratio $H/L \sim 1$ and $\mathrm{Ro} = 10$. Retention of the left-hand side acceleration term is now implied; this is consistent with the existence – especially in deep convective storms – of vertical accelerations of air driven in part by the force due to buoyancy (see Chapters 7 and 8).

Such a buoyancy force owes to local density variations acted on by gravity, but does not appear explicitly in (2.18). To reveal buoyancy in our equations, we assume that these density variations are small compared with some surrounding, spatially averaged value (the *base state*), and then proceed as follows.

Let us write the atmospheric variables for density, pressure, and temperature as the sum of a static base state and a variation or deviation from that state,

$$\begin{aligned} p &= p' + \overline{p}(z) \\ \rho &= \rho' + \overline{\rho}(z) \\ T &= T' + \overline{T}(z), \end{aligned} \tag{2.21}$$

where the deviation is denoted by primes, and where the base state, denoted by overbars, is chosen to be the hydrostatic balance

$$\frac{d\overline{p}}{dz} = -\overline{\rho} g. \tag{2.22}$$

Using (2.21) and then (2.22) in a simplified version of (2.18), we have

$$\frac{Dw}{Dt} = -\frac{1}{(\overline{\rho} + \rho')} \frac{\partial p'}{\partial z} + g \left[\frac{\overline{\rho}}{(\overline{\rho} + \rho')} - 1 \right]. \tag{2.23}$$

Two possible approximations may be introduced in (2.23) at this point. From the binomial series expansion

$$\left(\frac{1}{1+x} \right) = 1 - x + \frac{x^2}{2} - \frac{x^3}{3} + \ldots \text{ for all } |x| < 1, \tag{2.24}$$

we let $x = \rho'/\overline{\rho}$, assume it is small, and thus approximate

$$\frac{1}{(\rho' + \overline{\rho})} = \frac{1}{\overline{\rho}} \frac{1}{\left(1 + \frac{\rho'}{\overline{\rho}}\right)} \simeq \frac{1}{\overline{\rho}} \left(1 - \frac{\rho'}{\overline{\rho}}\right), \tag{2.25}$$

which thereby reduces the second term on the right-hand side of (2.23) to

$$-g\frac{\rho'}{\overline{\rho}}. \tag{2.26}$$

This is the buoyancy term, often written in a compact form as $B = -g\rho'/\overline{\rho}$. Equation (2.25) can also be applied to the vertical PGF term to yield

$$-\frac{1}{\overline{\rho}} \frac{\partial p'}{\partial z} + \frac{\rho'}{\overline{\rho}^2} \frac{\partial p'}{\partial z}, \tag{2.27}$$

which, on assuming that products of deviation state variables are negligibly small relative to other terms in the equation, reduces to

$$-\frac{1}{\overline{\rho}} \frac{\partial p'}{\partial z}. \tag{2.28}$$

The resultant equation

$$\frac{Dw}{Dt} = -\frac{1}{\overline{\rho}}\frac{\partial p'}{\partial z} + B \tag{2.29}$$

is common in mesoscale studies. As will be shown in Chapter 4, (2.16), (2.17), and (2.29) are often rewritten in terms of a normalized, nondimensional pressure Π in place of p. When the thermodynamic energy equation (2.2) is similarly rewritten in terms of Π and potential temperature θ, the density variable is eliminated from the governing equations.

2.1.3 Approximations to the Mass Continuity Equation

The mass continuity equation can also be reduced with the aid of scale analysis. The following approximations are relevant for problems in which the buoyancy force is a nonnegligible contributor to vertical accelerations.[5] We begin by using (2.21) in (2.3) to yield an expanded continuity equation,

$$\frac{1}{\overline{\rho}}\frac{D\overline{\rho}}{Dt} + \frac{1}{\overline{\rho}}\frac{D\rho'}{Dt} = -\left(1 + \frac{\rho'}{\overline{\rho}}\right)\nabla \cdot \vec{V}. \tag{2.30}$$

Dividing through by the first left-hand term, and then using (2.12), we arrive at a nondimensionalized equation

$$1 + \frac{\frac{W}{H}\frac{1}{\overline{\rho}}\frac{D\rho'}{D\tilde{t}}}{\frac{W}{H_\rho}} = -\frac{\frac{W}{H}\frac{\partial \tilde{w}}{\partial \tilde{z}}}{\frac{W}{H_\rho}} - \frac{\frac{W}{H}\frac{\partial \tilde{w}}{\partial \tilde{z}}\frac{\rho'}{\overline{\rho}}}{\frac{W}{H_\rho}}$$

or,

$$1 + \frac{H_\rho}{H}\frac{1}{\overline{\rho}}\frac{D\rho'}{D\tilde{t}} = -\frac{H_\rho}{H}\frac{\partial \tilde{w}}{\partial \tilde{z}} - \frac{H_\rho}{H}\frac{\partial \tilde{w}}{\partial \tilde{z}}\frac{\rho'}{\overline{\rho}}, \tag{2.31}$$

We have allowed the time scale to be described by vertical motions (i.e., $\tilde{t} = t/(H/W)$) and assumed an equivalency in the scaling of the horizontal and vertical derivatives in the divergence term. H_ρ is the *density scale height*

$$H_\rho = \left(-\frac{1}{\overline{\rho}}\frac{d\overline{\rho}}{dz}\right)^{-1}. \tag{2.32}$$

It can be vertically integrated to obtain $\overline{\rho}(z) = \overline{\rho}(0)\exp\left(-z/H_\rho\right)$, and thus to show that H_ρ is the height over which base-state density is reduced from its surface value to 1/exp of that value (also known as the *e*-folding distance).

The use of (2.31) requires two additional assumptions, namely: (1) that deviation quantities at some height are small compared with their base state counterpart, and in particular that $\rho'/\overline{\rho} \ll 1$; and (2) that the vertical length scale – in this case, the vertical excursion of convective parcels – is comparable to the density-scale height.[6] The latter serves, in effect, to put a thermodynamic depth to the troposphere, and thus implies convective motions that occupy most of this depth. The consequence of these assumptions in (2.31) is that the second terms on both sides are negligibly small, allowing (2.30) to be reduced to

$$\frac{1}{\overline{\rho}}\frac{D\overline{\rho}}{Dt} = -\nabla \cdot \vec{V} \tag{2.33}$$

and thus to

$$\nabla \cdot (\overline{\rho}\vec{V}) = 0. \tag{2.34}$$

This is the *anelastic approximation* to the continuity equation. Anelastic refers to the fact that this form of the continuity equation (and associated equations) does not permit the existence of acoustic waves (see Section 2.2.1).

Another form of the continuity equation follows from the *Boussinesq approximation*, in which density changes in the fluid are neglected – hence density is assumed constant – except when coupled to gravity in the buoyancy force term.[7] In the context of atmospheric convection, the Boussinesq approximation requires that the convective motions be confined to a relatively shallow layer: $H \ll H_\rho$. The implications of this approximation on the continuity equation are revealed by expanding (2.34):

$$\frac{w}{\overline{\rho}}\frac{d\overline{\rho}}{dz} + \nabla \cdot \vec{V} = 0, \tag{2.35}$$

then applying (2.12) and (2.32), and then dividing through by W/H, leaving

$$\tilde{w}\frac{H}{H_\rho} + \frac{\partial \tilde{w}}{\partial \tilde{z}} = 0. \tag{2.36}$$

Upon application of the condition $H \ll H_\rho$ to (2.36), it should be clear that (2.35) reduces to

$$\nabla \cdot \vec{V} = 0. \tag{2.37}$$

We shall refer to this as the *incompressibility form* of the continuity equation; it is equivalent to

$$\nabla \cdot (\rho_0 \vec{V}) = 0, \tag{2.38}$$

where ρ_0 is a constant reference density.

2.1.4 Inclusion of Moisture

The full and simplified set of equations considered so far applies to dry air. To include the effects of water (vapor, liquid, or solid), we must first rewrite the equation of state for an ideal gas:

$$p = \rho_a R_d T + \rho_v R_v T = \rho_a R_d T \left(1 + \frac{q_v}{\varepsilon}\right), \tag{2.39}$$

where ρ_a is the density of dry air, ρ_v is the density of water vapor, q_v is water vapor mixing ratio (ratio of mass of vapor to mass of dry air; kg kg^{-1}), and $\varepsilon = R_d/R_v = 0.6220$ is the ratio of the gas constant of dry air to that of water vapor (461 J kg^{-1} K^{-1}), or equivalently, the ratio of the mean molecular weight of water vapor to that of dry air.[8] It is assumed that the temperature of the vapor (and eventually, of all water substance) and the temperature of the dry air are equivalent. In (2.39), pressure is the sum of the partial pressure due to dry air and that to vapor; that is,

$$p = p_a + e, \tag{2.40}$$

and it follows that

$$q_v = \frac{\rho_v}{\rho_a} = \frac{e/R_v T}{p_a/R_d T} = \varepsilon \frac{e}{p - e}. \tag{2.41}$$

Both equations are useful in writing an alternative form of the equation of state. Note that

$$\alpha = \frac{\alpha_a}{1 + q_v} = \frac{R_d T}{p_a} \frac{1}{(1 + q_v)} = \frac{R_d T}{p} \frac{(p_a + e)}{p_a} \frac{1}{(1 + q_v)}. \tag{2.42}$$

If we solve (2.41) for $p - e\ (= p_a)$, and then substitute the result into (2.42), we have, after some algebraic manipulation,

$$p = \rho R_d T \left[\frac{1 + q_v/\varepsilon}{1 + q_v}\right]. \tag{2.43}$$

Virtual temperature is defined from (2.43):

$$T_v = T \left[\frac{1 + q_v/\varepsilon}{1 + q_v}\right], \tag{2.44}$$

but can be approximated based on the relation

$$s = \frac{q_v}{1 + q_v}, \tag{2.45}$$

where specific humidity s is the ratio of the mass of vapor to the mass of moist air. Substitution of (2.45) into (2.44) gives

$$T_v = T \left[1 + \left(\frac{1}{\varepsilon} - 1\right) s\right] \simeq T(1 + 0.61s). \tag{2.46}$$

This is approximated further by letting $s \simeq q_v$, a reasonable assumption for most situations, and yields the oft-used result:

$$T_v \simeq T(1 + 0.61 q_v). \tag{2.47}$$

By extension, the equation of state for moist air becomes

$$p \simeq \rho R_d T\,(1 + 0.61 q_v) = \rho R_d T_v; \tag{2.48}$$

the corresponding state variables should be understood to apply to moist, unsaturated flow, unless otherwise specified (or indicated with subscript a).

We may now proceed with the time-dependent equations. The thermodynamic energy equation for moist air can be written as

$$c_{vm}\frac{DT}{Dt} = -\frac{p}{\rho}\nabla\cdot\vec{V} + \dot{Q}, \tag{2.49}$$

where $c_{vm} = c_v + c_{vv}q_v + c_l q_l$ is the specific heat of moist air at constant volume, c_{vv} is the specific heat of water vapor at constant volume (1424 J kg^{-1} K^{-1}), c_l is the specific heat of liquid water (4186 J kg^{-1} K^{-1}), and q_l is the total (liquid) water mixing ratio;[9] it is often assumed that $c_{vm} \simeq c_v$, although its use in numerical models leads to inaccuracies in total energy conservation.[10] The thermodynamic energy equation can be expressed in terms of *potential temperature*:

$$\frac{D\ln\theta}{Dt} = \frac{\dot{Q}}{c_{pm}T}, \tag{2.50}$$

where

$$\theta = T\left(\frac{p_0}{p}\right)^{R_d/c_p}. \tag{2.51}$$

p_0 is a reference pressure usually equal to 1000 hPa, and $c_p = R_d + c_v = 1{,}004$ J kg^{-1} K^{-1} is the specific heat of dry air at constant pressure. The specific heat of moist air at constant pressure is $c_{pm} = c_p + c_{pv}q_v + c_l\, q_l$, where c_{pv} is the specific heat of water vapor at constant pressure (1885 J kg^{-1} K^{-1}); as with c_{vm} in (2.49), it is also often assumed that $c_{pm} \simeq c_p$. In the absence of water (in any phase), potential temperature is conserved. However, in our studies of moist convective processes, seldom can we justify this, especially given the diabatic heating owing to liquid $\leftrightarrow$ vapor phase changes:

$$\dot{Q} = -L_v Dq_v/Dt, \tag{2.52}$$

where L_v is the latent heat of vaporization ($\simeq$2.50 $\times$ 10^6 J kg^{-1}).

It is convenient at this point to introduce two additional thermodynamic variables: *virtual potential temperature* (θ_v), and *equivalent potential temperature* (θ_e). Virtual potential temperature is the potential temperature that a parcel of dry air would have if its pressure and density were the same as those of a parcel of moist air. From (2.47), this is simply

$$\theta_v \simeq \theta\,(1 + 0.61 q_v)\,. \tag{2.53}$$

Equivalent potential temperature is the temperature a parcel would have if all of its water vapor was condensed and the resulting latent heat was converted into sensible heat.[11] It is related to (reversible) moist entropy, and has the exact mathematical formulation

$$\theta_e = T\left(\frac{p_0}{p_a}\right)^{R_d/(c_p+c_l q_l)} \mathrm{RH}^{R_v q_v/(c_p+c_l q_l)} \exp\left[\frac{L_v q_v}{\left(c_p + c_l q_l\right) T}\right], \tag{2.54}$$

where $\mathrm{RH} = e/e_s$ is relative humidity and $e_s(T)$ is the saturation vapor pressure.[12] In practice, θ_e (following a pseudoadiabatic process) is usually computed using an approximation to (2.54), such as

$$\theta_e \simeq T\left(\frac{p_0}{p_a}\right)^{R_d/c_p} \mathrm{RH}^{-R_v q_v/c_p} \exp\left[\frac{L_0 q_v}{c_p T}\right], \tag{2.55}$$

where $L_0 = 2.555 \times 10^6$ J kg^{-1} ($\simeq L_v$).[13]

Of the dynamical equations, the vertical equation of motion is the one most directly affected by the presence of water. This effect is largely through the buoyancy term, $-g\rho'/\overline{\rho}$, which is exposed by applying the decompositions of the state variables to (2.48) and then expanding:

$$(\overline{p} + p') = (\overline{\rho} + \rho')R_d(\overline{T}_v + T_v')$$

or

$$\overline{p}\left(1 + \frac{p'}{\overline{p}}\right) = \overline{\rho}\left(1 + \frac{\rho'}{\overline{\rho}}\right) R_d \overline{T}_v \left(1 + \frac{T_v'}{\overline{T}_v}\right), \tag{2.56}$$

which assumes that water vapor mixing ratio can be decomposed as $q_v = \overline{q}_v(z) + q_v'$.[14] After taking the natural log of both sides of (2.56), recalling that $\ln xy = \ln x + \ln y$, and noting that $\ln \overline{p} = \ln \overline{\rho} + R_d \overline{T}_v$, we have

$$\ln\left(1 + \frac{p'}{\overline{p}}\right) = \ln\left(1 + \frac{\rho'}{\overline{\rho}}\right) + \ln\left(1 + \frac{T_v'}{\overline{T}_v}\right). \tag{2.57}$$

We now make use of the Maclaurin series expansion:

$$\ln(1 + x) = x - \frac{x^2}{2} + \frac{x^3}{3} - \frac{x^4}{4} + \ldots \text{ for } -1 < x \le 1. \tag{2.58}$$

For small $|x|$, which is generally the case for $x = p'/\overline{p}$, and so on, in (2.57), the series can be truncated, thus letting us approximate (2.57) as

$$-\frac{\rho'}{\overline{\rho}} \simeq \frac{T_v'}{\overline{T}_v} - \frac{p'}{\overline{p}}. \tag{2.59}$$

For consistency between variables in the buoyancy term and in the thermodynamic energy equation, (2.59) can be expressed in terms of virtual potential temperature rather than virtual temperature. To do so, first use (2.51) to rewrite (2.53) as:

$$\theta_v = T_v \left(\frac{p_0}{p}\right)^{R_d/c_p},$$

then apply the state variable decompositions with the addition of $\theta_v = \overline{\theta}_v(z) + \theta_v'$, take the natural log of this result, and follow the steps taken to reach (2.59). This yields

$$\frac{\theta_v'}{\overline{\theta}_v} \simeq \frac{T_v'}{\overline{T}_v} - \frac{R_d}{c_p}\frac{p'}{\overline{p}}, \tag{2.60}$$

which, when combined with (2.59), allows the buoyancy term to be rewritten as

$$-g\frac{\rho'}{\overline{\rho}} \simeq g\left[\frac{\theta_v'}{\overline{\theta}_v} - \frac{c_v}{c_p}\frac{p'}{\overline{p}}\right]. \tag{2.61}$$

This form of the buoyancy term applies strictly to a cloud-free, nonprecipitating atmosphere. The effect on buoyancy owing to water in the liquid and solid phase – the drag exerted by hydrometeors – is revealed through derivation of separate momentum equations for the liquid/solid hydrometeors, which are then combined with the moist-air momentum equations.[15] The result is a term that accounts for momentum exchange associated with change in phase, plus a term proportional to $q_T g$, where q_T is the sum of the mixing ratios of liquid and solid hydrometeors. Typically, only the latter is retained explicitly; its contribution to the buoyancy term is often referred to as "precipitation loading," and has consequences on convective updrafts and downdrafts, as will be discussed in Chapter 5. The resultant vertical equation of motion is written as

$$\frac{Dw}{Dt} = -\frac{1}{\overline{\rho}}\frac{\partial p'}{\partial z} + g\left[\frac{\theta_v'}{\overline{\theta}_v} - \frac{c_v}{c_p}\frac{p'}{\overline{p}} - q_T\right]. \tag{2.62}$$

2.1.5 Equations for Water Substance

It should be evident from the thermodynamic energy equation, and from the buoyancy term just developed, that the equations for moist air must be coupled with equations that govern the rate change of water substance. These equations, which express conservation of water substance, can be written in a variety of ways, depending on the complexity of the problem. At minimum, we expect a separate equation governing the rate change of water vapor

$$\frac{Dq_v}{Dt} = S_E - S_C, \tag{2.63}$$

where subscript E denotes evaporation (but could encompass other sources of water vapor), and subscript C denotes condensation (but could encompass other sinks of water vapor). The relevant j species of liquid and ice cloud and precipitation particles are then governed by

$$\frac{Dq_j}{Dt} = S_j, \tag{2.64}$$

where S_j represents the sources and sinks of specie mixing ratio q_j, including hydrometeor fallout or sedimentation. The sources and sinks generally provide a coupling between the species equations, as implied by water substance conservation. For example, evaporation is a source of water vapor but a sink of cloud water (q_c), and collection of cloud droplets by raindrops is a q_c sink but rain water (q_r) source. As discussed more in Chapters 4 and 6, many of these sources and sinks involve complex and often very-small-scale processes that must be approximated using empirical formulae.

2.2 Atmospheric Waves and Oscillations

In this section and the two that follow, we show how various forms of the equations just described can permit solutions that relate to phenomena described in later chapters.

Here we develop the theoretical means to examine well-known examples of waves and oscillations. Waves are an important means of propagating energy from one point in space to another, and are featured prominently in Chapter 5 and elsewhere because their associated displacements are often implicated in the initiation of deep convective clouds.

Let us begin[16] with linear, second-order partial differential equations of the general form

$$\frac{\partial^2 A}{\partial t^2} = c^2 \nabla^2 A. \tag{2.65}$$

Such a *wave equation* is formed herein by reducing the appropriate governing equations to one equation in one relevant unknown. We typically assume solutions to (2.65) of the form

$$A(x, y, z, t) = \mathsf{A} e^{i(kx+ly+mz-\sigma t)}, \tag{2.66}$$

where A represents a displacement with amplitude A and frequency σ; $i = \sqrt{-1}$; $k = 2\pi/\lambda_x$, $l = 2\pi/\lambda_y$, and $m = 2\pi/\lambda_z$ are wavenumbers in the x, y, and z directions, respectively; and λ_x, λ_y, and λ_z are the corresponding wavelengths. Equation (2.66)

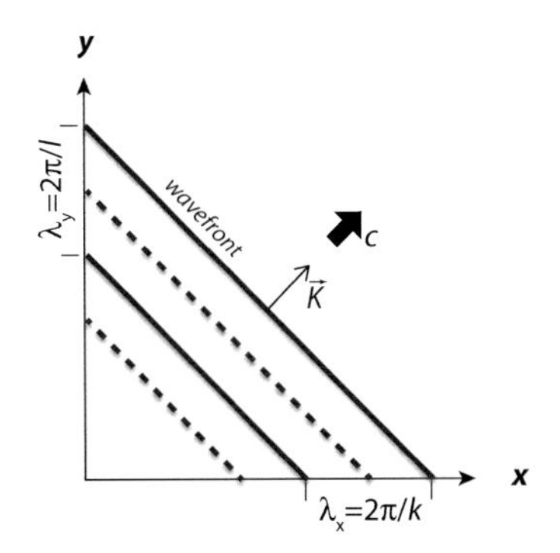

Figure 2.1 Idealized wave propagation in the *x*-*y* plane. The solid (dashed) lines are wave troughs (crests). Phase propagation is indicated by the bold arrow, speed is equal to C, and $\vec{K}$ is the horizontal wavenumber vector.

can be used to model a *plane wave*, or a sinusoidal wave with a constant frequency, and with wave fronts (lines or surfaces of constant phase) that are normal to wavenumber vector $\vec{K} = (k, l, m)$ and move with speed

$$c = \sigma/|\vec{K}| \tag{2.67}$$

(see Figure 2.1).

Frequency and thus phase speed of a given wave type will be shown to depend on some property of the medium, such as the base-state temperature in the case of the acoustic waves (Section 2.2.1). The frequency relationship will also reveal whether frequency is a function of wavenumber, and hence whether the waves are *dispersive*, such that different wavenumbers propagate at different speeds.

We assume that the displacement of the medium by a group of waves with different characteristics, such as wavelength and possibly phase speed, is the just the sum of the displacements of the individual waves.[17] Such a *superposition principle* equates, in turn, to an assumption of linearity of the system. This is embodied in the wave equation (2.65), which, as shown later, derives from a set of equations with nonlinear terms (such as $u\partial u/\partial x$) removed.

Linearization thus follows from a decomposition of all variables into a base state plus a deviation or perturbation, as was done above. Hence, we let

$$\begin{aligned} p &= \overline{p} + p' \quad & u &= \overline{u} + u' \\ \rho &= \overline{\rho} + \rho' \quad & v &= \overline{v} + v' \\ \theta &= \overline{\theta} + \theta' \quad & w &= \overline{w} + w', \end{aligned} \tag{2.68}$$

where, for now, we will leave open the details of the base state except to say that the base state must also satisfy the governing equations. In (2.68) the perturbations are assumed to be small, so that terms involving products of perturbations are negligible.

Three examples of pure waves are now presented. As just alluded to, these often coexist in the real atmosphere, forming mixtures of wave types.

2.2.1 Acoustic Waves

Acoustic or sound waves are omnipresent in the atmosphere. Although their displacements are of relatively low meteorological significance at the mesoscale, they do have consequences on the design and implementation of numerical prediction models. Thus, the following serves to motivate modeling decisions made later in Chapter 4, and otherwise is a good primer to the general procedure of theoretical analysis of atmospheric waves.

Let us assume a 2D (x, y) flow, and omit moisture, diabatic heating ($\dot{Q} = 0$), and planetary rotation ($f = 0$) because they are unnecessary for acoustic wave existence, and also because their omission simplifies the analysis. Let us also assume that the base state is at rest ($\overline{u} = \overline{v} = \overline{w} = 0$) and isentropic ($d\overline{\theta}/dz = 0$), but in hydrostatic balance. Using (2.68) in Eqs. (2.16), (2.17), (2.29), (2.3), (2.50), and (2.9) and then applying these assumptions yields the following set of equations that govern the perturbation variables:

$$\frac{\partial u'}{\partial t} = -\frac{1}{\overline{\rho}}\frac{\partial p'}{\partial x}, \tag{2.69}$$

$$\frac{\partial v'}{\partial t} = -\frac{1}{\overline{\rho}}\frac{\partial p'}{\partial y}, \tag{2.70}$$

$$\frac{\partial \rho'}{\partial t} = -\overline{\rho}\left(\frac{\partial u'}{\partial x} + \frac{\partial v'}{\partial y}\right), \tag{2.71}$$

$$\frac{\partial \theta'}{\partial t} = 0, \tag{2.72}$$

and

$$\frac{\rho'}{\overline{\rho}} = \frac{c_v}{c_p}\frac{p'}{\overline{p}} = \frac{\theta'}{\overline{\theta}}, \tag{2.73}$$

where the perturbation form of the equation of state makes use of Eq. (2.61).

For this problem, pressure is the most relevant variable (sound waves are oscillations in pressure), and thus we seek one equation in unknown p'. Take $\partial/\partial t$ of (2.73) and make use of (2.72) to write (2.71) in terms of pressure perturbation,

$$\frac{\partial p'}{\partial t} = -\overline{p}\frac{c_p}{c_v}\left(\frac{\partial u'}{\partial x} + \frac{\partial v'}{\partial y}\right). \tag{2.74}$$

Now, take $\partial/\partial x$ of (2.69), $\partial/\partial y$ of (2.70), and then use in $\partial/\partial t$ of (2.74) to eliminate u' and v' and yield a linear, second-order PDE in p':

$$\frac{\partial^2 p'}{\partial t^2} = \frac{\overline{p}}{\overline{\rho}} \frac{c_p}{c_v} \left(\frac{\partial^2 p'}{\partial x^2} + \frac{\partial^2 p'}{\partial y^2} \right). \tag{2.75}$$

Assume a solution of the form

$$p'(x, y, t) = P_0 e^{i(kx+ly-\sigma t)}, \tag{2.76}$$

where P_0 is amplitude, taken here to be constant, but need not be the case in general, and where it should be understood that only the real, physically relevant part of this complex solution is considered. Using this solution in (2.75), and dividing through by $P_0 e^{i(kx+ly-\sigma t)}$, we find a frequency-dispersion relation

$$\sigma^2 = c_s^2 K^2, \tag{2.77}$$

where $k^2 + l^2 = K^2$, and

$$c_s \equiv \sqrt{(c_p/c_v) R\overline{T}} = \sqrt{(c_p/c_v)\overline{p}/\overline{\rho}} \tag{2.78}$$

is the adiabatic speed of sound ($= 337$ m s^{-1} at $\overline{T} = 10°$C). The phase speed for propagation is

$$c = \frac{\sigma}{K} = \pm c_s \tag{2.79}$$

and the velocity of the wave group is

$$\vec{c}_g = \partial\sigma/\partial\vec{k} \tag{2.80}$$

with components $c_{g,x} = \partial\sigma/\partial k$, $c_{g,y} = \partial\sigma/\partial l$, and $c_{g,z} = \partial\sigma/\partial m$, and thus with a magnitude

$$|\vec{c}_g| = \sqrt{c_{g,x}^2 + c_{g,y}^2 + c_{g,z}^2}. \tag{2.81}$$

Using (2.77) in (2.81), we find that $|\vec{c}_g| = c_s = c$, or that individual waves move at the same speed as that of the group, and hence that acoustic waves are *nondispersive*. In addition, we find that the phase propagation, group velocity, and, consequently, the wave-energy propagation all are in the same direction.

The phase speed of acoustic waves is an order of magnitude larger than the typical wind speed at most atmospheric scales, including the mesoscale. Because the speeds of atmospheric motions determine how numerical models are integrated (Chapter 4), these waves must be accounted for, even though their effect on pressure (displacement) is small in comparison to typical mesoscale pressure perturbations. It is for this reason that acoustic waves have traditionally been filtered from, or otherwise treated in special ways in, numerical models.

2.2.2 Rossby Waves

Although a treatment of waves with planetary-scale lengths might seem out of place here, we will show in Chapter 9 that persistent mesoscale convective systems can excite *Rossby waves*. It is possible for these waves to propagate upstream (against the mean flow), such that an environment that favors subsequent deep convection is fostered.

In preparation for this discussion in Chapter 9, let us note first that barotropic Rossby waves owe their existence to the meridional gradient of planetary vorticity. Accordingly, the Coriolis force is necessary for this problem, whereas gravity, compressibility, moisture, and diabatic heating are not. We make the additional assumption of a 2D horizontal (x, y) flow that is inviscid and barotropic. In this flow, *absolute vertical vorticity* is conserved following the horizontal motion (denoted by the subscript H)

$$\frac{D_H(\zeta + f)}{Dt} = 0, \tag{2.82}$$

where $\zeta = \hat{k} \cdot \nabla \times \vec{V} = \partial v/\partial x - \partial u/\partial y$ is the vertical component of vorticity. Equation (2.82) is obtained by taking $\partial/\partial y$ of (2.16), subtracting from $\partial/\partial x$ of (2.17), and then applying the aforementioned assumptions.

(Barotropic) Rossby waves are a consequence of absolute-vorticity conservation, and thereby arise as a solution to (2.82). The solution procedure[18] begins with a linearization of (2.82) using

$$\begin{aligned} u &= \overline{u} + u' \\ v &= v' \\ \zeta &= \zeta', \end{aligned} \tag{2.83}$$

which assumes a base state characterized by westerly winds. Introduction of a perturbation streamfunction ψ'

$$\begin{aligned} u' &= -\partial \psi'/\partial y \\ v' &= \partial \psi'/\partial x \end{aligned} \tag{2.84}$$

allows (2.82) to be rewritten in terms of $\beta \equiv df/dy$ and ψ':

$$\left(\frac{\partial}{\partial t} + \overline{u}\frac{\partial}{\partial x}\right)\nabla^2\psi' + \frac{\partial \psi'}{\partial x}\beta = 0. \tag{2.85}$$

Equation (2.85) admits a solution of the form

$$\psi' = \Psi e^{i(kx+ly-\sigma t)}, \tag{2.86}$$

where Ψ is a constant amplitude; it is again understood that only the real part of (2.86) is physically relevant here. On substituting (2.86) into (2.85), the frequency-dispersion relation is found as

$$\sigma = \overline{u}k - \frac{k\beta}{K^2} \tag{2.87}$$

and the *zonal* phase speed, as

$$c = \overline{u} - \frac{\beta}{K^2}, \tag{2.88}$$

where $K = \sqrt{k^2 + l^2}$ is again the horizontal wavenumber magnitude. Recalling that $\vec{c}_g = \partial\sigma/\partial\vec{k}$, it can be shown that $c \neq |\vec{c}_g|$, and therefore that Rossby waves are dispersive. That is, long-wavelength (small-wavenumber) Rossby waves generally are retrogressive with respect to the base state westerly wind, and short-wavelength (large-wavenumber) Rossby waves are progressive.

2.2.3 Gravity Waves

Within a stably stratified fluid ($d\overline{\theta}/dz > 0$), the existence of the restoring force of gravity acts to generate propagating motions known as internal gravity or buoyancy waves.[19] The associated displacements yield temperature and wind perturbations that are meteorologically relevant on the mesoscale, thus providing extra incentive for a treatment of internal gravity waves herein.

To determine their theoretical characteristics, we make many of the same assumptions as before: 2D (x, z) flow, and omission of moisture, diabatic heating ($\dot{Q} = 0$), and planetary rotation ($f = 0$). Compressibility is unnecessary and the wave motions are relatively shallow, and therefore we make the incompressibility approximation to the mass continuity equation. Although a base-state wind is also unnecessary, we now, for the sake of illustration, consider the case with constant wind in the x-direction ($\overline{u} = U_0$; $\overline{v} = \overline{w} = 0$). As mentioned, a base-state stratification is required, and it is assumed to be in hydrostatic balance. The following set of equations result:

$$\frac{\partial u'}{\partial t} + U_0\frac{\partial u'}{\partial x} = -\frac{1}{\rho_0}\frac{\partial p'}{\partial x}, \tag{2.89}$$

$$\frac{\partial w'}{\partial t} + U_0\frac{\partial w'}{\partial x} = -\frac{1}{\rho_0}\frac{\partial p'}{\partial z} - g\frac{\rho'}{\rho_0}, \tag{2.90}$$

$$\left(\frac{\partial u'}{\partial x} + \frac{\partial w'}{\partial z}\right) = 0, \tag{2.91}$$

$$\frac{\partial \theta'}{\partial t} + U_0 \frac{\partial \theta'}{\partial x} + w' \frac{d\overline{\theta}}{dz} = 0, \tag{2.92}$$

and
$$\frac{\rho'}{\rho_0} = \frac{c_v}{c_p} \frac{p'}{\overline{p}} - \frac{\theta'}{\overline{\theta}}. \tag{2.93}$$

Because the displacement and restoring force act in the vertical direction in the pure problem, it would seem reasonable to seek a wave equation in vertical velocity, as is done traditionally.[20] Let us begin with the elimination of density in the buoyancy term in favor of potential temperature, a logical step if the background stratification is posed in terms of potential temperature. Equation (2.93) is helpful in this regard, made apparent by dividing through by ρ_0 and then using (2.78):

$$\rho' = \frac{p'}{c_s^2} - \rho_0 \frac{\theta'}{\overline{\theta}}.$$

Based on the value of c_s given previously, it should be evident that $|p'/c_s^2| \ll |-\rho_0 \theta'/\overline{\theta}|$, and thus that density fluctuations arise mostly from potential temperature fluctuations. Next, eliminate pressure by subtracting $\partial/\partial z$ of (2.89) from $\partial/\partial x$ of (2.90):

$$\left(\frac{\partial}{\partial t} + U_0 \frac{\partial}{\partial x}\right)\left(\frac{\partial w'}{\partial x} - \frac{\partial u'}{\partial z}\right) - \frac{g}{\overline{\theta}} \frac{\partial \theta'}{\partial x} = 0. \tag{2.94}$$

Take $\partial/\partial t$ of this equation, insert (2.92) into the result, and then substitute (2.94) solved for $\partial \theta'/\partial x$ to eliminate θ':

$$\begin{aligned} &\frac{\partial}{\partial t}\left[\left(\frac{\partial}{\partial t} + U_0 \frac{\partial}{\partial x}\right)\left(\frac{\partial w'}{\partial x} - \frac{\partial u'}{\partial z}\right)\right] \\ &\quad + U_0 \frac{\partial}{\partial x}\left[\left(\frac{\partial}{\partial t} + U_0 \frac{\partial}{\partial x}\right)\left(\frac{\partial w'}{\partial x} - \frac{\partial u'}{\partial z}\right)\right] + N^2 \frac{\partial w'}{\partial x} = 0. \end{aligned} \tag{2.95}$$

Finally, take $\partial/\partial x$ of (2.95) and use (2.91) to eliminate u':

$$\frac{\partial^2 \Psi}{\partial t^2} + 2U_0 \frac{\partial^2 \Psi}{\partial x \partial t} + U_0^2 \frac{\partial^2 \Psi}{\partial x^2} = -N^2 \frac{\partial^2 w'}{\partial x^2}, \tag{2.96}$$

where $\Psi = (\partial^2 w'/\partial x^2 + \partial^2 w'/\partial z^2)$. The quantity

$$N^2 = \frac{g}{\overline{\theta}} \frac{d\overline{\theta}}{dz} \tag{2.97}$$

is the square of the buoyancy frequency, also known as the *Brunt-Väisälä frequency*. This is the frequency of nonpropagating buoyancy oscillations, and has a value

$$N \sim [(10/300)(6.5/1000)]^{1/2} = 0.015\,\mathrm{s}^{-1}$$

for standard atmospheric conditions, and therefore an oscillation period of ~7 min.

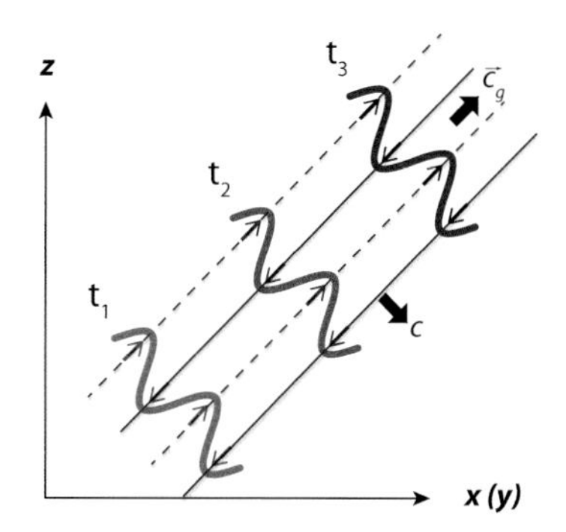

Figure 2.2 Schematic of gravity wave propagation. The thin solid (dashed) lines are wave troughs (crests), bold lines show the amplitude and phase of a segment of the wave at different times ($t_1 < t_2 < t_3$), and thin arrows indicate the corresponding air motion. It should be assumed that the wave was generated (at time t_0) impulsively, at a location below the plot origin. The group, and hence energy propagation, is in a direction perpendicular to the phase propagation. Based on Hooke (1986).

Assume again a solution of the form

$$w'(x, z, t) = W_0 e^{i(kx+mz-\sigma t)}, \tag{2.98}$$

where W_0 is a constant amplitude. After using this in (2.96) and manipulating the result, the frequency-dispersion relationship for internal gravity waves is found to be

$$\sigma - U_0 k = \pm \frac{Nk}{(k^2 + m^2)^{1/2}}. \tag{2.99}$$

The right-hand side is the *intrinsic frequency*, which in this case is *Doppler-shifted* as a consequence of the mean flow. The phase speed relative to the mean flow follows from (2.99):

$$\pm \frac{Nk}{k^2 + m^2}, \tag{2.100}$$

as does the group speed

$$\pm \frac{Nm}{k^2 + m^2}. \tag{2.101}$$

Because the phase and group speeds are unequal, we can conclude that internal gravity waves are dispersive, with small wave numbers propagating faster relative to large wave numbers. We also can conclude that the group velocity is parallel to the phase lines, and hence that phase propagation is perpendicular to group velocity propagation and accordingly to the propagation of wave energy (Figure 2.2). An interesting consequence is that gravity-wave energy can propagate upward over considerable vertical distances (approximately tens of km), even though the vertical displacements themselves occur over relatively small distances (~1 km).[21]

As will be discussed more in Chapter 5, such propagation is affected by the existence of a vertically sheared base-state wind. Let us prepare for this discussion by considering a slightly different base state wind: $\overline{u} = U(z)$ and $\overline{v} = \overline{w} = 0$. Following the same procedure as before and then assuming a solution

$$w'(x, z, t) = W(z)e^{i(kx-\sigma t)} \tag{2.102}$$

yields the Taylor-Goldstein equation

$$\frac{d^2W}{dz^2} + M^2W = 0, \tag{2.103}$$

where $W(z)$ is the vertically dependent amplitude, and

$$M^2 = \left[\frac{N^2}{(U-c)^2} - \frac{d^2U/dz^2}{(U-c)}\right] - k^2. \tag{2.104}$$

The relevant physical behavior is associated with the bracketed quantity in (2.104), which is known as the *Scorer parameter* ($= \ell^2$).[22] In essence, where $M^2 = \ell^2 - k^2 > 0$, vertical wave propagation is allowed, and where $M^2 < 0$, vertically exponential decay of vertical propagation (or *trapping*) is expected (see Chapter 5).

Gravity waves are excited when air is displaced vertically in the presence of a stable density stratification. One means of displacement is provided through convective clouds, either by the convective motions themselves, or by a precipitation-driven, surface-based cold pool that moves outward from the precipitating storm, into a stable environment. The dynamics for this particular mechanism are established in Section 2.4.

2.3 Instabilities

In this section we develop a theoretical background for our subsequent applications of instabilities. The basic methodology is similar to that used in Section 2.2 because of an underlying assumption of linearity.

2.3.1 Gravitational Instability

The preceding treatment of internal gravity waves provides a good segue to the first of three discussions on instability. We begin with *gravitational instability*,[23] and thus a consideration of the perturbation form of the inviscid, vertical equation of motion

$$\frac{\partial w'}{\partial t} = B. \tag{2.105}$$

In (2.105) a quiescent base state is assumed, $B = g\theta'/\overline{\theta}(z)$ as justified in the previous section, and perturbation pressure and its vertical gradient are assumed to be negligible; this later condition is discussed in Chapters 5 and 6 in the context of parcel

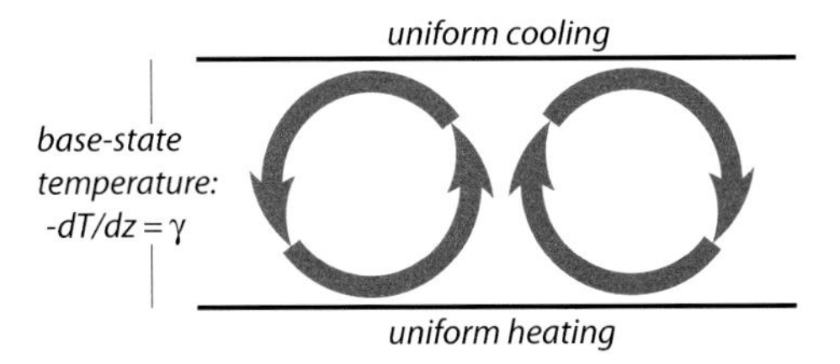

Figure 2.3 Idealized development of Rayleigh-Bénard convection.

theory. The perturbation form of the adiabatic thermodynamic energy equation can be written as

$$\frac{\partial B}{\partial t} = -w' N^2.$$

Combining this with $\partial/\partial t$ of (2.105) yields

$$\frac{\partial^2 w'}{\partial t^2} = -w' N^2. \tag{2.106}$$

If N^2 is constant (as given, e.g., by linear profiles in $\overline{\theta}$), (2.106) is a linear, second-order differential equation, with solutions of the form $w' = W_0 \exp(-i\sigma t)$. The time-dependent behavior of the vertical motion perturbation thus depends on $\sigma^2 = N^2$:

(1) If $N^2 > 0$, the perturbation oscillates with frequency $\sigma = \pm N$. This is the *stable* case, representing buoyancy oscillations.
(2) If $N^2 = 0$, then $\sigma = 0$, and the perturbation is constant in time. This is the *marginal* or *neutral* case.
(3) If $N^2 < 0$, then $\sigma = iN$, and the perturbation grows in time. This is the *unstable* case that, as implied earlier, will be examined more in the context of parcel theory in Chapter 5.

2.3.2 Rayleigh-Bénard Instability

What follows is a classic fluid dynamics problem that will help us understand the behavior of a heated atmospheric boundary layer, with implications on the development of shallow and deep cumuli. In essence, the problem is that of the convective overturning that results from a heated lower plate and/or cooled upper plate; the overturning is known as *Rayleigh-Bénard convection* (Figure 2.3). The uniform heating/cooling is assumed to maintain a base-state temperature distribution $\gamma = -dT/dz$. The base state otherwise is at rest. The incompressibility approximation is made, and the Coriolis force is neglected. This problem requires the inclusion of viscosity in the equations of motion and heat conduction in the thermodynamic energy equation. The linearized equations are

$$\frac{\partial u'}{\partial t} = -\frac{1}{\rho_0}\frac{\partial p'}{\partial x} + \mu\nabla^2 u', \tag{2.107}$$

$$\frac{\partial v'}{\partial t} = -\frac{1}{\rho_0}\frac{\partial p'}{\partial y} + \mu\nabla^2 v', \tag{2.108}$$

$$\frac{\partial w'}{\partial t} = -\frac{1}{\rho_0}\frac{\partial p'}{\partial z} + B + \mu\nabla^2 w', \tag{2.109}$$

$$\frac{\partial u'}{\partial x} + \frac{\partial v'}{\partial x} + \frac{\partial w'}{\partial z} = 0, \tag{2.110}$$

and $$\frac{\partial T'}{\partial t} + w'\gamma = \kappa\nabla^2 T', \tag{2.111}$$

where the thermodynamic energy equation is a reduced form of (2.2); μ and κ are constant viscosity and diffusivity coefficients, respectively; $B = g\Upsilon T'$ is buoyancy; and Υ is the thermal expansion coefficient.[24]

Because of the emphasis of this problem on overturning motions, we reduce the set of equations to a single equation in unknown w'. Begin by eliminating u' and v' with $\partial/\partial x$ of (2.107) and $\partial/\partial y$ of (2.108) and using these in (2.110) to yield

$$\left(\frac{\partial}{\partial t} - \mu\nabla^2\right)\frac{\partial w'}{\partial z} = \frac{1}{\rho_o}\left(\frac{\partial^2}{\partial x^2} + \frac{\partial^2}{\partial y^2}\right)p'. \tag{2.112}$$

Next, eliminate p' by taking $\partial/\partial z$ of (2.112) and using this in $\left(\partial^2/\partial x^2 + \partial^2/\partial y^2\right)$ of (2.109), giving

$$\left(\frac{\partial}{\partial t} - \mu\nabla^2\right)\nabla^2 w' = g\Upsilon\left(\frac{\partial^2}{\partial x^2} + \frac{\partial^2}{\partial y^2}\right)T'. \tag{2.113}$$

Finally, use (2.111) to eliminate T' from (2.113):

$$\left(\frac{\partial}{\partial t} - \kappa\nabla^2\right)\left(\frac{\partial}{\partial t} - \mu\nabla^2\right)\nabla^2 w' = g\Upsilon\gamma\left(\frac{\partial^2}{\partial x^2} + \frac{\partial^2}{\partial y^2}\right)w'. \tag{2.114}$$

Following the *method of normal modes*, we pose the following solution to (2.114):

$$w' = W(z)e^{i(kx+ly)+\sigma t}, \tag{2.115}$$

where $W(z)$ is determined by boundary conditions along the top and bottom plates. For example, an impermeability condition on w' (no normal flow at the boundaries, i.e., $z = 0$ and $z = 1$ for $n = 1$) is satisfied by $W(z) = W_0 \sin(n\pi z)$. Using this condition and (2.115) in (2.114) gives the following:

$$\mathcal{M}\sigma^2 + \mathcal{M}^2(\kappa + \mu)\sigma + [\kappa\mu\mathcal{M}^3 - K^2\gamma g \Upsilon] = 0, \tag{2.116}$$

where $\mathcal{M} = K^2 + n^2\pi^2$, and $K^2 = k^2 + l^2$ as before.

The quadratic equation in σ has roots

$$\sigma = -\mathcal{M}(\kappa + \mu) \pm [\mathcal{M}^4(\kappa + \mu)^2 - 4\mathcal{M}(\kappa\mu\mathcal{M}^3 - K^2\gamma g\Upsilon)]^{1/2}/2\mathcal{M} = 0. \tag{2.117}$$

Inspection of the radicand shows that a necessary condition for σ to be real, and thus for w' to undergo unstable growth, is that

$$\kappa\mu\mathcal{M}^3 - K^2\gamma g\Upsilon < 0 \tag{2.118}$$

or that

$$\mathrm{Ra} > \frac{\mathcal{M}^3 h^4}{K^2}, \tag{2.119}$$

where

$$\mathrm{Ra} \equiv \frac{\gamma g \Upsilon h^4}{\kappa\mu} \tag{2.120}$$

is the *Rayleigh number*, and h is the distance between the top and bottom plates (and hence the flow depth). This result can also be deduced from (2.116), which shows that the condition for marginal stability, $\sigma = 0$, is

$$\mathrm{Ra} = \frac{\mathcal{M}^3 h^4}{K^2}, \tag{2.121}$$

and thus constrains the conditions necessary for the onset of instability. We wish to find the *critical Rayleigh number* $\mathrm{Ra_c}$, which is smallest possible Rayleigh number for such onset, and express it in terms of the right-hand side parameters. Let the flow depth be $h = 1/n$, and then let $n = 1$. In this case, the K that minimizes Ra in (2.121) (and, thus, that satisfies $\partial \mathrm{Ra}/\partial K = 0$) is $K^2 = K_c^2 = \pi^2/2$, and accordingly:

$$\mathrm{Ra}_c = \frac{27\pi^4}{4} \tag{2.122}$$

(Figure 2.4). The onset of convective overturning will be exhibited when the Rayleigh number is gradually increased to a value $\mathrm{Ra} = \mathrm{Ra_c}$. The overturning, which is the manifestation of the instability, will occupy the depth of the fluid, and have horizontal wavenumber $K = \sqrt{k^2 + l^2} = \pi/\sqrt{2}$. As suggested by (2.119), illustrated in Figure 2.4, and to be shown in observational data in Chapter 5, Rayleigh-Bénard instability can occur for a limitless number of horizontal wavenumbers, and can take the form of convective cells ($k = l$) as well as rolls (e.g., $k \neq 0, l = 0$).

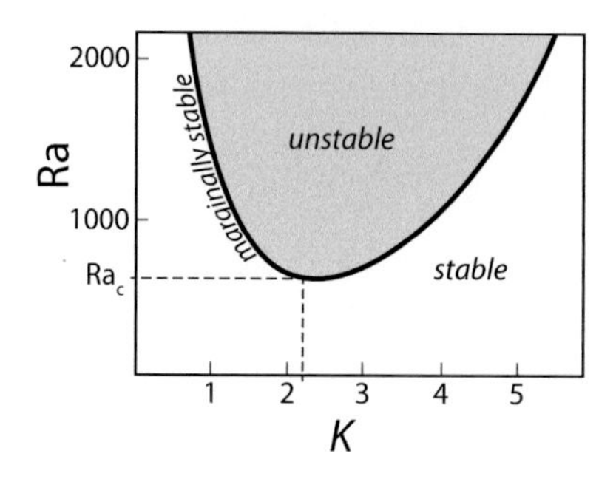

Figure 2.4 Rayleigh number as a function of horizontal wavenumber, for the case of vertical wavenumber 1. Adapted from Houze (1993), and Kundu (1990).

2.3.2 Kelvin-Helmholtz Instability

To conclude this section, we consider the instability that results at the infinitesimally thin interface between two fluids of differing density and wind speed (Figure 2.5). The interface itself is considered a *vortex sheet*, owing to the shear of the two flows. Credited to Lord Kelvin and Herman von Helmholtz,[25] the instability is realized in natural and laboratory flows. For example, it is known to occur atop the potentially cold surface-outflow of thunderstorms (Chapter 6). A variant of *Kelvin-Helmholtz* (KH) *instability* arises from wind shear (vertical as well as horizontal) in the absence of the density change. We will see in subsequent chapters (Chapters 5, 6, and 8) that such *shearing instability* develops in numerous mesoscale flows, and can play roles in convection initiation and even tornadogenesis.

Let us begin with a base state that consists of two overlying fluids with uniform velocity

$$\overline{v} = \overline{w} = 0; \quad \overline{u}(z) = \begin{cases} U_2, \ z > 0 \\ U_1, \ z < 0 \end{cases} \tag{2.123}$$

and density

$$\overline{\rho}(z) = \begin{cases} \rho_2, \ z > 0 \\ \rho_1, \ z < 0 \end{cases}. \tag{2.124}$$

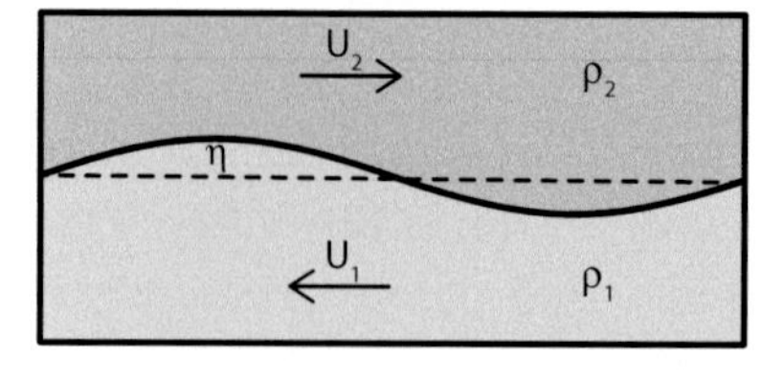

Figure 2.5 Base state configuration of wind (U) and density (ρ) in two overlying fluids. Kelvin-Helmholtz instability is realized as the disturbed height (η) of the interface between the two fluids. The undisturbed height of the interface is $z = 0$. Arrows in the two fluids show a possible flow example.

The base state is also hydrostatically balanced,

$$\frac{d\overline{p}}{dz} = \begin{cases} -\rho_2 g, & z > 0 \\ -\rho_1 g, & z < 0 \end{cases}, \tag{2.125}$$

and isentropic. As implied in Figure 2.5, we assume two-dimensionality (x,z), although this need not be the case in general. The classic problem treats the flow as incompressible and inviscid; the Coriolis force, diabatic heating, and moist processes are also neglected.

The undisturbed height of the interface is $z = 0$ (see Figure 2.5); here, we define a new dependent variable $\eta(x, t)$ as the height of the disturbed interface relative to $z = 0$. The interface itself is a *material surface*, composed at all times of the same fluid particles, and thus moving with the particles. We do not permit the two fluids to occupy the same point at the interface simultaneously, and also do not permit the existence of a cavity between the two fluids. Finally, although the interface itself is a vortex sheet, the disturbed flow on both sides of the interface is assumed to be *irrotational*, or free of vorticity.

The irrotational assumption allows us to define a velocity potential

$$\varphi = \begin{cases} \varphi_2, & z > \eta \\ \varphi_1, & z < \eta \end{cases}, \tag{2.126}$$

such that

$$\vec{V} = \nabla\varphi. \tag{2.127}$$

(Because the curl of a gradient is identically zero, it is easily verified that the vorticity in this case is also identically zero.) The incompressibility assumption $\nabla \cdot \vec{V} = 0$ with (2.126) and (2.127) gives

$$\begin{aligned} \nabla^2\varphi_2 &= 0, \quad z > \eta \\ \nabla^2\varphi_1 &= 0, \quad z < \eta, \end{aligned} \tag{2.128}$$

which serves as a boundary condition, as does the assumption that far from the interface, the disturbed flow becomes the base state flow as defined in (2.123).

We are now ready to introduce the governing equations. The first two represent the Lagrangian change of the interface on both sides, implicitly defining the vertical component of motion on both sides:

$$\left.\begin{aligned} \frac{D\eta}{Dt} &= \frac{\partial\eta}{\partial t} + \frac{\partial\varphi_2}{\partial x}\frac{\partial\eta}{\partial x} = w_2 = \frac{\partial\varphi_2}{\partial z} \\ \frac{D\eta}{Dt} &= \frac{\partial\eta}{\partial t} + \frac{\partial\varphi_1}{\partial x}\frac{\partial\eta}{\partial x} = w_1 = \frac{\partial\varphi_1}{\partial z} \end{aligned}\right\} \quad z = \eta. \tag{2.129}$$

The other equation is a consequence of a requirement that pressure be continuous across the interface, $p_1(\eta) = p_2(\eta)$, and comes from the vector equation of motion (2.1) with the imposed assumptions:

$$-\frac{1}{\rho}\nabla p = \frac{\partial \vec{V}}{\partial t} + \frac{1}{2}\nabla(\vec{V}\cdot\vec{V}) + \nabla(gz). \tag{2.130}$$

We can integrate (2.130) after making use of (2.127), and thereafter apply the continuous pressure requirement to give

$$\left.\begin{aligned}\rho_1\left[\frac{\partial\varphi_1}{\partial t} + \frac{1}{2}(\nabla\varphi_1)^2 + gz - C_1\right] \\ = \rho_2\left[\frac{\partial\varphi_2}{\partial t} + \frac{1}{2}(\nabla\varphi_2)^2 + gz - C_2\right]\end{aligned}\right\} \quad z = \eta. \tag{2.131}$$

C_1 and C_2 are constants of integration, ultimately ensuring that the base state satisfies (2.131), and thus are equivalent to $U_1^2/2$ and $U_2^2/2$, respectively.

To linearize these governing equations, we decompose the velocity potentials into a base state and perturbation as follows:

$$\begin{aligned}\varphi_2 &= U_2 x + \varphi_2', \quad z > \eta \\ \varphi_1 &= U_1 x + \varphi_1', \quad z < \eta.\end{aligned} \tag{2.132}$$

Note that η is, in essence, already a perturbation quantity; therefore, (2.129) and (2.131) become

$$\begin{aligned}\frac{\partial\eta}{\partial t} + U_2\frac{\partial\eta}{\partial x} &= \frac{\partial\varphi_2'}{\partial z} \\ \frac{\partial\eta}{\partial t} + U_1\frac{\partial\eta}{\partial x} &= \frac{\partial\varphi_1'}{\partial z}\end{aligned} \tag{2.133}$$

and

$$\rho_1\left[\frac{\partial\varphi_1'}{\partial t} + U_1\frac{\partial\varphi_1'}{\partial x} + g\eta\right] = \rho_2\left[\frac{\partial\varphi_2'}{\partial t} + U_2\frac{\partial\varphi_2'}{\partial x} + g\eta\right], \tag{2.134}$$

respectively. For the linear stability analysis we again pose general solutions of the form

$$\begin{aligned}\varphi_1' &= \hat{\varphi}_1(z)e^{ikx+\sigma t} \\ \varphi_2' &= \hat{\varphi}_2(z)e^{ikx+\sigma t} \\ \eta &= \hat{\eta}e^{ikx+\sigma t},\end{aligned} \tag{2.135}$$

where $\hat{\eta}$ is a constant amplitude. Using (2.135) in (2.133) and in the boundary conditions with (2.132) yields

$$\begin{aligned}\hat{\varphi}_2 &= -(\sigma + ikU_2)\frac{\hat{\eta}}{k}e^{-kz} \\ \hat{\varphi}_1 &= (\sigma + ikU_1)\frac{\hat{\eta}}{k}e^{kz}.\end{aligned} \tag{2.136}$$

Figure 2.6 Pictorial depiction of the development of Kelvin-Helmholtz instability over some interval $t_1 \leq t \leq t_n$. Lines represent the fluid interface, such as that defined in Figure 2.5.

After substituting the particular solution formed by (2.135) and (2.136) into Eq. (2.134), and then manipulating the result, we find a quadratic equation in σ that has roots

$$\sigma = -ik\frac{(\rho_1 U_1 + \rho_2 U_2)}{(\rho_1 + \rho_2)} \pm \left[\frac{k^2 \rho_1 \rho_2 (U_1 - U_2)^2}{(\rho_1 + \rho_2)^2} - gk\frac{(\rho_1 - \rho_2)}{(\rho_1 + \rho_2)}\right]^{1/2}. \tag{2.137}$$

The radicand in (2.137) is positive and thus allows for unstable growth of the interfacial disturbance η (as well as of velocity potentials ϕ_1 and ϕ_2) when

$$g\left(\rho_1^2 - \rho_2^2\right) < k\rho_1\rho_2 (U_1 - U_2)^2, \tag{2.138}$$

which occurs for sufficiently large wavenumbers (small wave lengths):

$$k > \frac{g\left(\rho_1^2 - \rho_2^2\right)}{\rho_1\rho_2 (U_1 - U_2)^2}. \tag{2.139}$$

It is always possible to satisfy this condition provided that $U_1 \neq U_2$.

The reader is reminded that the results of this theoretical analysis explain *only the initial behavior of the disturbances*. In real atmospheric occurrences of KH instability, as in association with thunderstorm outflow, the instability is known to amplify and even "break" (Figure 2.6).

Variants of the Kelvin-Helmholtz problem result in instabilities that undergo similar evolutions. Of particular relevance is the case of a vortex sheet within a homogeneous fluid – that is, one in which $U_1 \neq U_2$ but $\rho_1 = \rho_2$. Equation (2.137) becomes

$$\sigma = -\frac{1}{2}ik(U_1 + U_2) \pm \frac{1}{2}k(U_1 - U_2). \tag{2.140}$$

In this case, one of the roots will always lead to exponential growth of the interfacial disturbance, with short wavelengths growing the fastest.

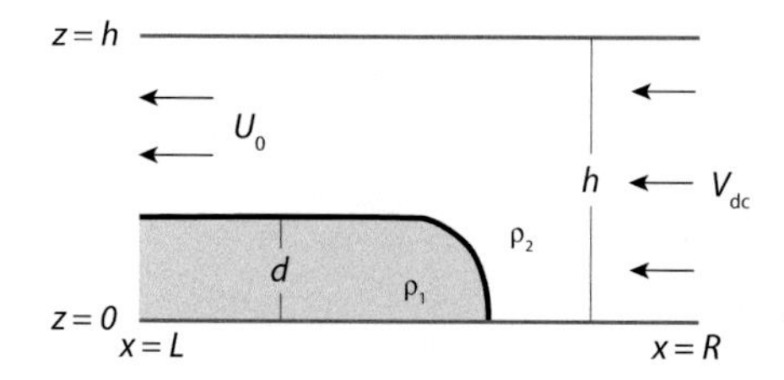

Figure 2.7 Depiction of the domain and relevant variables in the density current problem. Adapted from Bryan and Rotunno (2008) and Benjamin (1968).

Section 2.3.3 has served to illustrate the inherent instability of a vortex sheet – alone, and in the presence of buoyancy. In the next section, we focus only on the buoyancy, and examine the evolution of the dense fluid when it is initially beside the light fluid.

2.4 Density Currents

As an introduction to the basic dynamics of density, or gravity, currents, the reader is invited to imagine the following experiment: A long, rectangular channel contains a reservoir of motionless fluid of density ρ_2, separated by a divider from a motionless fluid of density $\rho_1(<\rho_2)$. After the divider is removed, the dense fluid flows beneath and displaces the light fluid. The horizontal advance of the dense fluid would continue indefinitely if the channel is infinitely long, the supply of dense fluid is unlimited, friction is zero, and the two fluids do not mix. This idealized, steady-state motion of a density current is what we now seek to mathematically model.[26]

For consistency with the idealization, assume a flow that is steady, inviscid, and in a 2D (x, z) domain as illustrated in Figure 2.7. The Coriolis force and diabatic heating are unnecessary for this problem, as is moisture. The density current depth d is generally assumed to be small relative to the density scale height H_ρ; thus, we make the Boussinesq approximation and also assume incompressibility.[27] The base state is hydrostatic and isentropic, and lacks vertical shear of the horizontal wind.

The resultant and relevant equations are

$$u\frac{\partial u}{\partial x} + w\frac{\partial u}{\partial z} = -\frac{\partial}{\partial x}\left(\frac{p'}{\rho_0}\right), \tag{2.141}$$

$$u\frac{\partial w}{\partial x} + w\frac{\partial w}{\partial z} = -\frac{\partial}{\partial z}\left(\frac{p'}{\rho_0}\right) + B, \tag{2.142}$$

$$u\frac{\partial \theta}{\partial x} + w\frac{\partial \theta}{\partial z} = 0, \tag{2.143}$$

$$\text{and} \quad \frac{\partial}{\partial x}(\rho_0 u) + \frac{\partial}{\partial z}(\rho_0 w) = 0, \tag{2.144}$$

where, respectively, the equations of motion follow from (2.16) and (2.29), the thermodynamic energy equation follows from (2.50) assuming dry air, and the continuity equation follows from (2.38). The buoyancy term in (2.142) is $B = -g\left(\rho'/\rho_0\right)$, but in this idealization, we let $B = -g' =$ const throughout the uniform ρ_2 fluid region comprising the density current, and $B = 0$ (and hence $\rho' = 0$) in the ρ_1 fluid region into which the current advances.

As shown in Figure 2.7, the equations are integrated over a *control volume* fixed with a current that moves at uniform speed V_{dc}. Control-volume depth h is chosen such that it is sufficiently far from the top of the current, and therefore that $w = 0$ along $z = h$. This means that the top boundary is impermeable, as also is the bottom boundary ($w = 0$ at $z = 0$). Both are also *free slip*, thus allowing flow along the boundary.[28] The right (R) and left (L) lateral boundaries have respective conditions of uniform horizontal velocity

$$\left.\begin{aligned} u(x = R, z) &= V_{dc} \\ w(x = R, z) &= 0 \end{aligned}\right\} \quad \text{for } z \le h \tag{2.145}$$

and

$$\begin{aligned} &\left.\begin{aligned} u(x = L, z) &= U_0 \\ w(x = L, z) &= 0 \end{aligned}\right\} \quad \text{for } d < z \le h \\ &\left.\begin{aligned} u(x = L, z) &= 0 \\ w(x = L, z) &= 0 \end{aligned}\right\} \quad \text{for } z \le d \end{aligned} . \tag{2.146}$$

There is no motion within the current itself in this reference frame. Given these conditions on u and w, (2.141) and (2.142) reveal that pressure at the lateral boundaries reduces to $\partial(p'/\rho_0)/\partial z = -g'$. Integrating, this is satisfied by

$$p'(x = R, z) = 0 \tag{2.147}$$

and by

$$p'(x = L, z) = \begin{cases} \rho'(z = h) + \rho_0 g'(d - z) & \text{for } z < d \\ \rho'(z = h) & \text{for } d \le z \le h \end{cases}. \tag{2.148}$$

The pressure at the upper boundary is unknown and can be eliminated as follows. Let us first obtain an equation governing energy conservation by taking $u \times$ (2.141) and adding this to $w \times$ (2.142):

$$u\frac{\partial E}{\partial x} + w\frac{\partial E}{\partial z} = 0, \tag{2.149}$$

where

$$E = \frac{u^2}{2} + \frac{w^2}{2} + \frac{p'}{\rho_0} - Bz, \tag{2.150}$$

which has the form of a *Bernoulli function*. With the free-slip condition, energy is conserved (thus, $E = \text{const}$) in this flow along surface ($z = 0$) streamlines. Let us choose a surface streamline that begins at the right boundary and then terminates at the leading edge or front of the density current. This termination point is also a *stagnation point* (SP), at which u must vanish. According to (2.150), and given conditions (2.145) and (2.147), it must be true that along this streamline,

$$\left(\frac{V_{dc}^2}{2}\right)_{x=R,z=0} = \left(\frac{p'}{\rho_0}\right)_{x=SP,z=0}.$$

Because there is no flow within the density current, it must also be true that

$$\left\{\frac{1}{\rho_0}\left[p'(z=h) + \rho_0 g' d\right]\right\}_{x=L,\,z=0} = \left(\frac{p'}{\rho_0}\right)_{x=SP,\,z=0}.$$

It follows that

$$p'(z=h) = \rho_0 \frac{V_{dc}^2}{2} - \rho_0 g' d,$$

and hence that condition (2.148) becomes

$$p'(x=L,z) = \begin{cases} \rho_0 \dfrac{V_{dc}^2}{2} - \rho_0 g' z & \text{for } z < d \\ \rho_0 \dfrac{V_{dc}^2}{2} - \rho_0 g' d & \text{for } d \le z \le h \end{cases}. \tag{2.151}$$

The control-volume approach, which we also will employ in Chapter 8, involves integration of the relevant equations over the domain indicated in Figure 2.7. Consider (2.141) written in flux form with the aid of (2.144):

$$\int_0^h \int_L^R \frac{\partial}{\partial x}(\rho_0 u u) dx\, dz + \int_L^R \int_0^h \frac{\partial}{\partial z}(\rho_0 u w)\, dz\, dx = -\int_0^h \int_L^R \frac{\partial p'}{\partial x} dx\, dz.$$

Recalling the conditions on w, this can be reduced immediately to

$$\int_0^h \rho u^2(x=R,z)\, dz - \int_0^h \rho_0 u^2(x=L,z) = \int_0^h p'(x=L,z)\, dz$$

and after substitution of the conditions on u and p',

$$\rho_0 V_{dc}^2 - \rho U_0^2 (h-d) = \left[\rho_0 \frac{V_{dc}^2}{2} d - \rho_0 g' \frac{d^2}{2}\right] + \left[\rho_0 \frac{V_{dc}^2}{2}(h-d) - \rho_0 g' d (h-d)\right]. \tag{2.152}$$

We seek a relationship between V_{dc} and the known quantities. In the preceding equation, U_0 is unknown, but we can eliminate it with the aid of the mass continuity equation, which is also integrated over the control volume

$$\int_0^h \int_L^R \frac{\partial}{\partial x}(\rho_0 u)\, dx\, dz + \int_L^R \int_0^h \frac{\partial}{\partial x}(\rho_0 w)\, dz\, dx = 0.$$

Owing again to the conditions on w, the second integral vanishes, leaving

$$\rho_0 V_{dc} h = \rho_0 U_0 (h - d).$$

Substituting this into (2.152), and then solving the result for V_{dc}, we find, after algebraic manipulation, that

$$V_{dc}^2 = g'd \left[\frac{(h-d)(2h-d)}{h(h+d)} \right] = g'd \left[\frac{(1-d')(2-d')}{(1+d')} \right], \tag{2.153}$$

where $d' = d/h$. The solution

$$V_{dc} = k\sqrt{g'd} \tag{2.154}$$

therefore depends on the depth of the density current relative to the domain (channel) depth. The theoretical upper limit in an incompressible flow is $d' \to 0$ (an infinitely deep channel), in which case $k = \sqrt{2}$.[29] The theoretical lower limit is $d' = 0.5$ and is known as the *energy-conserving case*, in which $k = 1/\sqrt{2}$.

Supplementary Information

For exercises and problem sets, please see www.cambridge.org/trapp/chapter2.

Notes

1 The derivations of these equations from "first principles" can be found elsewhere, such as in Holton (2004), Kundu (1990), and Batchelor (1967).
2 The complete form in the Navier-Stokes equations is given by the divergence of the deviatoric stress tensor; see, e.g., Batchelor (1967).
3 The Greek letter mu (μ) is often used to symbolize *dynamic* viscosity. For notational convenience, we use μ herein to denote *kinematic* viscosity, which is the dynamic viscosity divided by density.
4 Holton (2004).
5 The following analysis draws primarily from Bannon (1996).
6 See Bannon (1996) for a further discussion.
7 Thus, in the Boussinesq momentum equations, buoyancy is $-g\,(\rho'/\rho_0)$, the PGF becomes $-(1/\rho_0)\nabla p'$, etc., where ρ_0 is a constant reference density.
8 Bannon (2002).
9 To include the contribution of ice to c_{vm}, a term $c_i\, q_i$ is added to the right-hand side of (2.49), where c_i is the specific heat of ice water (=2100 J K^{-1} kg^{-1}). An analogous statement applies to c_{pm} in (2.50); Tripoli and Cotton (1981).
10 Bryan and Fritsch (2002).
11 Houze (1993).
12 Bryan (2008), Emanuel (1994).
13 See Bryan (2008), who offers an alternative to the formula given by Bolton (1980).
14 This generally follows the development of Cotton and Anthes (1989).

15 Bannon (2002).
16 This follows from Beer (1974).
17 Jacobson (2005) summarizes this concept nicely.
18 The mathematical development and application is based on Holton (2004).
19 An internal wave has its maximum amplitude *within* the fluid, in contrast to an external wave, which has its maximum amplitude at an external boundary, such as a free surface.
20 Much of this analysis is adapted from Holton (2004).
21 Holton (2004).
22 Crook (1988).
23 Following Houze (1993), although alternatively this may be referred to as *buoyant instability*.
24 Houze (1993), Emanuel (1994).
25 Lord Kelvin first posed and solved the mathematical problem in 1871, but Herman von Helmholtz put forward the physical problem in 1868; see Drazin (2002), whose mathematical development is followed herein.
26 This problem has long historical roots, with one of the more authoritative papers presented by Benjamin (1968). The analysis procedure used here, however, is that of Bryan and Rotunno (2008).
27 For density currents with depths of several kilometers, the anelastic approximation is more appropriate; see Bryan and Rotunno (2008).
28 The free slip condition is also a stress-free condition, such that the normal derivative of the tangential flow is zero. In the current problem, this equates to $\partial u/\partial z = 0$ at the upper and lower boundaries, for example.
29 See Bryan and Rotunno (2008), Klemp et al. (1994), and Benjamin (1968) to gain a sense for the range of possible solutions and associated comments.

3

Observations and Mesoscale Data Analysis

Synopsis: Chapter 3 introduces the different instrumentation for in-situ and remotely sensed observations, for operational/routine uses as well as special field collection. The placement of such instruments in observation networks is justified using sampling theory, and then is illustrated by example networks from past field programs. Finally, the spatial analysis of the data is motivated and described.

3.1 Introduction

This chapter provides an introduction to the data collection and analysis that underlie much of the discussion in subsequent chapters. For example, in Chapter 4 we will describe how observational data are used to provide the initial and boundary conditions for numerical weather prediction and simulation models. The characteristics of phenomena such as thunderstorm gust fronts (Chapters 5–6), supercells (Chapter 7), squall lines (Chapter 8), and mesoscale convective vortices (Chapter 9) are revealed by analysis techniques that extract the scale-relevant information obtained from appropriately configured observing systems.

The observations themselves are *discrete* samples of time- and space-*continuous* atmospheric variables. To illustrate this concept, consider the hypothetical time variation of a scalar such as pressure at a single observing station (Figure 3.1). Let this continuous signal in pressure, which in reality would be unknown, be instantaneously sampled at times $t = n\Delta$, where $n = 0, 1, 2, \ldots$; Δ is a uniform sampling interval determined in part by the capabilities of the instrument but otherwise is user-specified based on the process of interest.[1] Assume for the sake of this illustration that we unknowingly chose a Δ that results in three samples of a full oscillation of the pressure signal. It is possible to reconstruct the actual signal from these three discrete samples, suggesting that we were successful in observing the signal. Unfortunately, an infinite number of other signals can likewise be reconstructed from our samples,

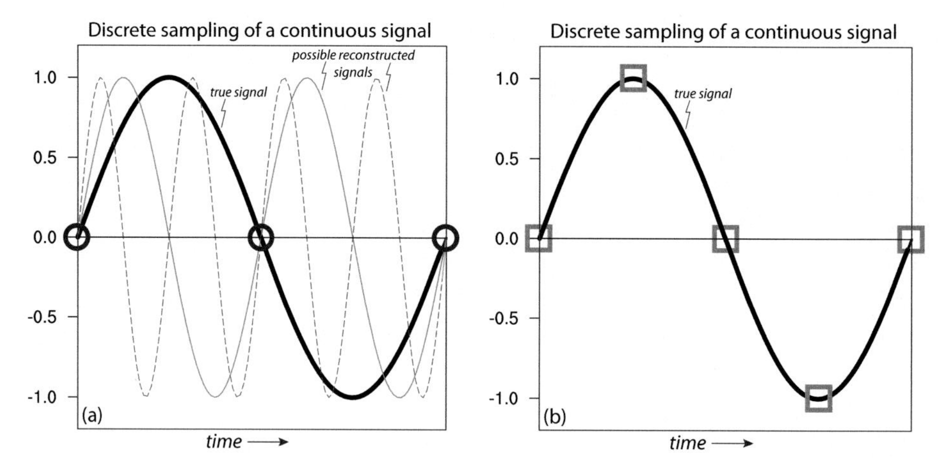

Figure 3.1 Illustration of the temporal sampling of a continuous scalar variable, such as pressure. Assume that the variable is observed at a single station, and discretely sampled at some interval Δ, as indicated in (a) by the three open circles. The solid bold curve represents the true signal. In (a), the other curves represent a few of the possible signals that could be reconstructed from the three samples. In (b), the open squares demonstrate how discrete sampling at interval $\Delta/2$ improves the reconstruction of the true signal.

suggesting instead that our choice for Δ was not necessarily the best one for this particular pressure signal.

Sampling theory provides additional insight into our illustrative example: The highest signal frequency that can be observed discretely without the deleterious effects of *aliasing* is limited by the *Nyquist frequency*, σ_N, where

$$\sigma_N = 1/2\Delta. \tag{3.1}$$

When spatial processes are sampled, the smallest wavelength is similarly limited by the *Nyquist wavelength*, λ_N, where

$$\lambda_N = 2\Delta. \tag{3.2}$$

Attempts to observe frequencies higher than σ_N (wavelengths smaller than λ_N) will result in their erroneous representation at lower frequencies (longer wavelengths) and hence in distortion of the true signal. Thus, (3.1) and (3.2) tell us a priori what we can, and cannot, be expected to resolve with a choice of Δ.

Returning to Figure 3.1 it should be evident that to more faithfully characterize a signal with a known frequency near σ_N, a better interval is $\Delta/2$, which yields five samples of the complete oscillation. Said a slightly different way, to observe a process or phenomenon with characteristic time T (or length L), we prefer an observing network or system that can sample at a rate

$$\Delta t \leq T/4 \tag{3.3}$$

or that has data spacing

$$\Delta x \leq L/4. \quad (3.4)$$

Hence, a datapoint spacing of 1 km would nominally resolve wavelengths of 4 km or larger; a sampling interval of 1 min would nominally resolve a temporal process of 4 min or larger (frequency of 0.25 min^{-1} or lower).

The hypothetical pressure signal in Figure 3.1 is simple in construction, but pressure and other variables in the real atmosphere are likely to be more complicated. Indeed, we may have only a vague understanding of the signal (or phenomenon) and its scale before observing it. Sampling at a relatively high rate is one strategy that could address this issue, although doing so within the constraints of limited observational resources can be challenging (see Section 3.5).

Measurements made at a high rate in space or time still do not always reflect the conditions immediately surrounding the measurement sites or times, however. Imagine how some highly localized difference in surface vegetation, or the disturbance caused by a moving vehicle, might lead to a locally anomalous measurement of temperature or wind, and hence skew the characterization of temperature or wind over a broader area. These *errors of representativeness* are pervasive in all types of measurements, as we will see later in this chapter.

The *precision* and *accuracy* of the measurements are additional considerations. Measurements made repeatedly in uniform conditions and yield the same result are precise. However, such measurements may or may not be accurate with respect to an accepted standard, as determined for example through tests of the instrument in a controlled laboratory setting. Accuracy is particularly important when a measurement from one instrument is used in tandem with similar measurements from other instruments. This is often the case in observing networks deployed during field campaigns or for other experimental purposes (see Section 3.5).

A final consideration is measurement *sensitivity*, which regards the measured change of output from a device given a measured change of input. A sensitive instrument would be able to detect and measure a relatively weak signal. Weather radars capable of making "clear-air" measurements have high sensitivity (Section 3.4.1).

With these characteristics in mind, let us now explore the means through which measurements are made.

3.2 In-Situ Observations

3.2.1 Surface

Atmospheric variables $\vec{V}_H (= u\hat{i} + v\hat{j})$, T, p, and relative humidity (RH) are measured routinely near the ground with instruments that have sensors in direct contact with the atmosphere. The instrumentation for such in-situ observations typically is mounted

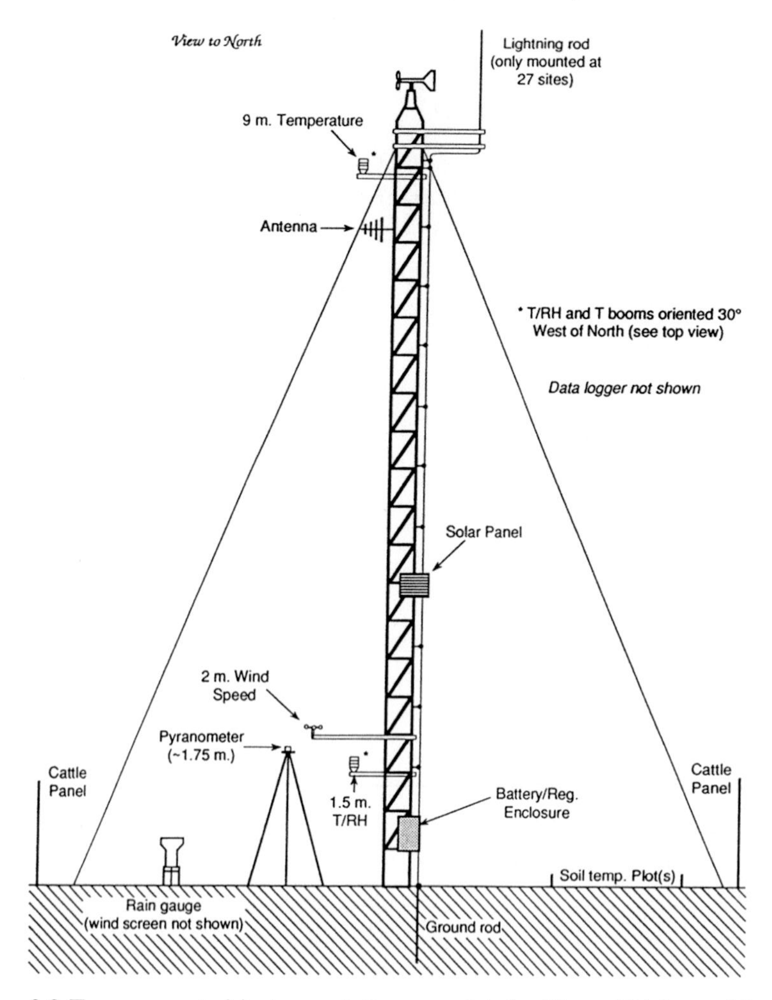

Figure 3.2 Tower-mounted instrumentation associated with an Oklahoma Mesonet surface observation station. The anemometer at the tower summit measures wind speed and direction at 10 m. Not shown in this schematic is the shelter that houses the barometer and other electronics. From Brock et al. (1995).

on a tower (Figure 3.2).[2] Optimally, the tower site is flat, locally homogeneous, and free of obstructions (trees, buildings, etc.) and anything else that could potentially cause representativeness errors. Meteorological observation towers have a standard height of 10 m, with wind measurements taken at the tower summit, and the remaining measurements made at ~1 to 2 m above ground level (AGL).[3] Rainfall (and perhaps snowfall) is measured using a gauge that is separated from the tower. Depending on the sophistication and purpose of the station, additional in-situ surface measurements may include incoming solar radiation, outgoing infrared (IR) radiation, soil temperature, and soil moisture.

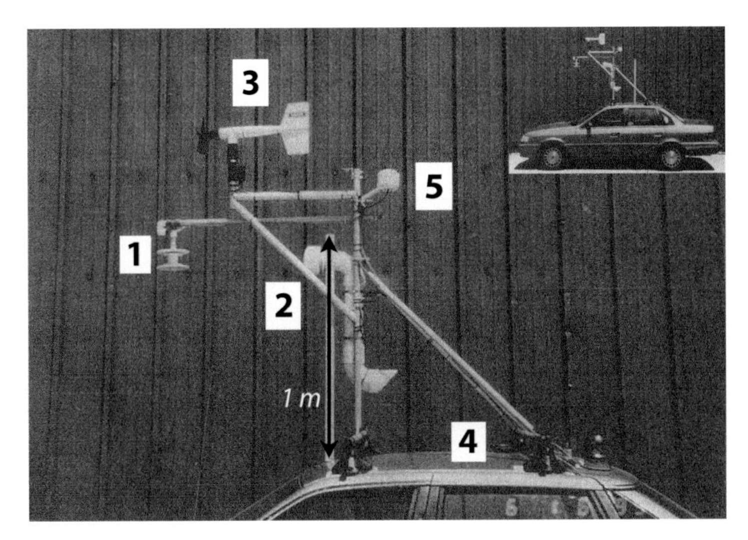

Figure 3.3 Mobile observing platform and associated instrumentation. Inset shows a complete view of the vehicle and instrument mast. In main photograph, labels correspond to: (1) pressure port, (2) aspirated weather shield for temperature and relative humidity sensors, (3) anemometer and vane, (4) GPS, and (5) flux-gate compass. A meter stick in both photographs gives a length reference. Relevant instruments are powered by a battery stored in the trunk of the vehicle, where the data logger also is stored. Real-time data are displayed via a laptop computer. From Straka et al. (1996).

Meteorological surface observations are now usually fully automated, with data collected and logged at intervals of seconds to minutes, and then transmitted to a remote storage/access computer via radio or some other form of communication. Observations of daily rainfall and daily maximum/minimum temperature are, however, still made manually across the United States and the world by volunteers as part of "cooperative" networks.

Although the platform shown in Figure 3.2 is permanent, portable and even mobile platforms have been developed for surface observations. Examples of portable platforms include a $\sim$1.5-m-tall surveyor tripod and a "turtle"-like housing that rests directly on the ground.[4,5] These platforms and their associated instrumentation are meant to comprise systems that are durable and easy to deploy, especially in adverse weather conditions. Their relatively low cost also allows for the construction of multiple systems that can then be used to form observing networks (see Section 3.5).

Similar design principles apply to mobile platforms, such as the automobile roof-mounted mast (Figure 3.3) that debuted during the Verification of the Origins of Rotation in Tornadoes Experiment (VORTEX) in 1994.[6] One advantage of this type of platform is that redeployment means simply driving to a new location. However, a corresponding disadvantage is that the observation sites are limited to roads or accessible pavements. If data are collected while the vehicle is actually moving, the vehicle motion vector, obtained using the Global Positioning System (GPS), must be removed from the wind measurement. Unfortunately, air displacements owing to nearby vehicles are difficult to remove, as are related effects on the other variables. Indeed, measurements made using mobile and portable platforms are more susceptible to these and other exposure errors as compared with measurements made on permanent (and presumably well-sited) platforms. Hence, these errors would need to be weighed against the benefits of nonpermanent platforms, which can be used to form novel, high-resolution observing networks (Section 3.5).

3.2.2 Upper-Air Observations

In-situ observations of the state of the atmosphere above the ground are most commonly made using a balloon-borne *radiosonde*. This instrument package contains sensors that measure pressure, temperature, and humidity.[7] The balloon and sonde are assumed to move with the wind; thus, $\vec{V}_H$ is determined from the sonde locations with respect to time during its ascent. Most sondes now are equipped with an inexpensive GPS receiver to provide 3D location. The GPS location includes sonde height, which also can be (and formerly was) computed using the hypsometric equation. The vertical sampling, and effectively the vertical resolution of the profile or *sounding*, depends partly on the measurement interval ($\sim$1 s or less) and on the balloon ascent rate. A typical ascent rate is 4 to 5 m s^{-1}, as controlled by the balloon mass and amount of helium inflation. A small radio transmitter is used to communicate the data to a ground receiving station.

The transmitter, GPS receiver, and other sonde components must operate on battery power for approximately three hours, to provide ample time for balloon ascent through the troposphere and into the stratosphere. Additional operational considerations include size, weight, durability, and cost: the sonde is assumed to have a single use, although retrieval of units is now better enabled with GPS location. Implicit in the consideration of durability is the fact that sondes experience a wide range of atmospheric conditions. Wetting and icing of sensors are known to lead to errors in RH, although improvements to radiosonde design are helping to mitigate such exposure issues.[8]

Radiosonde observations (or *raobs*) are particularly susceptible to errors of representativeness. Isolated clouds and areas of precipitation through which the radiosonde penetrate are one source. Another source is related to the fact that over the course of its ascent, the balloon (and sonde) can be displaced hundreds of kilometers from the launch location. The raob, therefore, does not yield a profile at and directly above the launch site, but rather describes the vertical structure of the atmosphere within some distance downwind of that site. This statement applies mostly to interpretation of a sounding as displayed on a thermodynamic diagram. When raobs are analyzed numerically, such as in the production of 3D gridded datasets (see Section 3.6) for numerical weather prediction, the geo-referenced sonde locations are used.

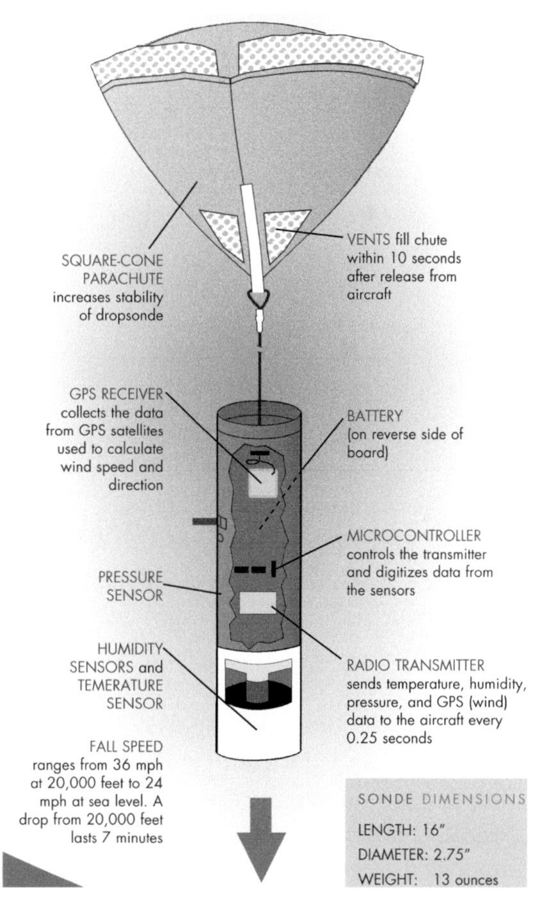

Figure 3.4 Schematic of the NCAR GPS dropsonde. Copyright © University Corporation for Atmospheric Research. Used with permission.

Mesoscale field campaigns often require high-frequency and/or finely spaced raobs, and this can be accommodated through launches from portable platforms (vehicle such as trucks or vans). The vehicles are typically: equipped with the electronics necessary to act as the ground receiving station; capable of transporting the expendables, such as the balloons, helium, and sondes; and instrumented so that the sensors can be calibrated before launch. *Dropsondes*, which essentially are radiosondes deployed from an aircraft flying at relatively high altitude, also serve this purpose. The basic operating principles are the same, except that the dropsondes collect data as they descend to the ground (or ocean surface) with the aid of a parachute (Figure 3.4).[9]

The rapid deployment of a succession of dropsondes from an airborne platform allows for in-situ sampling of a volume of the atmosphere over a relatively short time. This sampling time is partly a function of the sonde deployment rate, which is limited by the time required to prepare and then drop each sonde through the launch tube, as well as by the sonde descent rate and the number of sondes that

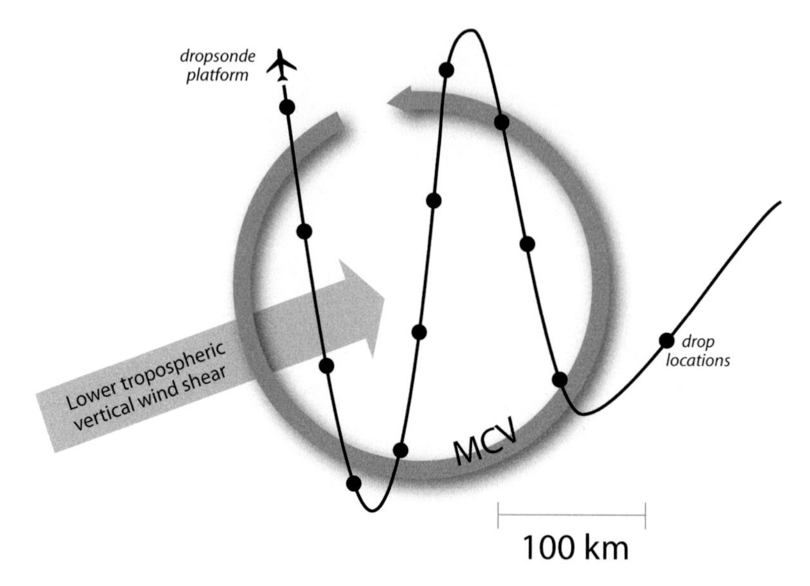

Figure 3.5 An example of a dropsonde deployment strategy employed during the Bow Echo and MCV Experiment (BAMEX). Adapted from Davis et al. (2004).

can be tracked simultaneously by the receiving system.[10] The aircraft speed is an additional factor because it is combined with the deployment rate to yield the drop spacing. As demonstrated in Figure 3.5, the drop spacing and the aircraft flight track itself are chosen based on the size and movement of the phenomenon of interest. For example, a "sawtooth" flight track was used during the Bow Echo and MCV Experiment (BAMEX) to observe the near environment and vertical structure of mesoscale convective vortices; this pattern offered logistical flexibility and spatial scalability.[11]

Naturally, there are issues related to airborne platforms that complicate atmospheric sampling, such as airspace restrictions, mission duration, cost, and safety. Some of these are at least eased with the use of unmanned aerial vehicles (UAVs). UAVs are piloted remotely, or programmed for autonomous flight, and can accomplish many of the same sorts of scientific missions as manned aircraft. A mission of particular relevance to this section is a spiral ascent or descent by the aircraft, which produces a vertical profile of the measured variables. The types of measurements depend on the payload capability. A UAV with a large capability ($\sim$1000 kg) is the Global Hawk (Figure 3.6). In addition to standard meteorological sensors, it can carry instrumentation for in-situ measurements of atmospheric chemistry and cloud microphysical properties; the Global Hawk can even serve as a dropsonde deployment platform.[12,13] The Aerosonde has a much smaller payload (5 kg), and currently carries the same meteorological sensors as in typical radiosonde units, although miniature instrumentation for other measurements could be accommodated.[14]

Figure 3.6 Image of the Northrop Grumman RQ-4 Global Hawk UAV operated by NASA. NASA photograph by Carla Thomas.

In-situ observations of the state of the atmosphere are not constrained to research platforms. Commercial aircraft, which are equipped with meteorological sensors to support their flight operations, also collect observations along the flight track and during takeoff and landing.[15] The takeoff/landing profiles are necessarily limited to near airports, and flight-track observations are concentrated along major commercial flight routes, although the coverage is nearly complete over the contiguous United States at typical flight levels.[16] After ingest and processing by the Aircraft Communications Addressing and Reporting System in the United States, these data become a critical supplement to the routine raobs.[17] As we will learn in Section 3.4, remote sensing provides another source of data that supplements, as well as complements, the in-situ datasets.

3.3 Report-Based Observations of Hazardous Convective Weather

Observations of severe and hazardous weather phenomena comprise a dataset that has been indispensible for mesoscale meteorological research, yet are associated with a unique set of issues deserving of the following discussion.[18] Our specific interest here is in occurrences of hail, convectively generated "straight-line" winds, and tornadoes. Because of the lack of an alternative and widespread means to unambiguously record these phenomena, their occurrence is established, with few exceptions, by witnesses through voluntarily issued reports.[19] One immediate concern with such observations is that they depend on the availability of potential witnesses, and thus are susceptible to changes in the density and geographical distribution of human population. Indeed, this is one reason that long-term trends in report-based observations are regarded as generally unreliable.[20] The recruitment and training of volunteer "storm spotters," education of the general public, and changes in reporting procedures and criteria have

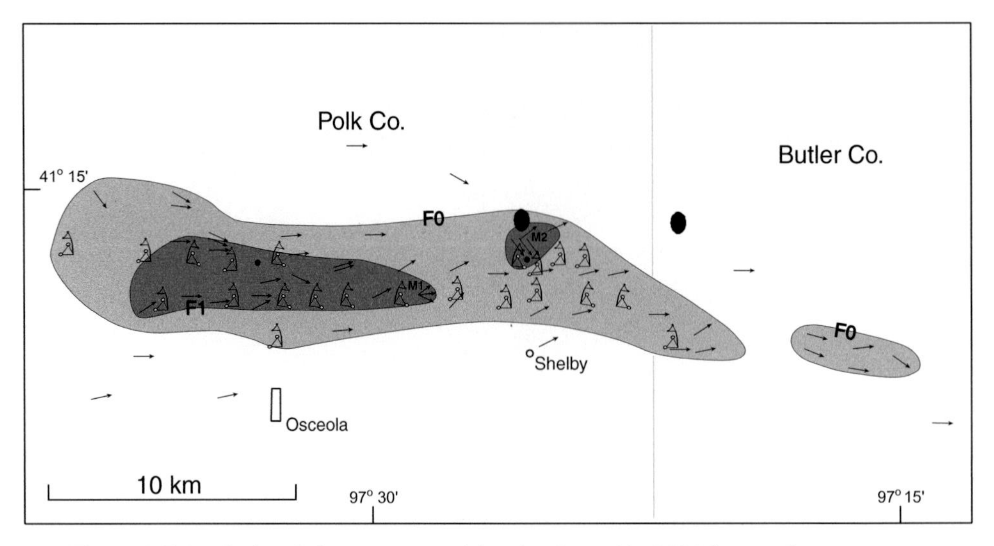

Figure 3.7 Analysis of damage caused by the June 10, 2003 bow echo event near Shelby, NE. The black ovals show the locations of severe wind reports in the official database. Shading and symbols are from an independent survey conducted by BAMEX personnel; areas of F0 (F1) damage are lightly (heavily) shaded. From Trapp et al. (2006).

all led to improvements in the observations during the recent decades, but consequently have also affected the long-term trends.

The reporting itself has different levels of active participation. In the United States, the observations serve the important purpose of verifying severe-weather warnings issued by the National Weather Service (NWS). Thus, there is effort by NWS personnel to solicit reports from presumed affected citizens, especially those known to provide reliable information; this does not preclude unsolicited reports, however. Both under- and overreporting are possible outcomes owing to this practice, and may misrepresent the actual scope of events (e.g., Figure 3.7).[21] In Europe, reports are contributed voluntarily to the European Severe Weather Database (ESWD).[22] In contrast to the U.S. system, the ESWD is supported outside the European national meteorological and hydrological services, the issuers of severe weather forecasts and warnings. This system is still subject to under- and overreporting, as can be imagined in the difference in contributed reports for a well-observed daytime event versus an overnight event.

Report-based observations are prone to errors not only in occurrence numbers, but also in geolocation and intensity. Consider the case of hail. The intensity metric is hailstone diameter, which is perhaps the most reliable of the human-estimated metrics, because the hailstone diameter can readily be measured, or at least compared to a common reference object (e.g., the size of a pea or a golf ball). Errors are possible, however, if the hailstone melts significantly before it is measured, or if it breaks apart on impact with the ground. Moreover, some fortuity is needed for an observer to locate the largest hailstone and thus properly assess the event by its maximum intensity. To begin to address these issues, dense networks of *hailpads* have been deployed around the world by various agencies and volunteer groups. Hailpads typically are squares of (foil-) covered, deformable material such as Styrofoam.[23] With proper calibration, indentations made on hailpads can be related to hailstone size and used to create size distributions.[24]

Damage to the natural and built environments is used most commonly to quantify the geolocation and intensity of straight-line wind and tornado events. Damage is related to wind speed through the Fujita (F) or "enhanced" Fujita (EF) scale and others like it;[25] these are empirical, but largely unconstrained by measurements. In the United States, the EF rating of an event is assigned by the NWS, based on information obtained either directly by NWS personnel from postevent damage surveys or from information supplied by event witnesses. Subjectivity is an inevitable part of the data-gathering and rating-assignment process. For instance the quality of construction must somehow be factored into the assessment of a damaged house, as also must the health of a damaged tree. Intensity assessment becomes complicated when windblown missiles cause damage not possible by wind loading itself. Finally, the damage-based intensity metric becomes ill posed, and obviously not representative, when the affected environment is devoid of vegetation or structures.

Fortunately, the potential exists to enhance the quantitative documentation of severe weather events with data from Doppler weather radar and weather satellites.[26] In Section 3.4, we explore the nature of the information provided through such remote sensing. As with the observations considered thus far, remotely sensed observations have unique advantages as well as limitations, but are used most effectively when combined with other data.

3.4 Remotely Sensed Observations

Remotely sensed observations are obtained either through active or passive sensing techniques. Weather radars and weather satellites, respectively, are the particularly relevant means of active and passive remote sensing.

3.4.1 Weather Radar

3.4.1.1 Overview of Basic Operations

Let us begin with a discussion of weather radar and its active remote sensing.[27] With a typical weather radar, short pulses of radiant energy are focused into a narrow beam and then transmitted through space in the form of electromagnetic waves with ~centimeter wavelengths. The *microwaves* travel at the speed of light ($c = 3 \times 10$ m s^{-1} in a vacuum) and are intercepted by meteorological targets. Only a small amount of the radiant energy incident on the targets is scattered back toward the radar, whereupon it is received during a "listening period" and then processed. As expressed through the radar equation, the (mean) microwave power P_r received at the radar is

$$P_r = \left[\frac{\pi^3 P_t G^2 \theta_b^2 c\tau \, |K|^2}{1024 \ln(2) \lambda^2} \right] \frac{z}{r^2} \tag{3.5}$$

and depends jointly on the radar-system characteristics (transmitted power P_t, radar antenna gain G, radar beam width θ_b, duration of transmitted pulse τ, and radar wavelength λ), radial distance or slant range r to the targets, and target properties $|K|^2$ and z.[28] K is the complex dielectric constant, and is a function of the composition and temperature of the targets as well as the radar wavelength; for typical weather radars and operating conditions, $|K|^2$ is 0.93 and 0.197 for water and ice targets, respectively.[29] z is the *radar reflectivity factor* in units of mm^6 m^{-3}, and depends on the size and concentration of the targets: if n perfect spheres of known diameters D fill a volume V sampled (or "illuminated") by the radar, then

$$z = \frac{1}{V} \sum_{i=1}^{n} D_i^6. \tag{3.6}$$

This is valid strictly for targets that are small compared with the radar wavelength, under which the *Rayleigh scattering approximation* applies. An important consequence of (3.6) is that the returned power is dominated by the largest scatterers, as in the case of a few large hailstones among a population of smaller raindrops.

Generally, the numbers, sizes, and even types of targets are not known, and therefore (3.5) is solved for z in terms of the power received, which is known. z is then expressed as a logarithmic radar reflectivity factor,

$$\mathrm{Z} = 10 \log_{10}(z), \tag{3.7}$$

with units of dBZ (decibels of reflectivity). For the sake of brevity, Z is often referred to as *radar reflectivity*, or simply *reflectivity*, but its origin from (3.6) should be understood.

Radar scans in azimuth and elevation produce fields of radar reflectivity factor that can be used qualitatively to infer precipitation structure. For example, the shape of the radar echo in scans at constant elevation angles (*plan position indicators* [PPIs]; Figure 3.8a) helps define the convective morphology (Chapter 6). Vertical scans at constant azimuth angles (*range-height indicators* [RHIs]; Figure 3.8b) are especially useful at revealing the depth and vertical structure of the precipitation regions. Even nonprecipitating echo structure can provide information on mesoscale-convective processes. Consider the radar *fine line*, which often depicts a mesoscale surface boundary (Figure 3.8c). The fine line is visible because of the existence of insects that tend to collect in the rising air currents, and/or to air density gradients and the associated changes in the index of refraction.

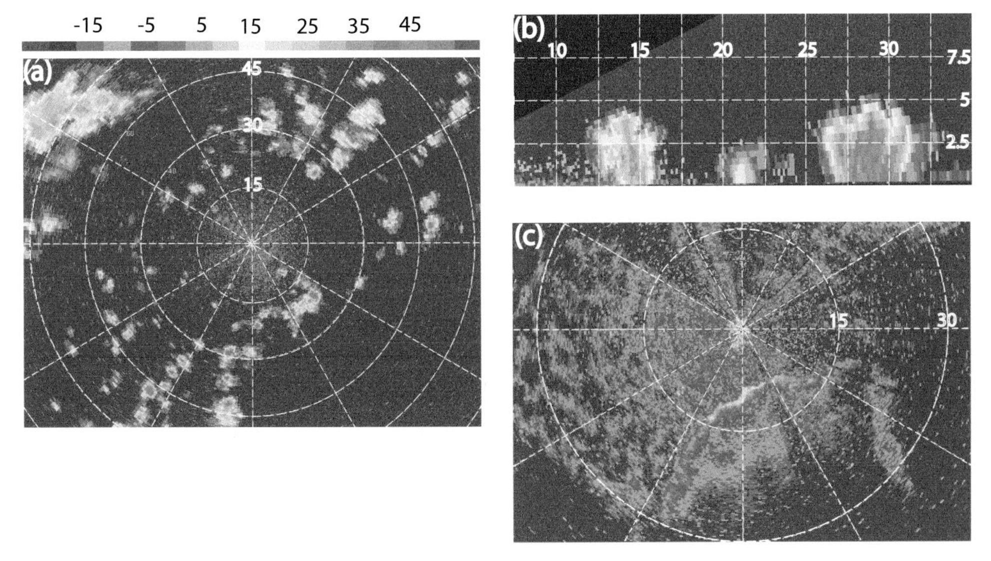

Figure 3.8 (a), (b) Example of radar reflectivity factor in a field a precipitating convective clouds, as displayed in (a) a PPI scan at 1° elevation, and (b) a corresponding RHI scan at 130° azimuth. (c) A fine line in radar reflectivity factor, associated with the sea-breeze front. As in (a), this PPI scan is at 1° elevation. All scans were collected in Florida in March 2012, using a Doppler on Wheels radar. (For a color version of this figure, please see the color plate section.)

Weather radar observations also provide quantitative information, the most notable of which is *rainfall rate* (R; mm h^{-1}). A vast number of empirical formulae have been introduced to relate radar reflectivity factor to rainfall rate, through power laws of the general form

$$z = a\, R^b. \tag{3.8}$$

An example is $z = 200R^{1.6}$, which was developed, and hence applies best, for estimation of stratiform-type rainfall.[30] Rainfall rate is overestimated via (3.8) in the presence of large hail (recall (3.6)) and radar "bright bands," which contain unrepresentatively high radar reflectivity due to melting hydrometeors (see Chapter 8). R is underestimated when relatively shallow clouds precipitate below the lowest radar beam; this is especially problematic for radars at installations in complex terrain.[31] Such effects are at least partially mitigated by automated algorithms, which also are designed to locally constrain the radar estimates with rain gauge data (rain gauge measurements themselves are susceptible to representativeness errors, however).[32]

Quantitative information about the (mean) motion of illuminated scatterers accompanies the backscattered power received by a Doppler weather radar. Specifically, pulsed Doppler radar transmits microwave pulses with a known phase, then compares this phase with the phase of the backscattered signal to determine shifts in frequency, and hence a Doppler velocity. The *pulse repetition time* (PRT), or its inverse the *pulse repetition frequency* (PRF) and radar wavelength combine to constrain the maximum Doppler velocity that can be determined unambiguously:

$$V_{\max} = \pm\frac{PRF\ \lambda}{4}. \tag{3.9}$$

Similarly, the PRF constrains the range or radial distance within which weather echoes can be detected unambiguously:

$$r_{\max} = \frac{c}{2\,PRF}. \tag{3.10}$$

These two equations show that a smaller (larger) PRF allows for a larger (smaller) maximum unambiguous range, but to the detriment (benefit) of the maximum unambiguous velocity. Because of this "Doppler dilemma," the PRF is a radar-system parameter that is usually adjusted according to the meteorological situation.

The Doppler velocity (V_r) is the component of the 3D scatterer motion in the direction of a radar beam. With assumptions about the motion orthogonal to the beam, fields of V_r can be used to characterize the precipitating flow. In the specific examples shown in Figure 3.9, the existence of rotation and divergence is deduced with an assumption of axisymmetry. The magnitudes of these respective kinematic measures are quantified through velocity differences across beams and along beams. For example, consider *differential velocity*,

$$\Delta V_r = V_{r,\max} - V_{r,\min}, \tag{3.11}$$

which forms the basis for mesocyclone and tornado detection using Doppler radar.[33] Here, $V_{\mathrm{r,max}}$ and $V_{\mathrm{r,min}}$ are the maximum and minimum Doppler velocities, respectively, in the V_r couplet associated with the vortex (see Figure 3.9a). An analogous measure for radial divergence (Figure 3.9b) forms the basis for microburst detection.

3.4.1.2 Retrievals from Multiple Doppler Radars

Nearly simultaneous scans by two or more (non-colocated) Doppler radars allow for the retrieval of the complete 3D winds in a spatial domain. The multiple-Doppler wind retrieval technique makes use of spherical geometry to relate the Doppler velocities of the radars to Cartesian wind components: The Doppler velocity measured at some "point" in space by radar i is

$$V_{r,i} = u\left(\frac{x - x_i}{r_i}\right) + v\left(\frac{y - y_i}{r_i}\right) + W_p\left(\frac{z - z_i}{r_i}\right), \tag{3.12}$$

where (x, y, z) is the Cartesian location of the point, (x_i, y_i, z_i) is the Cartesian location of the radar, and $r_i = [(x - x_i)^2 + (y - y_i)^2 + (z - z_i)^2]^{1/2}$ is the slant range. Equation (3.12) makes use of the coordinate transformations

$$\begin{aligned} x &= x_i + r_i \sin(az_i)\cos(el_i) \\ y &= y_i + r_i \cos(az_i)\cos(el_i) \\ z &= z_i + r_i \sin(el_i), \end{aligned} \tag{3.13}$$

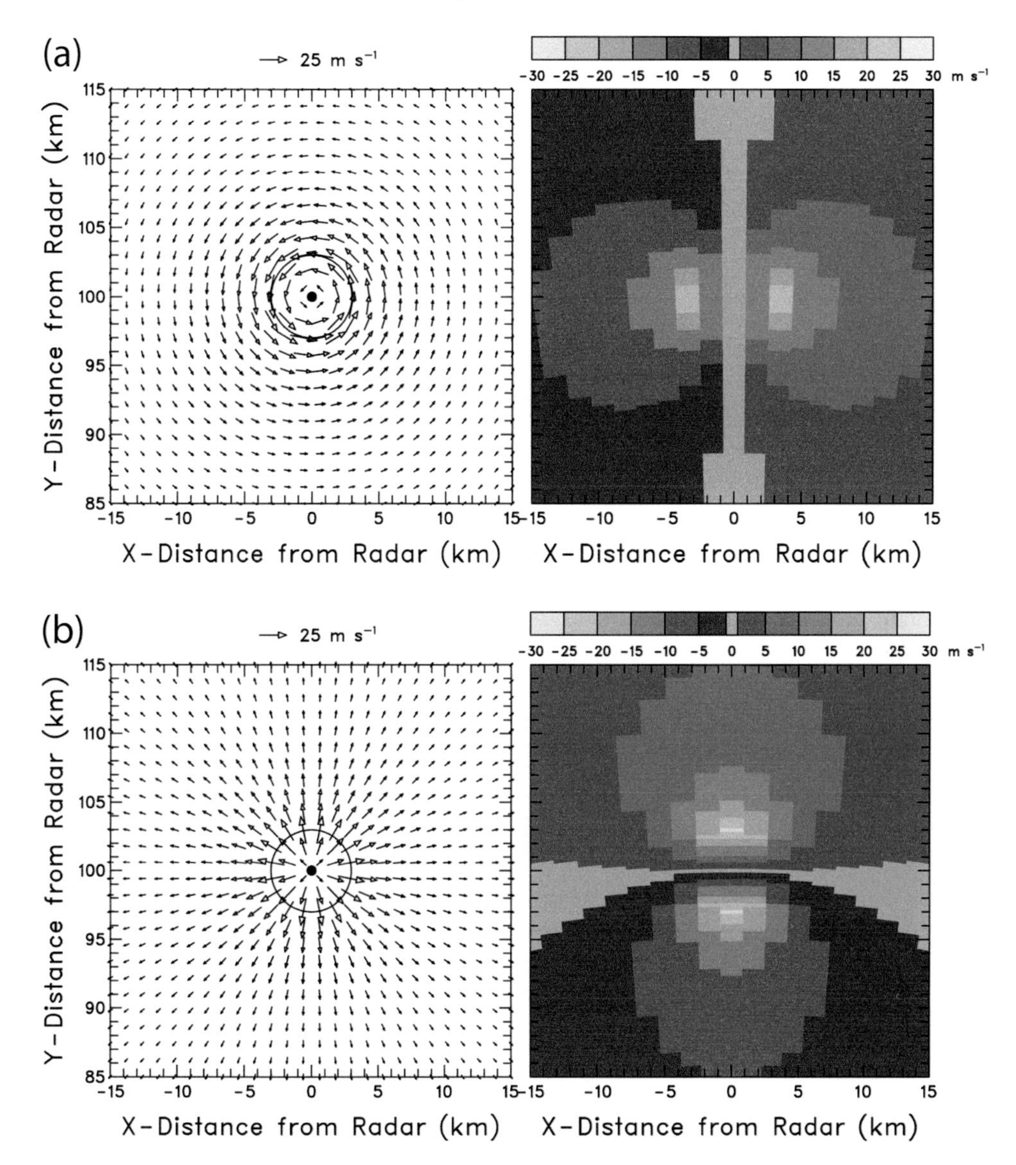

Figure 3.9 Low-level patterns of axisymmetric (a) vertical rotation and (b) horizontal divergence, in a vector wind field and corresponding field of Doppler velocity V_r. The circles in the vector wind fields indicate the relative location of the maximum winds. The black dot shows the center of the vortex in (a), and center of divergence in (b). The (simulated) radar is located 100 km to the south of the rotation and divergence centers. From Brown and Wood (2007). (For a color version of this figure, please see the color plate section.)

where *el* is the radar elevation angle (with respect to a plane tangent to the radar location) to a point in space, and *az* is the radar azimuth angle (with respect to north) (see Figure 3.10). In a precipitating atmosphere, the radar senses $W_p = w + V_f$, or the vertical airspeed plus the particle fallspeed V_f; the contribution from fallspeed can be estimated using empirical, reflectivity-based relationships.[34]

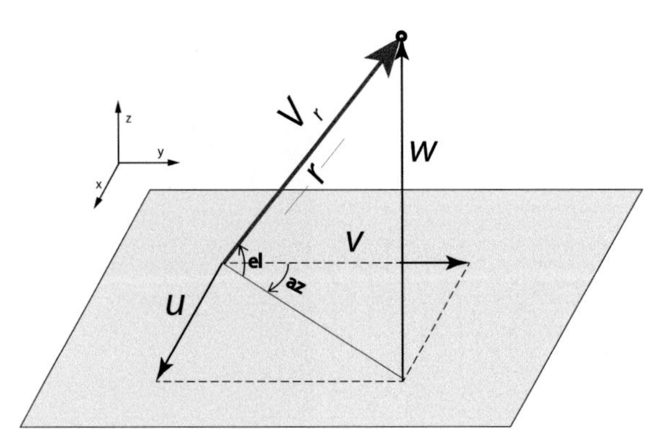

Figure 3.10 Illustration of the geometric relationship between Doppler velocity V_r and Cartesian velocity components u, v, and w. Indicated are the elevation angle (el), azimuth angle (az), and slant range (r).

A separate equation (3.12) for each radar allows us to solve for unknowns u, v, and w in terms of the known quantities (V_r, r, el, az, or V_r, x, y, z). However, because the beams from the i radars tend not to be mutually orthogonal, the $V_{\mathrm{r,i}}$ measured by the radars cannot uniquely determine the 3D wind vector. This is necessarily the case with a system of four or more radars, which results in an *overdetermined* problem. Thus, we can at best estimate the 3D wind vector.

Consider an estimation that minimizes error E in the least-squares sense:[35]

$$J = \sum_i \mathrm{E}^2 = \sum_i (a_i u + b_i v + c_i W_p - d_i)^2, \tag{3.14}$$

where a_i, b_i, c_i denote the respective geometric coefficients on u, v, and W_p in (3.12), and $d_\mathrm{i} = V_{r,i}$. Following this *least-squares minimization* procedure, we set

$$\partial J/\partial u = 0, \quad \partial J/\partial v = 0, \quad \partial J/\partial W_p = 0 \tag{3.15}$$

to yield the "normal equations"

$$\begin{aligned} u\sum_i a_i^2 + v\sum_i a_i b_i + W_p \sum_i a_i c_i &= \sum_i a_i d_i \\ u\sum_i a_i b_i + v\sum_i b_i^2 + W_p \sum_i b_i c_i &= \sum_i b_i d_i \\ u\sum_i a_i c_i + v\sum_i b_i c_i + W_p \sum_i c_i^2 &= \sum_i c_i d_i. \end{aligned} \tag{3.16}$$

Equations (3.16) can be rewritten using matrix notation

$$\mathbf{A}\vec{V} = \vec{D} \tag{3.17}$$

and then solved by inverting matrix $\mathbf{A}$ so that

$$\vec{V} = \mathbf{A}^{-1}\vec{D}. \tag{3.18}$$

A typical procedure is to interpolate coefficients a_i, b_i, c_i, and d_i to a uniform Cartesian grid (see Section 3.6) and then compute u, v, and W_p on this grid, following (3.18). The solution is constrained in various ways by the empirical estimate of V_f and/or by the anelastic continuity equation

$$\frac{\partial(\overline{\rho}w)}{\partial z} = -\overline{\rho}\left(\frac{\partial u}{\partial x} + \frac{\partial v}{\partial y}\right). \tag{3.19}$$

(see Chapter 2).

In the more common case of a two-radar, or "dual-Doppler," system we have fewer equations than variables, and thus an *underdetermined* system. Here we set

$$\partial J/\partial u = 0, \quad \partial J/\partial v = 0, \tag{3.20}$$

which yield the following normal equations:

$$\begin{aligned} u\sum_i a_i^2 + v\sum_i a_i b_i &= \sum_i a_i d_i - W_p \sum_i a_i c_i \\ u\sum_i a_i b_i + v\sum_i b_i^2 &= \sum_i b_i d_i - W_p \sum_i b_i c_i. \end{aligned} \tag{3.21}$$

In matrix form, these become

$$\begin{bmatrix} A_1 & B_1 \\ A_2 & B_2 \end{bmatrix} \begin{bmatrix} u \\ v \end{bmatrix} = \begin{bmatrix} D_1 - W_p C_1 \\ D_2 - W_p C_2 \end{bmatrix}. \tag{3.22}$$

When inverted, (3.22) is

$$\begin{aligned} u &= \frac{1}{\det}\left[B_2 D_1 - B_1 D_2 + W_p (B_1 C_2 - B_2 C_1)\right] \\ v &= \frac{1}{\det}\left[A_1 D_2 - A_2 D_1 + W_p (A_2 C_1 - A_1 C_2)\right], \end{aligned} \tag{3.23}$$

where $\det = A_1 B_2 - A_2 B_1$, $A_1 = a_1^2 + a_2^2$, $B_1 = a_1 b_1 + a_2 b_2$, and so on. Given an estimate of V_f, this system can be solved iteratively on a grid as follows: (1) compute u and v at some height from (3.23) using an initial estimate for W_p, (2) determine w at the current height from integration of anelastic continuity equation (3.19) over the previous height interval, and then (3) refine u and v using the new value of w. The cycle is repeated at the current height until the solutions for u, v, and w have sufficiently converged; thereafter, the velocity components are obtained at the next height. With this iterative procedure, the solution is especially sensitive to the direction of integration of (3.19) (i.e., "upward" or "downward"), and to other details such as the specification of integration limits.

Other procedures for multiple Doppler wind retrieval also exist, all motivated by goals of solution efficiency and retrieval error reduction, particularly in terms of vertical velocity.[36] The retrieval errors themselves owe to errors inherent in the original $V_{r,i}$ and associated with the geometric configuration of the radar network. Indeed, as discussed in Section 3.5, the design of a multiple radar network involves finding the acceptable compromise between the area of retrievable winds and the magnitude of the geometric error.

The resultant gridded 3D wind fields provide a means to obtain the corresponding 3D pressure and buoyancy fields.[37] To illustrate a basic approach of such *thermodynamic retrieval*, consider the x- and y-components of the equation of motion, written as

$$\partial p/\partial x = G_x$$
$$\partial p/\partial y = G_y, \tag{3.24}$$

where G_x and G_y contain the advection and tendency terms, in addition to other right-hand side terms (see (2.1)) as deemed appropriate for the particular application; each of these terms is evaluated at some height using the retrieved wind fields. Because the retrieved wind fields, and thus G_x and G_y, are not error free, Eqs. (3.24) are not solved directly, but rather in a least-squares sense:

$$J = \int\int [(\partial p/\partial x - G_x)^2 + (\partial p/\partial y - G_y)^2] dxdy, \tag{3.25}$$

in which we seek to minimize J. This is a standard problem in the calculus of variations, and yields an Euler-Lagrange equation

$$\partial^2 p/\partial x^2 + \partial^2 p/\partial y^2 = \partial G_x/\partial x + \partial G_y/\partial y, \tag{3.26}$$

which the reader may recognize as a Poisson equation in pressure. With appropriate boundary conditions (usually in the form of a Neumann condition), (3.26) is solved using a numerical method.[38] A unique solution for pressure is determined by subtracting off a horizontal average. Upon repeating the retrieval process at all grid levels, the vertical pressure gradient is then computed, so that buoyancy B can be retrieved using the vertical component of the equation of motion:

$$B = G_z, \tag{3.27}$$

where G_z contains the vertical pressure gradient force as well as the advection, tendency, and other relevant terms (e.g., see (2.18)). As for pressure, a unique solution for buoyancy is determined by subtracting off a horizontal average.

3.4.1.3 Mobile Radar Systems

The retrieval techniques and other radar applications discussed previously were developed largely for fixed-site weather radars, but apply equally well to data collected by *mobile radars*. Consider the mobile radar systems that are truck mounted and thus ground based.[39] These operate in the same basic way as do fixed-site systems, although they naturally have special design requirements, the foremost of which is physical size: To be mobile – specifically, for the platform and radar to meet typical road width and clearance restrictions – the antenna-reflector must have a relatively small ($\sim$2–3 m) diameter d_a (or otherwise be collapsible). The antenna diameter must be matched with beamwidth and wavelength, because all are interrelated through $\theta_b \sim \lambda/d_a$. Thus, if a narrow ($\sim 1^\circ$) beam is desired, as is often the case, transmission at a relatively short wavelength such as 3 cm is necessary.[40] Short wavelengths suffer relatively higher signal attenuation in precipitation, which means that the possible applications of the radar system are limited by the design requirements. For instance, a 3-cm mobile radar would not be very well suited to sample a large mesoscale convective system in its entirety, owing to the rapid loss of the transmitted signal through the system-generated rainfall. On the other hand, the mobility allows for the possibility of continuous data collection, at close range, on the phenomenon of interest. This is advantageous when high-resolution sampling of relatively rare and intermittent phenomena, such as tornadoes, is desired. The high-resolution sampling can even be extended to pairs of mobile radars that, if deployed appropriately, can gather data suitable for the application of dual-Doppler wind retrieval techniques. One of the challenges of this strategy, and to the successful use of mobile radars in general, is the availability of suitable roads, and moreover, of deployment sites that are relatively flat and free of trees, terrain, and other beam-blocking obstacles.

Although radar systems on airborne platforms are susceptible to airspace restrictions (see Section 3.2.2), they are immune to these road- and deployment-site challenges. The airborne Doppler radar shown in Figure 3.11a operates at 3-cm wavelength and consists of tail-mounted flat dish antennas that are configured to scan quasi-vertically, over a full 360° (or smaller-angle sectors).[41] Tail-mounted airborne radars collect data in beams at some angle to the aircraft flight track (Figure 3.11b), comprising scans like that given in Figure 3.12; this particularly striking example reveals the recirculation of hydrometeors within a developing hailstorm. As indicated in Figure 3.11b, the along-track distance Δ between such scans depends on the aircraft speed and antenna rotation rate. Figure 3.11b also indicates that tail-mounted radar data can be collected in beams directed fore and then aft relative to the flight track. The fore-aft data from these *pseudo* dual-Doppler radar scans can then be combined to retrieve the 3D wind vector in the same basic way as are data from separate radars (Figure 3.11b). Potential error in the retrieved wind arises from the fact that the fore-aft scans sample the same physical space at different times, implying the possibility of nonadvective evolution of the phenomenon during sampling. Other radar systems, such as *millimeter-wavelength* polarimetric radars, have also been used successfully on airborne platforms.[42] The short wavelength has particular applicability for studies of cloud-scale properties and processes.

3.4.1.4 Polarimetric Weather Radar

The polarimetry of this millimeter-wavelength radar, and of longer-wavelength radars including the Weather Surveillance Radar-1988 Doppler (WSR-88D) in the United States, regards the radar's capability to transmit and receive microwaves in orthogonal

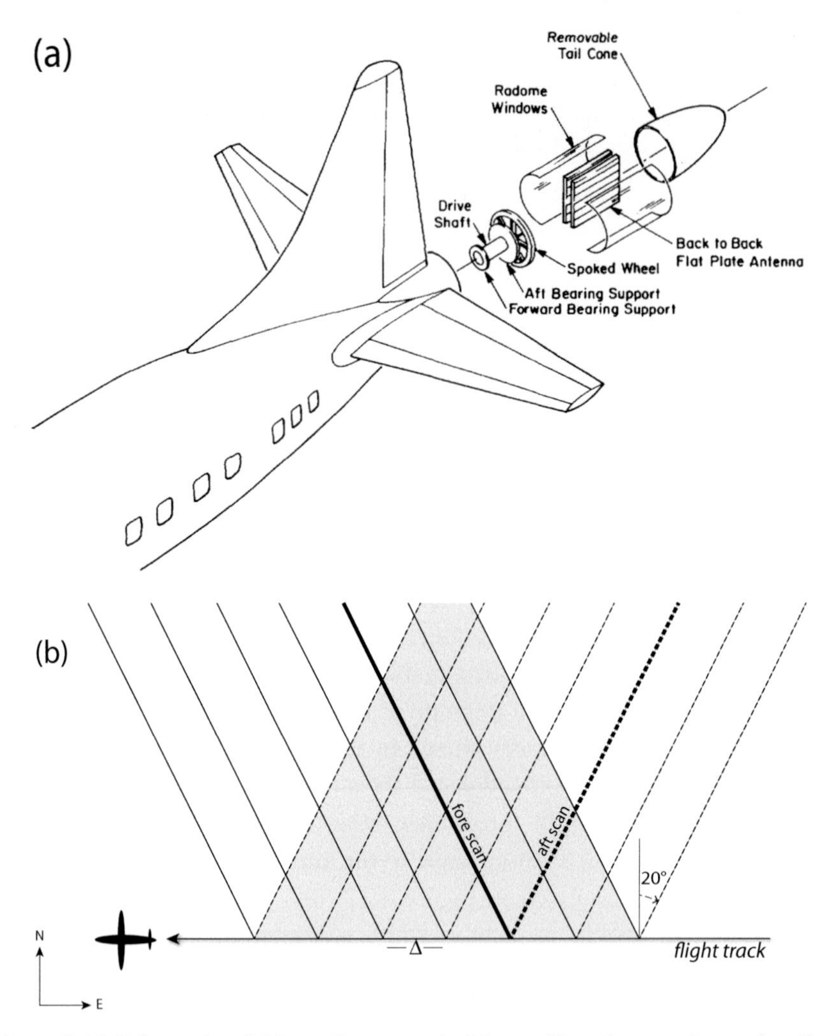

Figure 3.11 Schematic of (a) a tail-mounted airborne Doppler weather radar (from Hildebrand et al. 1994; used with permission), and (b) the strategy used to collect "pseudo-dual-Doppler" data with an airborne Doppler radar (adapted from Dowell et al. 1997). In (b), lines show the projection of the quasi-vertical scans in the horizontal plane of the flight track (denoted by the arrowed line). Solid (dashed) lines are scans aft (fore) of the track. In this particular application, the beams in these scans intersect at 40°; the area of intersections, and hence of pseudo-dual-Doppler coverage, is shaded.

planes of polarization. A key benefit of this capability, as demonstrated in Figure 3.13, is that it supplies information about the hydrometeor aspect ratio. Knowledge of the aspect ratio, in turn, can be used to discriminate (albeit with some ambiguity) between hydrometeor types; numerous possible applications follow, not the least of which is a means to improve rainfall estimation. Indeed, the issues of hail and bright-band contamination encountered with $Z - R$ relationships (Eq. (3.8)) are mitigated with the

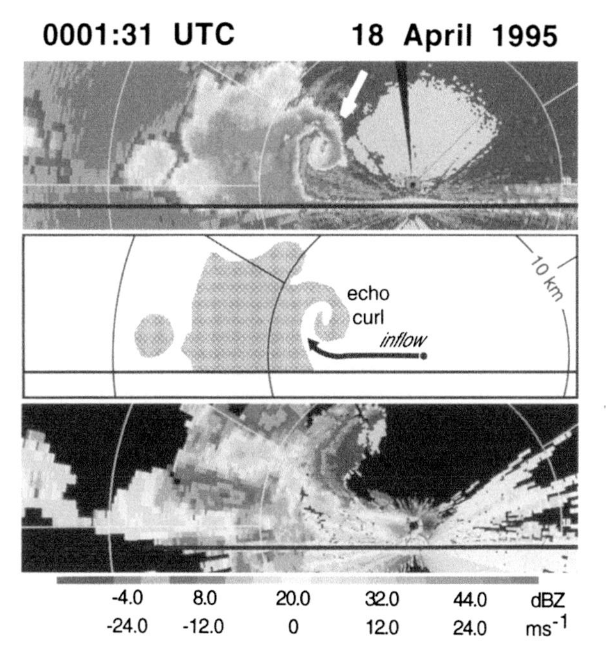

Figure 3.12 An example of airborne Doppler radar scan through a developing hailstorm. The top panel shows radar reflectivity factor, the bottom panel is of Doppler velocity, and the middle panel gives a physical interpretation. From Wakimoto et al. (1996). (For a color version of this figure, please see the color plate section.)

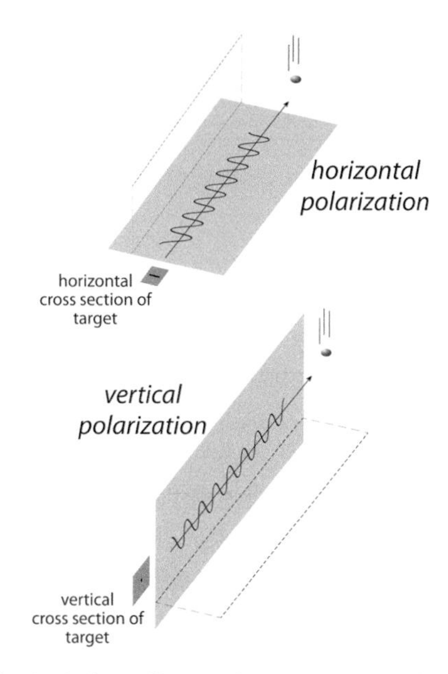

Figure 3.13 Schematic depiction of how microwaves at horizontal and vertical polarization would detect an oblate spheroidal raindrop. The plane of polarization (shaded) is that of the electric field.

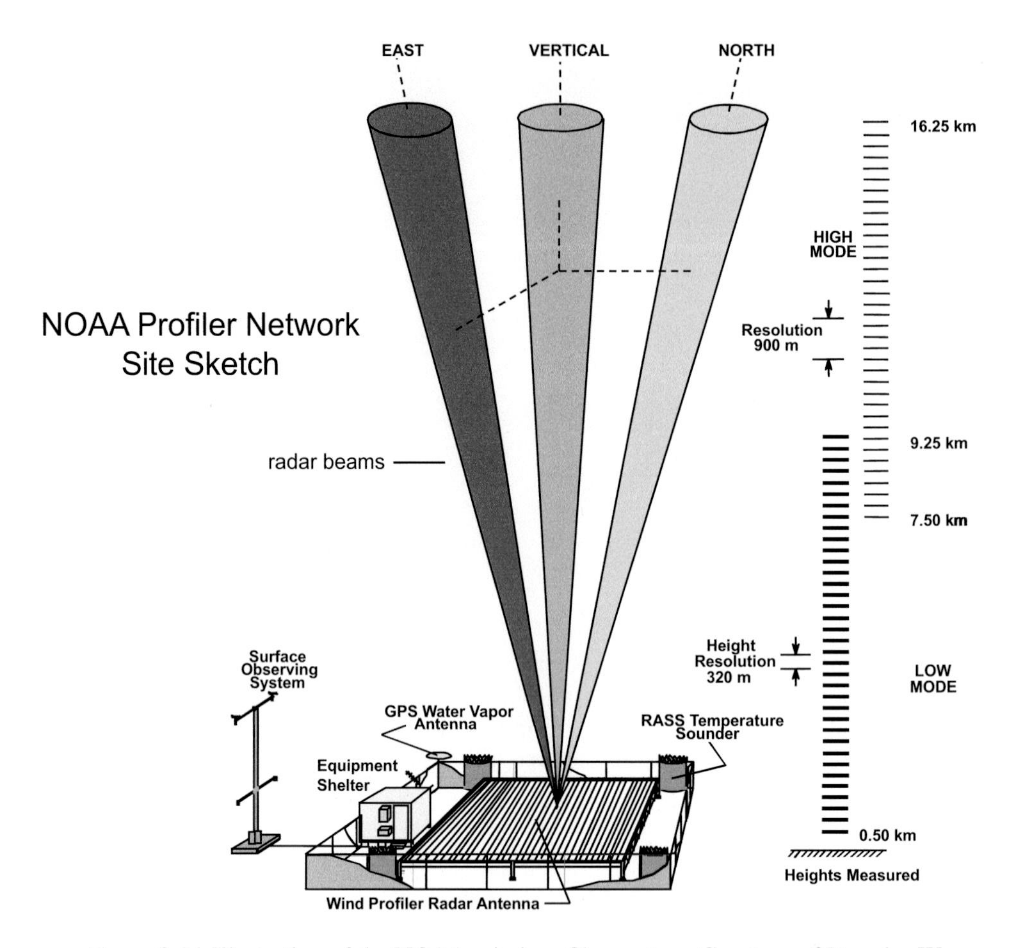

Figure 3.14 Illustration of the NOAA wind profiler system. Courtesy of Douglas W. van de Kamp, NOAA.

use of polarimetric radar measurands.[43] As will be seen in future chapters, knowledge of hydrometeor type and size is essential to a complete understanding of convective processes such as precipitating downdrafts and their associated pools of chilled air.

3.4.1.5 Wind Profilers

We conclude this section with a discussion of radar wind profilers. These are pulsed Doppler radars with fixed beams that are controlled by a coaxial antenna array rather than a parabolic reflector (or dish). Figure 3.14 shows an example of a three-beam configuration: a vertical beam, a beam at 73.7° elevation angle directed toward local north, and another at 73.7° elevation angle but directed toward local east. Power is transmitted in these relatively wide beams ($\theta_b \sim 4.5°$) at common operating frequencies of 915 MHz and 404 MHz, corresponding to wavelengths of 33 cm and 74 cm, respectively.[44,45] Relatively long wavelengths facilitate detection of gradients of

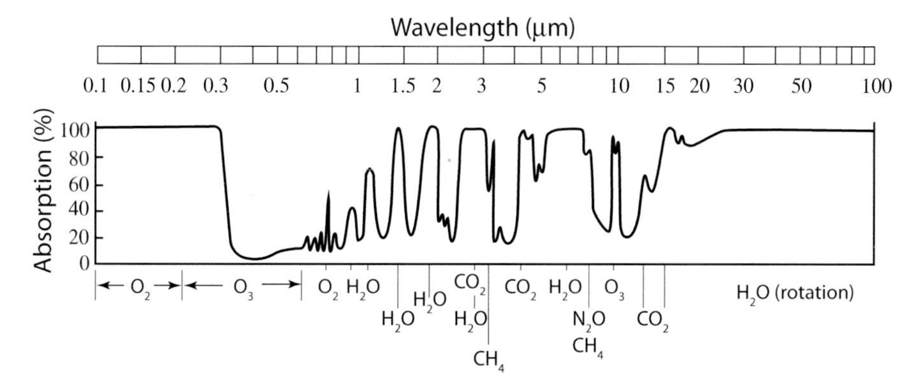

Figure 3.15 The spectrum of atmospheric absorption for a beam of solar radiation reaching the ground. After Goody and Yung (1989). Used by permission of Oxford University Press, Inc.

index of refraction in clear air, which produce the backscatter received by profilers. The refractive index gradients owe to turbulent eddies, assumed to move with the mean wind. Geometric relations such as (3.12) are used to convert the Doppler velocity in the fixed beams to Cartesian wind components. Thus, the wind profile is a retrieved horizontal wind at height increments controlled by the pulse duration; the NOAA Profiler Network in the United States provides hourly averaged tropospheric wind profiles, at 250-m height increments. For daily forecast operations, these data are considered a valuable supplement to the profiles determined from twice-daily in-situ measurements made with the radiosondes.[46] A similar comment can be made regarding the utility of portable profiler systems for experimental data collection.[47]

3.4.2 Weather Satellites

3.4.2.1 Overview of Basic Operations

Space-borne observing systems exploit the fact that electromagnetic radiation emitted at certain wavelengths by clouds and the Earth's surface suffers little absorption by the atmosphere (Figure 3.15). This is the case for, radiation at wavelengths in the visible (VIS) part of the electromagnetic spectrum (0.4–0.7 μm), which consequently provides information about visible cloud and surface attributes when sensed in space. Radiation sensed at infrared (IR) wavelength intervals (or *channels*) at ~3.5 to 4.5 μm and ~8.5 to 12.5 μm similarly helps quantify effective surface and cloud temperature. In contrast, the strong *absorption* of wavelengths in the IR channel at ~5 to 7.5 μm is exploited to help characterize atmospheric water vapor (WV).

The effective spatial and temporal resolution of this information depends in part on the characteristics of the satellite's orbit.[48] *Near-polar-orbiting satellites* are sunsynchronous and pass over the same location only twice per day. Their ~850-km altitude,

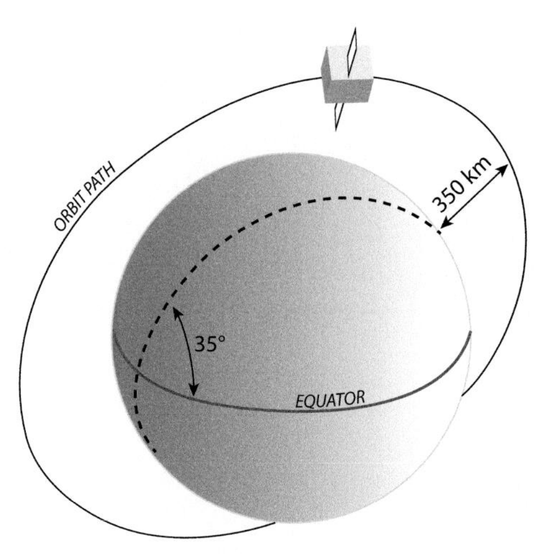

Figure 3.16 Orbital path and parameters of the Tropical Rainfall Measurement Mission (TRMM) satellite. Based on a NASA graphic.

or low Earth orbit, results in a relatively small geographic field of view at nadir. *Geostationary satellites* have altitudes of ~36,000 km, but their geosynchronous orbit above the equator allows continuous observations of all locations within a roughly hemispheric viewing area, and accordingly renders them well suited to monitor the time evolution of atmospheric phenomena including convective storms. The *Tropical Rainfall Measurement Mission (TRMM) satellite* has an altitude of 403 km, and a non-sunsynchronous orbit with an inclination of 35° to the equator (Figure 3.16). The path limits of ± 35° latitude allow the TRMM satellite to complete one orbit every 92.5 min, or 16 orbits per day.[49] A "precession" to the orbit further allows TRMM to sample or "visit" a given averaging box a different hour during the diurnal cycle, every ~23 days.

Although the payload of the TRMM satellite includes an instrument for active remote sensing (a precipitation radar [PR], discussed later), it and other weather satellite systems are designed largely for passive remote sensing, and thus for signal reception. An example of a passive sensing instrument is a *radiometer*, which receives and measures the (monochromatic) radiance. Satellite-borne radiometers such as the Advanced Very High Resolution Radiometer (AVHRR) are passive-sensing instruments, but do participate in the radiance data collection by physically scanning in lines across the subsatellite track (Figure 3.17). The result is that satellite images are comprised of sequences of scan lines, which have segments known as *scan spots* (or pixels).

The AVHRR is the type of imaging radiometer typically found on operational polar-orbiting satellites. The I-M and N/O/P series Geostationary Operational Environmental Satellites (GOES) carry the Imager, another imaging radiometer.[50] When obtained in a rapid-scan mode (as frequent as 1 min, over small geographic areas), the VIS data from Imager Channel 1 (0.55–0.75 μm) provide a captivating, high-resolution (1 km)

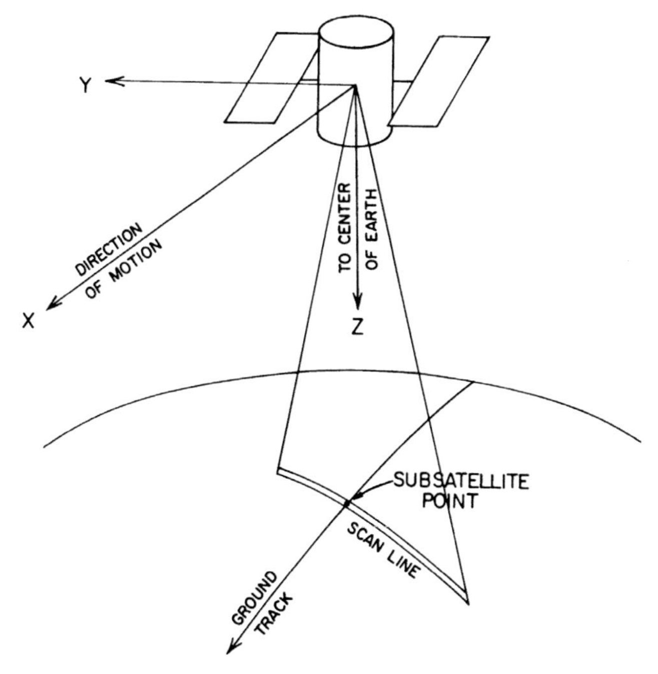

Figure 3.17 Illustration of data collection along weather-satellite scan lines, as a function of the Earth-relative motion. From Kidder and Vonder Haar (1995).

portrayal of cloud evolution and cloud-top structure. Imager Channels 4–5 are the IR channels, with 4-km resolution pixels, and Channel 3 is the WV channel, with 8-km resolution.

The GOES I-M and N/O/P also carry the Sounder, a separate radiometer designed to obtain measurements from which vertical profiles of temperature and moisture at scan spots can be retrieved. These measurements are still of radiance, but over multiple channels (19 for the Sounder, and 20 for the High Resolution Infrared Radiation Sounder found on polar orbiting satellites) with different central wavelengths. Retrieval techniques exploit this wavelength dependency.

3.4.2.2 Radiative Transfer Theory and Satellite Retrievals

To appreciate temperature retrievals and other quantitative applications of weather satellites, let us pause here to review basic theoretical aspects of how electromagnetic radiation is transferred through the atmosphere.[51] Encompassing much of the relevant theory is the *radiative transfer equation*,

$$\frac{dI_\lambda}{ds} = -\sigma_a I_\lambda - \sigma_s I_\lambda + \varepsilon_\lambda B_\lambda + \sigma_s \left\langle I'_\lambda \right\rangle, \tag{3.28}$$

which governs the change in radiance through a volume element (Figure 3.18) of the (atmospheric) medium. Here, I_λ is the intensity of electromagnetic radiation at a certain wavelength (monochromatic radiance), s is the slant path length of the volume

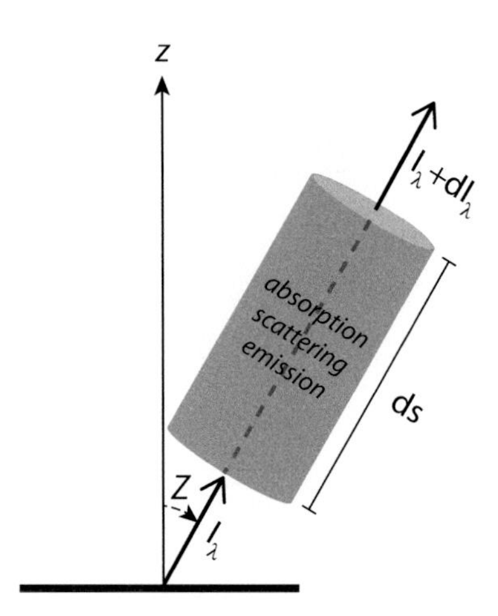

Figure 3.18 Change in monochromatic intensity I_λ over a slant path segment ds of a volume element. The slant path is relative to the satellite zenith angle Z and local vertical coordinate z. Adapted from Kidder and Vonder Haar (1995).

segment, σ_a is the absorption coefficient, σ_s is the scattering coefficient, and the quantity $\langle I'_\lambda \rangle$ represents the averaged radiation from other directions that is scattered into the beam. The remaining variables are expressed as follows: the Planck function B_λ is

$$B_\lambda = \frac{c_1 \lambda^{-5}}{\exp\left(\frac{c_2}{\lambda T}\right) - 1}, \tag{3.29}$$

where c_1 and c_2 are the first and second radiation constants, and emissivity ε_λ is, by definition,

$$\varepsilon_\lambda \equiv \frac{I_\lambda(emitted)}{B_\lambda}. \tag{3.30}$$

Notice that the Planck function can be inverted and solved for temperature, given the emissivity and the emitted radiance at a specified wavelength.[52]

The radiative transfer equation tells us that radiation is depleted owing to absorption by the medium, depleted through scattering out of the beam, added by emission of radiation by the medium, and added owing to out-of-beam radiation that is scattered into the beam. In the absence of clouds, and at IR wavelengths, the effects of scattering are commonly neglected. With this simplification, (3.28) reduces to *Schwarzchild's equation*

$$\frac{dI_\lambda}{ds} = (-I_\lambda + B_\lambda)\,\sigma_a, \tag{3.31}$$

where emissivity is equated to absorptivity through Kirchhoff's law. It is convenient to rewrite (3.31) with respect to vertical optical depth, $d\delta_\lambda = \cos(\mathrm{Z})\,\sigma_e ds$, where Z is the zenith angle of the satellite (see Figure 3.18), and σ_e is the extinction coefficient;

under the nonscattering assumption, $\sigma_a/\sigma_e = 1$. Integration of the resultant alternative form of (3.31) from the Earth's surface ($\delta_\lambda = 0$) to the satellite (and effectively, the top of the atmosphere; $\delta_\lambda = \delta_0$) yields

$$I_{\lambda,\delta_0} = I_{\lambda,0} \exp\left[-\delta_0 \sec(\mathrm{Z})\right] + \int_0^{\delta_0} \exp\left[-(\delta_0 - \delta_\lambda)\sec(\mathrm{Z})\right] B_\lambda \sec(\mathrm{Z})\, d\delta_\lambda, \tag{3.32}$$

where I_{λ,δ_0} is the monochromatic radiance measured at the satellite, and $I_{\lambda,0}$ is the monochromatic radiance upwelling the ground.

An approximation to (3.32) forms the basis for physical[53] retrieval of temperature from measured radiances:

$$I_i = W_s B_i(T_s) + \sum_n W_{i,n} B_i(T_n), \tag{3.33}$$

where subscript i indicates wavelength channel, subscript s denotes the surface, and the atmosphere is assumed to be divided into n isothermal layers of temperature T_n and depth $\Delta\delta$. The remaining variable W is a layer- and channel-dependent weight function, and is expressed as a vertical derivative of the transmittance $\tau_\lambda = \exp(-\Delta\delta)$. The channels of sounding radiometers are such that each has a weight-function peak in a unique layer. Thus, with known weight functions, an iterative technique is used to retrieve temperature, given a first guess of T_n: radiances I_i are computed from (3.33) with (3.29), compared to the measured radiances, and then used to adjust T_n based on the difference between the observed and computed I_i. The steps are repeated until the solution convergences. Retrieval of humidity follows a similar iterative procedure, but employs the retrieved temperature profile and a coupling with retrieval of column-integrated water vapor.

Assimilation of retrieved temperature and humidity profiles[54] into global and limited-area NWP models has resulted in significant gains in forecast skill, especially over oceanic and other regions with sparse in-situ observations.[55] This has relevance to the prediction of mesoscale processes in a number of ways (see Chapters 4 and 10), perhaps the most intuitive of which is that the retrieved data improve the characterization of the convective environment.

Environmental wind fields are also retrievable from satellite data and subsequently assimilated into NWP models. The essence of the wind retrieval is fairly simple: geographic locations of specific clouds/cloudy pixels are tracked in time using a sequence of geostationary satellite images, with wind vectors – commonly referred to as cloud-drift winds or atmospheric-motion vectors (AMVs) – computed from the location/time changes (Figure 3.19). Historically, this was implemented manually. However, despite the benefits associated with trained human analysts, the costs of human time and labor have led to the development of automated approaches. An automated algorithm searches an image subdomain for a pixel with a local maximum in

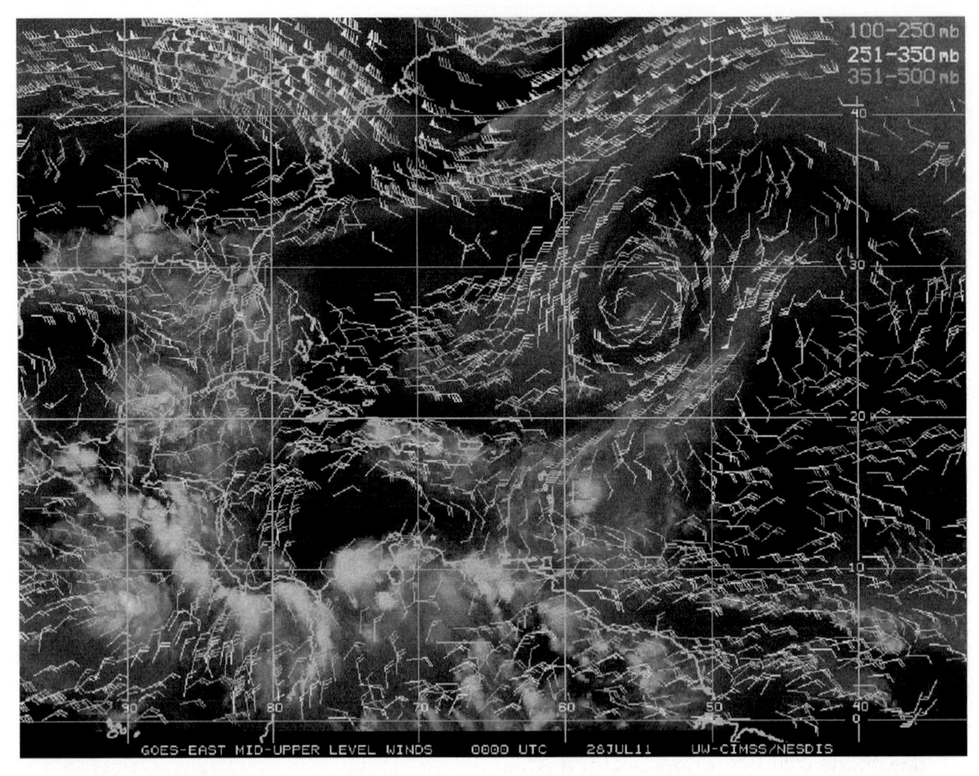

Figure 3.19 Atmospheric-motion vectors in three upper-level layers, derived from satellite WV/IR imagery. Courtesy of Dr. Christopher S. Velden, Cooperative Institute for Meteorological Satellite Studies, University of Wisconsin.

channel radiance, and then attempts to track this pixel in time by searching subsequent images and identifying similar pixel characteristics.[56] The heights of the motion vectors are then determined with the aid of the IR channel radiances at the tracked pixel locations using, for example, a weight-function method similar to that described earlier.[57] As evidenced by Figure 3.19, automated algorithms have been extended for use with data from WV (and IR) channels.[58] With this capability, satellite winds are now retrievable at any time of day, in cloudy as well as cloud-free conditions.

One source of error in the retrieved wind fields is rooted in the *satellite navigation*. Navigation regards the Earth-relative coordinates of the "scene" viewed by the satellite. Calculation of the coordinates, and thus of the geographical location of the tracked pixels, is based on similar geometrical relations as ground-based radar (Eq. (3.13)). It also requires knowledge of the orbital location of the satellite, the orientation (or attitude) of the satellite itself, and the scanning geometry of the instrument.[59] Another source of error follows from the implicit assumption that the clouds move with the environmental wind. This assumption is violated in cases of clouds with significant propagation, such as supercell thunderstorms (Chapter 7). The "perturbed"

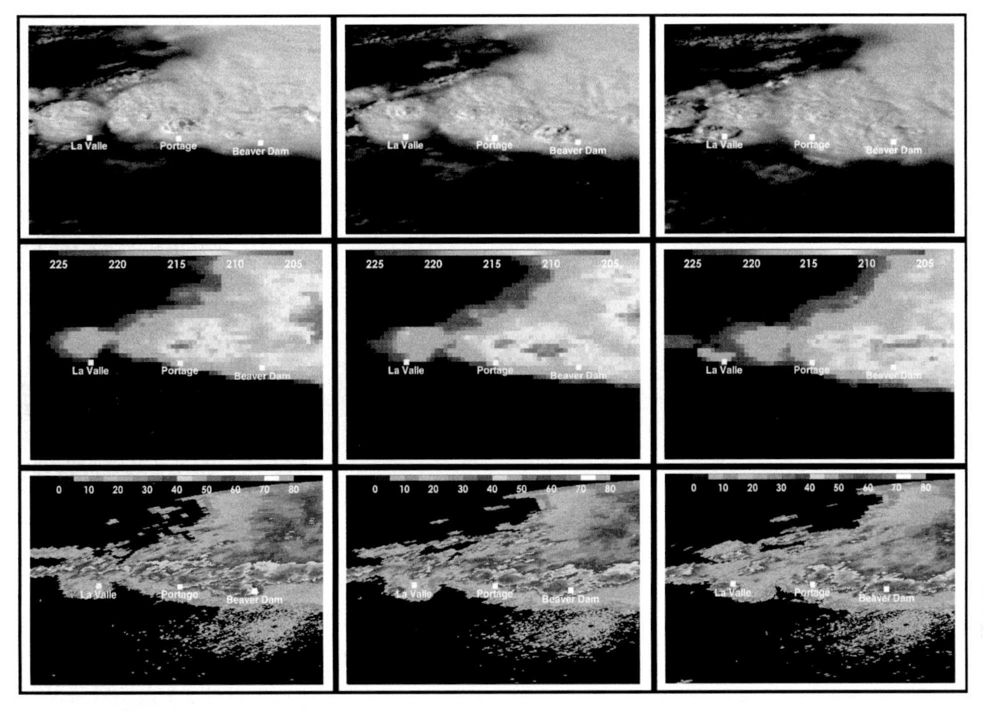

Figure 3.20 GOES-12 visible channel imagery with objective overshooting-top detections (red dots) (top), GOES-12 IR (10.7-μm) brightness temperatures (middle), and KMKX WSR-88D composite reflectivity (bottom). White dots show the locations of La Valle, Portage, and Beaver Dam, WI. From Dworak et al. (2012). (For a color version of this figure, please see the color plate section.)

motion vectors still contain useful information. For example, upper-level divergence calculated from these motion vectors provides a measure of updraft intensity.[60]

3.4.2.3 Satellite Applications

Diagnoses of cloud intensity, type, and amount follow directly from the measured radiances, as does information about (land and sea) surface characteristics. These can be deduced from VIS imagery, using attributes such as texture, opacity, and brightness (see Figure 3.20).[61] Cloud characteristics also can be diagnosed from IR imagery using *brightness temperature* T_b: under the assumption that the medium is a perfect emitter ($\varepsilon_\lambda = 1$) and therefore behaves as a blackbody, T_b is the temperature that results from (3.30) and inversion of the Planck function (3.29). "Cloud-top" brightness temperatures are used to help identify hazardous thunderstorms as demonstrated in Figure 3.20, and form criteria that define phenomena such as *mesoscale convective complexes* (see Chapter 8).[62] The total coverage of clouds and of particular cloud types in scenes can be quantified by summing pixels in brightness-temperature intervals, and/or by the use of advanced image processing techniques.[63] These have direct applicability to climatological studies.[64]

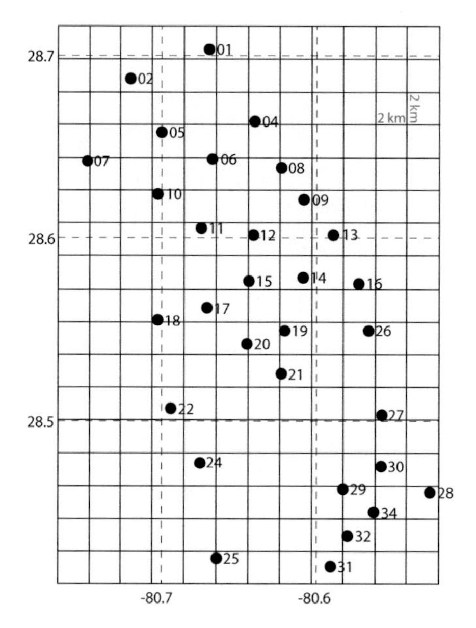

Figure 3.21 The NASA Kennedy Space Center rain gauge network (circles). Data from these gauges are used to validate TRMM rainfall estimates within the corresponding 2 × 2 km square areas. After Wang and Wolff (2010).

Finally, weather satellites provide a means to estimate surface rainfall. Empirical formulae have been developed to relate rainfall rate to T_b as well as to VIS channel brightness. Satellites with microwave radiometers allow additional formulae that take advantage of the interactions between microwaves and precipitation-sized particles. Application of $z - R$ relations is possible with the TRMM PR, a space-borne radar that has the same basic operating principles as described in Section 3.4.1 except that the PR beams are steered electronically.[65] As with many of the satellite applications, rainfall estimation is particularly valuable over oceanic and other data-sparse regions.

3.5 Observation Networks

Spatial arrays – or *networks* – of meteorological instruments are used extensively in the detection, prediction, and analysis of mesoscale processes. Consider the example of convective rainfall. Variations in rainfall amount over mesoscale areas are readily characterized by a network of rain gauges such as that deployed in the vicinity of the NASA Kennedy Space Center in Florida (Figure 3.21). The gauge data from this particular network serves the purpose of validating the TRMM-estimated rainfall over 2 × 2-km grid cells. Along similar lines, development of rainfall estimation techniques with research (and operational) radars in Oklahoma has been facilitated by the Little Washita Agricultural Research Service (ARS) Micronet, an array of 42 gauges with an average spacing of 5 km.[66] The ARS Micronet is embedded within the Oklahoma

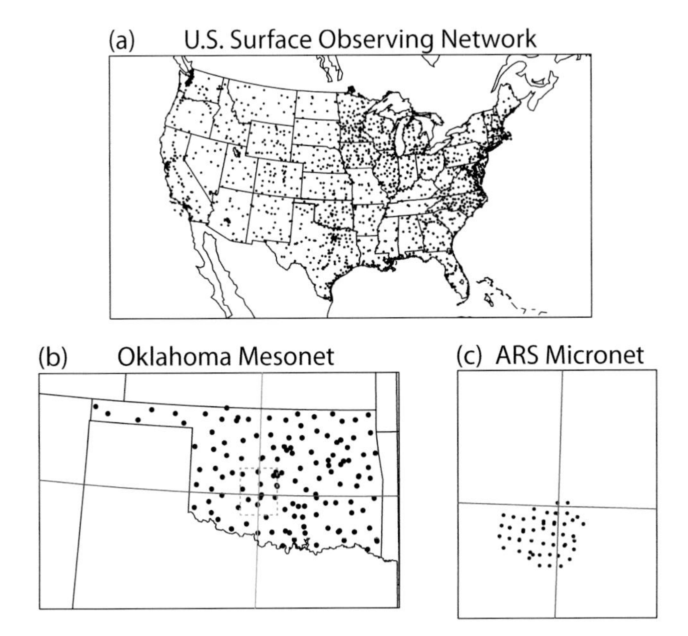

Figure 3.22 Comparison of three observing networks: (a) U.S. surface observing stations, (b) the Oklahoma Mesonet, and (c) the ARS Micronet. The location of the domain for (c) is indicated by the dashed box in (b).

Mesonet, a network of 119 surface observing stations with an average spacing of ~40 km;[67] the Oklahoma Mesonet provides regional information about rainfall and associated surface meteorological variables. The spatial distribution of rainfall over an even broader, yet coarser scale is revealed by the U.S. network of surface observing stations known as the Automated Surface Observing System (ASOS) and Automated Weather Observing System (AWOS) (see Figure 3.22).

These experimental and permanent network have respective station/instrument[68] spacings Δx capable of spatially resolving length scales of approximately $4\Delta x$ (see Section 3.1). An important caveat to this statement is that seldom is the station spacing exactly uniform, owing to the many logistics involved in siting the stations. This nonuniformity is especially likely in "opportunistic" networks, composed of observing stations operated by different agencies (or investigators) and perhaps with some variation in instrumentation type, error characteristics, and so forth. Some care is required, therefore, in the data analysis and interpretation. Analysis methods that account for data nonuniformities are described in Section 3.6. We will find that such methods also treat 3D data, as collected, for example, in radiosonde and Doppler radar networks, and are indispensible in merging data from networks of mixed platforms as often deployed during field campaigns.

The temporal sampling of the network instrumentation must also be considered in addition to the spatial characteristics of networks, because the mesoscale phenomena of interest herein evolve in time. As argued in Section 3.1, the temporal sampling rate

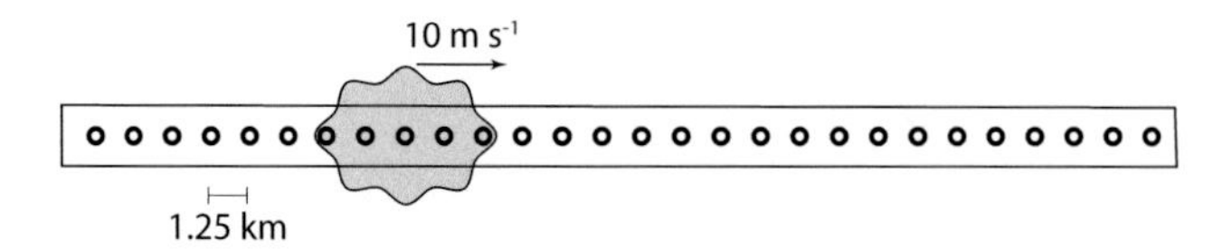

Figure 3.23 Linear array of observing stations, designed to sample a hypothetical "storm" (shaded region) of length scale 5 km, time scale 1 h, and horizontal speed 10 m s^{-1}. The stations have a separation distance of 1.25 km. Because the storm traverses this linear network in ~1 h, a station sampling rate of 0.1 h is chosen.

Δt is determined by the time scale of the phenomenon. However, as will be shown through the following exercise, Δt should also be consistent with the station spacing and domain size.[69]

Assume a "storm" with a length scale of 5 km, a time scale of 1 h, and a horizontal speed of 10 m s^{-1}. Let us propose a horizontal array of stations with $\Delta x = 1.25$ km, which adequately resolves the storm at some instant. If we distribute the stations uniformly over a linear distance of 35 km, we can expect the storm to traverse this domain in $35 \times 10^3\,\mathrm{m}/10\,\mathrm{m\,s}^{-1} \sim 1$ h (Figure 3.23). If we then choose a sampling rate equal to some fraction $1/n$ of this traversal time, (e.g. $\Delta t = 1$ h/10 $= 0.1$ h), we are able to collect n (=10) observations of the storm while it is within the domain. This particular sampling rate of $\Delta t = 0.1$ h is consistent with the assumed time scale, and also is within the capabilities of most modern instruments. Thus, our proposed network would appear to be well designed; the question we now face is whether the network design is *feasible*.

The cost and logistics (installation, maintenance, data management, etc.) of the instrumentation establish the feasibility of the network. Our linear array in Figure 3.23 has 28 stations; a comparable square array (35 × 35 km domain) would have 784 stations. It is reasonable to conclude that a 784-site radiosonde network configured according to the proposed square array would not be feasible; the same conclusion would probably be reached for an experimental network consisting of 784 modern surface observing systems (Section 3.2.1).We may choose to reduce the size of the domain, which results in fewer stations and thus lowers the cost, but also necessitates adjustments in the sampling rate. Unfortunately, a smaller domain also lowers the probability that the phenomenon of interest will move through or otherwise occur within the observing network. This is especially true for relatively infrequent and inherently transient phenomena such as tornadic storms. Hence, our network-design exercise ends in a dilemma between feasibility and scientific value.

In principle, this dilemma is at least partially resolved through the use of a *mobile network*. Consider the mobile surface network deployed during field campaigns such as VORTEX, VORTEX2, and the International H20 Program (IHOP).[70,71] The vehicles comprising this "mobile mesonet" had roof-mounted surface observing systems (Section 3.2.1) and were driven to relevant geographical locations;[72] the spacing and

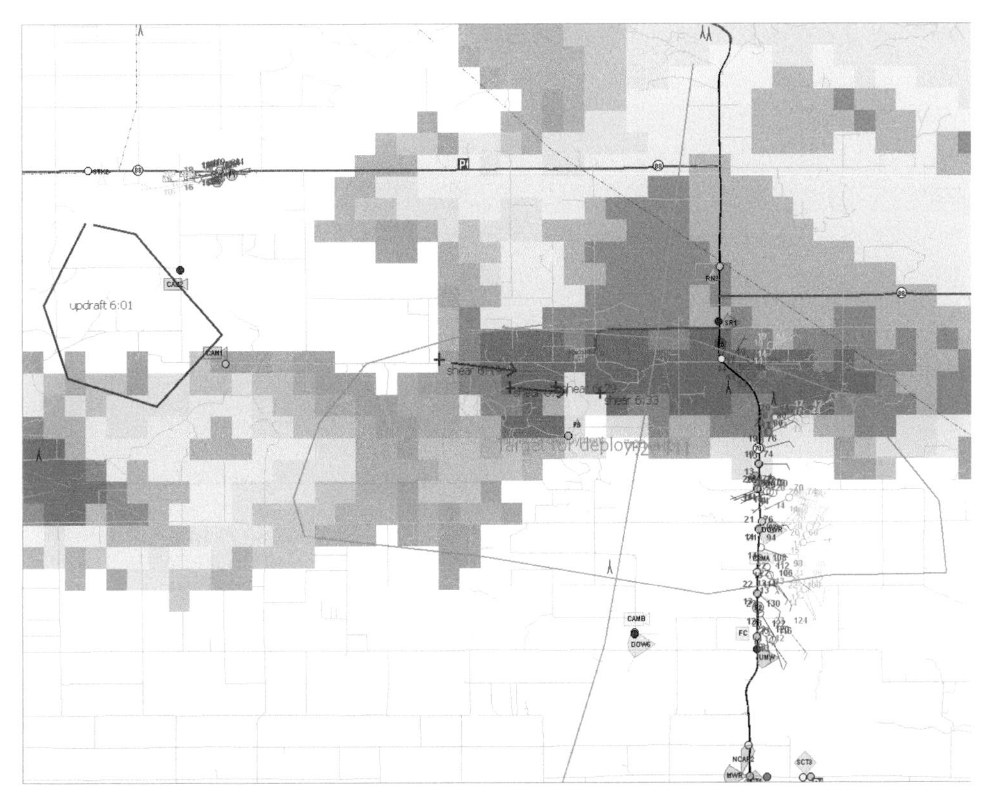

Figure 3.24 Display of Situational Awareness for Severe Storms Intercept (SASSI) software application used during VORTEX2 in 2009 and 2010. Color fill is a radar-reflectivity overlay, and icons represent various (mobile or portable) observing systems at a current as well as recent locations. The light-blue icon labeled FC shows the location of the field coordinator. Courtesy of Dr. Erik N. Rasmussen, Rasmussen Systems, LLC. (For a color version of this figure, please see the color plate section.)

uniformity of the resulting observing sites were dictated in part by the specific application (and, in part, by roads; discussed later). Coordination of the vehicle deployment was done semiremotely, based on GPS locations that were transmitted to a field command center (see Figure 3.24). Observations were then collected as the vehicles were moving (albeit with the potential for greater exposure errors; Section 3.2.1) or, when stationary, with vehicle redeployment as warranted.

One benefit of a mobile mesonet is that a moving phenomenon literally can be followed, and hence observed, during its entire lifetime. A challenge for such an observing strategy is the identification of suitable observing sites and suitable roads. For example, unpaved roads in rural areas often prove to be hazardous when wetted; urban areas usually have paved roads, but buildings, trees, and vehicular traffic may contaminate the observations. Deployment decisions, therefore, are made often with the aid of navigation–coordination software that utilizes a geographic information system (GIS). Suitable roads on uniform grids (e.g., the 1 mi $\times$ 1 mi road grids in

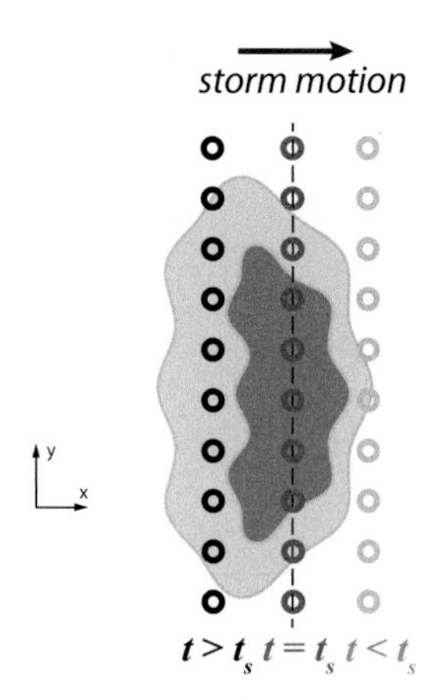

Figure 3.25 Depiction of an application of time-to-space conversion to a linear array of observing stations (open circles along dashed line). The shaded regions represent the radar echo of a moving "storm" at some time t. The fixed locations of the stations are converted to storm-relative locations using a representative storm motion at some reference time t_s. The shading of the open circles is proportional to the relative age of the observations.

many rural areas of the United States) are especially sought out, because they offer the potential for observing sites with spatial uniformity.

Mobile (and also portable; Section 3.2.1) observing stations that are placed along a single road segment can yield a set of 2D gridded observations of the moving phenomenon or "storm" with a *time-to-space conversion* (Figure 3.25): the geographic locations (x, y) of measurements taken at time t are converted to storm-relative locations (x_s, y_s) using a discrete form of (2.4):

$$x_s = x - u_s \Delta t$$
$$y_s = y - v_s \Delta t, \tag{3.34}$$

where (u_s, v_s) is the representative storm velocity at a reference time t_s. The underlying assumption of this technique is that the storm does not evolve substantially over the interval $\Delta t = t - t_s$. The Δt is limited in practice by the specific phenomenon; studies of fast-evolving tornadic supercells, for example, have imposed a limit of $|\Delta t| \leq$2–3 min.[73]

The mesoscale observational objectives often require that the surface network be complemented by networks of other instrumentation such as mobile Doppler radars, as was the particular case for VORTEX2. Mobile (and fixed) radar networks have many of the same design issues as do networks of in-situ instrumentation. However, an issue unique to radar networks is the relative geographical placement of each radar, because this affects the geometry of the wind retrieval (Section 3.4.1) as well as the data collection strategy. To illustrate, assume a two-radar network with respective radars 1 and 2 located at $(x = +d, y = 0)$ and $(x = -d, y = 0)$ (Figure 3.26).[74] The $2d$ separation

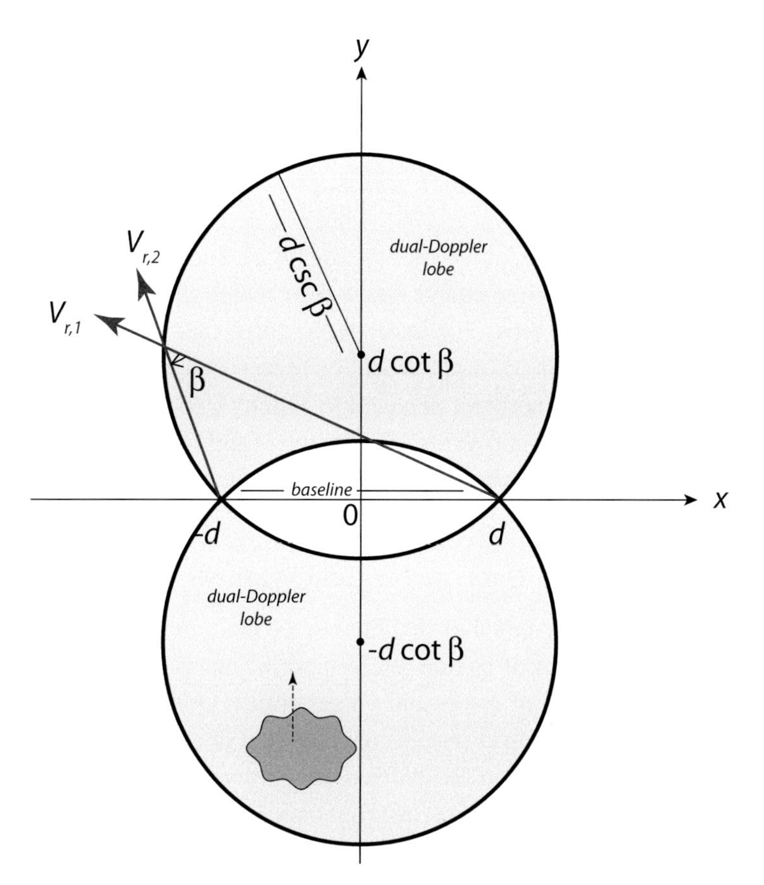

Figure 3.26 A two-Doppler radar network, with radars located at ($x = \pm\, d$, $y = 0$), and hence with a $2d$ baseline. Angle β is the minimum acceptable intersection angle between beams of the two radars. The stippling indicates the dual-Doppler "lobes," which correspond to the area (A_R; see text) of retrievable winds. The two lobes are centered at ($x = 0$, $y = \pm\, d$ cot β), and have radii equal to d csc β. Shown in the southern lobe is a "storm" moving from south to north. After Davies-Jones (1979).

distance, or *baseline*, is one of the user-defined network design parameters. The other is β, the minimum acceptable angle of intersection between beams of the two radars ($0 \le \beta \le \pi/2$): small beam intersections imply nearly dependent measurements of the true wind vector (scatterer motion) by the two radars, whereas beam intersections at angles approaching $\pi/2$ imply nearly independent measurements of the true wind vector. All acceptable beam intersections are constrained to lie within the interval $[\beta, \pi - \beta]$ and thus within two overlapping circles with radii d csc (β). As shown in Figure 3.26, these two circles, minus their overlap (which contains beam intersections outside the interval $[\beta, \pi - \beta]$), describe the so-called *dual-Doppler lobes*. This is the area of dual-Doppler wind retrieval, quantified as

$$A_R = [d \csc(\beta)]^2 \, [\pi - 2\beta + \sin(2\beta)] \, . \tag{3.35}$$

Decreases in β yield increases in area A_R but also increases in the geometric errors of the retrieved winds. Increases in the baseline distance $2d$ yield increases in A_R but also increases in the slant range r of beam intersections, and thus implicitly in decreases in the resolution of the radar data because of beam broadening. The latter follows from that fact that the linear distance across a beam, and therefore the azimuthal data spacing Δ_{az}, varies as $\sim r\,\theta_b(\pi/180)$, where θ_b is the half-power beamwidth (in degrees). Although the orientation of the radar network does not affect the magnitude of A_R, it does influence the time and/or space over which the phenomenon is within the dual-Doppler coverage area. For instance, a storm that moves along the y-axis from south to north (or north to south) would be observed in one lobe and then in the other lobe of the two-radar network depicted in Figure 3.26. Clearly, the design of a radar network is an exercise in maximizing the area (and time) of retrievable winds, while maximizing resolution and minimizing the geometric error.

3.6 Data Analysis and Synthesis

The previous sections have shown us that mesoscale meteorological observations are collected over a wide range of spatial and temporal intervals, and potentially have a corresponding wide range of errors and uncertainties. Observations yielded by the mobile surface and radar networks just discussed would provide ample support for this statement. Methods to synthesize these disparate (and voluminous) data are pursued in this section. With diagnostic as well as prognostic (or NWP) purposes in mind, our focus is on the class of spatial analysis methods wherein observations are remapped and/or interpolated to a common set of geographic locations, at a common time. Approaches that involve function fitting and empirically based, distant-dependent weighted averaging are treated here; statistical interpolation is included as part of the discussion on data assimilation in Chapter 4.

3.6.1 Function Fitting

The approach of fitting a mathematical function, such as a polynomial, to a set of observations is common in numerical analysis of scientific data. In mesoscale meteorological applications, the function typically is a 2D surface of, say, 850-hPa temperature. The true, continuous function $T(x, y)$ is unknown, but the premise is that $T(x, y)$ can be approximated from discrete temperature observations.

Let us, for ease of presentation, assume that the problem is 1D and choose an approximating function that is linear:

$$f(x) = a_0 + a_1 x. \tag{3.36}$$

Two (or, generally, $n + 1$) observations or data values are needed to determine the unknown coefficients a of the first- (or nth-) order polynomial. Evaluation of (3.36) at these two points, x_1 and x_2, gives

$$\begin{aligned} f(x_1) &= f_1 = a_0 + a_1 x_1 \\ f(x_2) &= f_2 = a_0 + a_1 x_2, \end{aligned} \tag{3.37}$$

which in matrix form is

$$\begin{bmatrix} 1 & x_1 \\ 1 & x_2 \end{bmatrix} \begin{bmatrix} a_0 \\ a_1 \end{bmatrix} = \begin{bmatrix} f_1 \\ f_2 \end{bmatrix} \tag{3.38}$$

or, more generally,

$$\mathbf{X}\vec{A} = \vec{F}. \tag{3.39}$$

The unknown coefficients a are found by inverting $\mathbf{X}$ such that:

$$\vec{A} = \mathbf{X}^{-1}\vec{F}. \tag{3.40}$$

For (3.38), we find that

$$\begin{aligned} a_0 &= \frac{1}{\Delta}\left[f_1 x_2 - f_2 x_1\right] \\ a_1 &= \frac{1}{\Delta}\left[f_2 - f_1\right], \end{aligned} \tag{3.41}$$

and hence that our approximating polynomial (3.36), applicable over the interval $[x_1, x_2]$, becomes

$$f(x) = f_1\left(1 - \frac{\delta}{\Delta}\right) + f_2\frac{\delta}{\Delta}, \tag{3.42}$$

where $\Delta = x_2 - x_1$, and $\delta = x - x_1$.

Equation (3.42) is an expression of the well-known technique of *linear interpolation*, which gives an estimate of the true function only in the local neighborhood of the data. It can be adapted for use in 2D (bilinear interpolation) or 3D (trilinear interpolation); bilinear interpolation of radar data from its native spherical coordinate system to a Cartesian coordinate grid is one relevant application.[75] This technique has the distinct advantages of being simple and computationally inexpensive. It does not, however, allow for the analyst to consider data values other than the two immediately surrounding the analysis or grid point. This disadvantage is especially relevant for meteorological data, as typically there are many more observations than the order of the desired function. This leads to an overdetermined problem, as encountered already in Section 3.4.1.

The method of least squares is used to solve the overdetermined problem. Here we seek to minimize the quadratic of the difference between a data distribution and its polynomial-based approximation:

$$J = \sum_{i=1}^{N} [f(x_i) - f_o(x_i)]^2, \tag{3.43}$$

where $f_o(x_i)$ is the data distribution or set of N observations, and $f(x_i)$ is an evaluation of the approximating function at the observation points x_i. We are again free to choose the order and form of the approximating function, so let us now, for the sake of variety, consider a second-order polynomial

$$f(x) = a_0 + a_1 x + a_2 x^2. \tag{3.44}$$

Equation (3.43) then becomes

$$J = \sum_{i=1}^{N} \left[\left(a_0 + a_1 x_i + a_2 x_i^2\right) - f_o(x_i)\right]^2. \tag{3.45}$$

To determine the unknown coefficients a_n, take $\partial J/\partial a_n$ in (3.45), set equal to zero, and then solve for the a_n. The resulting normal equations have the now-familiar general form

$$\mathbf{X}\vec{A} = \vec{F}. \tag{3.46}$$

Equation (3.46) is inverted to solve for the coefficients $\vec{A}$, exploiting the fact that $\mathbf{X}$ is a symmetric matrix. The matrix inversion is reasonable in 1D, but quickly becomes complicated in 2D, which is the usual case for meteorological data analysis. Indeed, to fit a quadratic polynomial surface in the least squares sense requires that a 6×6 matrix be solved at every grid point. This is a computationally expensive process, and has other disadvantages that are addressed by the class of methods described next.

3.6.2 Empirical Analysis: Successive Corrections Method

The *successive corrections method* (SCM) is an empirical, iterative data analysis approach. It has the general form[76]

$$f_k^{n+1} = f_k^n + \frac{\sum_{i=1}^{N} W_{i,k}^n \left(f_o - f_i^n\right)}{\sum_{i=1}^{N} W_{i,k}^n + e^2}, \tag{3.47}$$

where f_k^n is the analysis at grid point k and iteration n; f_o is the observation at data point i; f_i^n is the analysis after iteration n, evaluated at data point i; e is a measure of observation error, if known; and $W_{i,k}^n$ is the weight that observation i receives at grid

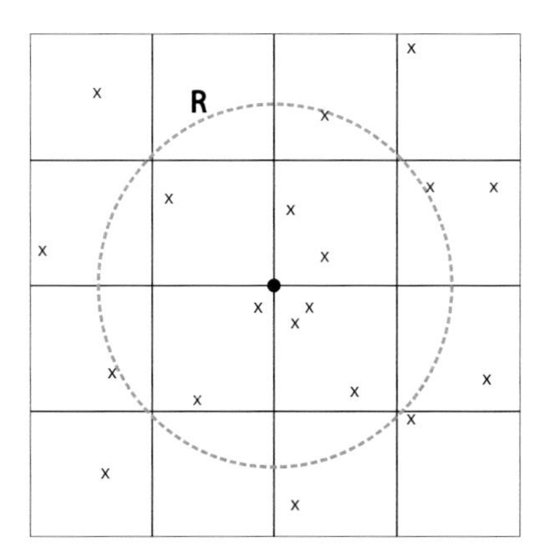

Figure 3.27 Illustration of data point locations (x) and the radius of influence R about a grid point. A 2D domain is shown. All data that fall within the radial distance R of a given grid point are allowed to affect the analysis at that grid point.

point k and iteration n. Unlike the function fitting approaches, the SCM allows for the use of "first guess" or "background" information, applied at the 0th iteration:

$$f_k^0 = f_k^b, \tag{3.48}$$

where, for example, f_k^b is a model forecast valid at the same time of the observations. The choice of the number of subsequent iterations, as well as the choice of *weight function*, is where empiricism enters the SCM. A survey of the literature will quickly reveal two weight function formulations that are commonly used in mesoscale data-analysis applications.

The first is attributed to G. P. Cressman[77] and is written as

$$W_{i,k}{}^n = \begin{cases} \dfrac{R_n^2 - r_{i,k}^2}{R_n^2 + r_{i,k}^2}, & r_{i,k}^2 \leq R_n^2 \\ 0, & r_{i,k}^2 > R_n^2, \end{cases} \tag{3.49}$$

where R_n is the *radius of influence* at iteration n, and $r_{i,k}$ is the Euclidean distance between a data point and grid point; in 2D, $r_{i,k}$ is

$$r_{i,k}^2 = \left[(x_i - x_k)^2 + (y_i - y_k)^2\right]. \tag{3.50}$$

Only the observational data points that lie within the distance R_n of a given grid point affect the analysis at that grid point (Figure 3.27). Owing to the functional form of (3.49), the weight of an observation falls off quickly with increases in the gridpoint – datapoint separation within R_n (Figure 3.28a). Thus, the magnitude of R_n controls the desired level of smoothness (or coarseness) of the analysis.

Implicit in the preceding statement is a fundamental purpose of an *objective analysis* approach such as SCM: to produce a gridded analysis that retains the adequately sampled spatial scales (or wavelengths) in the data, but is absent of under- or marginally

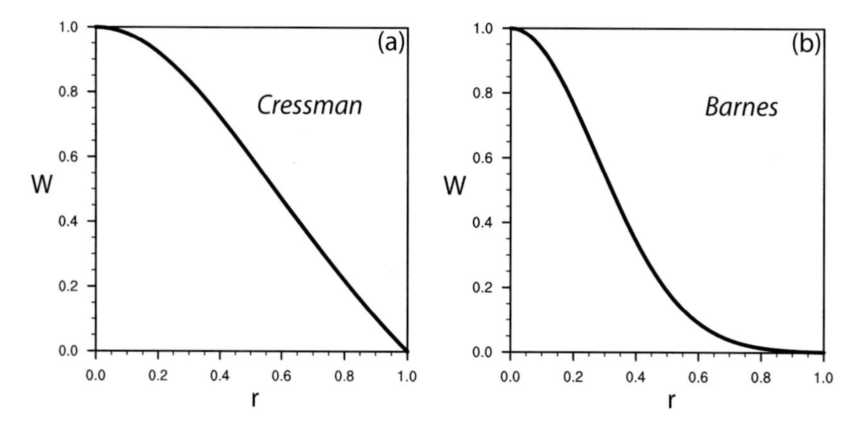

Figure 3.28 Example evaluations of the weight functions due to (a) Cressman (1959), and (b) Barnes (1964), over the interval $0 \geq r \geq 1$ km. In (a), the radius of influence R_n has a value of 1 km. In (b), the smoothing parameter κ_n has a value of 0.15 km^2.

sampled wavelengths. With the Cressman scheme, this is achieved through trial-and-error experimentation with different R_n (and n); multiple iterations (or passes) with successively smaller R_n are known to refine the range of wavelengths retained.[78] An analytical relationship between the radius of influence in (3.49) and the spectral content of the analysis does not exist, although it is possible to determine such a *spectral response* of the weight function using numerical approximations to the Fourier transform (discussed later).[79]

The other popularly used weight function is attributed to S. L. Barnes[80] and is written as

$$W_{i,k}^n = \exp\left(-r_{i,k}^2/\kappa_n\right), \tag{3.51}$$

where κ_n is a smoothing parameter. The smoothing parameter determines the rate at which the weight function asymptotes to zero as $r_{i,k} \rightarrow \infty$ (gridpoint–datapoint distance becomes large) (Figure 3.28b). In other words, κ_n controls the steepness of the weight function and thus the smoothness of the analysis.

As with R_n in (3.49), κ_n represents the subjective part of an otherwise objective analysis method. When κ_n is judiciously chosen, data scales smaller than the Nyquist wavelength are filtered and thus removed from the analysis. Toward this end, guidance is provided through the analytical representation of (3.51) in spectral space. Consider the Fourier transform pair associated with a 1D, continuous function:

$$\Psi(k) = \int_{-\infty}^{\infty} W(x)e^{-ikx}dx$$

$$W(x) = \frac{1}{2\pi}\int_{-\infty}^{\infty} \Psi(k)e^{ikx}dk, \tag{3.52}$$

where $k = 2\pi/\lambda$ is the wavenumber.[81] Assuming that $W(x)$ is symmetric (i.e., $W(x) = W(-x)$) (3.52) can be written as a Fourier cosine transform pair, and on substituting (3.51) for $W(x)$, we have:

$$\Psi(k) = 2\int_0^\infty e^{\left(-x^2/\kappa\right)} \cos(kx)dx, \tag{3.53}$$

where we have let $r = x$ in (3.51). The integral in (3.53) can be evaluated using standard integral tables, giving

$$\Psi(k) = \frac{\sqrt{\kappa}\sqrt{\pi}}{2} \exp\left(-k^2\kappa/4\right). \tag{3.54}$$

The spectral response is usually normalized as

$$D(k) = \frac{\Psi(k)}{\Psi(0)} = \exp\left(-k^2\kappa/4\right)$$

and then expressed in terms of wavelength

$$D(\lambda) = \exp\left(-\pi^2\kappa/\lambda^2\right). \tag{3.55}$$

In simple terms, the theoretical spectral response D quantifies which data or input scales will be filtered during application of the Barnes scheme, and which scales will be retained.

To better appreciate the information offered by a spectral response function, let us rewrite (3.55) is in terms of a dimensionless smoothing parameter and wavelength,

$$D(\tilde{\lambda}) = \exp(-\pi^2\tilde{\kappa}/\tilde{\lambda}^2), \tag{3.56}$$

where $\tilde{\lambda} = \lambda/L$, $\tilde{\kappa} = \kappa/L^2$, and length scale L is twice the mean datapoint spacing; that is, $L = 2\overline{\Delta}$. The evaluation of (3.56) in Figure 3.29 indicates that for a relatively large smoothing parameter, say $\tilde{\kappa} = 1.0$, input scales with $\tilde{\lambda} = 1 = 2\overline{\Delta}$ are completely filtered ($D = 0$), thus meeting our previously stated intentions. Figure 3.29 also indicates a dilemma: the benefit of the complete removal of the marginally (and under-) resolved scales is partially offset by the reduction in amplitude of the well-resolved scales. For example, the spectral response for a wavelength $\tilde{\lambda} = 4 = 8\overline{\Delta}$ is D ~0.5, meaning that only ~50 percent of the amplitude of this wavelength is retained. A smaller smoothing parameter value, such as $\tilde{\kappa} = 0.1$, has a sharper spectral response, and hence yields a higher retention of the amplitude of well-resolved scales; unfortunately, this comes with the cost of a retention of some fraction of the amplitudes of the underresolved scales.

This dilemma – which also is encountered with the Cressman weight function – is partially resolved with multiple iterations of the SCM equation (3.47).[82,83] In a multipass Barnes scheme, the successive κ_n values are chosen such that $\kappa_{n+1} = \gamma\kappa_n$, where the convergence parameter γ ($0 < \gamma < 1$) has the effect of steepening the

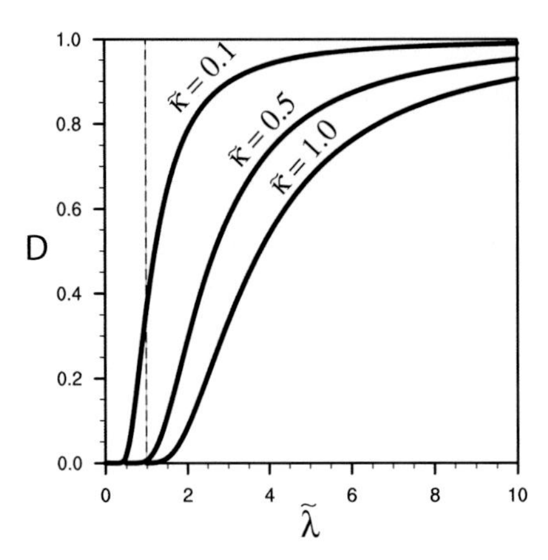

Figure 3.29 Evaluation of theoretical spectral response D of the (1D) Barnes weight function, for dimensionless smoothing parameter values of $\tilde{\kappa} = 0.1$, 0.5, and 1.0, where $\tilde{\kappa} = \kappa/L^2$, and L is equal to the mean datapoint spacing, $L = 2\overline{\Delta}$. The abscissa is dimensionless wavelength, $\tilde{\lambda} = \lambda/L$, and the ordinate represents the fraction of the amplitude of the input wavelength that is retained in the analysis. The dashed vertical line indicates $\tilde{\lambda} = 1$, the Nyquist wavelength.

weight function per iteration. A two-pass implementation is common in analyses of mesoscale data, with the resultant spectral response given by

$$D = D_1(1 + D_1^{\gamma-1} - D_1^{\gamma}), \tag{3.57}$$

where D_1 is the response associated with the first pass (i.e., 3.55) evaluated with κ_1) (see Figure 3.30a).[84] Note that: (1) a background field, as applied at the 0th iteration (see (3.48)) is not often available for mesoscale data analyses, nor is error information e, particularly when the data are used for diagnostic purposes; and (2) evaluation of f_k^2 from (3.47) during the second iteration requires "back-interpolation" of the current analysis f_k^1 to the datapoint locations, to allow evaluation of the term $f_0 - f_i^1$. A simple method, such as bilinear interpolation, is typically employed.

A variant of the multipass Barnes scheme is a scale-separation or *bandpass* objective analysis technique.[85] A bandpass analysis b_k is the difference between an analysis f_k produced with one κ and an analysis g_k produced with a judiciously chosen larger κ. In essence, the subtraction removes both "short" and "long" wavelength scales and leaves the "intermediate" scales. The bandpass technique would therefore allow one to extract mesoscale perturbations from the total field (Figure 3.30).

A closing remark is appropriate here on practical applications and interpretations of (3.56) based on a mean datapoint spacing $\overline{\Delta}$. As mentioned in Section 3.5, the spacing of observations in most networks tends not to be uniform. The relevant implication of such nonuniformity is that an underresolved scale in one part of the analysis domain may in fact be well resolved in another part of the domain. Indeed, this is the usual case with weather radar data, because the (azimuthal) spacing increases continuously with slant range. Hence, the guidance provided by the spectral response function

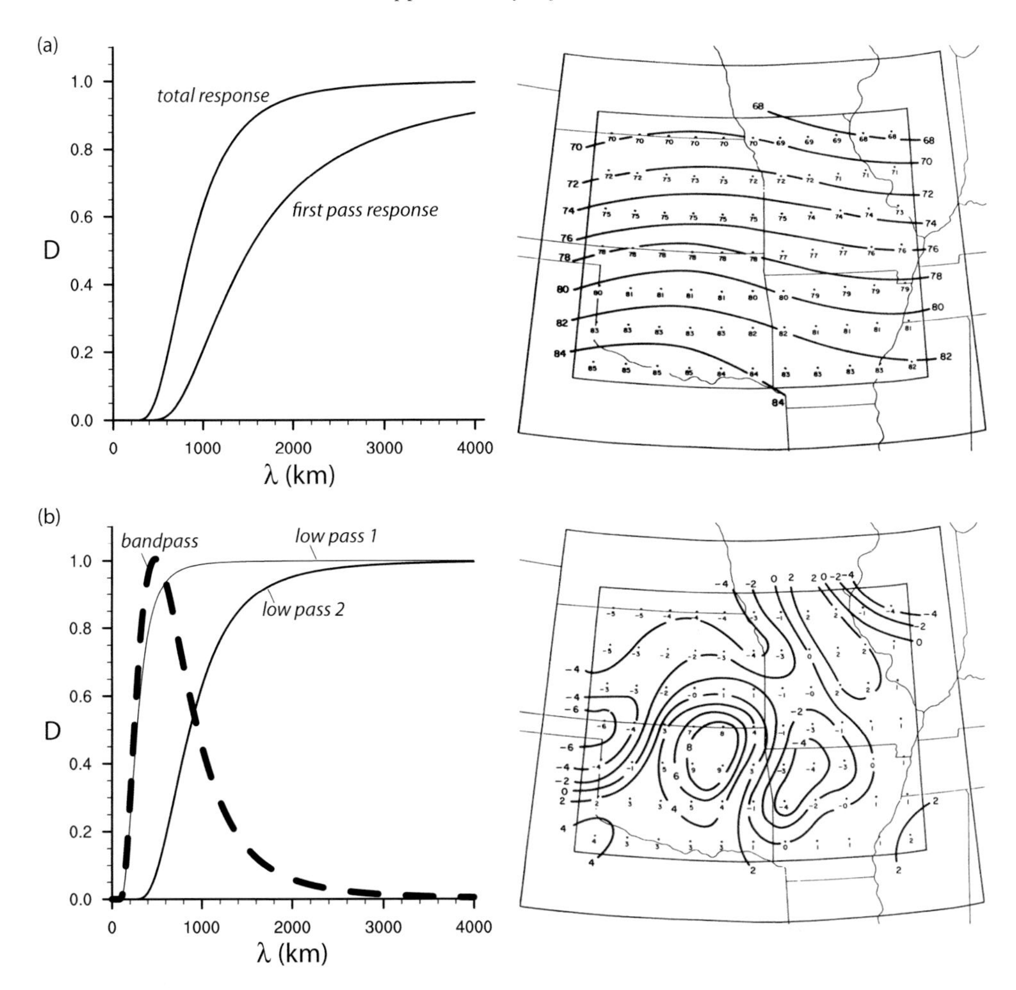

Figure 3.30 Macroscale and mesoscale-perturbation surface temperature (°F) analysis produced using objective analysis techniques. (a) The first-pass and total spectral response (left panel) for the two-pass scheme used to generate the macroscale temperature field (right panel). The smoothing parameter is 40,000 km^2, and convergence parameter is 0.4. (b) The spectral responses for the two low-pass analyses (both of which are two-pass schemes; left panel) used to generate the mesoscale perturbation temperature field (right panel). The "bandpass," or difference between the two spectral responses, is multiplied by a constant coefficient (1.25). "Low-pass 2" is shown in (a). "Low-pass 1" has a smoothing parameter of 5000 km^2, and convergence parameter of 0.3. The intermediate scales retained in this perturbation field have wavelengths in a band centered at 500 km. After Maddox (1980b).

should be applied with full knowledge of the underlying data distribution and then adjusted accordingly.

Supplementary Information

For exercises, problem sets, and suggested case studies, please see www.cambridge.org/trapp/chapter3.

Notes

1 Hence, the resolution of the measurements is distinct from the resolution of the instrument itself, which depends on the sensor, electronics, etc.; Thomson (1986).
2 See Brock and Richardson (2001) for a complete description of the various sensors and their error characteristics.
3 Brock and Richardson (2001).
4 Schroeder and Weiss (2008).
5 Brock et al. (1987), Winn et al. (1999).
6 Straka et al. (1996).
7 See Brock and Richardson (2001) for a complete description of the various sensors and their error characteristics.
8 Brock and Richardson (2001).
9 Hock and Franklin (1999).
10 Both tasks were done manually, but now tend to be automated.
11 Davis et al. (2004).
12 MacDonald (2005).
13 See www.eol.ucar.edu/development/avaps-iii/documentation/overall-global-hawk-dropsonde-system-description/.
14 Holland (2001).
15 Worldwide, these data are generally referred to as *aircraft meteorological data reports* (AMDAR); see Moninger et al. (2003).
16 Moninger et al. (2003).
17 Benjamin et al. (1991).
18 Trapp et al. (2006).
19 There are examples of surface observing networks that could nominally resolve large mesoscale convective systems and their associated straight-line winds. However, the existence of such networks tends not to be widespread. Furthermore, they would tend not to be able to resolve smaller-scale events, such as microbursts.
20 Diffenbaugh et al. (2008).
21 See Trapp et al. (2006) for examples.
22 Dotzek et al. (2009).
23 Long et al. (1980).
24 Palencia et al. (2011).
25 Fujita (1981), Doswell et al. (2009).
26 Toth et al. (2012).
27 The reader is encouraged to consult Doviak and Zrnic (1993), Rinehart (1997), and Battan (1973) for in-depth treatments of scattering theory, signal processing, and radar electronics and components.
28 See the development of this equation by Rinehart (1997).
29 Rinehart (1997), Battan (1973).
30 This particularly well-known and widely used relationship derives from the data that describe the Marshall-Palmer (1948) dropsize distribution; see Battan (1973), and Rinehart (1997).
31 Westrick et al. (1999).
32 Fulton et al. (1998).
33 Stumpf et al. (1999), Mitchell et al. (1999).
34 Joss and Waldvogel (1970).
35 Miller and Fredrick (1998).
36 Armijo (1969), Brandes (1977), Kessinger et al. (1987).
37 Gal-Chen (1978), Hane et al. (1981).
38 An example is the numerical method known as sequential overrelaxation (SOR).
39 The Doppler on Wheels (DOW) radar is an example of a mobile, ground-based system; see Wurman et al. (1997).
40 Mobile C-band (5-cm wavelength) radars have also been developed, e.g., Biggerstaff (2005).
41 Jorgensen et al. (1983), Hildebrand et al. (1994).
42 Leon et al. (2006), Galloway et al. (1997).
43 Zrnic and Ryzhkov (1999).

44 Parsons (1994).
45 Weber (1990).
46 Benjamin et al. (2004).
47 Parsons (1994).
48 Kidder and Vonder Haar (1995).
49 Simpson et al. (1988).
50 See, e.g., Kidder and Vonder Haar (1995) for tabulated information on the Imager and the Sounder.
51 The following material draws from Wallace and Hobbs (2006), Kidder and Vonder Haar (1995), and Liou (2002). The reader is invited to consult these references for more information on radiative transfer.
52 Kidder and Vonder Haar (1995).
53 As compared to statistical retrieval approaches, which do not make use of the radiative transfer equation.
54 Or of the radiances themselves, with proper operators in the data assimilation technique; e.g., Derber and Wu (1998).
55 Kelly et al. (2008).
56 Nieman et al. (1997).
57 Neiman et al. (1997).
58 Velden et al. (1997).
59 Kidder and Vonder Haar (1995).
60 Velden et al. (2005).
61 Kidder and Vonder Haar (1995).
62 Bedka et al. (2010).
63 Kidder and Vonder Haar (1995).
64 Schiffer and Rossow (1983).
65 Simpson et al. (1988).
66 Ryzhkov et al. (2005).
67 Brock et al. (1995).
68 We will use the terms "station" and "instrument" interchangeably in this section, although strictly speaking, a station is comprised of an instrument or instrument package. Most of the instruments are used for in-situ measurement, although weather radars and other remote-sensing instrumentation are also networked.
69 The following draws from Thomson (1986).
70 Wurman et al. (2012).
71 Weckwerth (2004).
72 Straka et al. (1996).
73 Markowski et al. (2002).
74 This discussion of a dual-Doppler network follows the example given by Davies-Jones (1979). See Kessinger et al. (1987) for a discussion of networks composed of three or more radars.
75 See Mohr and Vaughan (1979), but also note the concerns raised by Trapp and Doswell (2000).
76 The convention of Kalnay (2003) is followed throughout this section.
77 Cressman (1959).
78 Daley (1991).
79 Trapp and Doswell (2000).
80 Barnes (1964).
81 For 2D isotropic functions, (3.52) becomes the Hankel transform.
82 See Figure 2 in Trapp and Doswell (2000).
83 Koch et al. (1983).
84 From Barnes (1973), but see also Maddox (1980b) and Koch et al. (1983).
85 Following Maddox (1980b).

4
Mesoscale Numerical Modeling

Synopsis: This chapter provides information on the design and implementation of mesoscale numerical models. The governing equations of typical mesoscale models are given, as are their numerical approximations. Physical processes such as those involving cloud and precipitation microphysics are represented as simplified functions of model variables. Schemes for the parameterization of these and other relatively complex processes are provided to show basic formulations. Design and implementation issues, such as the size of the model domain, the use of nested grids, and model initialization, are also discussed.

4.1 Introduction

The objective of this chapter is to introduce the reader to the basic design and implementation of mesoscale numerical models. As will be demonstrated in the remaining chapters – and perhaps as the reader has already experienced – such models play dual roles as experimental and weather prediction tools. Indeed, some community models, such as the Weather Research and Forecasting (WRF) model, have a built-in functionality that allows for (1) *idealized modeling*, which employs simplified initial and boundary conditions (ICs, BCs), and for (2) *real-data modeling*, which employs observationally derived ICs and BCs such that real events can be simulated or predicted. Our treatment herein assumes a fairly generous definition of a mesoscale model so that both approaches can be discussed.

The discussion that follows is focused on nonhydrostatic models appropriate for the explicit representation of deep moist convection, but does acknowledge when (and generally how) the convective processes must be parameterized. Section 4.3 includes a description of this parameterization and others that are considered particularly relevant. The philosophical approach adopted in Section 4.3 pervades the entire chapter: the topics are covered in sufficient detail so the reader can appreciate the complexities of model design, and understand the need to make informed choices as a model user. Accordingly, this chapter is not meant to serve as a handbook or reference manual on model development. The interested reader is encouraged to consult entire technical reports or books devoted to: mesoscale models;[1] related subjects such as computational fluid dynamics, data assimilation, and model parameterizations;[2] and specific topics such as convective parameterizations and cloud and microphysical parameterizations.[3]

4.2 Equations and Numerical Approximations

4.2.1 Continuous Equations

We consider a slightly different form of the relevant continuous equations introduced in Chapter 2, beginning with the vector equation of motion in components:

$$\frac{Du}{Dt} = -\frac{1}{\rho}\frac{\partial p'}{\partial x} + F_u \tag{4.1}$$

$$\frac{Dv}{Dt} = -\frac{1}{\rho}\frac{\partial p'}{\partial y} + F_v \tag{4.2}$$

$$\frac{Dw}{Dt} = -\frac{1}{\rho}\frac{\partial p'}{\partial z} + B + F_w, \tag{4.3}$$

the thermodynamic energy equation

$$c_v\frac{DT}{Dt} = -\frac{p}{\rho}\nabla \cdot \vec{V} + \dot{Q}, \tag{4.4}$$

and a prognostic pressure equation that is formulated using the equations for mass continuity, thermodynamic energy, and state:

$$\frac{Dp}{Dt} = -\frac{c_p}{c_v}p\nabla \cdot \vec{V} + \frac{p}{c_v T}\dot{Q}. \tag{4.5}$$

The total derivative is $D/Dt = \partial/\partial t + \vec{V} \cdot \nabla$, as before. In (4.1) through (4.3), F includes the Coriolis force terms, curvature terms, and other effects represented explicitly or through some approximation; turbulent eddy mixing is one specific example (see Section 4.3.1). B in (4.3) is the buoyancy term. Pressure p' in (4.1) through (4.3) is expressed as the deviation (indicated by a prime) from a base state (indicated by an overbar) in hydrostatic balance, as is done in many, though not all, numerical models of nonhydrostatic flow.[4,5] The horizontal velocity components can also be expressed as the deviation from a horizontally homogenous, vertically stratified base state ($\overline{u} = U(z)$, $\overline{v} = V(z)$, $\overline{w} = 0$). This practice is often followed in idealized model applications. Finally, the diabatic heating rate $\dot{Q}$ in (4.4) and (4.5) is kept in its general form, with possible contributions from the latent heating due to water phase changes, radiative heating, and conduction, as relevant (see Section 4.3).

Models of moist mesoscale-convective processes include the equation of state for moist air:

$$p = \rho R_d T \left[\frac{1 + q_v/\varepsilon}{1 + q_v} \right] = \rho R_d T_v, \tag{4.6}$$

with moist density $\rho = p/R_d T_v$ used in (4.1) through (4.5), and with vapor and total cloud and precipitation mixing ratios q_v and q_T contributing to the buoyancy term in (4.3) (discussed later). Additionally, the specific heat c_v (c_p) is replaced with c_{vm} (c_{pm}) in (4.4) and (4.5), although it is often assumed, as it is here and with admission of some error, that $c_{vm} \simeq c_v$ and $c_{pm} \simeq c_p$.[6] A moist model also includes an equation governing vapor:

$$\frac{Dq_V}{Dt} = S_{q_v} + F_{q_v,\, turb}, \tag{4.7}$$

and equations for all j desired liquid and frozen species of cloud and precipitation particles:

$$\frac{Dq_j}{Dt} = S_{q_j} + F_{q_j,\, turb}. \tag{4.8}$$

Descriptions of microphysical sources and sinks S, and *subgrid scale* (SGS) turbulent mixing of water substance mixing ratio $F_{q,\, turb}$, are provided in Section 4.3.

The SGS contributions to F_u, F_v, and F_w are also discussed in Section 4.3, so let us consider here the remaining contributions, which were introduced in Chapter 2. The treatment of the curvature terms

$$F_{u,curv} = \frac{uv \tan\phi}{r_E} - \frac{uw}{r_E}$$
$$F_{v,curv} = -\frac{u^2 \tan\phi}{r_E} - \frac{vw}{r_E}$$
$$F_{w,curv} = \frac{u^2 + v^2}{r_E}$$

is guided by scale analysis similar to that developed in Chapter 2. For relatively small horizontal domains (L_x, L_y ~ a few hundred kilometers), as typically used in applications such as idealized thunderstorm modeling, the neglect of these terms is easily justified. In such small domains, the Coriolis force terms

$$F_{u,Cor} = v2\Omega \sin\phi - w2\Omega \cos\phi$$
$$F_{v,Cor} = -u2\Omega \sin\phi$$
$$F_{w,Cor} = u2\Omega \cos\phi$$

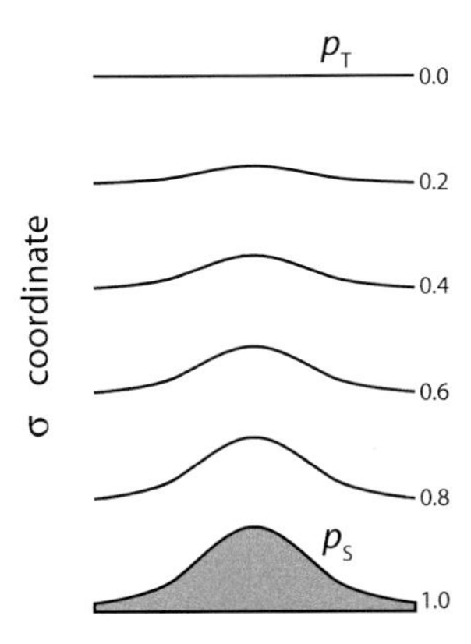

Figure 4.1 Example of a terrain-following vertical coordinate. Because the coordinate is normalized, it ranges from a value of 1 at the surface to 0 at the specified top of the computational domain.

are either similarly neglected,[7] or are reduced to

$$F_{u,Cor} = f_0 v, \quad F_{v,Cor} = -f_0 u, \quad F_{w,Cor} = 0. \tag{4.9}$$

Implicit in (4.9) is an f-plane assumption, with constant f assigned a value such as $f_0 = 10^{-4}\ \mathrm{s}^{-1}$, which assumes a midlatitude application. This reduced form of the Coriolis force is usually assumed to apply only to the *deviations* from a horizontally homogenous, vertically stratified, base-state velocity. Otherwise, the base-state velocity would need to be balanced with a base-state horizontal temperature gradient according to the thermal wind equation, and the base-state profile in the atmospheric boundary layer would need to satisfy a three-way balance among the Coriolis force, horizontal pressure gradient force (PGF), and shear stresses (see Section 4.3).[8]

The preceding equations are written explicitly in terms of height z as the vertical coordinate. However, mesoscale models, and NWP models in general, tend to employ some type of monotonic function of z as the vertical coordinate. The sigma coordinate:

$$\sigma = (p - p_T)/(p_S - p_T), \tag{4.10}$$

where $p_S = p_S(x, y, t)$ is the surface pressure and p_T is a specified domain-top pressure, belongs to the class of *terrain-following vertical coordinates*.[9] The advantage of such coordinates is that vertical model levels conform to the local terrain (Figure 4.1), thus allowing, for example, for a better representation of mountain and valley flows and their associated phenomena.

Pressure p is also replaced in some model formulations by a normalized, nondimensional pressure Π. This is done in part to reduce the computer round-off errors that accrue when finite differences in p are evaluated.[10] If T is also replaced by θ, the use of Π additionally eliminates density from the governing equations, thereby simplifying the model system. Thus, consider:

$$\Pi = \left(\frac{p}{p_0}\right)^{R_d/c_p}, \tag{4.11}$$

where $p_0 = 1000$ hPa, and Π is the *Exner function*. Substituting (4.11) in (4.5) and (4.4) gives, respectively,

$$\frac{D\Pi}{Dt} = -\Pi\frac{R_d}{c_v}\nabla\cdot\vec{V} + \frac{R_d}{c_p}\frac{1}{c_v\theta}\dot{Q} \tag{4.12}$$

and

$$\frac{D\theta}{Dt} = \frac{1}{\Pi c_p}\dot{Q}, \tag{4.13}$$

where

$$\theta = \frac{T}{\Pi}. \tag{4.14}$$

The use of (4.11), (4.14), (4.6), and an assumed decomposition $\Pi = \overline{\Pi}(z) + \Pi'$ in (4.1) through (4.3) also gives

$$\frac{Du}{Dt} = -c_p\theta_v\frac{\partial\Pi'}{\partial x} + F_u, \tag{4.15}$$

$$\frac{Dv}{Dt} = -c_p\theta_v\frac{\partial\Pi'}{\partial y} + F_v, \tag{4.16}$$

and

$$\frac{Dw}{Dt} = -c_p\theta_v\frac{\partial\Pi'}{\partial z} + B + F_w, \tag{4.17}$$

where θ_v is virtual potential temperature, and the base state of hydrostatic balance is now expressed as

$$\frac{d\overline{\Pi}}{dz} = -\frac{g}{c_p\overline{\theta}}. \tag{4.18}$$

Notice that in the Π-θ system, the buoyancy term

$$B = g\left[\frac{\theta_v}{\overline{\theta}_v} - 1 - q_T\right] \tag{4.19}$$

no longer contains pressure (cf. (2.62)). The equations governing water substance are unchanged, except for how the source/sink terms are expressed.

4.2.2 Numerical Approximations

The nonlinearity and general complexity of the continuous governing equations thwart exact solutions. It is possible, however, to form discrete approximations to the equations that then can be solved numerically. *Spectral* and *finite volume* methods are sometimes used towards this end, but herein we focus on the use of *finite difference* (FD) methods, which conceptually are relatively straightforward and enjoy widespread use in atmospheric mesoscale models.[11]

Let us begin some function $y = f(x)$. If the limit

$$\frac{dy}{dx} = \lim_{\Delta x \to 0} \frac{f(x + \Delta x) - f(x)}{\Delta x} \tag{4.20}$$

exists and is finite, the mean value theorem tells us that this limit is the derivative of y with respect to x. Equation (4.20) provides a way to understand the finite-differencing concept, but the FD approximations themselves stem from *Taylor series expansions*. For example,

$$f(x + \Delta x) = f(x) + \frac{df}{dx}\Delta x + \frac{d^2 f}{dx^2}\frac{(\Delta x)^2}{2!} + \cdots + \frac{d^n f}{dx^n}\frac{(\Delta x)^n}{n!} \tag{4.21}$$

is an approximation to the function f at a small positive distance Δx from x. If we truncate (4.21) after the df/dx term and hence neglect all higher-order terms, we have, on rearranging and dividing through by Δx,

$$\frac{df}{dx} = \frac{f(x + \Delta x) - f(x)}{\Delta x} + \mathrm{O}(\Delta x). \tag{4.22}$$

Equation (4.22) is known as a *first-order approximation* to the derivative df/dx, because the largest power of Δx common to all truncated terms is one; (4.22) is also known as a *one-sided* or *forward* FD approximation. The order notation O() is used to characterize the *truncation error* (TE), which is the difference between the complete Taylor series and the truncated approximation or, equivalently, the difference between a continuous derivative and its finite-difference approximation. In the case of (4.22), the truncation error is

$$|\mathrm{TE}| \leq a\,|\Delta x| \quad \text{as } \Delta x \to 0,$$

where a is some positive real constant. O() therefore communicates the behavior of the TE as $\Delta x \to 0$.[12]

Notice that a Taylor series for f can also be written in terms of a small negative distance from x:

$$f(x - \Delta x) = f(x) - \frac{df}{dx}\Delta x + \frac{d^2 f}{dx^2}\frac{(\Delta x)^2}{2!} + \cdots + \frac{d^n f}{dx^n}\frac{(-\Delta x)^n}{n!}. \tag{4.23}$$

If we truncate series (4.23) and (4.21) after the d^2f/dx^2 terms, subtract the results, and then divide through by Δx, we have

$$\frac{df}{dx} = \frac{f(x+\Delta x) - f(x-\Delta x)}{2\Delta x} + O(\Delta x)^2, \tag{4.24}$$

which is the "centered" FD approximation of the derivative df/dx. In this second-order or $O(\Delta x)^2$ approximation, the TE approaches zero at a relatively faster rate as $\Delta x \to 0$ than does the TE of the $O(\Delta x)$ approximation, thus connoting relatively more accuracy in the $O(\Delta x)^2$ approximation.

Higher-order approximations of the first derivative follow from the basic procedure of combining truncated Taylor series, as do approximations of higher-order derivatives. For example:

$$\frac{d^2f}{dx^2} = \frac{f(x+\Delta x) - 2f(x) + f(x-\Delta x)}{(\Delta x)^2} + O(\Delta x)^2 \tag{4.25}$$

is formed by truncating (4.21) and (4.23) after the d^2f/dx^2 terms and adding the results.

Each continuous derivative in an equation such as (4.15) is discretized using finite differencing. We can illustrate this simply with the aid of the 1D linear advection equation

$$\frac{\partial A}{\partial t} + c\frac{\partial A}{\partial x} = 0, \tag{4.26}$$

where $A = A(x,t)$ is some scalar function, and c is a constant wind speed. Of the myriad ways to discretize the terms in (4.26), one popular choice is:

$$\frac{A(x,t+\Delta t) - A(x,t-\Delta t)}{2\Delta t} + c\left[\frac{A(x+\Delta x,t) - A(x-\Delta x,t)}{2\Delta x}\right] = 0. \tag{4.27}$$

This centered-in-time, centered-in-space FD approximation is second-order accurate. It is often referred to as the *leapfrog scheme* , and can be expressed symbolically as

$$A_i^{n+1} = A_i^{n-1} - c\frac{\Delta t}{\Delta x}\left(A_{i+1}^n - A_{i-1}^n\right), \tag{4.28}$$

where index n corresponds to a time level, Δt is the time difference between successive time levels (or *time step*), index i corresponds to a grid point in space, and Δx is spatial difference between adjacent grid points (or *grid point spacing*), The actual implementation of the right-hand side of (4.28) on a grid and, more generally, of a centered-in-space FD approximation of $\partial A/\partial x$ ($\partial A/\partial y$) at grid point i (j), is shown graphically in Figure 4.2; an extension of (4.28) to advection in 2D would require the FD approximations to spatial derivatives in x and y.

Equation (4.28) represents a simple numerical model that, upon specification of an initial and boundary conditions, is "marched" forward in time. The numerical

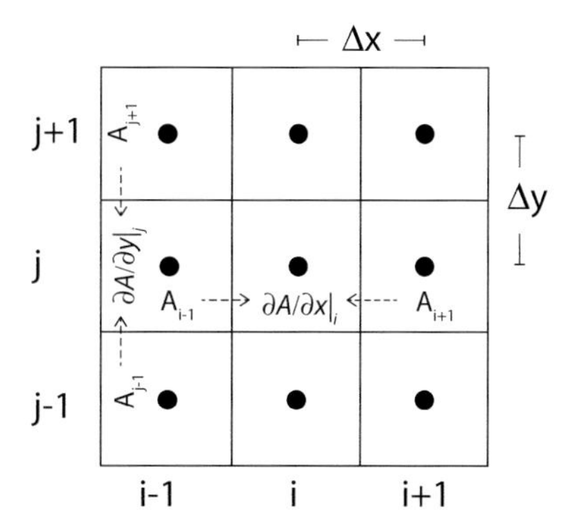

Figure 4.2 Implementation of a second-order, centered FD approximation of $\partial A/\partial x$ at grid level i, and as well as of a second-order, centered FD approximation of $\partial A/\partial y$ at grid level j. The closed circles are grid points, and Δx and Δy indicate the distances between the grid points in the respective Cartesian directions.

solution from this forward integration posseses small errors that have the potential for unbounded growth, in absence of the following imposed constraint:

$$|c\Delta t/\Delta x| \leq 1. \tag{4.29}$$

Equation (4.29) is the *Courant-Friedrichs-Lewy* (CFL) criterion for numerical stability of (4.28), as determined from theoretical analysis.[13] It is imposed through the time step Δt, computed from (4.29) using the problem-specific values of c and Δx. In applications of the leapfrog scheme to problems with non-constant advective winds (e.g., to the time tendency and advection terms on the left-hand side of equations such as those in Section 4.2.1), Δx is still specified, but Δt follows from an evaluation of the CFL criterion using an estimate (or actual determination) of the fastest advective speed (V_{max}) to be encountered in the particular problem.

Numerical stability is of critical importance in model design, but other – perhaps competing – factors must also be considered, including the following:[14]

- *Computational efficiency*, which regards the relative number of computations per time step required by the FD scheme, and has particular relevance for numerical weather prediction models
- *Accuracy*, which relates back to the FD approximation, but can be quantified upon comparison of the numerical solution to analytic solutions in idealized problems
- *Monotonicity*, or positive definiteness, which essentially is the extent to which the scheme precludes negative values of a physically positive quantity, such as water-substance mixing ratio
- *Numerical diffusiveness*, or the degree to which the scheme damps the solution, which can be desirable because a diffusive scheme also suppresses noise

Compromises are necessary in other aspects of the numerical model, such as in the treatment of compressibility,[15] or in the chosen complexity of parameterized processes, as we will see next.

4.3 Parameterizations

The goal of this section is to illustrate basic approaches to *parameterizations*, which represent complicated physical processes as simplified functions of model variables. Usually, the processes are sources/sinks in the continuous equations. When expressed in a complete, nonsimplified way, these sources/sinks often have their own sets of equations, but are intractable even when numerically approximated, or are just too computationally intensive. Thus, we will find that most parameterization schemes make use of empirical relationships to bypass additional equations and/or computations and, consequently, to "close" the basic set of governing equations (e.g., (4.1)–(4.8)). Examples of widely employed or particularly straightforward schemes will be used in the following sections to demonstrate basic formulations.[16] Some types of parameterizations, such as those involving atmospheric chemistry, are excluded here but are treated well elsewhere.[17] It is important for the reader – and potential model user – to recognize that presentation of a scheme herein does not necessarily represent an endorsement or recommendation: the user has a responsibility to make an informed decision on the capabilities and limitations of a scheme, its complexity (and, therefore, computational requirements), and its appropriateness for specific application.

4.3.1 Subgrid Scale Turbulence

Imagine a developing cumulonimbus cloud. Its visual appearance is partly a result of a broad range of turbulent eddies within the cloud and along its periphery. The eddies simultaneously entrain environmental air into the cloud's vertical drafts and detrain draft air into the environment, and in doing so contribute to the convective dynamics (see Chapter 6). It should be evident that an accurate model simulation of this cloud and its salient dynamics requires proper representation of the effects of turbulent eddies, over all sizes.

The largest of these eddies are well resolved on model grids with spacings (Δx) of several hundred meters.[18] Resolution of the progressively smaller turbulent eddies, down to the ones responsible for energy dissipation, requires grid point spacings of several millimeters or less (and of a correspondingly small time step; Eq. (4.29)). Unfortunately, a computational domain of sufficient size for convective-cloud simulation, but with millimeter grid point spacings, is currently (and foreseeably) unfeasible, owing to computer resource limitations. Hence, the generally accepted strategy is to parameterize the effects of the small eddies, using a grid point spacing that lends to a computationally feasible problem but still is capable of resolving the large, energetic eddies. This is the essence of *large-eddy simulation* (LES), although strict LES assumes that the grid point spacing is much smaller than the scale of the phenomenon of interest, much larger than the scale of the dissipative eddies, and well within the *inertial subrange*.[19]

A methodology for parameterizing the effects of SGS eddies is revealed through a simple application of *Reynolds averaging*.[20] Let us begin by expanding the advection term in (4.1) as

$$\vec{V} \cdot \nabla u = -u \nabla \cdot \vec{V} + \nabla \cdot \left(u \vec{V}\right)$$

and then using the incompressible form of the continuity equation to rewrite (4.1) in *flux form*:

$$\frac{\partial u}{\partial t} = -\frac{1}{\rho_0}\frac{\partial p}{\partial x} - \nabla \cdot \left(u \vec{V}\right), \tag{4.30}$$

where the F_u terms have been neglected (including internal friction), ρ_0 is a constant reference density consistent with the incompressibility assumption, and, for ease of presentation (but without loss of physical interpretation), the PGF has been written with a total rather than deviation pressure. Next, we decompose the dependent variables as

$$\begin{aligned} \vec{V} &= \left\langle \vec{V} \right\rangle + \vec{V}' \\ p &= \langle p \rangle + p' \end{aligned}, \tag{4.31}$$

where the brackets indicate application of a filter, which commonly is a grid-volume average, and primes indicate a deviation from that averaged quantity; in studies of turbulence, the primes would represent turbulent fluctuations from a time-averaged state. The averaged quantities are assumed to be well resolved on the grid; the deviations are unresolved and represent SGS processes.

For a volume average, the following relations hold for symbolic variables A and a:

$$\langle A' \rangle = 0, \quad \langle \langle A \rangle \rangle = \langle A \rangle, \quad \langle \langle A \rangle a' \rangle = 0, \quad \langle \langle A \rangle \langle a \rangle \rangle = \langle A \rangle \langle a \rangle. \tag{4.32}$$

Importantly, the volume average of the products of deviation quantities are not necessarily zero; hence $\langle A'a' \rangle \neq 0$. Substituting (4.31) into (4.30) and then applying the volume average to the entire equation yields

$$\begin{aligned} \frac{\partial \langle u \rangle}{\partial t} = &-\frac{1}{\rho_0}\frac{\partial \langle p \rangle}{\partial x} - \frac{\partial}{\partial x}(\langle u \rangle \langle u \rangle) - \frac{\partial}{\partial y}(\langle u \rangle \langle v \rangle) - \frac{\partial}{\partial z}(\langle u \rangle \langle w \rangle) \\ &- \frac{\partial}{\partial x}(\langle u'u' \rangle) - \frac{\partial}{\partial y}(\langle u'v' \rangle) - \frac{\partial}{\partial z}(\langle u'w' \rangle). \end{aligned} \tag{4.33}$$

The first four terms in (4.33) equate to the grid-resolved form of (4.30) and are integrated using the numerical techniques discussed earlier (see Section 4.2.3). The remaining three terms comprise the divergence of the unresolved *eddy momentum*

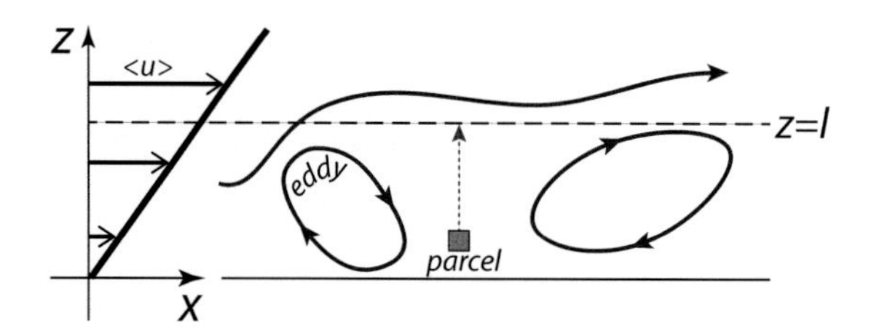

Figure 4.3 A basis for the representation for SGS turbulent mixing. This example depicts the vertical transfer of a parcel's momentum by turbulent eddies (thin arrowed lines) over a length l. Only vertical mixing of x-momentum, and a linear profile of (grid-resolved) wind $<u>$, is assumed. Through mixing-length theory, the magnitude of mixing coefficient K_m, and thus the momentum flux, increases with the grid-resolved shear and mixing length l. Based on Tennekes and Lumley (1972) and Stull (1988).

fluxes,[21] and effectively represent, in F_u, F_v, and F_w, a frictional-type force due to small eddies. Because these "eddy mixing" terms are unresolved, they cannot be numerically integrated directly, and thus require some approximation based on resolved variables (i.e., the parameterization).

It is convenient at this point to rewrite the eddy mixing terms in a form that makes use of *index notation*:

$$F_{u_i,turb} = \frac{\partial \tau_{i,j}}{\partial x_j}, \tag{4.34}$$

where $i = 1, 2, 3$, and summation over the other index ($j = 1, 2, 3$) is assumed. The quantity $\tau_{i,j} = -\langle u_i' u_j' \rangle$ is the *Reynolds stress tensor*, and can be interpreted as the rate of mean momentum transfer by turbulent eddies (see Figure 4.3).[22] Through the *K-theory* approach to the turbulence parameterization, the Reynolds stress tensor is expressed as

$$\tau_{i,j} = -\left\langle u_i' u_j' \right\rangle = K_m D_{i,j}, \tag{4.35}$$

where K_m is the coefficient of eddy diffusivity, also referred to as *eddy viscosity* or *eddy mixing coefficient*, and $D_{i,j}$ is the deformation tensor

$$D_{i,j} = \frac{\partial \langle u_i \rangle}{\partial x_j} + \frac{\partial \left\langle u_j \right\rangle}{\partial x_i}. \tag{4.36}$$

As indicated, $D_{i,j}$ is determined from the grid-resolved velocity. The task at hand, then, is to approximate K_m so the equation can be closed and numerically integrated.

In the simplest way, K_m is treated as a constant (positive) coefficient. An alternative, yet still first-order, approximation is through *mixing length theory*:[23]

$$K_m = C_m l^2 |D_{i,j}|, \tag{4.37}$$

where C_m is a dimensionless constant and l is a mixing-length scale set equal to a measure of the size of the grid volume such as,

$$l = (\Delta x \Delta y \Delta z)^{1/3}. \tag{4.38}$$

If the horizontal grid point spacing is much different from the vertical grid point spacing, separate horizontal and vertical mixing lengths are used, leading to separate horizontal and vertical eddy diffusivity coefficients and, thus, anisotropic mixing.[24] The degree of mixing is driven by the magnitude of the grid-resolved deformation – in effect, 3D wind shear – as modulated by the length scale (Figure 4.3). In some formulations, (4.37) is modified to include a Richardson number factor that accounts for buoyancy effects on turbulence and, thus, on mixing.

The 1.5-order turbulence kinetic energy (TKE) closure approximation allows for time-dependent production and dissipation of turbulence by buoyancy and shear.[25] The Reynolds stress tensor is expressed as

$$\tau_{i,j} = K_m D_{i,j} - \frac{2}{3}\delta_{ij} E, \tag{4.39}$$

where δ_{ij} is the Kronecker delta ($= 1$ for $i = j$, $= 0$ for $i \neq j$), K_m is related to the SGS TKE as

$$K_m = C_m E^{1/2} l, \tag{4.40}$$

C_m and l are as defined in (4.37), and kinetic energy $E = \frac{1}{2}\langle u_i' u_i' \rangle$ is computed through a separate prognostic equation of the general form

$$\frac{DE}{Dt} = -\langle u_i' u_j' \rangle \frac{\partial \langle u_i \rangle}{\partial x_j} + \frac{g}{\theta}\langle w' \theta' \rangle + \frac{\partial}{\partial x_j}\left(K_m \frac{\partial E}{\partial x_j}\right) - \varepsilon. \tag{4.41}$$

The respective terms in (4.41) represent the effects of shear, buoyancy, diffusion, and dissipation on E. Each has an SGS part that requires a separate approximation. The shear term follows from (4.35), the diffusion term uses (4.40), and the dissipation term

$$\varepsilon = \frac{C_\varepsilon}{l} E^{3/2}, \tag{4.42}$$

comes from the Kolmogorov hypothesis, where C_ε is a constant empirical coefficient. The buoyancy term – written with respect to dry air but expandable to include moist air processes – is the turbulent eddy heat flux, and originates from the Reynolds averaged thermodynamic energy equation. Thus, the thermodynamic energy equation has an SGS heat flux divergence

$$F_{\theta,turb} = \frac{\partial \tau_{\theta,i}}{\partial x_i} \tag{4.43}$$

and $\tau_{\theta,i}$ is approximated through

$$\tau_{\theta,i} = -\langle u_i'\theta' \rangle = K_H \frac{\partial \langle \theta \rangle}{\partial x_i}. \tag{4.44}$$

The mixing coefficient K_H follows from the relationship $K_H = K_m/Pr$, where Pr is the Prandtl number. This mixing coefficient is also used in SGS terms in the water substance equations, for instance

$$F_{q,turb} = \frac{\partial}{\partial x_i}\left(K_H \frac{\partial \langle q \rangle}{\partial x_i}\right) \tag{4.45}$$

for $q = q_v$, q_r, and so on.

4.3.2 The Land Surface and Atmospheric Boundary Layer

Vertical eddies within the atmospheric boundary layer (ABL) are one means through which the meteorological state at the Earth's surface is coupled to the free troposphere. These eddies develop in a heated (and vertically sheared) ABL, and serve to vertically mix heat, moisture, and momentum throughout the ABL; in some instances, their consequence is the initiation of convective clouds (see Chapter 5). Such mixing – which is still subgrid scale unless Δz is significantly less than a few hundred meters – is distinct from the isotropic SGS mixing and/or vertical SGS mixing in the free troposphere, thus meriting a separate parameterization.

An outcome common to ABL parameterization schemes is the representation of vertical SGS fluxes over grid columns within the ABL. The details of how this is accomplished depend mostly on whether a *local* or *nonlocal* approach is taken. In a local ABL scheme, the parameterized calculation of a SGS flux at some height is literally from model data at or near this height. The calculation itself typically follows from K-theory; for example,

$$-\frac{\partial}{\partial z}\langle w'\theta' \rangle = \frac{\partial}{\partial z}\left[K_H \frac{\partial \langle \theta \rangle}{\partial z}\right], \tag{4.46}$$

where K_H is again described by $K_H = K_m/Pr$, and K_m is determined from a relation similar to (4.37) or (4.40). Equation (4.46) approximates an SGS term implicit in the right-hand side of (4.13). Because the mixing coefficients K_H and K_m are necessarily positive, as defined previously, the eddy fluxes are constrained to be *downgradient*. Accordingly, the eddy transfers (of heat, etc.) will be in the direction opposite to that of the gradient, and thus from regions of higher values to regions of lower values. Under this constraint, the consequence of a local scheme is a well-mixed ABL.

This consequence can be viewed as a limitation of local schemes, however, because superadiabatic layers are known to exist in the ABL, as are vertical gradients in ABL moisture and momentum.[26] Hence, consider a widely used nonlocal formulation,

$$\frac{\partial}{\partial z}\left[K\left(\frac{\partial \langle\theta\rangle}{\partial z}-\gamma_\theta\right)\right], \tag{4.47}$$

where the eddy diffusivity K depends highly on the boundary-layer depth, which is quantified through the use of Richardson-number-like criteria.[27] The factor γ_θ corrects the local gradient for the effects of large eddies, and is a function of the vertical SGS flux at the surface. Perhaps more importantly, γ_θ allows for *upgradient* transfer of heat (and moisture and momentum).[28] In other words, the flux can be in the same direction as the gradient, effectively allowing for gradients to be increased rather than removed.

As just alluded to, fluxes from the Earth's surface provide the lower boundary conditions on ABL parameterizations. In the context of the discrete equations, these surface fluxes allow for an approximation of vertical derivatives at the lowest model level. In some types of idealized modeling experiments, surface fluxes are specified with constant or even random values. In real-data mesoscale model applications, they are usually determined using surface exchange coefficients and output from a *land-surface model* (LSM).[29]

LSMs parameterize the interaction of the land surface and its underlying soil properties with short- and long-wave radiation, sensible and latent heating, precipitation, and ultimately with the overlying atmosphere (Figure 4.4). LSMs are based in part on a *surface energy budget*:

$$F_{rad}^{sfc} = F_{SH}^{\uparrow} + F_{LH}^{\uparrow} + F_{G}^{\downarrow} + F_M, \tag{4.48}$$

which states that the net surface radiative flux[30] is balanced by the energy lost to the atmosphere as sensible heat (↑SH), the energy lost as latent heat from the evaporation of water at the surface (↑LH), the heat flux into the ground (↓G), and the heat flux (M) associated with the melting (freezing) of snow or ice (water).[31] Surface energy storage is neglected because of the assumed balance in (4.48). The net surface radiative flux,

$$F_{rad}^{sfc} = F_{SW}^{\downarrow}(1 - A_{sfc}) - F_{LW}^{\uparrow} + F_{LW}^{\downarrow}, \tag{4.49}$$

depends on incoming (↓SW) solar or shortwave radiation, and incoming (↓LW) and outgoing (↑LW) surface longwave radiation; these derive from a separate parameteriation (see Section 4.3.4). The surface albedo, A_{sfc}, depends on the specification of the land use type (urban, cropland, grassland, forest, etc.), which is an LSM input

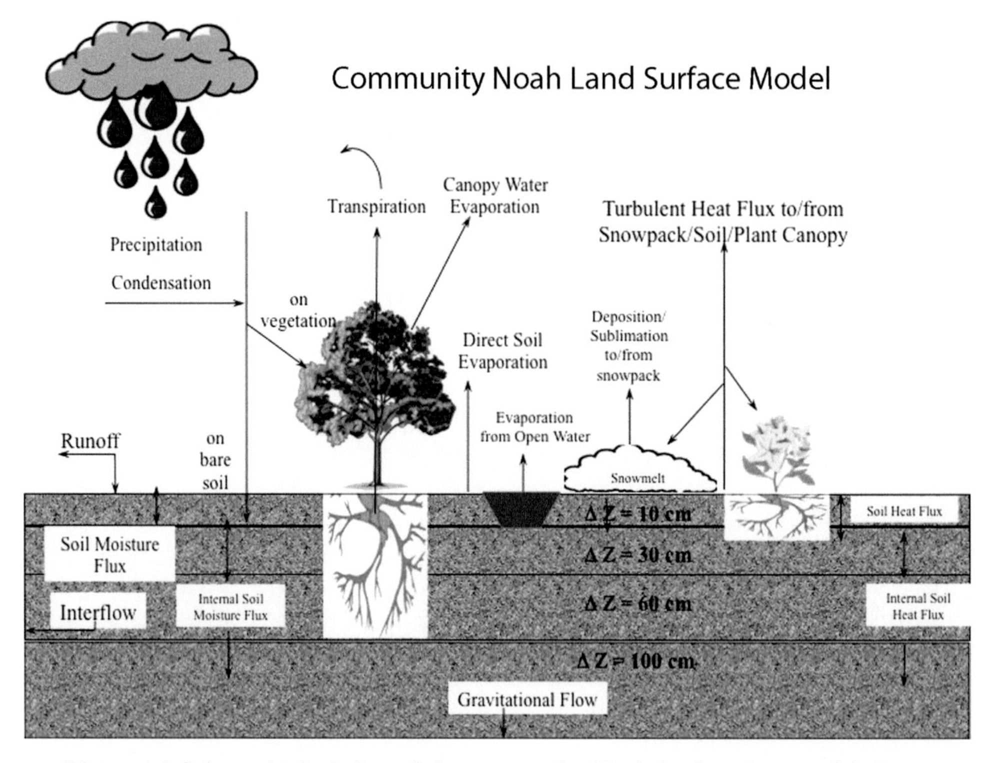

Figure 4.4 Schematic depiction of the community Noah land-surface model. From Chen and Dudhia (2001).

variable. An output variable common to LSMs is the temperature of the land surface (skin temperature), as provided through (4.48) via the:

- Surface sensible heat flux, $\sim c_p|\vec{V}|(T_{sfc} - T_a)$, where $|\vec{V}|$ and T_a are the wind and temperature at the lowest model level
- Outgoing longwave radiative flux, $\sim T_{sfc}^4$, by virtue of the Stefan-Boltzmann law
- Ground heat flux, $\sim T_{sfc} - T_{soil}$

The soil temperature is usually determined from a separate, time-dependent equation in the form of

$$c_g \frac{\partial T_{soil}}{\partial t} = \frac{\partial}{\partial z}\left(k_g \frac{\partial T_{soil}}{\partial z}\right), \tag{4.50}$$

where k_g are c_g are the thermal conductivity and heat capacity of the soil, respectively. Equation (4.50) has the form of the heat conduction equation, thus governing the (vertical) transfer of heat from warm to cool soil, and the corresponding local vertical profile of soil temperature. It is applied, in 1D, to two or more interacting soil layers, one of which is in contact with the ground surface ($z = 0$). If the soil is frozen, an additional term accounting for the latent heat of melting/freezing is required.[32]

In simple models, the thermal conductivity and heat capacity in (4.50) are treated as constants, but in actuality, both are dependent on variations – over time and depth – in the *soil volumetric water content* (Θ, or fraction of a unit of soil volume occupied by liquid water) and in other soil properties. The variation in Θ is governed by an equation of the form

$$\frac{\partial \Theta}{\partial t} = \frac{\partial}{\partial z}\left(K_{\Theta}\frac{\partial \Theta}{\partial z}\right) + \frac{\partial k_{\Theta}}{\partial z} + S_{\Theta}, \tag{4.51}$$

where K_{Θ} is the soil water diffusivity, k_{Θ} is hydraulic conductivity, and S_{Θ} contains soil-water sources and sinks, such as precipitation, evaporation, and runoff.[33] Analogous to the equation for soil temperature, this equation includes the effects of diffusion of water toward drier soil, and governs the local vertical profile of soil moisture. It is similarly applied in 1D to interacting soil layers, although the sources and sinks of each layer may differ and will depend on the complexity of the model. For example, precipitation would be a source for the layer in contact with the ground surface; uptake of water by the roots of plants would be a sink for this upper layer and perhaps for deeper layers, depending on the type of vegetation that is assumed at the grid point.

Evaporation is a soil moisture sink in the upper layer, and connects the subsurface/surface properties back to the surface energy balance through the surface latent heat flux term, $F_{LH} \sim L_v|\vec{V}|[q_{vs}(T_{sfc}) - q_{v,a}]$, where $q_{v,a}$ is the water vapor mixing ratio at the lowest model level, and q_{vs} is the saturation vapor mixing ratio as a function of the skin temperature; thus, evaporation ultimately contributes to the surface water-vapor mixing ratio and then to a surface moisture flux. Additional contributions come from transpiration from vegetation, and evaporation of liquid water from a plant canopy, depending again on the complexity of model (e.g., Figure 4.4). As perhaps will be appreciated even more following the discussion of cloud and precipitation microphysical parameterizations, the decision of when to use more or less complexity is directed by the model application, and by the available computational resources.

4.3.3 Cloud and Precipitation Microphysics

The guiding principle behind the water substance equations, and the source/sink terms therein, is a conservation of total water in absence of SGS mixing and diffusion. The terms themselves represent exchanges of water among the vapor, liquid, and solid phases. Because a by-product of these exchanges is diabatic heating, the microphysical processes influence the air motion. The air motion, in turn, influences the microphysics, because the wind transports the water substances throughout the cloud and from (to) the cloud interior to (from) its environment, thereby actuating further exchanges.

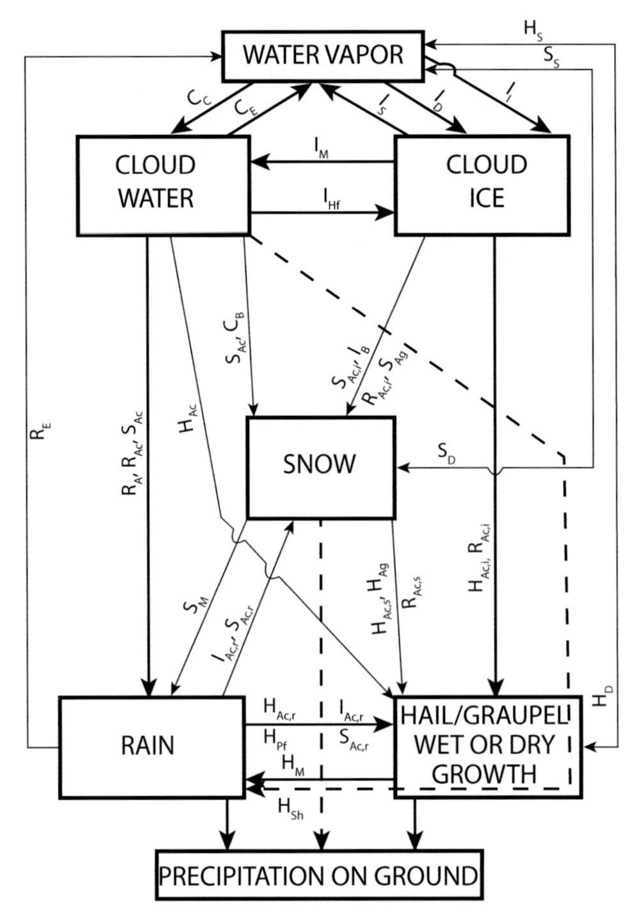

Figure 4.5 Flow chart illustration of the ice- and liquid-phase microphysics scheme presented by Gilmore et al. (2004).

A complete example of a microphysical-dynamical feedback is provided in Chapter 6 through the use of 1D cloud models. Here we simply introduce the exchanges in a generally symbolic manner. Although it is tempting to bypass the complexities of an ice–liquid microphysics scheme (see Figure 4.5) in favor of a simple, rain-only scheme, the inclusion of ice microphysics is considered paramount to accurate numerical simulation and prediction of most mesoscale-convective processes. Thus we resist this temptation and consider a set of coupled exchanges[34] that involve the mixing ratios of vapor, cloud water (q_c), cloud ice (q_i), rain (q_r), snow (q_s), and graupel/hail (q_h):

$$S_{q_v} = (C_E + R_E + I_S + S_S + H_S) - (C_C + I_I + I_D + S_D + H_D) \tag{4.52}$$

$$S_{q_c} = (C_C + I_M) - \left(C_E + C_B + R_A + R_{Ac} + I_{Hf} + S_{Ac} + H_{Ac}\right) \tag{4.53}$$

$$S_{q_i} = \left(I_{Hf} + I_I + I_D\right) - \left(R_{Ac,i} + I_M + I_S + I_B + S_{Ag} + S_{Ac,i} + H_{Ac,i}\right) \tag{4.54}$$

$$S_{q_r} = \left(R_A + S_{Ac} + R_{Ac} + S_M + H_M + H_{Sh}\right) - \left(R_E + I_{Ac,r} + S_{Ac,r} + H_{Ac,r} + H_{Pf}\right) \tag{4.55}$$

$$S_{q_s} = \left(C_B + R_{Ac,i} + I_{Ac,r} + I_B + S_{Ac} + S_{Ac,r} + S_{Ac,i} + S_D + S_{Ag}\right) - \left(R_{Ac,s} + S_M + S_S + H_{Ac,s} + H_{Ag}\right) \tag{4.56}$$

$$S_{q_h} = (R_{Ac,i} + R_{Ac,s} + I_{Ac,r} + S_{Ac,r} + H_{Ac} + H_{Ac,r} + H_{Ac,s} + H_{Ac,i} + H_{Ag} + H_D + H_{Pf}) - (H_{Sh} + H_M + H_S), \tag{4.57}$$

where, in order of their appearance in (4.52) through (4.57):

- C_E and R_E are the evaporation rates of cloud droplets and raindrops, determined mostly by the ambient humidity, and dependent on the drop size distribution. Note that C_E, R_E, and all other the exchanges directly involving a phase change contribute to diabatic heating $\dot{Q}$ in the thermodynamic energy equation.
- I_S, S_S, and H_S are respective sublimation rates of cloud ice, snow, and hail, and similarly are functions of the ambient humidity and size distributions, but are active only at subfreezing temperatures.
- C_C is the rate of condensation of vapor into cloud droplets. Essentially, for any occurrence of supersaturation ($q_v > q_{vs}$), the excess water vapor ($q_v - q_{vs}$) is converted into a (specified) distribution of cloud droplets.
- I_I is the initiation of cloud ice. Analogous to C_C, when the air is supersaturated *with respect to ice*, and temperatures are below freezing, excess vapor is converted into a (specified) distribution of ice crystals.
- I_D, S_D, and H_D are respective deposition rates of cloud ice, snow, and hail. Snow and hail deposition represent "dry growth." Deposition of vapor on cloud ice represents growth to snow.
- I_M is the melting rate of cloud ice, which occurs for temperatures above freezing, and converts into cloud water.
- C_B is the transfer of cloud water to snow at subfreezing temperatures via the *Bergeron process*.[35]
- R_A is autoconversion of cloud water to rain, a simple transfer that activates when q_c exceeds some threshold.
- R_{Ac}, S_{Ac}, and H_{Ac}, respectively, represent the accretion of cloud droplets by rain, snow, and hail. These hydrometeors collect cloud droplets as they fall through the cloud, with rates that depend mostly on the hydrometeor diameter. For temperatures above freezing, accretion by snow is a source of rain.

- I_{Hf} is the rate of homogeneous freezing of cloud droplets into ice crystals, which is assumed to occur at temperatures below $-40°$C.
- $R_{Ac,i}$, $S_{Ac,i}$, and $H_{Ac,i}$, respectively, represent the accretion of cloud ice by rain, snow, and hail, which occurs in way that is similar to the accretion of cloud droplets. The likelihood at which these hydrometeors collect cloud ice (collection efficiencies) can vary with temperature. When rain accretes ice, the result is either hail/graupel or snow, depending on q_r.
- I_B is the transfer of cloud ice to snow through deposition and riming in a supercooled cloud, when the Bergeron process is active.
- S_{Ag} is the aggregation of ice crystals to form snow. Physically, this rate would depend on temperature and humidity, which, in turn, controls habit type and how the ice crystals interact. In simple parameterization schemes, this is represented crudely by a transfer that activates when q_i exceeds some threshold.
- S_M and H_M are the respective rates at which snow and hail melt to form raindrops.
- H_{Sh} is the rate at which raindrops are shed from hail stones, and is distinct from H_M; both processes occur during "wet growth."
- $I_{Ac,r}$, $S_{Ac,r}$, and $H_{Ac,r}$ represent the accretion of rain by cloud ice, snow, and hail, respectively, and are treated in way that is similar to the accretions of other species. This is a form of "wet growth." In the case of $I_{Ac,r}$, the accretion of rain by cloud ice results in the creation of snow or hail rather than in an increase in cloud ice.
- H_{Pf} is the (probabilistic) rate at which raindrops freeze to become hail.
- $R_{Ac,s}$ and $H_{Ac,s}$, respectively, represent the accretion of snow by rain and hail.
- H_{Ag} is the aggregation of snow to form hail, which has a similar representation to S_{Ag}.

Although not listed explicitly, hydrometeor fallout or sedimentation also contributes to the rate change of water substance, and has the general form $1/\rho\partial/\partial z(\overline{V}_{f,j}\rho q_j)$, where $\overline{V}_{f,j}$ is the mean fallspeed of hydrometeor type j.

Equations (4.7) and (4.8), with contributions such as those in (4.52) through (4.57), comprise a *bulk microphysics scheme*. In bulk schemes, a distribution of cloud and precipitation particles is predicted within the volume associated with each grid point, rather than the individual particles themselves. The general form of the *drop size distribution* (DSD) is constrained by a function such as:

$$N_j(D) = N_{0j}e^{-\lambda_j D_j}, \tag{4.58}$$

where $N_j(D)$ is the concentration, per unit volume, of drops with a diameter in the interval $(D, D + dD)$; N_{0j} is a specified constant, known as the intercept (literally, the value of N_j when the distribution curve intercepts the y-axis; i.e., when D_j is zero); and $j = r$, s, or h denotes the species. Equation (4.58) is commonly referred to as a *Marshall-Palmer distribution*. The slope λ_j of the distribution is the inverse of the mean diameter, or

$$D_{Nj} = \lambda_j^{-1} = [\rho q_j/(\pi\rho_j N_{0j})]^{1/4}, \tag{4.59}$$

where ρ_j is the density of rain, snow, or hail, and is assumed to be constant. The total number concentration of particles is then

$$\mathbf{N}_j = N_{0j} D_{Nj}. \tag{4.60}$$

Hence, the species mixing ratio q_j predicted at each grid point determines the slope and mean diameter of the local drop-size distribution through (4.59). Both values, as well as total number concentration $\mathbf{N}_j$, are required in the evaluation of several of the exchange terms, such as the accretion rates. These rates contribute to the local change in mixing ratio (4.8), with the new q_j value then used to reevaluate the new mean diameter, and so on.

In this example, the cloud droplet size distribution is assumed to be *monodisperse*, or, in other words, the droplet population has a uniform diameter

$$D_c = \left(\frac{m_c}{\rho_w} \frac{6}{\pi} \right)^{1/3}, \tag{4.61}$$

and a uniform droplet mass

$$m_c = q_c \rho / N_{ccn}, \tag{4.62}$$

that is determined by the cloud water mixing ratio, and by the specified number of cloud condensation nuclei (N_{ccn}); all nuclei are assumed to be activated. Thus, the cloud-water mixing ratio at each grid point determines the mass and (size-limited) diameter of the local cloud-droplet distribution. This information is used to evaluate exchange terms, such as accretion rates, which contribute to local changes in q_c through (4.8), a new q_c, and then a new diameter and mass. Relations among D_i, m_i, and q_i for an assumed monodisperse distribution of cloud ice are similarly used to predict q_i.

An alternative to the preceding *single-moment bulk scheme* is one in which mixing ratio as well as number concentration are predicted. In a *double-moment bulk scheme*, an additional equation governs the rate change of number concentration

$$\frac{DN_j}{Dt} = S_{N_j} + F_{N_j}, \tag{4.63}$$

where F_{Nj} is SGS mixing, and S_{Nj} are the source/sink terms.[36] Although the underlying DSD is still constrained to a function such as that due to Marshall-Palmer, (4.63) contributes to a time-varying DSD slope and intercept value,

$$\begin{aligned} \lambda_j &= \left[\frac{\Gamma(1+d_j) c_j N_j}{\rho q_j} \right]^{1/d_j}, \\ N_{0j} &= N_j \lambda_j \end{aligned} \tag{4.64}$$

where Γ is the gamma function, and parameters c_j and d_j follow from how the mass is related to diameter, $m_j = c_j D^{d_j}$, for the particular hydrometeor.[37] The benefit of a double-moment scheme is a more physically realistic evolution of the DSD: it allows, for example, for depletion of the small sizes in the distribution that result from collection of small raindrops by large raindrops, and thus for the corresponding upward shift in the lower range of sizes in the distribution. Such an improved treatment of the DSD influences the exchange terms as well as the precipitation rate, amount of diabatic heating, and microphysical-dynamical feedback. The cost, of course, is additional computations.

Another alternative, albeit with an even higher computational cost, is an explicit microphysics parameterization. Explicit schemes require additional equations that govern the evolution of drops in diameter bin sizes dD, and thus are not constrained by a specified distribution function.[38] As with higher-moment bulk schemes, the benefits of an explicit microphysics approach would need to be weighed against its computational expense and the expenses inherent in other parameterized processes, the domain size, gridpoint spacing, and so on. This practical need to compromise poses an ever-present challenge to users of numerical models.

Irrespective of the choice of parameterization scheme, a means to compare the model output with observations allows for model evaluation and also facilitates interpretation. Although observations of DSDs are collected from airborne and ground-based platforms during field campaigns, a more readily available source of quantitative information about precipitation structure is weather radar (see Chapter 3). It is possible to the convert the predicted distributions to a radar reflectivity factor as follows:

$$Z_{er} = 720 N_{0r} \lambda_r^{-7} \times 10^{18}, \tag{4.65}$$

$$Z_{es} = 161.3 N_{0s} \lambda_s^{-7} \left(\frac{\rho_s}{\rho_w} \right)^2 \times 10^{18}, \tag{4.66}$$

$$\text{and } Z_{eh} = 161.3 N_{0h} \lambda_h^{-7} \left(\frac{\rho_h}{\rho_w} \right)^2 \times 10^{18}, \tag{4.67}$$

where Z_e indicates *equivalent reflectivity*.[39] Summing over the hydrometeor categories

$$Z_e = Z_{er} + Z_{es} + Z_{eh} \tag{4.68}$$

and then converting to dBZ gives

$$\text{SRF} = 10 \log (Z_e), \tag{4.69}$$

which is referred to as simulated radar reflectivity factor (SRF). Examples of applications of SRF from high-resolution numerical weather prediction models are given in Chapter 10.

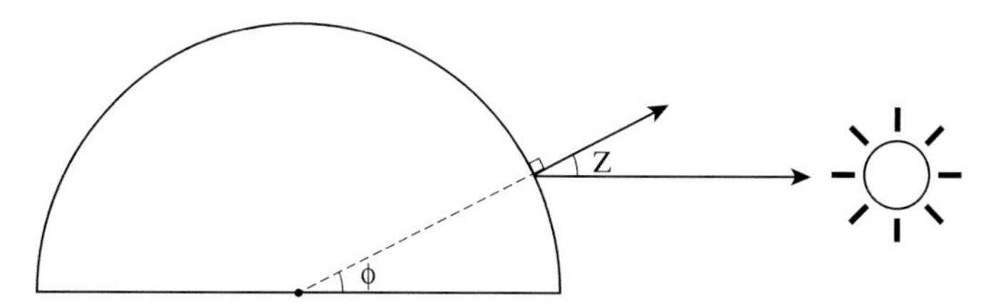

Figure 4.6 Graphical depiction of solar zenith angle.

4.3.4 Atmospheric Radiation

Short- and long-wave radiative fluxes at the surface are critical components to the surface energy budget and hence to LSMs; more generally, F_{SW} and F_{LW} contribute to diabatic heating over the entire atmosphere. Although the sun is the source of the atmospheric radiation, the radiative fluxes are not constants, even over short-time scales: they depend on time-varying details of clouds and precipitation, the chemical composition of the atmosphere, geographical location, season of year, time of day, and so forth. This section provides an example of how some of these factors are accounted in a parameterization of the radiative fluxes.

We begin with incoming short-wave flux,

$$F_{SW}^{\downarrow}(z) = S_0 \cos(\mathrm{Z}) - \int_z^{top} \left(dF_{SW}^{cs} + dF_{SW}^{ca} + dF_{SW}^{s} + dF_{SW}^{a}\right), \tag{4.70}$$

where S_0 is the solar constant, and dF_{SW}^{cs}, dF_{SW}^{ca}, dF_{SW}^{s}, and dF_{SW}^{a} are the differential changes in the short-wave fluxes due to scattering by clouds, absorption by clouds, clear-sky scattering, and clear-sky absorption, respectively.[40] The solar zenith angle Z (see Figure 4.6) is determined through

$$\cos(\mathrm{Z}) = \cos(\phi)\sin(\delta_s) + \cos(\phi)\cos(\delta_s)\cos(t_s), \tag{4.71}$$

where ϕ is latitude, δ_s is the solar declination angle, which is a function of the day of the year, and t_s is the local hour of the sun and depends on longitude.[41] The scattering and absorption by clouds are a function of cos (Z), the vertically integrated liquid water path as modulated by cos (Z), and whether or not the grid box is occupied by clouds (i.e., the cloud fraction is either 0 or 1). Increases in the zenith angle beget increases in the radiative path through the cloudy atmosphere and, consequently, in the absorption and scattering. Clear-sky scattering is proportional to the path-integrated atmospheric mass. Finally, clear-sky absorption is assumed to be due only to water vapor, and hence is a function of the path-integrated water vapor. Clear-sky and cloud absorption are the only short-wave contributors to diabatic heating:

$$\dot{Q}_{SW} = \frac{1}{\rho c_p}\frac{\partial}{\partial z}\left[\int \left(dF_{SW}^{ca} + dF_{SW}^{a}\right)\right]. \tag{4.72}$$

As has been the case with each parameterization discussed, complexity can be added to this representation of the short-wave radiative flux by including, for example, absorption by other clear-sky constituents, scattering by atmospheric aerosol, and cloud scattering that depends effectively on the type of cloud properties (ice versus liquid, concentration, etc.).

The parameterization of upwelling long-wave radiation from the Earth's surface is relatively straightforward:

$$F_{LW}^{\uparrow} = \varepsilon_{sfc}\sigma_{SB}T_{sfc}^{4}, \tag{4.73}$$

where T_{sfc} (K) is the ground surface or skin temperature as stated previously, $\sigma_{SB} = 5.67 \times 10^{-8}$ W m^{-2} K^{-4} is the Stefan-Boltzmann constant, and ε_{sfc} is the emissivity of the ground surface.[42] The emissivity depends on the land surface; for most surfaces on the Earth, ε_{sfc} ranges from 0.9 to 0.99.

Clouds and atmospheric constituents such as carbon dioxide and water vapor also affect long-wave emission. Hence, the surface radiative flux is only part of the radiative flux integrated over an entire atmospheric column. Conceptually,[43] we have

$$\begin{aligned} F_{LW}^{\uparrow} &= \int \mathrm{B}(T)d\varepsilon_u \\ F_{LW}^{\downarrow} &= \int \mathrm{B}(T)d\varepsilon_d \end{aligned}, \tag{4.74}$$

where B is the Planck function, which, when integrated over all wavelengths, has the temperature dependence in (4.73) (see Chapter 3). The upwelling (u) and downwelling (d) emissivities in (4.74) are less straightforward because they have a number of dependencies, including height, path-integrated atmospheric mass (especially accounting for liquid water and vapor), cloud properties, and the amount of atmospheric absorption within wavelength intervals (or bands) in the long-wave part of the electromagnetic radiation spectrum. The absorption itself is a function of the specific atmospheric constituent. In essence, radiation parameterization schemes differ in the details of how integrals such as those in (4.74) are treated. The output of the schemes allow for calculation of the long-wave contribution to diabatic heating:

$$\dot{Q}_{LW} = \frac{1}{\rho c_p}\frac{\partial}{\partial z}\left[F_{LW}^{\downarrow} - F_{LW}^{\uparrow}\right]. \tag{4.75}$$

4.3.5 Convective Clouds

The updrafts and downdrafts of deep convective clouds transport heat, moisture, and and momentum throughout the troposphere. In a bulk sense, such transports act to warm and moisten the local environment in the upper troposphere, and cool and dry the environment in the lower troposphere. Releases of the latent heats of condensation and freezing due to cloud and precipitation processes provide additional warming to the environment aloft; latent-heat releases from the melting and/or evaporation of precipitation further cool the low-level environment, especially below cloud base. When the horizontal grid point spacing exceeds several kilometers, the horizontal extent of the overturning motions of cumulus convection generally becomes SGS, as do these associated diabatic processes.[44] Accordingly, representation of the SGS convective mixing and heating in terms of resolved variables becomes necessary.

At a basic level, convective parameterization schemes seek to estimate the following vertically varying quantities:

$$Q_{1c} = -\frac{1}{\rho}\frac{\partial}{\partial z}(\langle \rho w' s' \rangle) + L_v\left(\langle C \rangle - \langle E \rangle\right) + \dot{Q}_R \tag{4.76}$$

$$\text{and } Q_{2c} = -\frac{1}{\rho}\frac{\partial}{\partial z}(\langle \rho w' q_v' \rangle) - (\langle C \rangle - \langle E \rangle), \tag{4.77}$$

where subscript c indicates application to the cumulus cloud region, $s = c_p T + gz$ is the *dry static energy*, $\dot{Q}_R$ is the heating rate due to radiation (as determined, for example, through (4.72) and (4.75)), and C (E) is the condensation (evaporation) rate. Equations (4.76) and (4.77) follow from Reynolds averaging applied to heat energy and moisture conservation equations, with horizontal eddy transport neglected.[45] Q_{1c} is usually referred to as an *apparent heat source*, and contributes to the right-hand side of the thermodynamic energy equation. Q_{2c} is known as an *apparent moisture sink*, and contributes as a source/sink term to the equation governing water vapor mixing ratio. Adjustments to grid-resolved thermodynamic variables as a consequence of these terms may result in a conversion of vapor to rainwater; this is considered convective precipitation, P_c.

Available schemes differ largely in how Q_{1c}, Q_{2c}, and P_c are parameterized, and in what activates or triggers the parameterization. For example, the Kain-Fritsch scheme[46] represents the vertical fluxes in (4.76) and (4.77) through updraft (and downdraft) mass fluxes, in an amount sufficient to mitigate (most of) the convective available potential energy (CAPE):

$$\text{CAPE} = \int_{z_1}^{z_2} B dz, \tag{4.78}$$

where B is buoyancy (see Chapter 5). Activation of this scheme is based on an evaluation of the thermodynamic stability of candidate "air parcels" (Chapter 5), in a way that basically simulates the initiation of convective clouds in the real atmosphere.

4.4 Model Design and Implementation

4.4.1 Initial and Boundary Conditions

Numerical solutions to the coupled set of equations (e.g., (4.1)–(4.8)) require initial and boundary conditions (ICs, BCs). The ICs used in *idealized simulations* of convective storms typically assume a vertically stratified but horizontally homogeneous atmosphere. This initial state is described by a sounding, or an idealization thereof, and is meant to represent the convective-storm environment. The areal extent of the environment is defined by the model horizontal domain, $0 \leq x \leq L_x$, $0 \leq y \leq L_y$; in most studies, L_x and L_y range from several tens to several hundreds of kilometers. The BCs are then based on a specification of the initial state at the lateral boundaries ($x = 0$, L_x) and ($y = 0, L_y$). Specifically, if inflow is diagnosed just inside the lateral boundary (e.g., $u|_{x=\Delta x} > 0$), then variables at this boundary are assigned environmental values; for example,

$$T|_{x=0} = \overline{T}, \quad q_v|_{x=0} = \overline{q}_v, \text{ etc.}$$

If outflow is diagnosed, a *radiation boundary condition* is imposed in the form of the 1D advection equation (4.26), such that the variable information is allowed to pass freely out of the domain. Idealized model applications that require a domain with closed lateral "walls" would assume an impermeability condition on the flow normal to the wall, and perhaps free-slip conditions on the flow tangential to the wall:

$$u|_{x=0} = 0 \text{ and } \partial v/\partial x|_{x=0} = 0.$$

Such conditions are also appropriate for the upper ($z = L_z$) and lower ($z = 0$) domain boundaries. Finally, periodic boundary conditions,

$$u|_{x=0} = u|_{x=L_x},$$

are the BCs of choice for idealized simulations of channel flow, especially when these conditions are coupled with free-slip conditions along the channel walls,

$$\partial u/\partial y|_{y=0} = 0 = \partial u/\partial y|_{y=L_y}.$$

In real-data applications of mesoscale models, the ICs/BCs come either directly from observations or from output of a larger-scale model (henceforth referred to as the larger-scale "driver") (Figure 4.7). The basic implementation involves interpolation of this information to the model grid. If data are used, the actual implementation involves analysis techniques such as those discussed in Chapter 3 to remove poorly resolved scales.[47] As noted in Section 4.4.2, additional techniques are required if the model-predicted variables need to be retrieved or otherwise converted from the data.

After model initialization (see also Section 4.4.2), the evolution of the larger scale is communicated to the domain interior through the BCs. Typically, data from observations or from the larger-scale driver are interpolated at fixed time intervals to

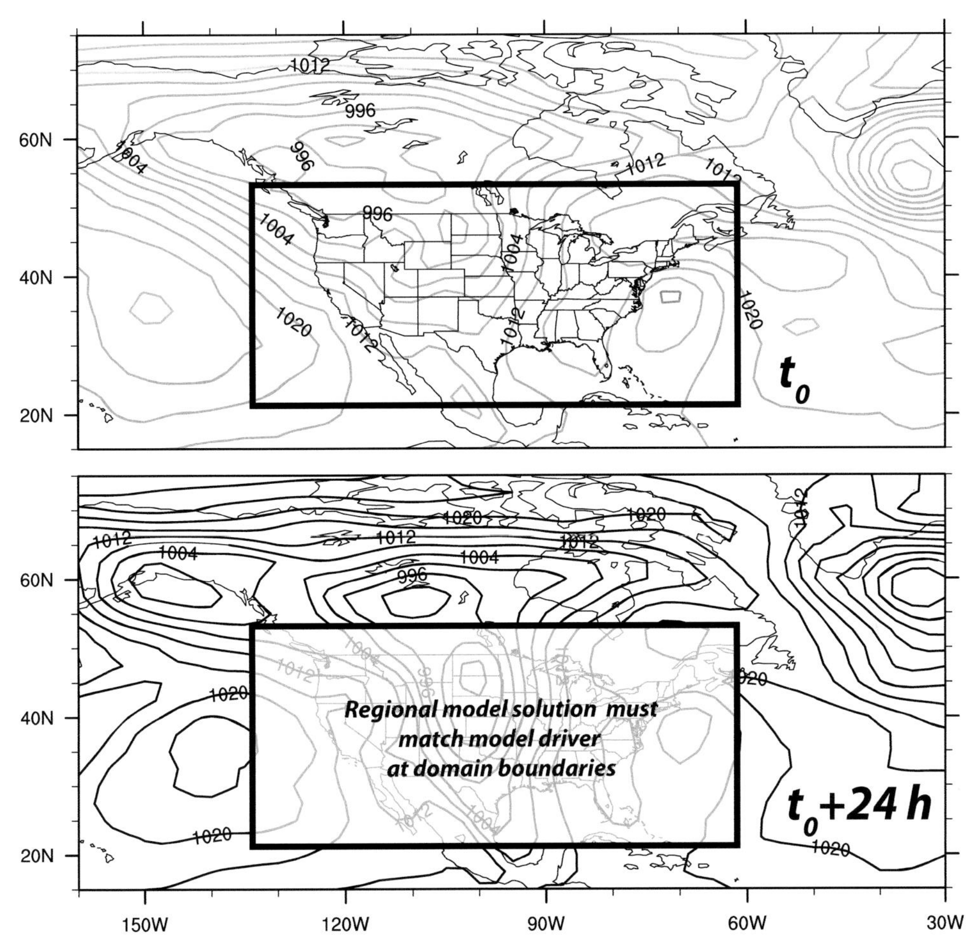

Figure 4.7 Example of initial and boundary conditions for a regional model domain. The contours (black, gray) are of mean sea-level pressure (hPa) at two times (t_0, t_0 + 24 h), from the model driver; the gray lines would represent the initial condition (t_0). The regional model solution in the domain interior is free to evolve, but must match the model driver at the boundaries (bold line) at all times.

the model domain boundaries to form the BCs. A consequence worthy of model-user cognizance is that the interior solution is at all times constrained by the specified boundary condition. Thus, for example, the eastward progression of some feature in the domain interior must be matched with its progression in the larger-scale data at the time the feature encounters a domain boundary (e.g., Figure 4.7).

4.4.2 Data Assimilation

It is possible to introduce updates of the observed state of the atmosphere into the model solution, even after the forward integration has begun. The means by which this is done falls in the broad category of *data assimilation* (DA). Keeping with the philosophy of this chapter, this Section will provide only a basic overview of DA methods and capabilities.

Well-recognized DA techniques include the three- (or four-) dimensional variational data assimilation (3DVAR; 4DVAR), and the ensemble Kalman filter (EnKF). Their primary objective is to realign the solution with the observed state, thus producing an updated estimate of the state variables; the expectation is that the subsequent model integration will yield a more accurate solution. In essence, the assimilation acts to reinitialize the model; indeed, DA techniques are used to generate the ICs as well as provide updates.

The following simple example illustrates the basic components of DA.[48] Let us consider the estimate, or analysis, of a model-predicted variable such as temperature:

$$T_a = T_t + \varepsilon_a. \tag{4.79}$$

Although we will generalize this to a 3D gridded field, assume for now that T_a is a scalar, and represents a single estimate at a point. We recognize first that T_a will always contain some error ε_{a} relative to the truth T_t. DA methods are designed to minimize this error, as accomplished through a least-squares approach, and given by an expression of the form:

$$T_a = T_b + W\left[T_o - T_b\right]. \tag{4.80}$$

Equation (4.80) states that the analysis is the linear combination of a background value T_b and the weighted (W) difference between the background and the observation T_o. This difference is known as the *observational increment* or *innovation*. The background, or "first guess," usually derives from a prior model forecast, valid at the analysis time. Conceptually, (4.80) is the same as the data analysis equation (3.47) presented in Chapter 3.

It is assumed in (4.80) that the observed variable is the same as that analyzed/predicted. However, there are numerous observing systems that yield data that must be transformed into a predicted variable: an example from Chapter 3 is satellite radiances, from which temperature is retrieved. To allow for this, the assimilation equation is written as:

$$T_a = T_b + W\left[O - H\left(T_b\right)\right], \tag{4.81}$$

where H is the *observational operator*, acting here to transform the background to the form of the observation (O). H can also account for the fact that the observation often is not collocated with the background.

The optimal weighting in (4.80) or (4.81) is

$$W = \frac{\sigma_b^2}{\sigma_b^2 + \sigma_o^2}, \tag{4.82}$$

where σ_o^2 is the variance of the error associated with the observation (i.e., with $\varepsilon_o = (T_o - T_t)$), and σ_b^2 is the variance of the error associated with the background (i.e., with $\varepsilon_b = (T_b - T_t)$). In our current example, it is assumed that the squared errors ε_o^2 and ε_b^2 are consistent with the variances of the distributions of errors associated with variable T.[49] The observation and background error variances are related to the analysis error variance as follows:

$$\frac{1}{\sigma_a^2} = \frac{1}{\sigma_b^2} + \frac{1}{\sigma_o^2}, \tag{4.83}$$

where the individual terms represent the accuracy of the background and observation, respectively.

The primary difference between EnKF and the simple scalar versions of 3DVAR is in the treatment of the background error variance, and more generally in the implementation of the assimilation equation. Both might initialize the forecast model with the current analysis and integrate forward in time to obtain a new forecast and, hence, background:

$$T_b^{n+1} = M\,(T_a^n), \tag{4.84}$$

where M denotes application of the forecast model. Per its name, EnKF requires an ensemble of many model integrations, each of which is initialized at time level n with the observational analysis plus a small random perturbation (see Chapter 10). The perturbed solutions are used toward the calculation of a new background error variance σ_b^2 at time level $n + 1$. Assuming a constant observation error variance σ_o^2 that is estimated from knowledge of the observational system the new background error variance is combined with the new observation and background at $n + 1$ to compute the new analysis, which is then used in another forward model integration. This assimilation cycle is repeated multiple times, at time increments consistent with observational data availability. Eventually, though, the model is integrated forward to completion without further assimilation.

In the various implementations of 3DVAR, σ_b^2 is often estimated from different forecasts that are valid at the same time.[50] Otherwise, the assimilation cycle is essentially the same as that just described. 4DVAR, which is an extension of 3DVAR, is designed to incorporate observations at varying times within the assimilation cycle time interval.

When the assimilation equation is written for a gridded 3D field, differences arise between EnKF and 3DVAR in the respective formulations of the optimal weight: the weight is now represented as a matrix, and involves operations between the observation and background error covariance matrices rather than error variances.[51] The assimilation cycle itself, however, is basically the same as that outlined previously.

A data assimilation application that is particularly relevant herein involves the use of radar data. The predicted water-species variables are retrieved from radar reflectivity, and the 3D winds are retrieved from Doppler velocity (and radar reflectivity). These observations are then assimilated over a number of cycles before a commencement of a forward integration.[52] The radar DA is not necessarily concerned with the initial formation of a convective storm, especially because initial storms have weak reflectivity, but it does allow for representation of the precipitating and dynamical structure once the storm is ongoing.

4.4.3 Other Design Issues

In high-resolution, real-data model simulations, the initiation of convective storms is attributable to the processes (e.g., a synoptic-scale front) present in the IC/BC, although parameterized processes (e.g., land-surface energy exchanges and SGS mixing) also affect the initiation. When convective storms are numerically simulated through idealized approaches, an experimental design issue arises in how to trigger cumulus convection within the quiescent, horizontally homogeneous environment. Despite the intent to represent processes in the real atmosphere (see Chapter 5), initiation procedures used to date still have some artificial aspects.[53] One common procedure is to specify an impulsive "warm bubble." The bubble typically is a 3D, spheroidal distribution of (perturbation) temperature, introduced instantaneously at $t = 0$ (Figure 4.8). The free parameters, such as bubble size, maximum temperature excess at bubble center, and distance above the ground, depend on the model application and on the homogeneous environment. As we will learn in Chapters 5 and 6, the size of the bubble influences the size of the storm. This has motivated the development of alternative initiation procedures, an example of which involves the specification of random perturbations in temperature and water vapor mixing ratio at the model surface, meant to simulate daytime heating and aid in the generation of boundary-layer circulations.[54]

Another issue regards how to reconcile available computer resources with parameterization-scheme complexity, integration length, domain size, and grid point spacing. The method of *grid nesting* provides some reconciliation of this issue. In analogy with the Russian matryoshka doll, nested grids are computational subdomains that are located – though not necessarily centered – inside one another (Figure 4.9). The incrementally smaller subdomains have correspondingly smaller gridpoint spacings and associated timesteps:

$$\Delta x_n = \Delta x_p / n c_n \text{ and } \Delta t_n = \Delta t_p / n c_n, \tag{4.85}$$

where Δx_p and Δt_p are the gridpoint spacing and timestep, respectively, of the primary or "parent" domain, n denotes the subdomain increment, and c_n is the nesting factor.[55] A significant benefit of grid nesting is that a relatively fine grid can be applied only to

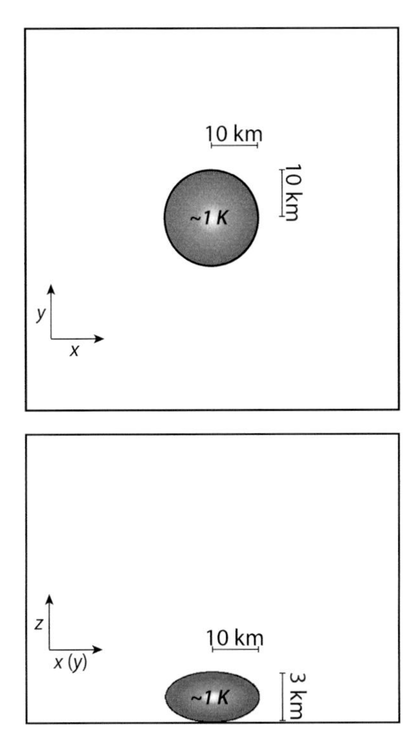

Figure 4.8 Example specification of a warm-bubble for convection initiation in an idealized numerical simulation. The gray shading represents the temperature gradation, with a maximum temperature excess of $\sim$1 K at the bubble center. In this example, the bubble is circular in horizontal cross section, with a radius of 10 km; in vertical cross section, the bubble is elliptical, with a minor-axis radius of 1.5 km.

the physical region that requires it, and only that region is limited by the associated smaller timestep.

Consider the example in Figure 4.9, which is motivated by a desire to explicitly resolve convective storms that initiate in advance of the sea-breeze front in Florida.[56]

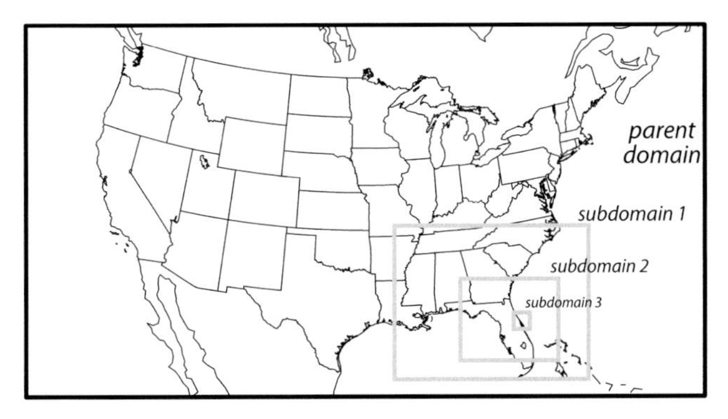

Figure 4.9 Example of grid nesting within a parent domain. The gray boxes represent the subdomains.

The configuration consists of a high-resolution ($\Delta x \sim 250$ m) inner subdomain at and west of the coast; an intermediate, yet convective-permitting ($\Delta x \sim 1$ km) domain that includes most of peninsular Florida; another intermediate, convective-permitting ($\Delta x \sim 4$ km) domain that includes the southeastern United States; and then a parent domain encompassing most of the United States. Because convection is parameterized in the relatively coarse ($\Delta x \sim 16$ km) parent domain, it is possible for an area of convective precipitation to form via this parameterization and then move across subdomain boundaries until the convective processes are explicitly represented. The implicit-to-explicit transition of this and other processes may lead to an unphysical evolution, which is one liability of grid nesting.[57,58]

Grid nesting can operate so that the flow of information is one-way only, from a coarse to a fine grid. Alternatively, it can allow for two-way interaction between grids: coarse-grid data are used to integrate the model equations on the fine grid over a time increment, and then these data are upscaled for integration on the coarse grid. Such two-way interactivity is even the basis for a design feature in climate models, wherein a 2D convective-permitting model is embedded within a climate-model grid cell, and functions to vertically redistribute heat and moisture in place of a typical convective parameterization scheme.[59,60] This *superparameterization* is yet another example of the resource and design compromises that pervade all numerical modeling.

Supplementary Information

For exercises, problem sets, and suggested case studies, please see www.cambridge.org/trapp/chapter4.

Notes

1 Skamarock et al. (2008), Pielke (2002).
2 Anderson et al. (1984), Lewis et al. (2006), Stensrud (2007).
3 Straka (2009), Kessler (1969).
4 Klemp and Wilhelmson (1978), Bryan and Fritsch (2002).
5 Xue et al. (2000).
6 For an example of model equations without this approximation, see Bryan and Fritsch (2002).
7 Klemp and Wilhelmson (1978).
8 Skamarock et al. (1994).
9 See Haltiner and Williams (1980) and Kalnay (2003) for a discussion of other vertical coordinates.
10 Roundoff errors are the inevitable consequence of the limited memory of all computers: no matter what the memory, real numbers will always be rounded off at some decimal place.
11 Kalnay (2003).
12 Anderson et al. (1984), Jacobson (2005).
13 The analysis, known as a *von Neumann* or *Fourier stability analysis*, involves substitution of a hypothetical error distribution of the general form $\exp(ikx + at)$ into the finite difference equation, and then an evaluation of the conditions, if any, under which the error grows in time. As noted, these conditions are cast in terms of the Courant number; see Anderson et al. (1984).
14 The interested reader should consult reference material on computational fluid dynamics, such as Anderson et al. (1984), and also chapters in Haltiner and Williams (1980), Kalnay (2003), and Jacobson (2005).

15 An alternative to an assumption of incompressibility or the use of the anelastic approximation is to maintain the compressibility of the atmosphere but employ a *time-splitting method*; see Klemp and Wilhelmson (1978).
16 Additional schemes, and their complete developments, are provided by Stensrud (2007), Straka (2009), Jacobson (2005), and others.
17 Jacobson (2005).
18 Bryan et al. (2003).
19 See Bryan et al. (2003) and references therein.
20 Anderson et al. (1984).
21 *Flux* is the movement of some quantity across a unit area, per unit time. In this context, flux is the rate of transfer of momentum across a unit area.
22 Kundu (1990).
23 The following draws from Emanuel (1994).
24 Xue et al. (2000).
25 Klemp and Wilhelmson (1978).
26 Stensrud (2007).
27 Hong and Pan (1996).
28 Pielke (2002).
29 Much of the following discussion is based on Chen and Dudhia (2001).
30 In this context, the flux is the rate of transfer of energy across a unit area.
31 Peixoto and Oort (1998).
32 Stensrud (2007).
33 Chen and Dudhia (2001), Stensrud (2007).
34 This particular scheme is described in the supplement to Gilmore et al. (2004). A overview and more detailed information about cloud and precipitation microphysical parameterization schemes can be found in Straka (2009).
35 In subfreezing clouds that still contain liquid water (supercooled clouds), ice crystals will grow by diffusion of vapor instead of, and at the expense of, supercooled cloud droplets. This process will even consume the water droplets, resulting in a cloud composed completely of ice. See Pruppacher and Klett (1978).
36 Ferrier (1994).
37 See Ferrier (1994) for a general expression of the distribution function.
38 Kogan (1991).
39 Kain et al. (2008), Koch et al. (2005).
40 Dudhia (1989).
41 Stensrud (2007).
42 Ibid.
43 Dudhia (1989).
44 Although this is a somewhat contentious issue, it is generally assumed that the critical convective processes are sufficiently resolved on the grid scale when the grid point spacing is several kilometers or less (Weisman et al. 1997). These processes are not well resolved, however, until the grid point spacing is $\sim$100 m or less (Bryan and Fritsch 2002).
45 Stensrud (2007).
46 Kain and Fritsch (1990).
47 See Haltiner and Williams (1980) for further discussion.
48 This follows Kalnay (2003).
49 As discussed by Kalnay (2003), this follows from the fact that $E[\epsilon^2] = \sigma^2$, where E is the *expected value*.
50 See Kalnay (2003) for further discussion.
51 The interested reader should refer to Kalnay (2003) and elsewhere for more details on the complete form of these expressions.
52 Dowell et al. (2011).
53 Loftus et al. (2008).
54 Balaji and Clark (1988).
55 The nesting factor is typically set to 3 or 4, but is not constrained physically to these particular values.
56 Conceptually, this domain configuration applies to many regions around the world.

57 Warner and Hsu (2000).

58 The alternative approach of dynamic grid adaption removes this transition issue, because the grid is everywhere smoothly varying. However, one limitation of this approach is that the same time step, as constrained by the smallest grid increment, applies to the entire domain; see Fiedler and Trapp (1993).

59 In this context, such models are usually referred to as *cloud-system resolving models*.

60 Randall et al. (2003).

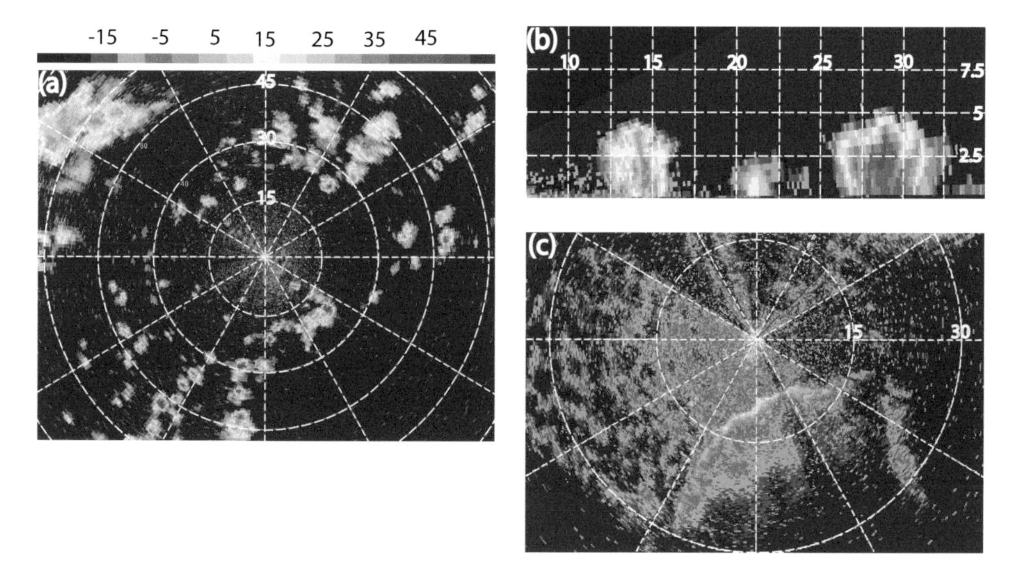

Figure 3.8 Example of radar reflectivity factor in a field a precipitating convective clouds, as displayed in (a) a PPI scan at 1° elevation, and (b) a corresponding RHI scan at 130° azimuth. (c) A fine line in radar reflectivity factor, associated with the sea-breeze front. As in (a), this PPI scan is at 1° elevation. All scans were collected in Florida in March 2012, using a Doppler on Wheels radar.

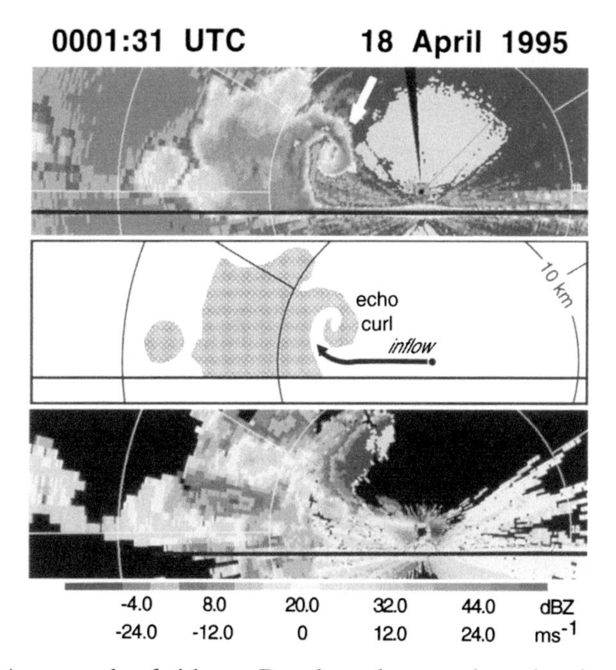

Figure 3.12 An example of airborne Doppler radar scan through a developing hail-storm. The top panel shows radar reflectivity factor, the bottom panel is of Doppler velocity, and the middle panel gives a physical interpretation. From Wakimoto et al. (1996).

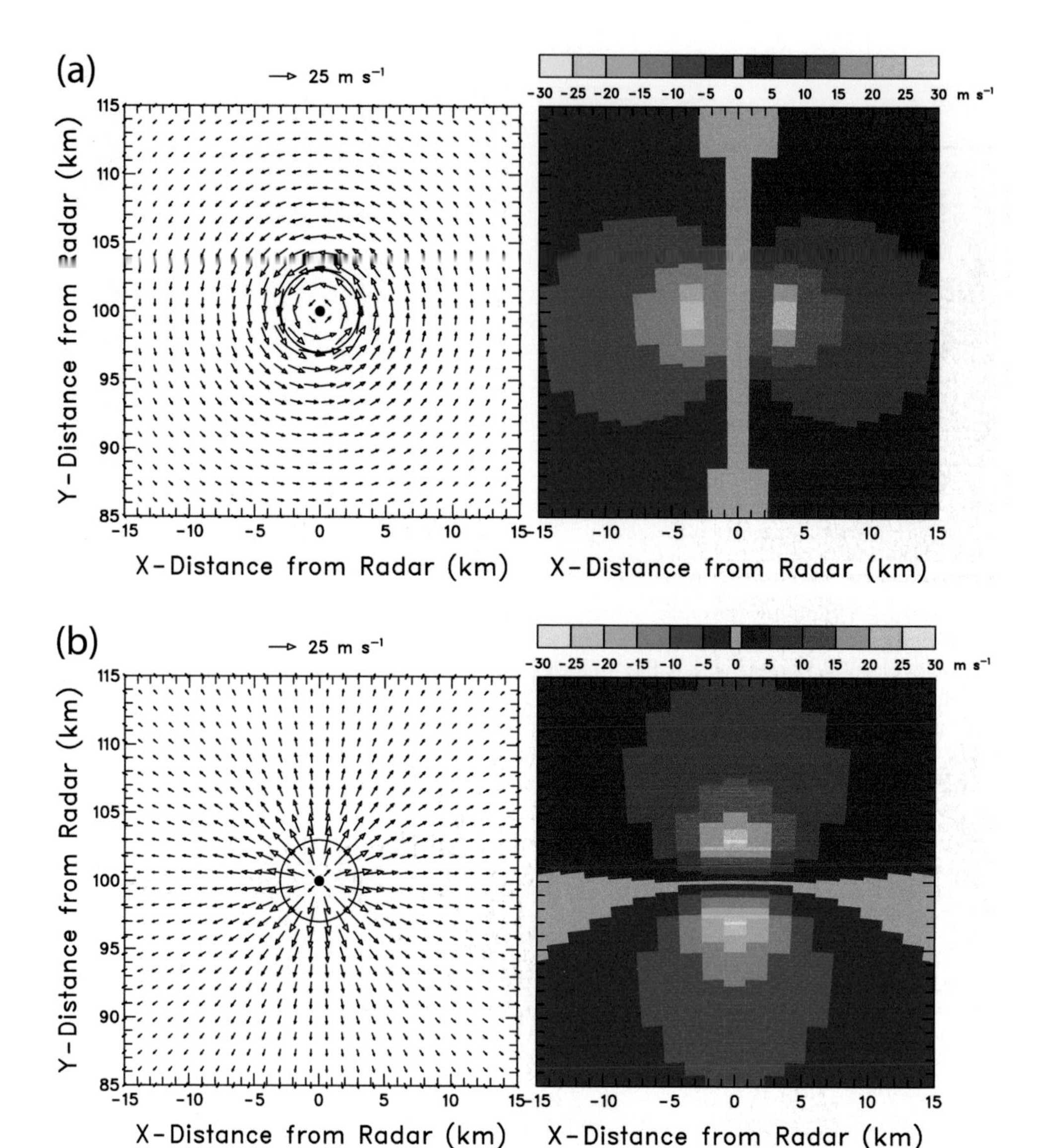

Figure 3.9 Low-level patterns of axisymmetric (a) vertical rotation and (b) horizontal divergence, in a vector wind field and corresponding field of Doppler velocity V_r. The circles in the vector wind fields indicate the relative location of the maximum winds. The black dot shows the center of the vortex in (a), and center of divergence in (b). The (simulated) radar is located 100 km to the south of the rotation and divergence centers. From Brown and Wood (2007).

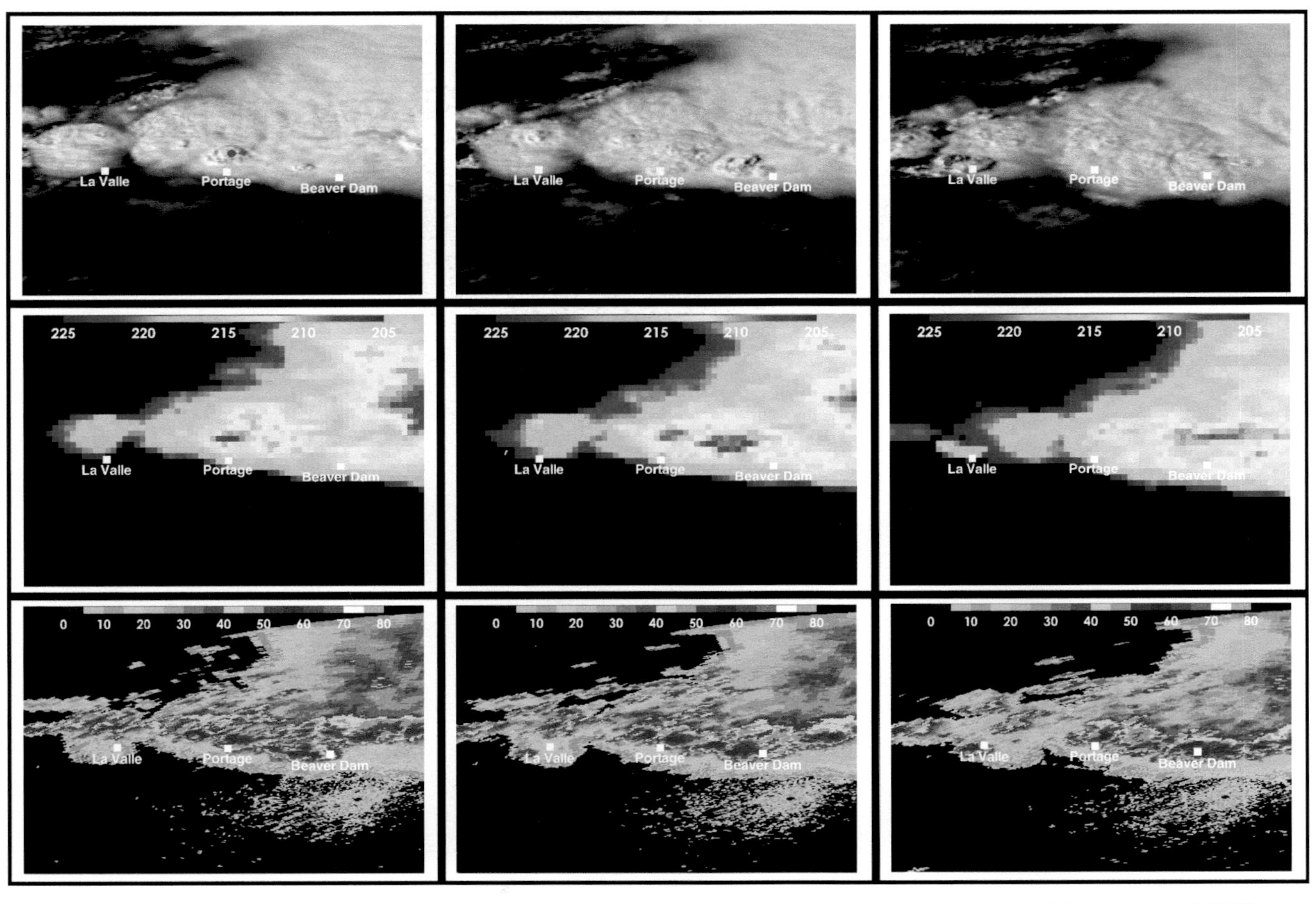

Figure 3.20 GOES-12 visible channel imagery with objective overshooting-top detections (red dots) (top), GOES-12 IR (10.7-μm) brightness temperatures (middle), and KMKX WSR-88D composite reflectivity (bottom). White dots show the locations of La Valle, Portage, and Beaver Dam, WI. From Dworak et al. (2012).

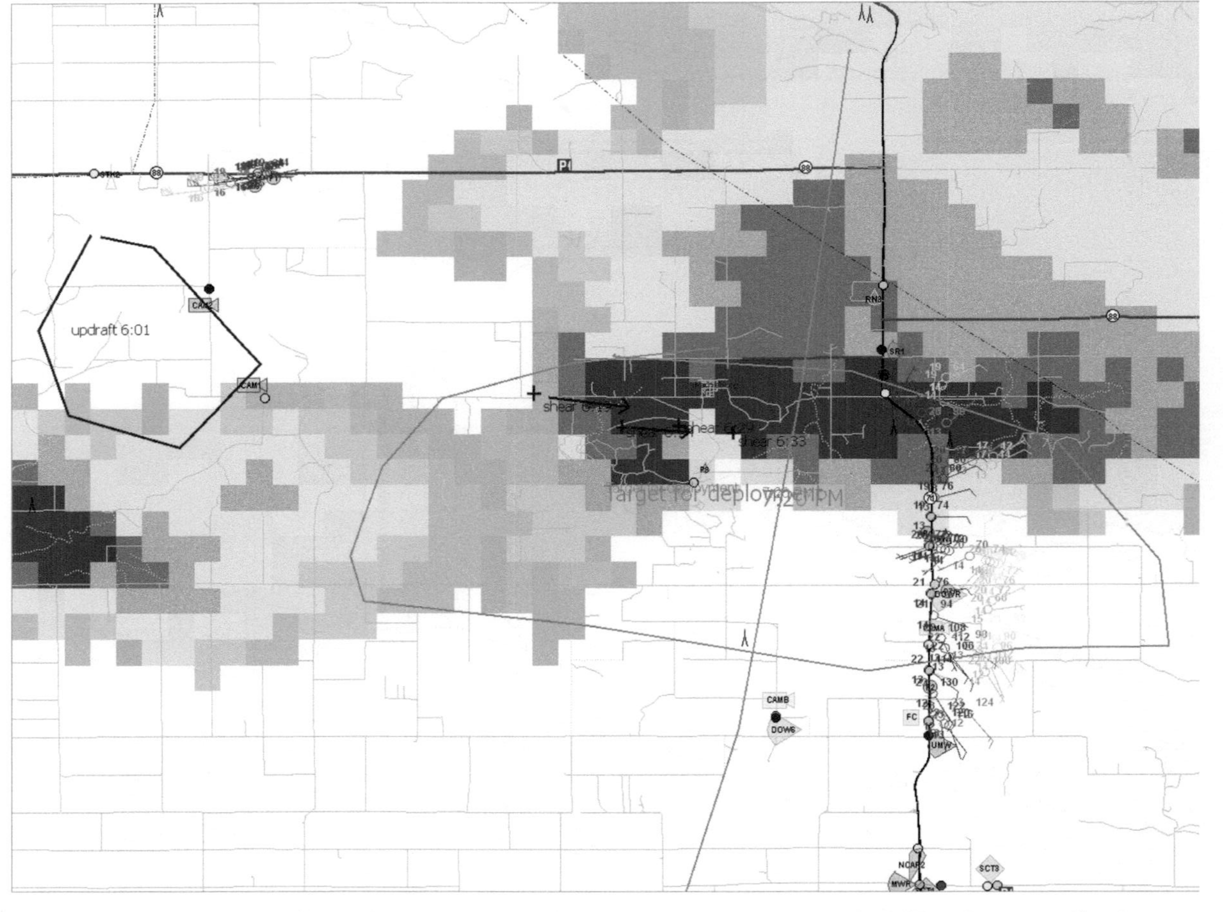

Figure 3.24 Display of Situational Awareness for Severe Storms Intercept (SASSI) software application used during VORTEX2 in 2009 and 2010. Color fill is a radar-reflectivity overlay, and icons represent various (mobile or portable) observing systems at a current as well as recent locations. The light-blue icon labeled FC shows the location of the field coordinator. Courtesy of Dr. Erik N. Rasmussen, Rasmussen Systems, LLC.

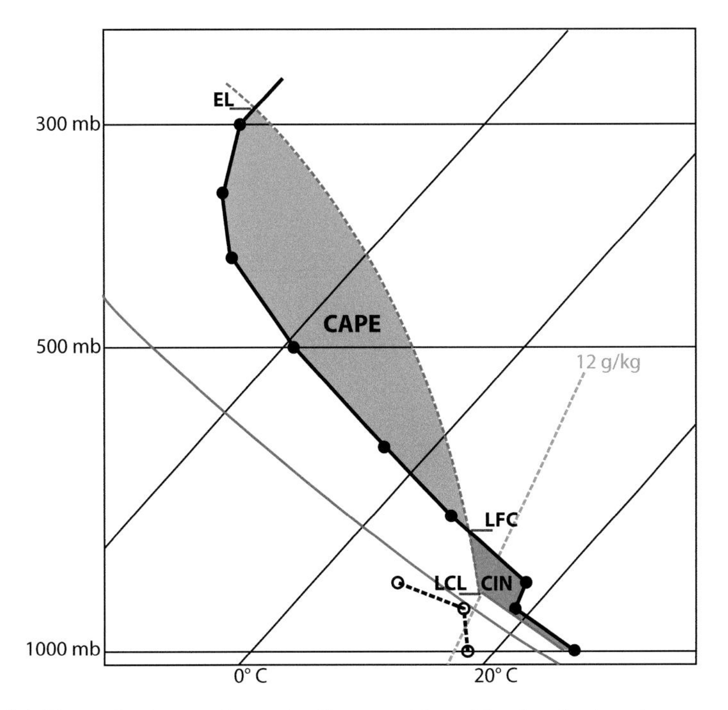

Figure 5.1 Example thermodynamic diagram (skew T – ln p), showing a sounding with an LCL, LFC, and EL. Positive and negative areas corresponding to CAPE and CIN, are shaded and red and blue, respectively. Bold black line is temperature sounding, and dashed line is segment of dewpoint sounding. Solid orange line is a dry adiabat, dashed orange line is a relevant moist adiabat, and dashed green line is a mixing ratio line.

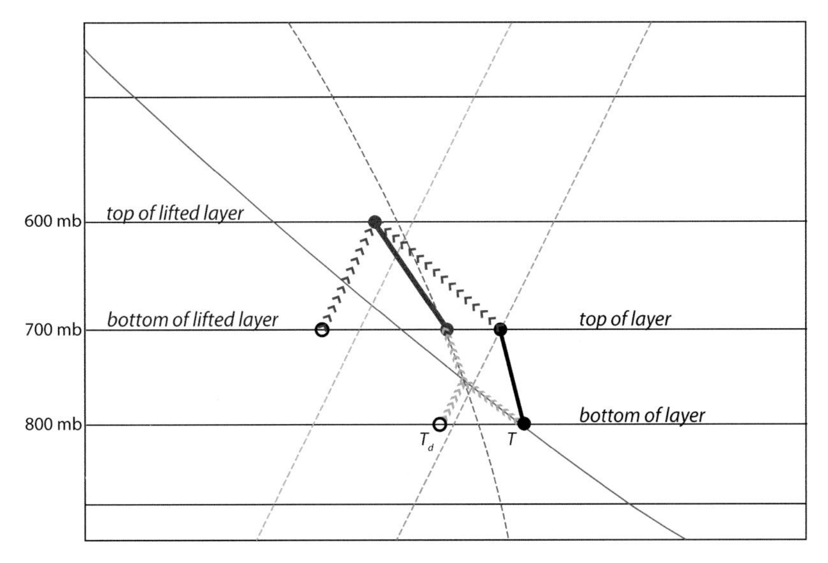

Figure 5.2 Illustration of thermodynamic destabilization of a layer through ascent. Bold black line is segment of original temperature sounding, and bold gray line represents the subsequent modification. Solid orange line is dry adiabat, and dashed orange line is relevant moist adiabat. The parcel process associated with the lifting of the bottom (top) of the layer is indicated by the light blue (dark blue) arrowed line.

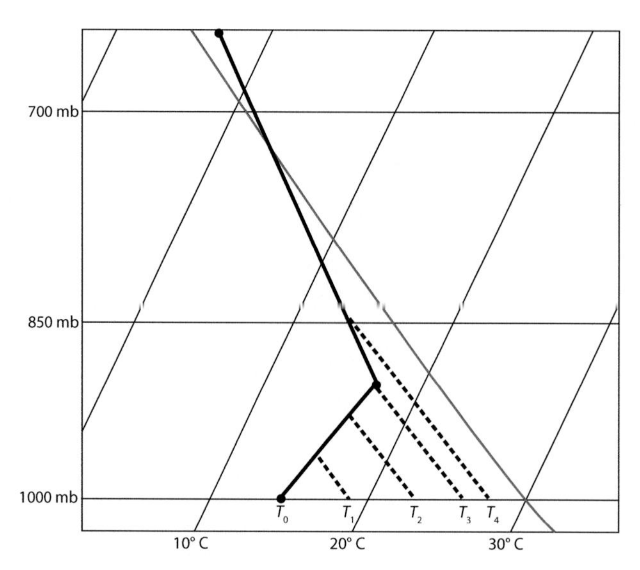

Figure 5.3 Illustration of thermodynamic destabilization of the atmospheric boundary layer through solar heating. Bold black line is original temperature sounding, and dashed lines represent subsequent modifications, as coupled to changes in surface temperature T. Solid orange line is dry adiabat.

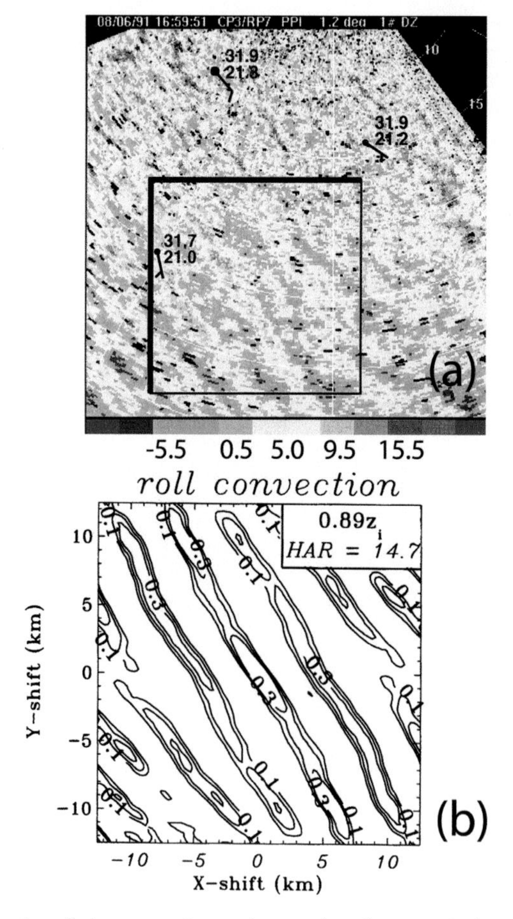

Figure 5.10 Example of the use of weather radar data to deduce HCR structure. (a) Equivalent radar reflectivity at low elevation angle in a PPI. (b) Field of spatial autocorrelation based on (a), and for the subdomain indicated in (a). From Weckwerth et al. (1997).

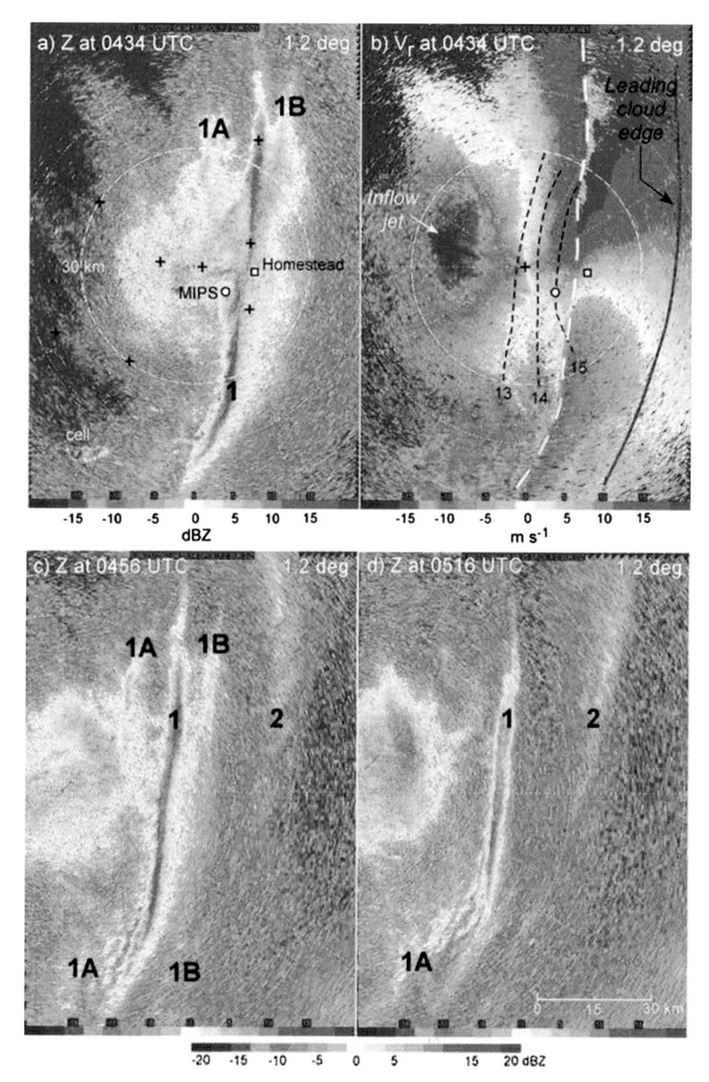

Figure 5.25 Example atmospheric bore evolution as observed by Doppler weather radar. Radar reflectivity factor (Z) is shown in (a), (c), and (d), and radial velocity in (b). From Knupp (2006).

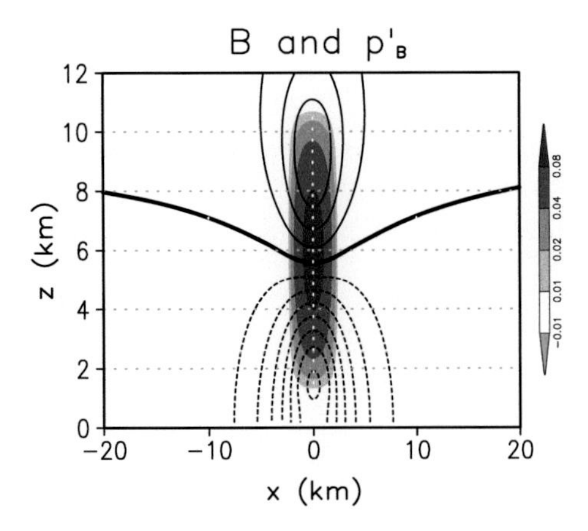

Figure 6.10 Vertical cross section of an idealized two-dimensional element of buoyancy (color-filled as shown; m s^{-2}) and the corresponding buoyancy pressure perturbation (contour interval 10 Pa, with $p'_B = 0$ indicated by the bold contour, and $p'_B < 0$ indicated by dashed contours). After Parker (2010).

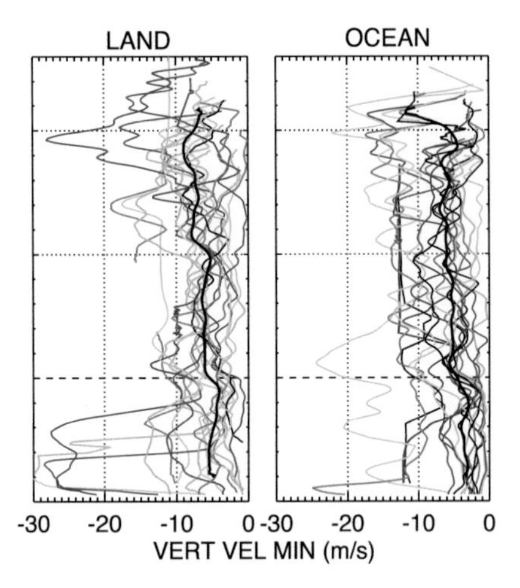

Figure 6.13 Vertical profiles of minimum vertical velocity in a variety of convective storms, over continental versus oceanic domains. The profiles are derived from a high-altitude airborne Doppler radar (EDOP). After Heymsfield et al. (2010).

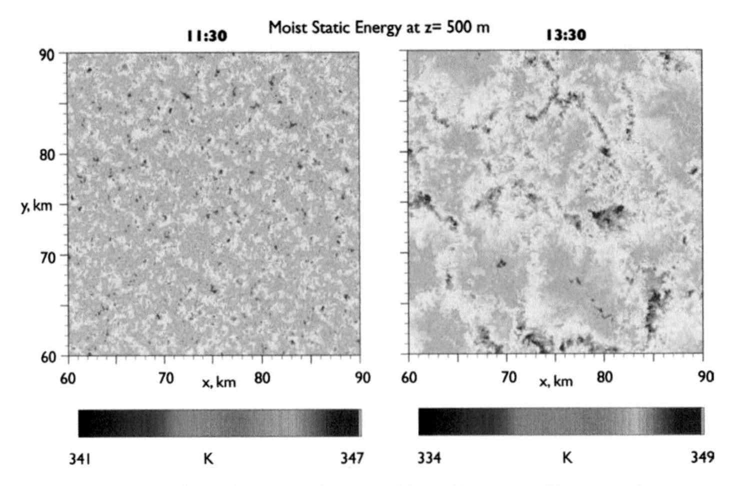

Figure 6.17 Example of a (simulated) transition from shallow to deep convective clouds. Shown are horizontal sections (at 500 m) of moist static energy. From Khairoutdinov and Randall (2006).

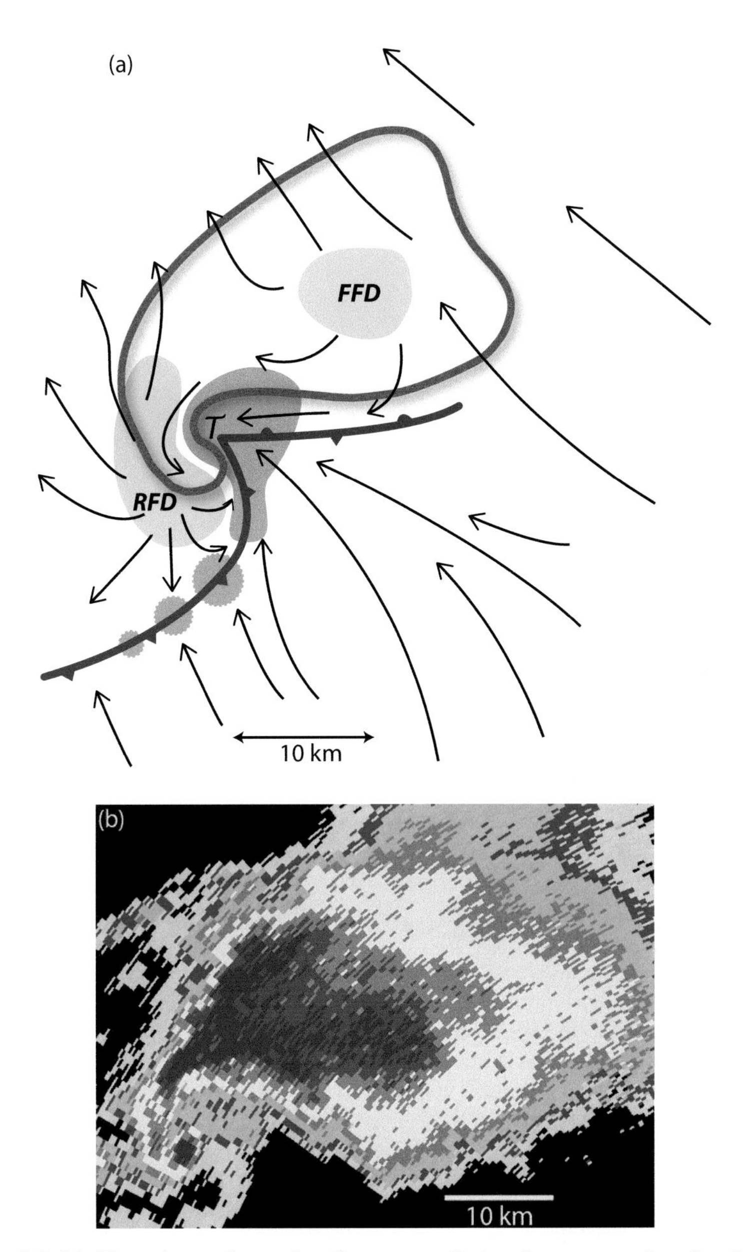

Figure 7.3 (a) Plan-view schematic of a supercell thunderstorm at low levels. Red (blue) shading shows regions of updrafts (downdrafts). Streamlines are of (storm-relative) near-surface flow. Bold contour shows ~40 dBZ contour. Adapted from Lemon and Doswell (1979). (b) Weather radar image of an actual (tornadic) supercell thunderstorm.

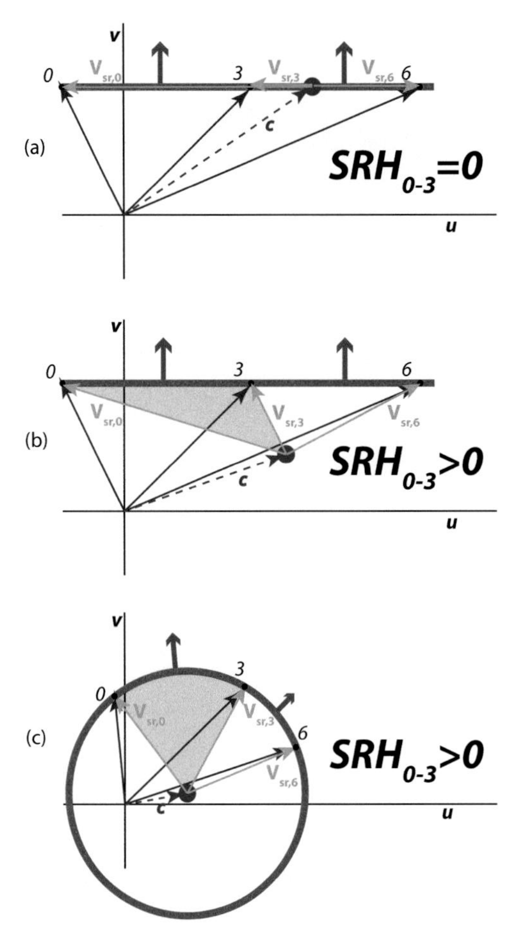

Figure 7.6 The hodograph (bold gray line) of an environmental wind (purple arrows). The environmental horizontal vorticity vector (direction indicated by red arrows) is everywhere normal to the shear vector, and hence to the hodograph curve. In (a) and (b), the hodograph shape is a straight line, and thus the environmental horizontal vorticity vector does not change with height. (a) The storm motion (blue dashed arrow) lies on the hodograph, and using this as a hodograph origin (blue circle), the storm-relative winds (orange vectors) are everywhere perpendicular to the horizontal vorticity vector. In this case, $SRH_{0\text{–}3} = 0$. (b) The storm motion lies off the hodograph, and the resultant storm-relative winds are not perpendicular to the horizontal vorticity vector. In this case, $SRH_{0\text{–}3} > 0$, with a magnitude equal to $-2 \times$ signed area (shaded region) swept out by the storm-relative wind vector between 0 and 3 km. (c) A circular hodograph with storm motion at the circle center. In this case, storm-relative winds (orange vectors) are everywhere parallel to the horizontal vorticity vector, thus maximizing $SRH_{0\text{–}3} > 0$.

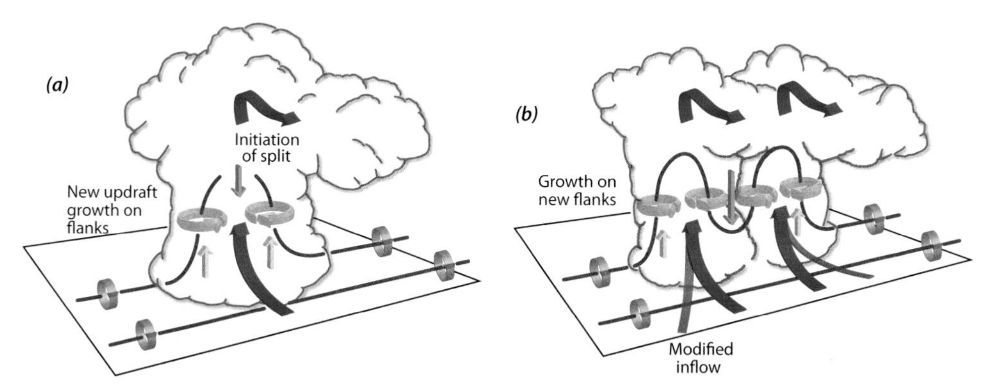

Figure 7.7 Schematic showing the location and effects of the dynamically (rotationally) induced vertical pressure gradient forcing on supercell evolution. In (a), the initial, midlevel vortex pair (gray ribbons) results in positive vertical pressure gradient forcing and subsequent vertical accelerations (yellow arrows) on the flanks of the initial updraft (red ribbons). As facilitated by a precipitation downdraft (blue arrow), a split of the initial cell ensues. In (b), the cell-splitting process bears two new cells, each of which generates a new midlevel vortex pair, and is accompanied by a modified low-level inflow (red ribbons). Positive vertical pressure gradient forcing is again found beneath each of the vortices; lifting with the inner vortices is impeded by precipitating downdrafts, and thus new updraft growth is favored beneath the outer vortices. Adapted from Klemp (1987).

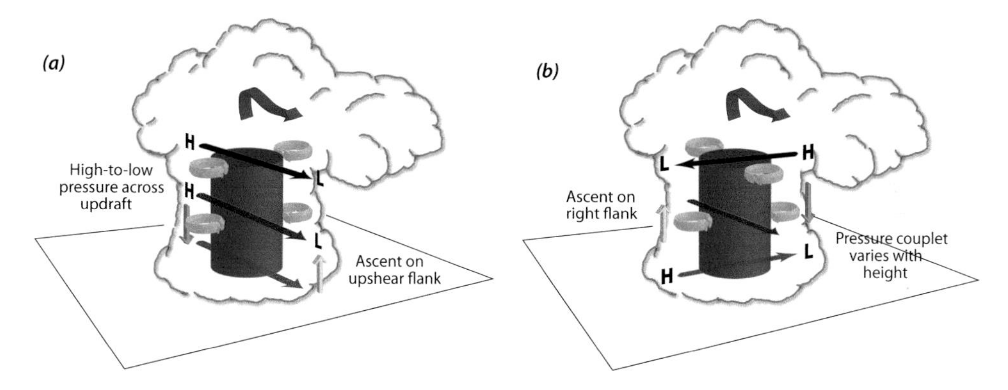

Figure 7.12 Qualitative depiction of the location and effects of the linear, dynamically induced pressure (H for high pressure, L for low pressure), as well as the accelerations responding to the associated vertical pressure gradient. Red cylinder represents the updraft, and the flat arrows show the shear vector orientation at the corresponding level. For reference, the location of vertical vortices at these levels is also shown (gray ribbons), which depict the nonlinear dynamic pressure effects. In (a), the supercell has formed in an environment with a straight hodograph. The linear-dynamics forcing results in upward acceleration and ascent (downward acceleration and descent) on the downshear (upshear) flank of the updraft. In (b), the supercell has formed in an environment with hodograph curvature (half-circle) over the lowest half of the troposphere. The linear dynamics forcing results in ascent (descent) on the right (left) flank of the updraft. Adapted from Klemp (1987).

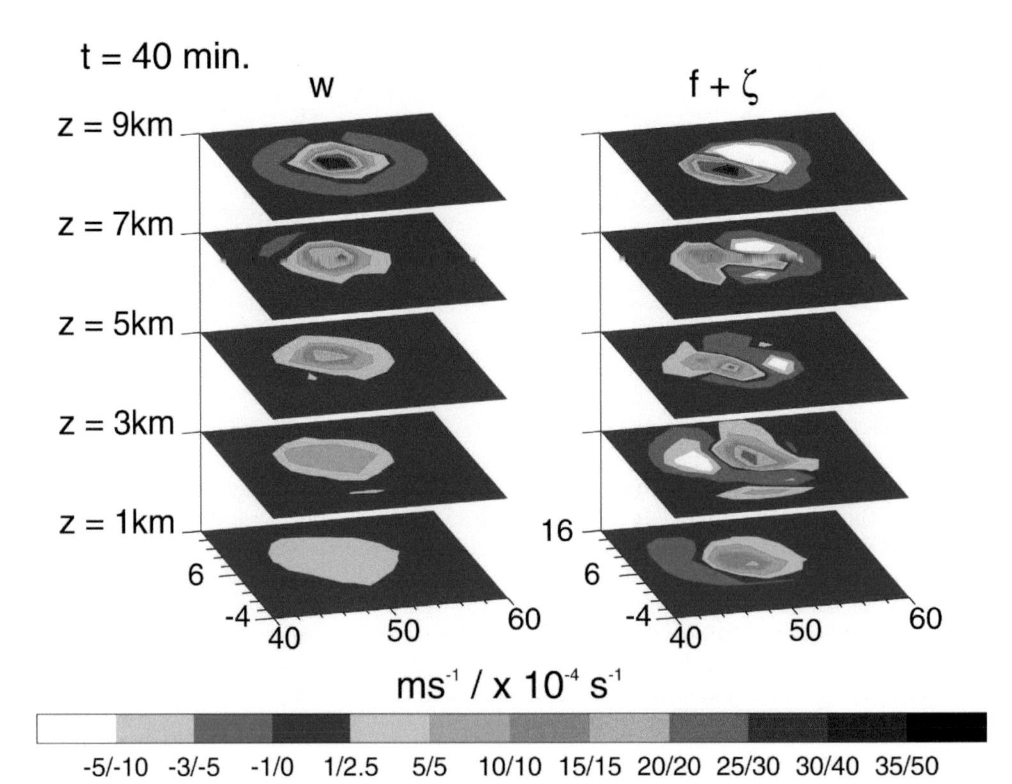

Figure 7.24 Vertical velocity and absolute vertical vorticity associated with a numerically simulated vortical hot tower. From Montgomery et al. (2006).

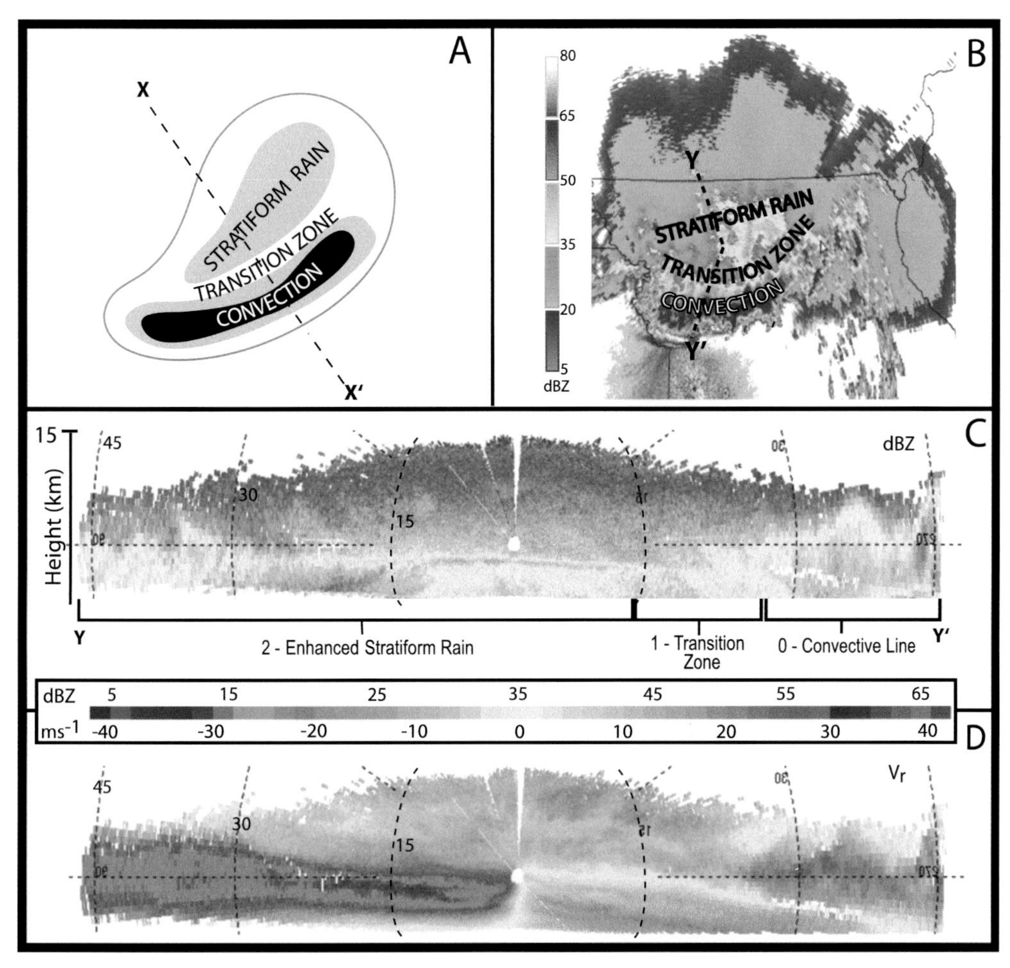

Figure 8.2 Schematic depiction (A), and corresponding Doppler radar observations (B)–(D), of a mature squall line with a trailing stratiform region. Radar reflectivity in (B) is from a WSR-88D scan at 0.5° elevation. Radar reflectivity and Doppler velocity in (C) and (D), respectively, are from quasi-vertical scans from an airborne Doppler radar (NOAA P-3). After Smith et al. (2009).

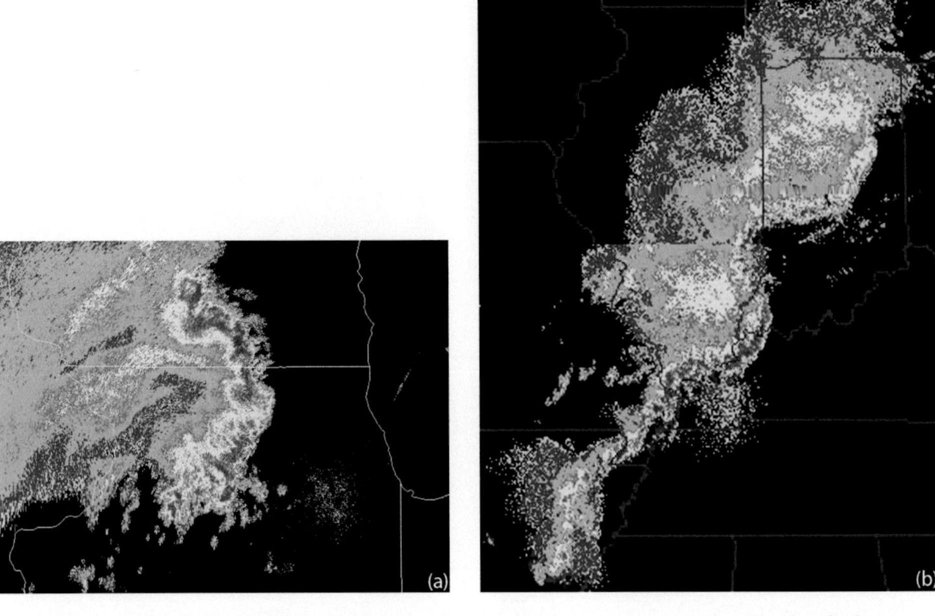

Figure 8.5 Radar reflectivity imagery of example bow-echo events: (a) classic bow echo, July 11, 2011, and (b) squall-line bow echo, or LEWP, April 19, 2011. Both are from scans at 0.5° elevation angle, and (b) is a composite of data from two radars.

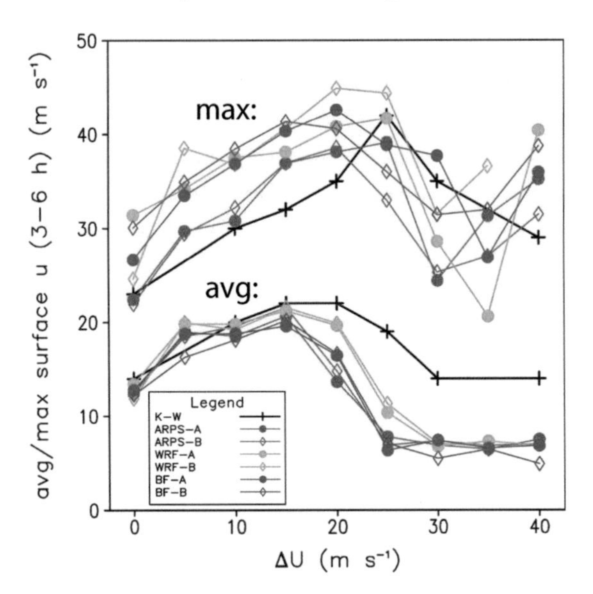

Figure 8.16 Maximum surface winds generated by numerically simulated squall lines as a function of unidirectional environmental shear over the 0–5-km layer. All other parameters, including an environmental CAPE $\sim$ 2200 J kg^{-1}, are held constant. The simulations are performed using four different numerical models (see Bryan et al. (2006) for details). The maximum surface winds are relatively weaker when the environmental shear exceeds $\sim$25 m s^{-1} because the convective cells become downshear tilted and also organized into more 3D, supercell-like entities rather than MCSs. From Bryan et al. (2006).

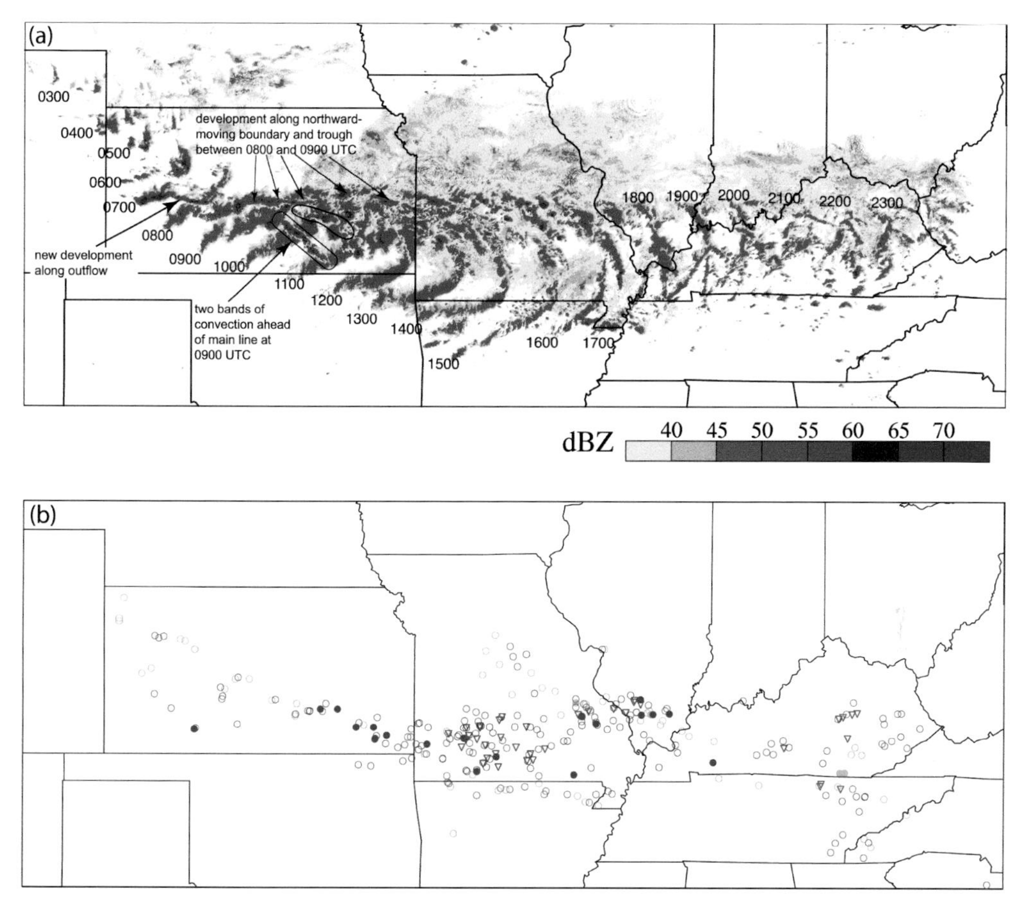

Figure 8.18 Chronology of the derecho that occurred from 03 UTC to 23 UTC, May 8, 2009. (a) Hourly composite radar reflectivity (dBZ). (b) Location of severe weather reports: open (closed) green circles indicate hail ≥ 0.75 in (≥ 2.0 in), open blue (closed) circles indicate wind damage or wind gusts ≥ 26 m s^{-1} (wind gusts measured or estimated ≥ 33.5 m s^{-1}), and red triangles indicate tornado reports. From Coniglio et al. (2011).

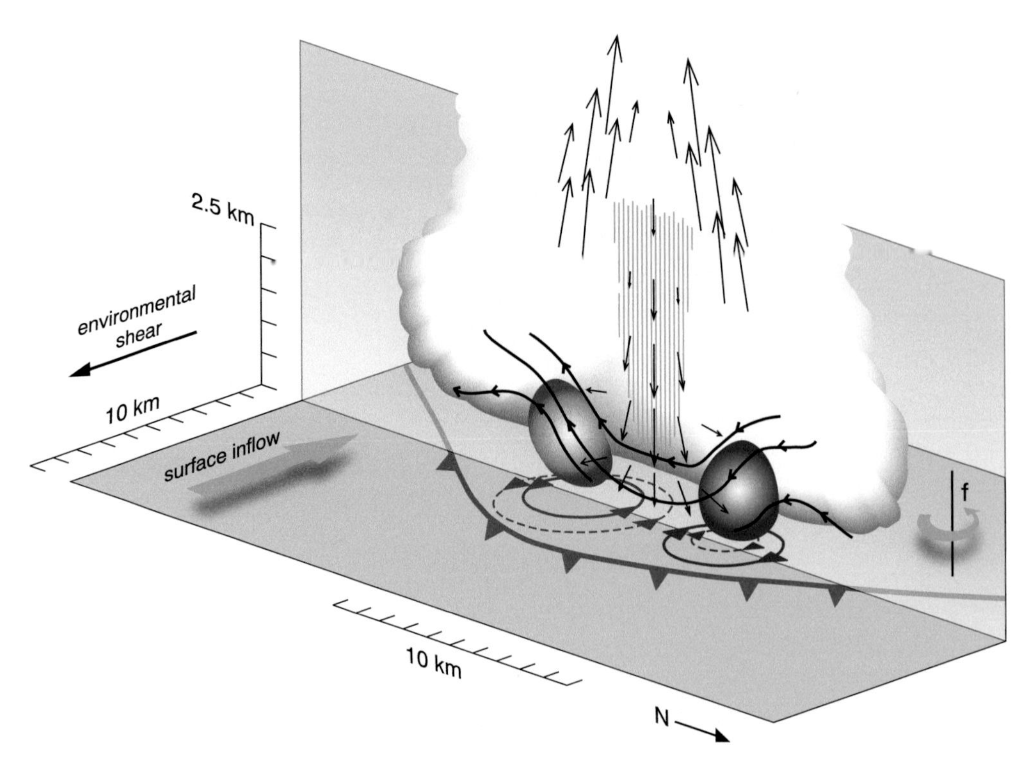

Figure 8.20 Schematic of mesovortexgenesis in the early stages of a quasilinear MCS. Vortex lines (black) are tilted vertically by the downdraft (vectors, and blue hatching), to result in a surface vortex couplet (cyclonic vertical vorticity is red; anticyclonic is purple). The dashed red and purple circles represent the future state of the vortex couplet, which is due in part to the stretching of planetary vorticity (*f*) as shown. During the mature stage, relevant vortex lines would have opposite orientation, and thus the resultant vortex-couplet orientation would be reversed. From Trapp and Weisman (2003).

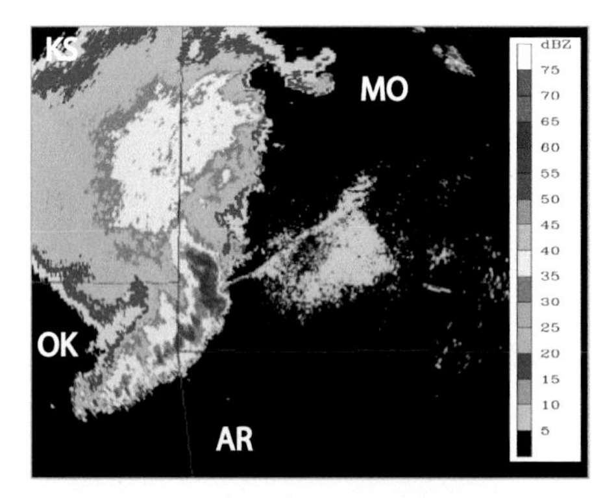

Figure 9.2 Radar reflectivity factor from the 0.5° scan of the Springfield, Missouri WSR-88D, at 1230 UTC on July 4, 2003. The SW-NE–oriented thin line shows the boundary. The point of interaction with the asymmetric bow echo corresponds to a low-level mesovortex that was associated with wind damage and a tornado. State boundaries (and state abbreviations) provide scale.

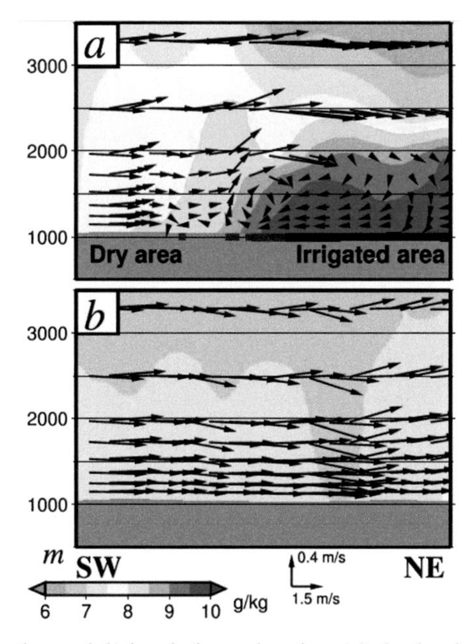

Figure 9.13 Mesoscale model simulations showing (a) the landscape-induced circulation that arises between an irrigated and nonirrigated land area, and (b) the lack of such a circulation when the land areas are not irrigated. Vectors indicate wind in a vertical cross section, and red and blue shadings are of water vapor mixing ratio. The light gray shows topography. From Kawase et al. (2008).

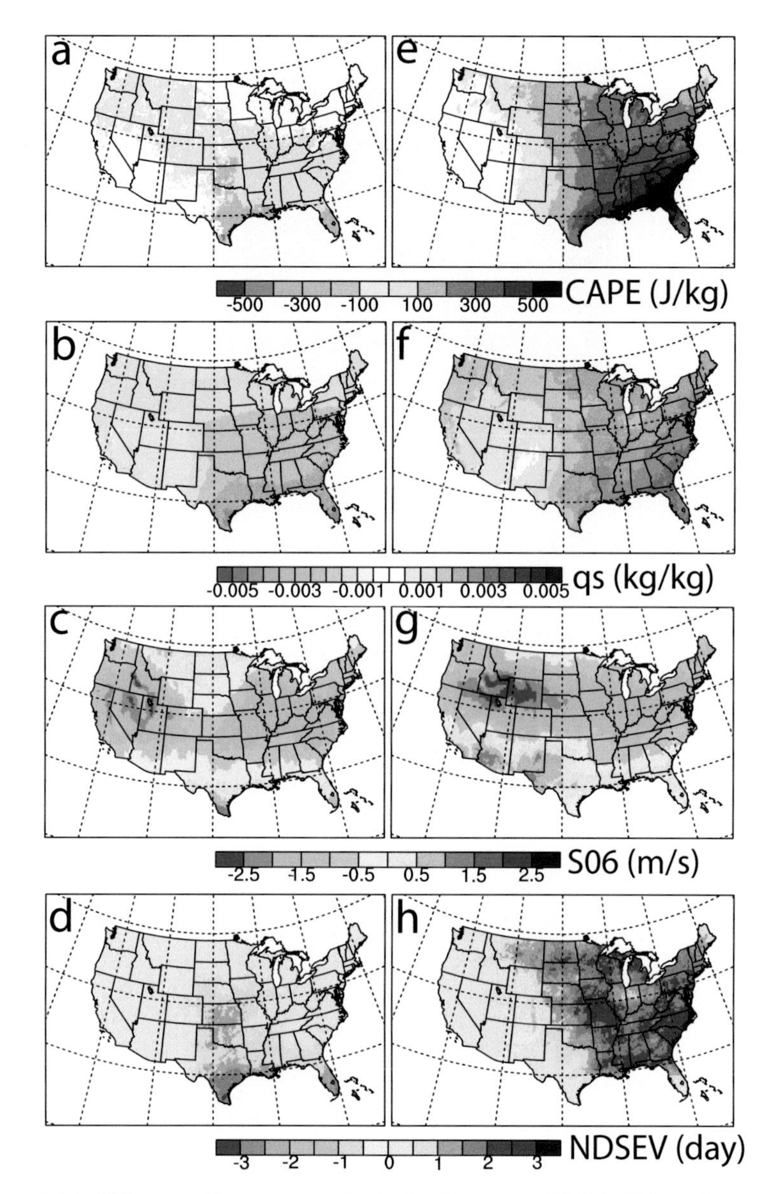

Figure 9.14 Difference (future minus historical) in mean CAPE, S06, surface specific humidity, and occurrence of the product CAPE $\times$ S06 $\geq$ 10,000. The latter is treated as the frequency of a severe convective storm environment. The future integration period is 2072–2099, the historical integration period is 1962–1989, and the analyses are valid for March-April-May (a–d), and June-July-August (e–h). From Trapp et al. (2007a). Copyright 2007 National Academy of Sciences, USA.

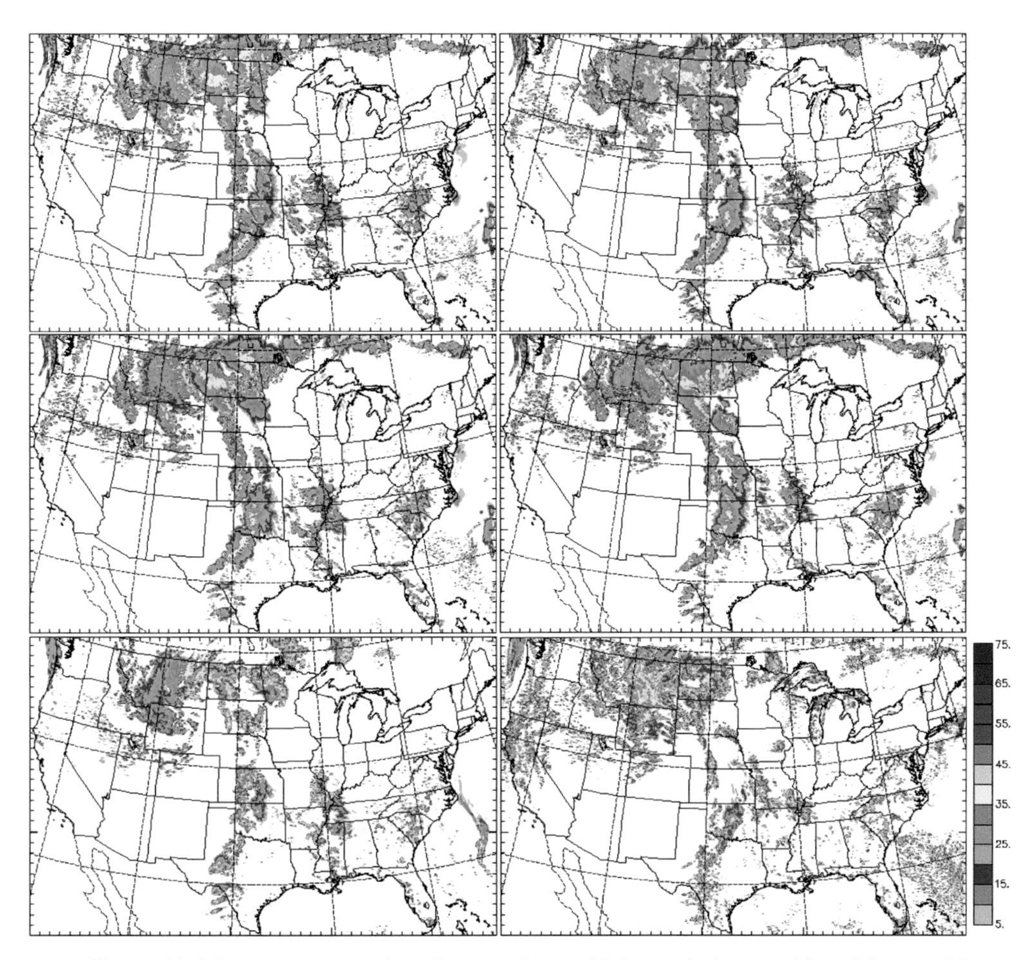

Figure 10.8 Postage-stamp plot of output from a high-resolution, multimodel ensemble system composed of 26 members. Shown is simulated radar reflectivity factor, from six members, at the forecast hour valid 0000 UTC on May 25, 2010. The system was initialized on May 24, 2010 at 0000 UTC. Courtesy of Dr. Fanyou Kong and the Center for Analysis and Prediction of Storms, University of Oklahoma.

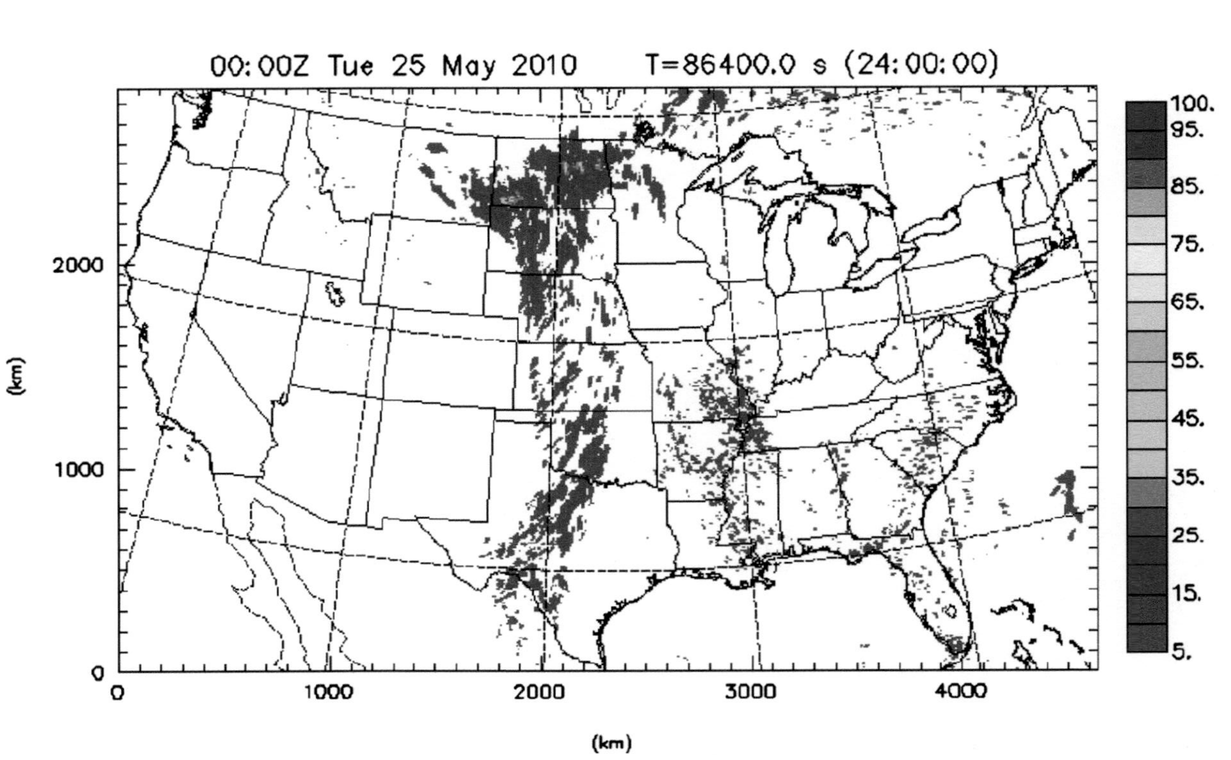

Figure 10.9 Probability of 1-h accumulated precipitation exceeding 0.5 in., at the forecast hour valid 0000 UTC on May 25, 2010. This is derived from output of a high-resolution, 26-member, multimodel ensemble system, initialized on May 24, 2010 at 0000 UTC (see Fig. 10.8). Courtesy of Dr. Fanyou Kong and the Center for Analysis and Prediction of Storms, University of Oklahoma.

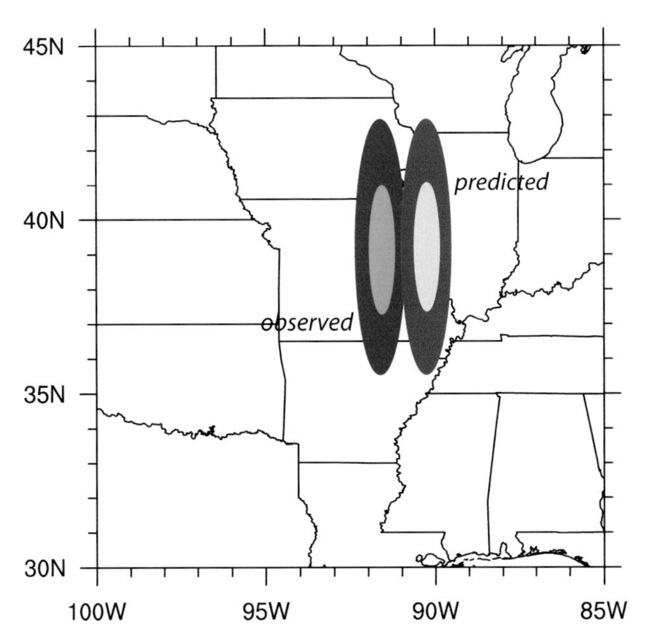

Figure 10.11 Hypothetical example of a forecast field of 1-h precipitation (red/yellow), with the verifying observations (blue/cyan). Assume that the red and blue (yellow and cyan) contours correspond to the same amounts. According to traditional measures (such as MSE), this forecast would have little to no skill, owing to a lack of local correspondence between the forecast and observation. Based on Gilleland et al. (2010).

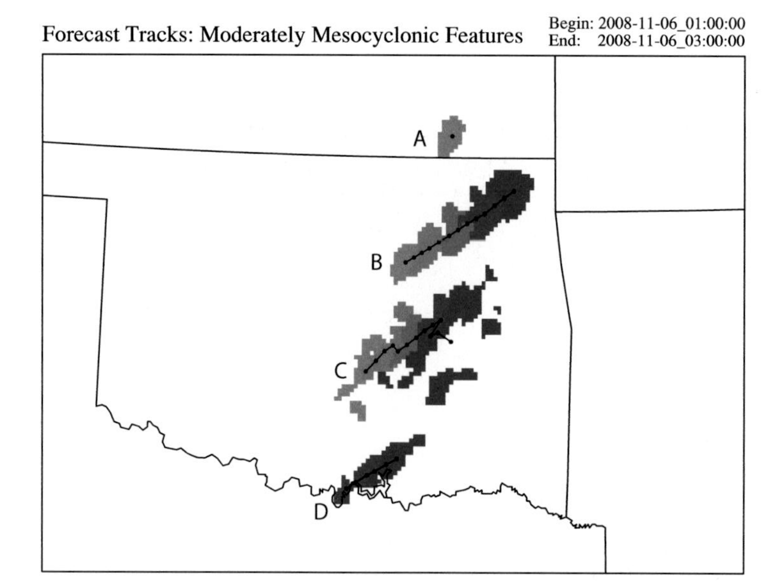

Feature ID	Start Time (UTC)	End Time (UTC)	Temporal Persistence (minutes)	Max dBZ (time UTC)
A	01:00:00	01:10:00	10	54 (0100)
B	01:00:00	03:00:00	120	59 (0140)
C	01:00:00	03:00:00	120	59 (0140)
D	02:10:00	03:00:00	50	57 (0210)

Figure 10.15 Example a feature-specific system applied to supercell prediction. Tracks and shadings indicate locations of supercells, as objectively determined from high-resolution model output. The table provides attributes of the individual supercells. From Carley et al. (2011).

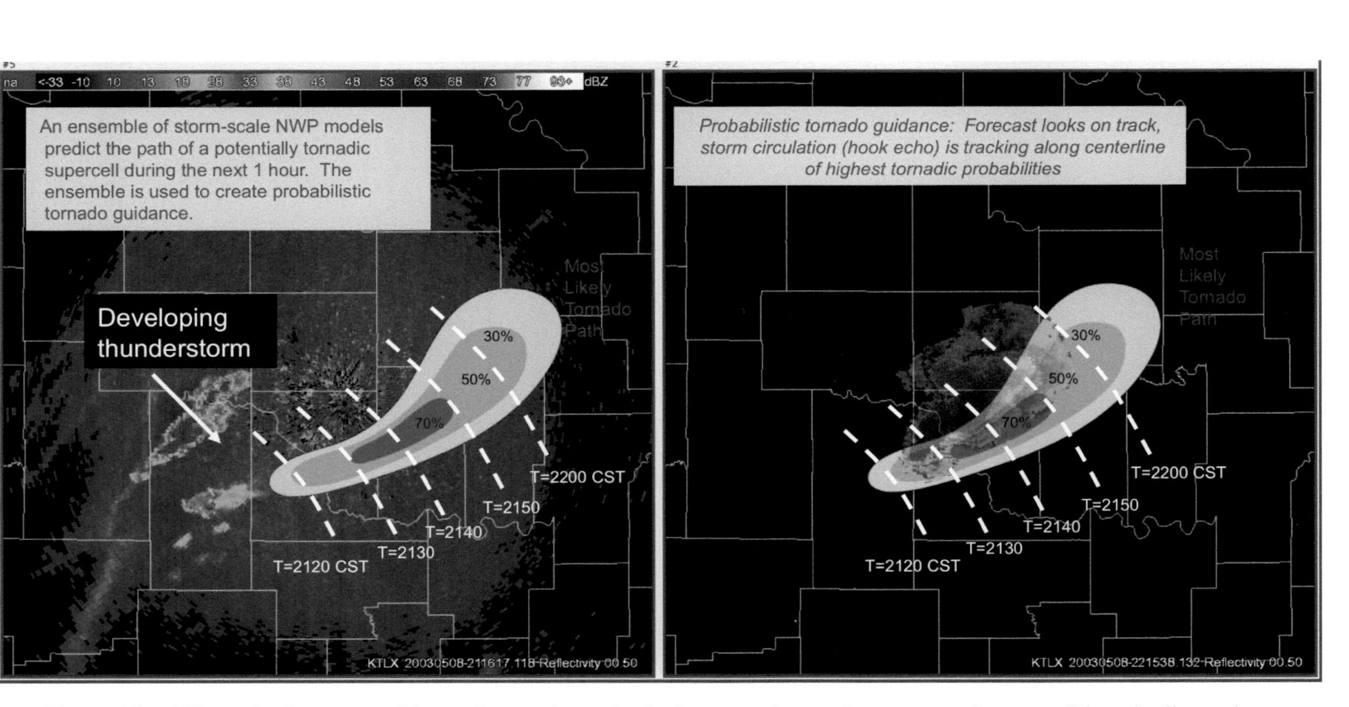

Figure 10.16 Tornado forecast guidance from a hypothetical convective-scale warn-on-forecast. Blue shadings show areal probabilities of tornado occurrence. White dashed lines indicate predicated storm locations. Color fill is of radar reflectivity factor. From Stensrud et al. (2009).

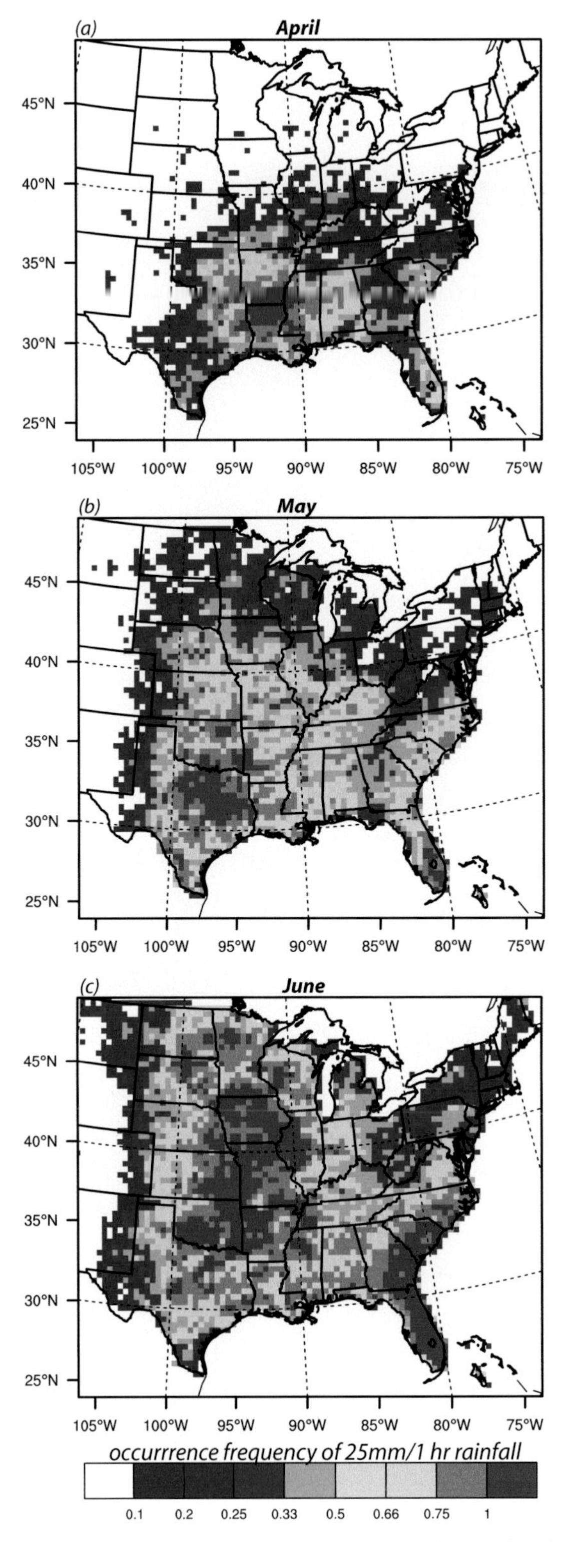

Figure 10.17 Mean frequency of 1-h rainfall exceeding 1 in., during the warm seasons of April–June, 1991–2000, from a dynamical downscaling method. From Trapp et al. (2010). Used with kind permission from Springer Science and Business Media.

5

The Initiation of Deep Convective Clouds

Synopsis: This chapter considers the basic problem of how moist air becomes positively buoyant and thereafter rises freely in the form of a deep convective cloud. Following a review of parcel theory, much of the discussion in this chapter regards the means by which air parcels are "lifted" some vertical distance so that they become positively buoyant. Synoptic-scale processes provide weak lifting, but mostly serve to precondition the thermodynamic environment. Orographic lifting is the canonical example, whereby air parcels are forced to rise as they encounter sloped terrain. Other lifting mechanisms include horizontal convective rolls, gravity waves, horizontal outflow due to other convective storms, and relatively larger-scale fronts, drylines, and sea-breeze fronts. As shown, these mechanisms may operate individually or in tandem.

5.1 Parcel Theory

Paramount to studies of convective processes is the origin of the deep cumuli that subsequently organize into convective storms. Such *convection initiation* (CI) is a topic that is treated separately here, even though some of the concepts will be used in later chapters to explain the sustenance and longevity of storms.

In simplest terms, the convection initiation problem is one of determining how moist air becomes positively buoyant and thereafter rises freely. Buoyancy is often assessed by comparing the thermodynamic properties of a hypothetical air "parcel" to those properties of the surrounding, or environmental, air. Given the two restrictions that (1) the parcel does not mix with the environment and retains its identity and (2) the environment does not generate motions to compensate for the parcel motion, *parcel theory* provides a theoretical means for such an assessment.[1]

Despite the recognized weaknesses of this theoretical construct, it is still a worthwhile exercise to understand how parcel motion can be represented through a simplified version of the vertical equation of motion (2.18):

$$\frac{Dw}{Dt} = \frac{d^2z}{dt^2} = -\frac{1}{\rho}\frac{dp}{dz} - g, \tag{5.1}$$

which neglects viscous forces, the vertical component of the Coriolis force, and the effects of water substance in the buoyancy term. The motion, which is confined to the vertical axis, occurs within an environment (denoted by an overbar) that is assumed to be in hydrostatic balance,

$$d\overline{p}/dz = -\overline{\rho}g. \tag{5.2}$$

It is also assumed is that the pressure of the parcel quickly adjusts to that of the environment, and it follows that the pressure gradient force (PGF) in (5.1) can be rewritten in terms of the hydrostatic pressure

$$\frac{d^2z}{dt^2} = -\frac{1}{\rho}\frac{d\overline{p}}{dz} - g,$$

which is then eliminated with the use of (5.2), leaving

$$\frac{d^2z}{dt^2} = g\frac{(\overline{\rho} - \rho)}{\rho}. \tag{5.3}$$

Upon substitution of the equation of state (2.9) into (5.3) for ρ and $\overline{\rho}$, and then noting that $p = \overline{p}$ per the aforementioned assumption, (5.3) can be expressed as

$$\frac{d^2z}{dt^2} = g\frac{(T - \overline{T})}{\overline{T}}. \tag{5.4}$$

Equation (5.4) shows us that a parcel will experience a vertical acceleration locally when its temperature exceeds that of the environment. This equation also allows for considerations of the (hydro)static stability of *parcel displacements* in the context of temperature lapse rates.[2] Thus let us expand parcel temperature in a Taylor series as

$$T = T_0 + \left.\frac{dT}{dz}\right|_0 (z - z_0) + \frac{1}{2}\left.\frac{d^2T}{dz^2}\right|_0 (z - z_0)^2 + \cdots, \tag{5.5}$$

where the parcel's initial state is denoted by subscript 0 and is assumed to be at level z_0. The environmental temperature is similarly expanded as

$$\overline{T} = \overline{T}_0 + \left.\frac{d\overline{T}}{dz}\right|_0 (z - z_0) + \frac{1}{2}\left.\frac{d^2\overline{T}}{dz^2}\right|_0 (z - z_0)^2 + \cdots. \tag{5.6}$$

Second-order and higher terms can be neglected in (5.5) and (5.6) if the displacements, which we now indicate as $\delta z = z - z_0$, are required to be small. Upon substituting these truncations of (5.5) and (5.6) into (5.4), we have

$$\frac{d^2(\delta z)}{dt^2} \simeq \frac{g(\gamma - \Gamma)\delta z}{\overline{T}_0 - \gamma \delta z}, \tag{5.7}$$

where $\gamma = -d\overline{T}/dz$ is the *environmental lapse rate*, $\Gamma = -dT/dz$ is the (dry, adiabatic) *parcel lapse rate*, and $T_0 = \overline{T}_0$, which is due to the parcel theory assumption that the parcel conditions initially equal those of the environment. Because δz is required to be small, the quantity $\gamma \delta z / \overline{T}_0$ will also be small, and thus the numerator in (5.7) can be approximated using the binomial series expansion (2.24) to yield

$$\frac{d^2(\delta z)}{dt^2} \simeq \frac{1}{\overline{T}_0}\left(1 + \frac{\gamma \delta z}{\overline{T}_0}\right)(\gamma - \Gamma) g \delta z = \frac{g}{\overline{T}_0}\left[(\gamma - \Gamma)\delta z + \frac{(\gamma - \Gamma)\gamma(\delta z)^2}{\overline{T}_0}\right]$$

or,

$$\frac{d^2(\delta z)}{dt^2} + \frac{g}{\overline{T}_0}(\Gamma - \gamma)\delta z = 0, \tag{5.8}$$

which again neglects the second-order terms in δz. When the coefficient is constant (i.e., constant parcel and environmental lapse rates), (5.8) is a linear, second-order differential equation. We recognize, given our work in Chapter 2, that this differential equation has a solution of the form

$$\delta z = z_0 \exp(-i\sigma t), \tag{5.9}$$

where $\sigma^2 = (\Gamma - \gamma)g/\overline{T}_0$. Thus, from (5.9) we find:

(1) If $\Gamma > \gamma$, the parcel displacement is *stable*, and the parcel oscillates about level z_0 with a frequency $\sigma = \pm\sqrt{(\Gamma - \gamma)g/\overline{T}_0}$.
(2) If $\Gamma = \gamma$, the parcel displacement is *neutral* and does not change in time.
(3) If $\Gamma < \gamma$, the parcel displacement grows in time and is *unstable*. The parcel rises freely when this condition is satisfied, illustrating why deep convection is considered to arise out of an instability.

Application of parcel theory is aided by thermodynamic diagrams (Figure 5.1) on which parcels are assumed to cool dry-adiabatically when initially lifted or vertically displaced, and then cool moist-adiabatically once this initial lifting has brought about condensation of water vapor within the parcel. Thus, the preceding static stability conditions actually depend on whether the parcel has a *dry adiabatic lapse rate*

$$\Gamma = \Gamma_d = \frac{g}{c_p}, \tag{5.10}$$

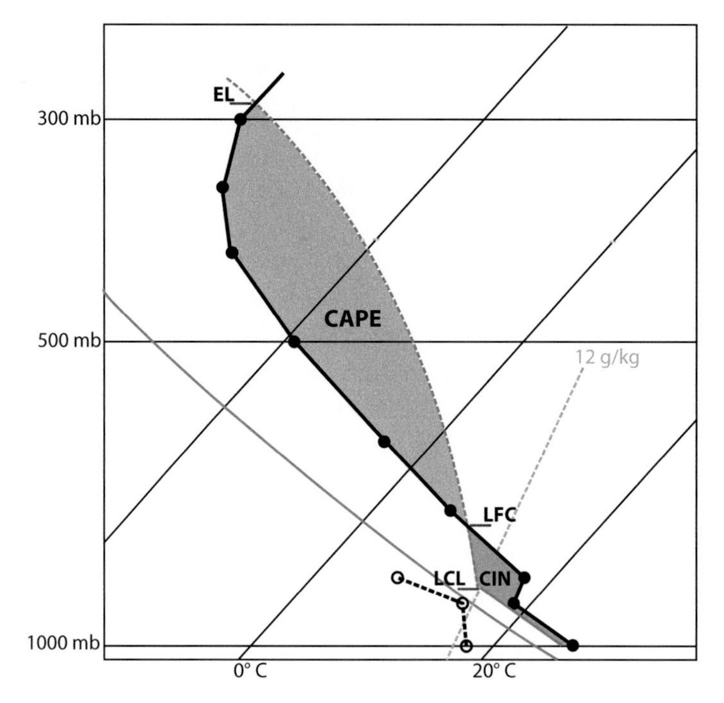

Figure 5.1 Example thermodynamic diagram (skew T – ln p), showing a sounding with an LCL, LFC, and EL. Positive and negative areas corresponding to CAPE and CIN, are shaded and red and blue, respectively. Bold black line is temperature sounding, and dashed line is segment of dewpoint sounding. Solid orange line is a dry adiabat, dashed orange line is a relevant moist adiabat, and dashed green line is a mixing ratio line. (For a color version of this figure, please see the color plate section.)

which follows from a form of the first law of thermodynamics (2.10) for dry air under adiabatic conditions, or *a moist (saturated) adiabatic lapse rate*,

$$\Gamma = \Gamma_s = \frac{g}{c_p}\left[\frac{1+\dfrac{L_v}{R_d}\dfrac{q_{v,s}}{T}}{1+\dfrac{\varepsilon L_v^2}{c_p R_d}\dfrac{q_{v,s}}{T^2}}\right], \tag{5.11}$$

which follows from a form of the first law of thermodynamics for saturated air, where $q_{v,s}$ is the water vapor mixing ratio at saturation, $\varepsilon = R_d/R_v$, and all other variables are as defined in Chapter 2.[3] The cooling rate during the moist ascent is less than that during dry ascent because of the release of latent heat of condensation; the process itself assumes that the parcel stays just saturated. The stability criteria[4] become:

$$\begin{aligned} \gamma &< \Gamma_s, \text{ absolutely stable} \\ \gamma &= \Gamma_s, \text{ saturated neutral} \\ \Gamma_s < \gamma &< \Gamma_d, \text{ conditionally unstable}. \\ \gamma &= \Gamma_d, \text{ dry neutral} \\ \gamma &> \Gamma_d, \text{ absolutely unstable} \end{aligned} \tag{5.12}$$

An additional criterion applies to a saturated atmosphere:

$$\gamma_s > \Gamma_s, \text{ absolutely unstable,} \tag{5.13}$$

where γ_s is the lapse rate of a saturated environment. The moist absolutely unstable layers (MAULs) that describe this state are formed by nonbuoyant lifting (see Section 5.2) over mesoscale areas, as is known to occur within certain mesoscale-convective systems.[5] In contrast to the moist absolute instability, which requires saturation of the parcel and environment, the state of conditional instability literally is conditional on whether the parcel is saturated; it may be argued therefore that conditional instability is not a true instability.[6]

It is appropriate to be reminded here that, as strictly defined, an instability requires existence of stored energy in the base state and a perturbation to which the base state is unstable and thus that can access the stored energy.[7] In the specific case of gravitational (or buoyant) instability, potential energy in an atmospheric column or slab is drawn by perturbations that are manifest and grow as vertically overturning motions. The stored energy is *convective available potential energy* (CAPE; J kg^{-1}),

$$\text{CAPE} = \int_{z_{LFC}}^{z_{EL}} B dz, \tag{5.14}$$

where the limits of integration z_{LFC} and z_{EL} are, respectively, the level of free convection (LFC) and the equilibrium level (EL). On a thermodynamic diagram, this is the "positive area" bounded by the environmental and parcel curves between the LFC and EL (see Figure 5.1). It is convenient at this point to introduce *downdraft CAPE* (DCAPE; J kg^{-1}), which quantifies the potential (negative) buoyant energy during saturated descent,

$$\text{DCAPE} = \int_{z_p}^{z_0} -B dz, \tag{5.15}$$

where z_p is the source height of the parcel and z_0 is the ground.[8] The exact form of buoyancy B in (5.14) and (5.15) has purposely been left vague, although it is common for B to be expressed as a thermal buoyancy in terms of virtual temperature:

$$B = g\frac{T_v - \overline{T}_v}{\overline{T}_v}, \tag{5.16}$$

which follows from (5.3) with a substitution from the equation of state for moist air (2.48).[9]

Existence of nonzero CAPE can be considered a necessary, though not sufficient, condition for instability.[10] The insufficiency is because access of perturbations to the stored energy is not guaranteed. Literally inhibiting this access is a layer or layers below the LFC in which buoyancy is negative. Such *convective inhibition* (CIN; J kg^{-1}) can be quantified as:

$$\mathrm{CIN} = \int_{z_0}^{z_{LFC}} B dz, \tag{5.17}$$

which on a thermodynamic diagram is the "negative area" bounded by the environmental and parcel curves, and hence is relevant for layers in which $B < 0$ (Figure 5.1). CIN may, in some sense, be a predictive parameter for the likelihood of convection initiation: the larger the CIN, the more large-scale destabilization and/or parcel lifting is required for the CAPE to be realized. Mechanisms for both will be discussed in the sections that follow.

Though arguably still useful in a pedagogical sense, parcel theory is often criticized as an oversimplification of cumulus convection.[11] One objection to parcel theory is the neglect of the parcel's pressure and its vertical gradients in (5.4) (see Chapter 2). A related objection is the application of (5.4) without explicit representation of mass continuity. As explored more in Chapter 6, the implication is an overestimate in parcel ascent or descent.

5.2 Synoptic-Scale Conditioning of the Convective Environment

In this section we focus on how synoptic-scale processes condition the convective environment thermodynamically; effects of the synoptic-scale wind field on convective organization are considered in later chapters. Let us begin with a simple example. Assume the existence of typical quasi-geostrophic (QG) vertical motions of 1 cm s^{-1}, in an atmosphere with low static stability. Under this forced ascent only, a surface-based parcel would reach a hypothetical 1500-m LFC in roughly 42 hours. This is clearly an implausible time interval for parcel lifting and convection initiation, especially in light of the fact that the synoptic-scale thermodynamic environment would be expected to evolve over this time, with likely changes to the static stability; parcel buoyancy would also be diluted significantly, if not completely, during a 42-hr ascent (Chapter 6). Much stronger synoptic-scale vertical motions of 10 cm s^{-1} would allow a parcel to reach a 1500-m LFC in about four hours. This is a more plausible time interval for parcel lifting and convection initiation. However, because QG vertical motions are usually maximized in the middle troposphere, it is uncertain whether lifting of this magnitude could actually occur within the lowest levels of the troposphere.[12]

These simple calculations demonstrate that QG vertical motions unlikely serve as what is often referred to as a convective "trigger."[13] The weak yet persistent ascent does, however, modify the convective environment through adiabatic cooling. Figure 5.2 illustrates how vertical layers are destabilized, and effectively moistened,

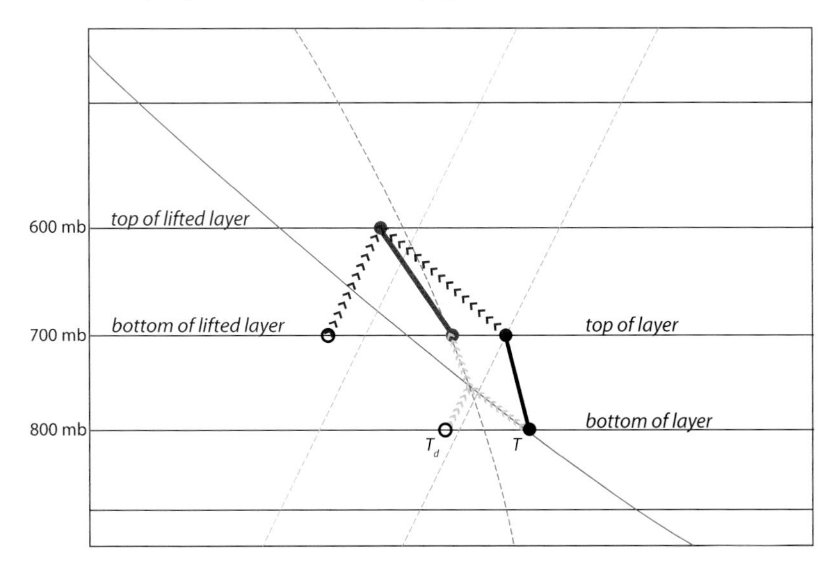

Figure 5.2 Illustration of thermodynamic destabilization of a layer through ascent. Bold black line is segment of original temperature sounding, and bold gray line represents the subsequent modification. Solid orange line is dry adiabat, and dashed orange line is relevant moist adiabat. The parcel process associated with the lifting of the bottom (top) of the layer is indicated by the light blue (dark blue) arrowed line. (For a color version of this figure, please see the color plate section.)

through ascent. The *potential* for instability upon layer lifting can be assessed from calculations of equivalent potential temperature of the environment. The criteria are

$$\begin{array}{c} \partial\theta_e/\partial z < 0, \text{ potentially unstable} \\ \partial\theta_e/\partial z > 0, \text{ potentially stable} \end{array} . \tag{5.18}$$

The synoptic-scale destabilization is often coupled with solar radiative heating at the ground, with sensible heat transferred to the air immediately above the ground via conduction and shallow convection (Figure 5.3). The latter is the well-recognized process by which the nocturnal temperature inversion is gradually eroded during the day.

The traditional view of synoptic-scale conditioning is from QG ascent (descent) downstream of a mid- to upper-tropospheric trough (ridge) axis. An alternative perspective is provided through the use of potential vorticity (PV), defined as

$$\mathrm{P} = g(\zeta_\theta + f)(-\partial\theta/\partial p), \tag{5.19}$$

where

$$\zeta_\theta = \left(\frac{\partial v}{\partial x} - \frac{\partial u}{\partial y}\right)_\theta . \tag{5.20}$$

ζ_θ is the vertical component of the vorticity vector evaluated on an isentropic surface, as denoted by the subscript θ, and $-\partial\theta/\partial p$ quantifies static stability. Locally large (hence anomalously positive) PV requires an upward (downward) displacement of

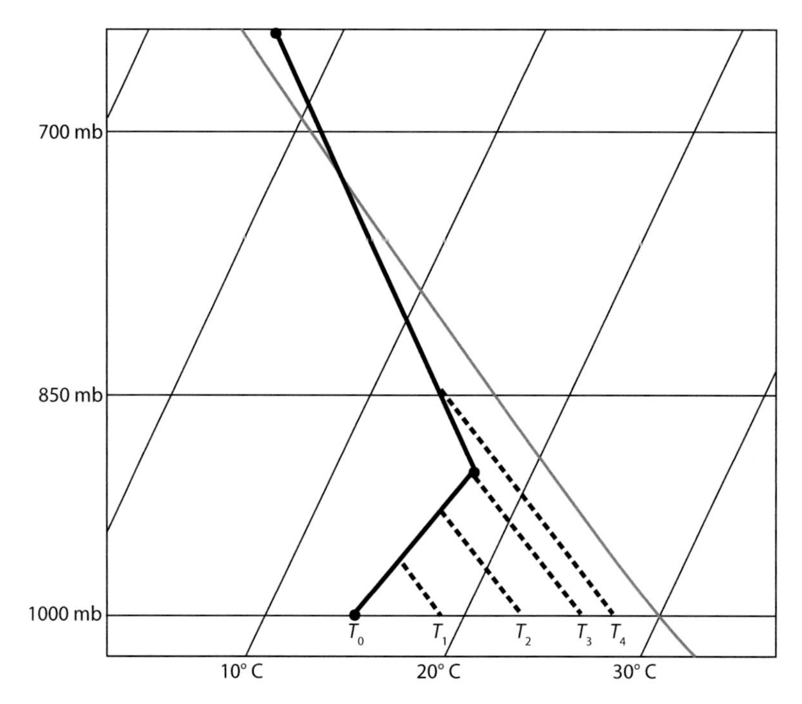

Figure 5.3 Illustration of thermodynamic destabilization of the atmospheric boundary layer through solar heating. Bold black line is original temperature sounding, and dashed lines represent subsequent modifications, as coupled to changes in surface temperature *T*. Solid orange line is dry adiabat. (For a color version of this figure, please see the color plate section.)

isentropes beneath (above) the anomaly (Figure 5.4; see also (5.19)). In a migratory anomaly that conserves its PV (see Chapter 10), this requirement translates into ascent (descent) along isentropic surfaces in advance of (behind) the anomaly, and accordingly, into destabilization (stabilization) in the ascent upstream (descent downstream). As the reader may have recognized, this is consistent with the QG perspective. Knowledge of the PV perspective, however, allows us to examine cases of synoptic-scale preconditioning that have been cast in terms of PV and, hence, observed relationships between PV and thunderstorm activity.[14]

An example is the infamous May 3, 1999 tornado outbreak within the U.S. Great Plains.[15] During this event, low- to mid-tropospheric ascent in advance of an upper-tropospheric PV anomaly (jet streak) had the destabilizing effect just illustrated. Concurrently, upper-tropospheric ascent associated with the same anomaly resulted in the development of cirrus clouds. The interesting consequence of the cirrus shield was that it reduced solar insolation and surface heating, and thus limited the destabilization of the low levels of the atmosphere. Widespread CI was prevented by the cirrus, but gaps in the shield did allow for localized CI. Indeed, aided by mesoscale lift (see Section 5.3), isolated convective storms ultimately initiated beneath these gaps, and some of these storms quickly became supercell thunderstorms (Chapter 7).

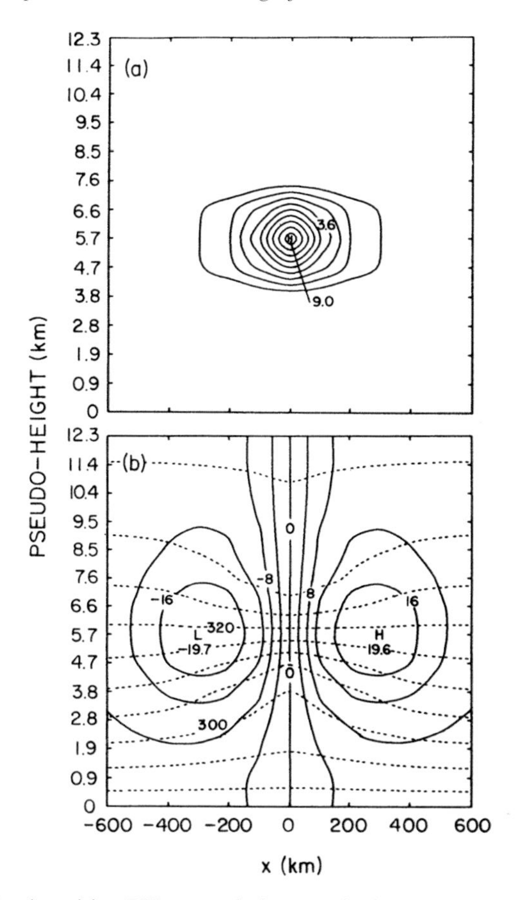

Figure 5.4 Idealized positive PV anomaly in a vertical plane (top), and the associated fields of potential temperature (dashed contours) and tangential horizontal wind (solid contours) (bottom). Negative tangential wind values represent horizontal flow out of the plane. From Davis (1992).

A general conclusion from this case study is that synoptic-scale processes have a multitude of effects that directly or even indirectly relate to CI. Not yet considered are the positive temperature and moisture advections in the warm sector of an archetypal synoptic-scale extratropical cyclone (ETC) (see Figure 5.5). Temperature and moisture advections that warm and moisten the lower troposphere have a destabilizing effect, as similarly does the advection of drier, cooler air aloft. These effects are particularly strong and rapid when the advecting winds are concentrated in a jet (Figure 5.5), as in the case of the low-level jet (LLJ), which develops in association with synoptic-scale forcing and/or diurnally varying force balances within the atmospheric boundary layer (ABL).[16] LLJs and other advecting airstreams are not entirely horizontal, and may, at times, have nonnegligible vertical components. One consequence of these sloping streams is a partial offset of the warming due to advection by the cooling from adiabatic ascent. Depending on the slope and the wind speed, another possible consequence is parcel lifting to free convection. This could occur, for example, when

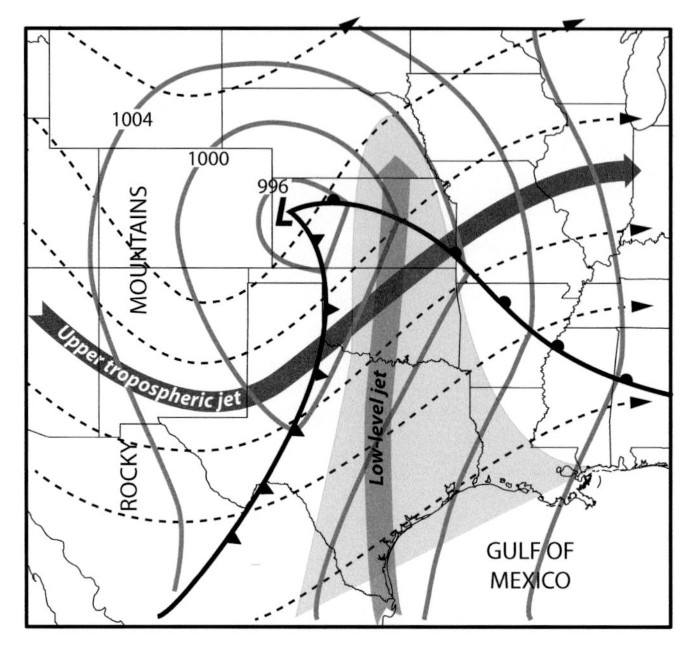

Figure 5.5 Characteristic synoptic-scale airflow in association with severe convective storms. The synoptic-scale cyclone at the surface is identified by surface frontal locations (bold black lines) and isobars of mean sea-level pressure (bold gray lines). The associated upper level trough is indicated in streamlines of upper-level flow (dashed lines). The upper-level jet (black ribbon) represents a concentration of this flow. Similarly, the low-level jet (LLJ; gray ribbon) shows where lower-tropospheric flow is particularly strong and concentrated, and the light gray shading indicates the tongue of low-level moisture that has resulted from the strong low-level advection. Adapted from Newton (1976).

an LLJ is oriented perpendicular to a warm- or quasi-stationary-frontal boundary, and parcel motion on isentropic (or equivalent isentropic) surfaces is conserved. *Isentropic upglide*[17] is one potential way that synoptic-scale fronts contribute to CI; others are considered in Section 5.4.1.

The net effect of advection depends in part on the geographical setting. In the United States, ETCs moving east of the Rocky Mountains have access to an abundant source of low-level moisture from the Gulf of Mexico (Figure 5.5); particularly intense systems can transport moisture poleward from the Gulf of Mexico to considerable northern latitudes. The convective environment in the Great Plains is also affected by air originating in the semi-arid region of the Central Mexican Plateau. Owing to the aridity, solar insolation results in high sensible heating of the elevated surface (average elevation is ~1800 m). As boundary-layer air from this elevated terrain is transported eastward, it overruns the relatively cooler, moister boundary layer of the lower terrain, and thus contributes to a "capping inversion." The elevated layer of air often has a near dry-adiabatic lapse rate and a constant mixing ratio (Figure 5.6),

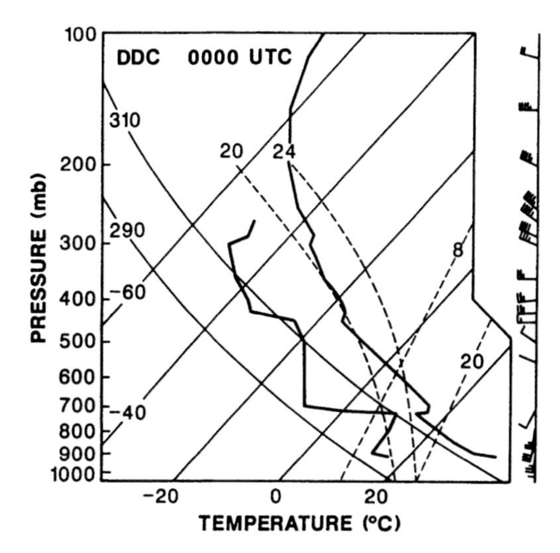

Figure 5.6 An elevated mixed layer (EML) and associated "lid" or "capping inversion" in a sounding at Dodge City, Kansas. The EML is between 700 and 450 hPa. From Stensrud (1993).

consistent with the structure of well-mixed boundary layer and hence deserving of the term *elevated mixed layer* (EML), although such layers may not always be well mixed.[18]

EMLs are a class of "caps" or "lids" found in environments supportive of deep moist convection around the world. Some have geographical links similar to that found in the United States, as in the EMLs from the hot, dry air generated on the Spanish plateau (the "Spanish plume"), which then inhibit convection in southern France.[19] Lids can also originate when stable and dry stratospheric air is introduced through folds in the tropopause.[20] Regardless of their origin, all serve to prevent widespread and relatively shallow convection, in favor of allowing the local environment to accumulate CAPE over time through the larger-scale processes just described. The release of the CAPE then requires a sufficiently deep and strong lifting mechanism, as discussed in the following sections.

5.3 Mechanisms of Local or Near-Field Mesoscale Lift

Mechanisms whereby CI occurs in general proximity to the source of parcel lifting are described in this section. These often explain the first radar echoes, observed early in the diurnal cycle of solar radiative heating. The resultant convective cells are advected (and propagate) away from their initiating source, and then provide a means for the initiation of subsequent, or secondary convection later in the diurnal cycle. Such *far-field* CI is discussed in Section 5.5.

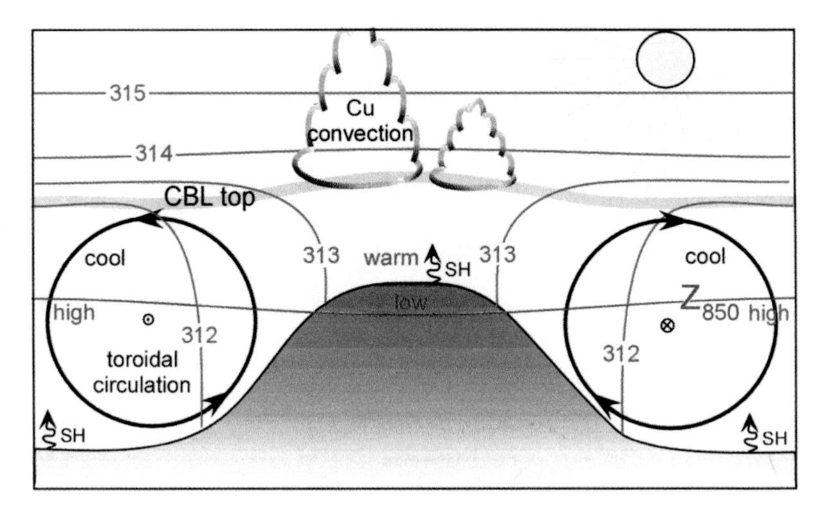

Figure 5.7 Depiction of convection initiation on an idealized heated mountain. Thin lines are isentropes or the height of the 850-hPa pressure surface (as labeled Z_{850}), the thick gray line is the top of the convective boundary layer, and arrowed circles indicate the baroclinically generated circulations. This is a case of quiescent flow; with nonzero flow, CI is more likely on the lee side of the mountain, where thermally forced upslope flow converge with ridgetop winds. From Geerts et al. (2008).

5.3.1 Orographic Lifting

Several of the canonical lifting mechanisms are associated with regional physiogeography. *Orographic lifting*, or the ascent of air parcels in association with sloped terrain, is one example; other examples are provided in Section 5.4. A typical portrayal of orographic lifting is of mechanically forced upslope flow, with CI on the windward side. Valleys and other complex terrain also contribute to mechanical forcing, and lead to correspondingly complex patterns in CI occurrence.[21]

CI near the mountain peak often results from upslope flow that is thermally forced: owing to the greater (radiative) heating of the elevated land surface relative to that of the nearby atmosphere at the same altitude, a horizontal gradient in density forms and serves to drive a solenoidal or baroclinic circulation with an upslope branch (Figure 5.7). This vertical circulation – and others analogous to it that will be encountered in this chapter – is explained by an equation that governs time-dependent, inviscid changes in the component of vorticity normal to this plane:

$$\frac{D\xi}{Dt} = -\xi\left(\frac{\partial u}{\partial x} + \frac{\partial w}{\partial z}\right) - \frac{\partial B}{\partial x}, \tag{5.21}$$

where $\xi = \hat{j} \cdot \nabla \times \vec{V} = \partial u/\partial z - \partial w/\partial x$, and $B = -g\rho'/\overline{\rho}$. Equation (5.21) assumes an idealized 2D mountain and flow restricted to the x–z plane, and derives from scaled versions of the component equations of motion introduced in Chapter 2, with $\partial/\partial x\,(Dw/Dt)$ subtracted from $\partial/\partial z\,(Du/Dt)$. The first right-hand term in (5.21) is a horizontal stretching term, and accounts for amplification of existing y-component relative vorticity; the second right-hand term is the baroclinic generation term. Notice that when the environment is quiescent, as assumed in Figure 5.7, the heated mountain results in baroclinically generated circulations of equal strength but opposite orientation across the mountain peak. This favors the deep lifting of boundary layer air that is drawn up the mountain slopes, with CI over the peak.[22] When the environmental flow is nonzero, CI is more likely to occur on the lee side of the mountain, where thermally forced upslope flow converges with ridgetop winds.

There is the suggestion in the preceding text that orographically initiated convection is a consequence of a singular instance of mechanically and/or thermally forced ascent. Figure 5.8 shows, instead, the emergence of deep convection from a series of buoyant "thermals" that are aided by the upslope flow.[23] Each thermal rises through a predecessor's wake: because the successive thermals ingest moist wake air rather than drier environmental air, they are more buoyant, rise more freely, and reach successively higher levels (see Chapter 6). This form of atmospheric conditioning can be localized or widespread, and applies to lifting aided by other mechanisms, such as through *horizontal convective roll* (HCR) circulations organized in the atmospheric boundary layer.

5.3.2 Horizontal Convective Rolls

HCRs are longitudinal vortices (Figure 5.9), often made visible by bands of shallow, roll-topped cumuli or "cloud streets," and known as a means by which heat and moisture are vertically mixed within the ABL. HCRs form in the presence of environmental vertical wind shear and thermal buoyancy. The ubiquity of a vertically sheared ABL relative to the less frequent occurrence of cloud streets might lead one to argue that HCRs are primarily a manifestation of Rayleigh-Bénard instability, with radiative heating at the ground (and/or cold-air advection above the ground) providing for the base-state temperature profile $\gamma = -d\overline{T}/dz$. We learned in Chapter 2 that the theoretical onset of Rayleigh-Bénard convection (in the specific case of free-slip conditions on the tangential velocity components, and a resting base state) occurs when the Rayleigh number exceeds the critical Rayleigh number $Ra_c = 27\pi^4/4$. Recall, however, that this theoretical value does not reference a specific horizontal direction (i.e., it is based on horizontal wavenumber $K = \sqrt{k^2 + l^2}$), and therefore provides no explicit information about whether the instability should take the form of hexagonal cells ($k = l$) or rolls ($k = 0$ or $l = 0$) at onset.

Geometrical shape is influenced by wind shear. A parameter that accounts for such dynamical effects as well as thermal effects on boundary-layer convective eddies is the *Monin-Obukhov length L*,

$$L = -\frac{(\overline{u'w'})^{3/2}}{\overline{w'B'}}, \tag{5.22}$$

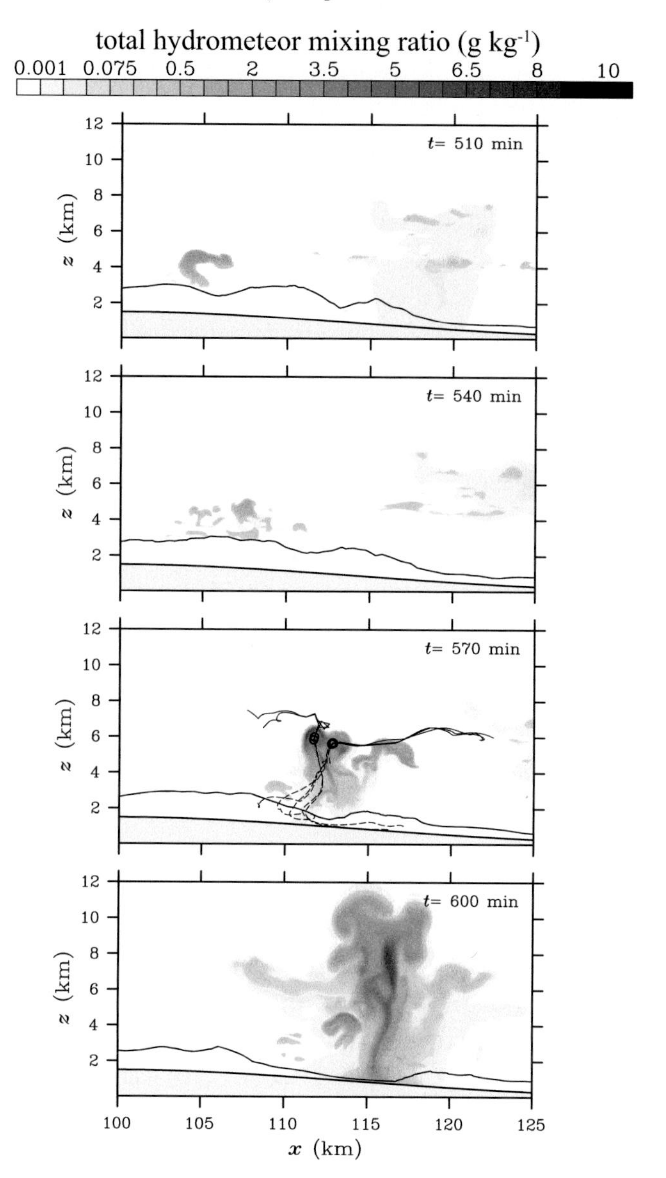

Figure 5.8 High-resolution simulation of convection initiation over an idealized, heated mountain, in a weakly sheared environment. Gray shading is of total hydrometeor (cloud and precipitation) mixing ratio. Single solid line indicated top of boundary layer. Series of lines at t = 570 min show parcel trajectories. After Kirshbaum (2011).

where $u'w'$ is the near-surface eddy momentum flux, $w'B'$ is the near-surface eddy buoyancy flux, and an overbar indicates a horizontal average.[24] By convention, L is negative when the surface buoyancy flux is positive, which will be the case in a heated boundary layer. Physically, $|L|$ represents the height above the surface where the production of turbulence by buoyancy is greater than that by vertical shear, but here we compare it to the boundary-layer depth z_i to address HCR occurrence. Specifically, the

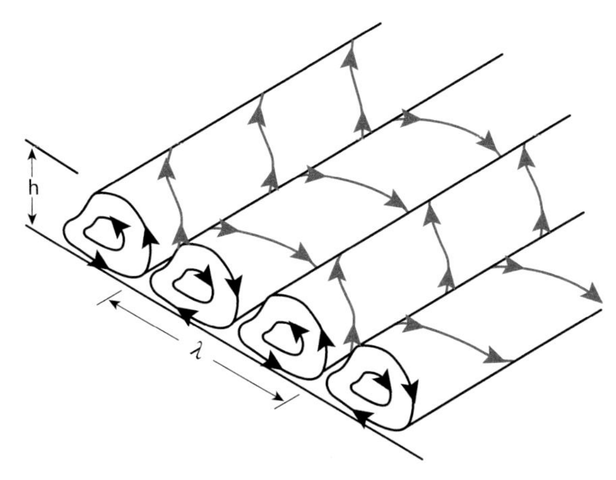

Figure 5.9 Depiction of horizontal convective rolls (HCRs). The roll wavelength is given as λ, and role depth, as h. After Weckwerth et al. (1997).

nondimensional ratio $-z_i/L$ is treated as an ABL stability parameter, with instability realized in the form of (observed) roll convection generally when $0 < -z_i/L < 21$.[25] The aspect ratio (horizontal/vertical scale) of the rolls is also proportional to $-z_i/L$, with observed and modeled values generally between 2 and 4. Thus, the nominal roll wavelength for a 1.5-km-deep ABL varies from 3 to 6 km. We will see in Section 5.3.2 that interactions between HCRs and internal gravity waves could modulate this wavelength en route to convection initiation. However, let us consider here the question of whether HCRs alone can provide sufficiently strong and deep lifting to initiate cumulus convection in absence of these interactions.

Although an affirmative answer to this question is implied in analyses of events such as the May 3, 1999 tornado outbreak mentioned previously,[26] quantification of observed HCR structure provides more convincing evidence. For example, single-Doppler radar data have been used to establish the existence of HCRs and, through analysis of spatial autocorrelation fields, also used to estimate roll wavelength (Figure 5.10).[27] Proximity soundings have then provided corresponding evaluations of z_i and the LFC. When combined, these data show that convection initiation via HCRs is indeed possible but, not surprisingly, only when the roll depth is nearly equal to the LFC (Figure 5.11).

As mentioned previously, HCRs serve the more general purpose of boundary-layer mixing, which in the context of the current chapter equates to a modification of the local environment. Indeed, an effect of the upward transport of moist surface air is an enhancement of water vapor mixing ratio ($\sim$1–2 g kg^{-1}) in the roll updraft regions.[28] It is thus within the roll updrafts that the potential for cumulus convection should be evaluated. For precise forecasts of CI, this represents a significant challenge, because the short temporal and small spatial scales of HCRs suggest limited predictability of roll updrafts.

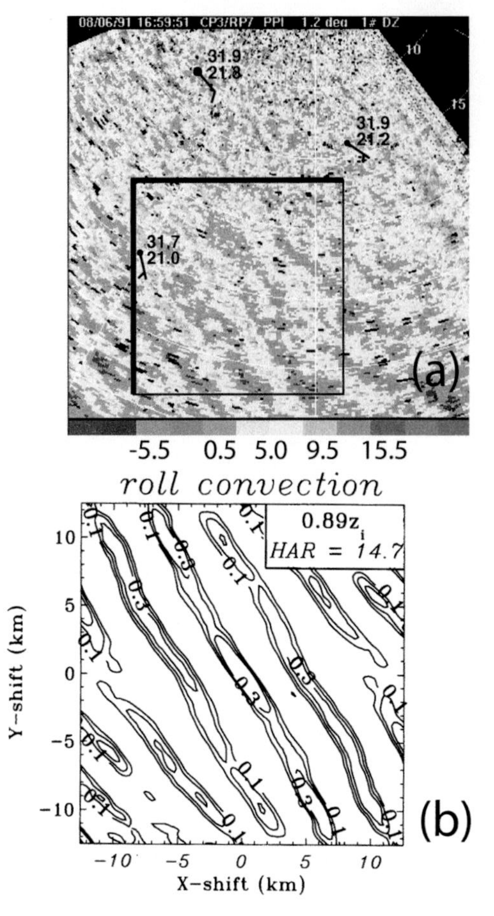

Figure 5.10 Example of the use of weather radar data to deduce HCR structure. (a) Equivalent radar reflectivity at low elevation angle in a PPI. (b) Field of spatial autocorrelation based on (a), and for the subdomain indicated in (a). From Weckwerth et al. (1997). (For a color version of this figure, please see the color plate section.)

5.3.3 Internal Gravity Wave–HCR Interaction

A convectively active day often features the development of a field of shallow cumuli, followed by its subsequent evolution into fewer, but larger cumulus congestus clouds. One might infer from this evolution that the atmosphere favors – or *selects* – a certain scale of cumulus for further growth into congestus. From a theoretical perspective, the selected scale would represent the fastest growing mode of an instability.

Let us assume that on this hypothetical day, the cumuli form atop HCRs. As just discussed, the HCRs are controlled in scale by $-z_i/L$ and fundamentally by the environmental temperature and wind. For typical conditions in midlatitudes, the roll wavelength is 3 to 6 km, and hence the boundary-layer-topped cumuli would initially have updrafts of roughly half this length (i.e., within the rising branches of adjacent rolls that comprise the HCR wavelength; see Figure 5.9). Severe midlatitude convective storms, on the other hand, are known to have updraft diameters on the

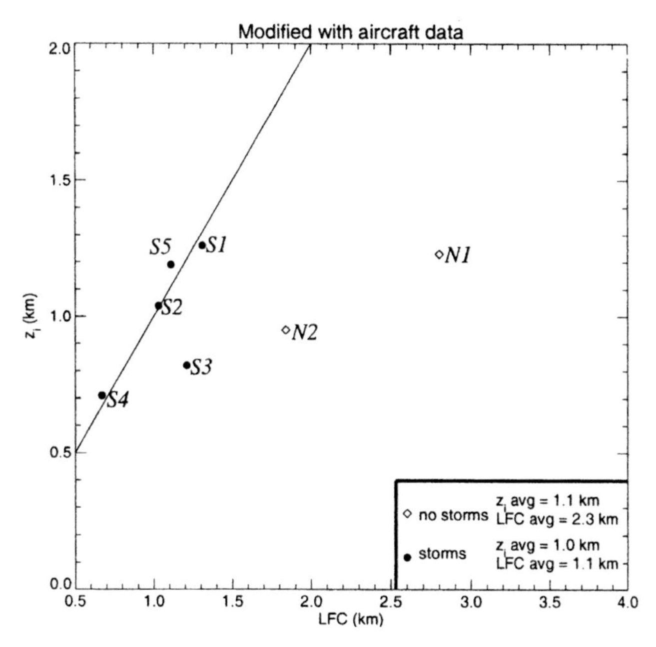

Figure 5.11 Comparison of HCR cases with and without convection initiation. Here, z_i is the depth of the ABL. The LFC has been computed using sounding data modified with aircraft and surface data, such that the sounding represents conditions within a roll updraft. For reference, the case shown in Figure 5.10 is a "no-storm" case (N1). Line is the one-to-one agreement between LFC and z_i. Note that convection initiation (or the lack thereof) was confirmed using the criterion that $Z > 35$ dBZ within a radar echo. From Weckwerth (2000).

order of 10 km.[29] One could argue that the mechanism of mesoscale lift must, at minimum, be of this same order, because much narrower deep cloud growth twould be overly susceptible to the deleterious effects of entrainment by small, turbulent eddies (see Chapter 6).[30] Updraft size depends of course on other factors, such as the environmental profiles of temperature, humidity, and wind shear. Nonetheless, a basic relationship between the scale of lift and the scale of cloud growth is routinely realized in idealized thunderstorm models (and consistent with the established practice of using warm bubbles of ~10-km radius to initiate convective storms in such models; see Chapter 4).

For insight on how this relationship may be brought about more naturally, consider the following idealization of the environmental (or base state) temperature $\overline{\theta}(z)$, in which the associated (dry) static stability is divided into three layers: a boundary layer in which $S \equiv \partial \ln \overline{\theta} / \partial z = 0$, a free troposphere where S is some constant value $S_T > 0$, and a stratosphere where $S = S_S > S_T$. Internal gravity waves are supported in the free tropospheric layer. They can be excited by HCRs when the roll updrafts perturb or vertically displace the top of the boundary layer. The dominant waves (or fastest growing modes) have been shown both in modeling studies and in observations to have wavelengths $\geq$10 km.[31,32] As illustrated in Figure 5.12, subsequent nonlinear interaction between these relatively larger-scale gravity waves and the HCRs results

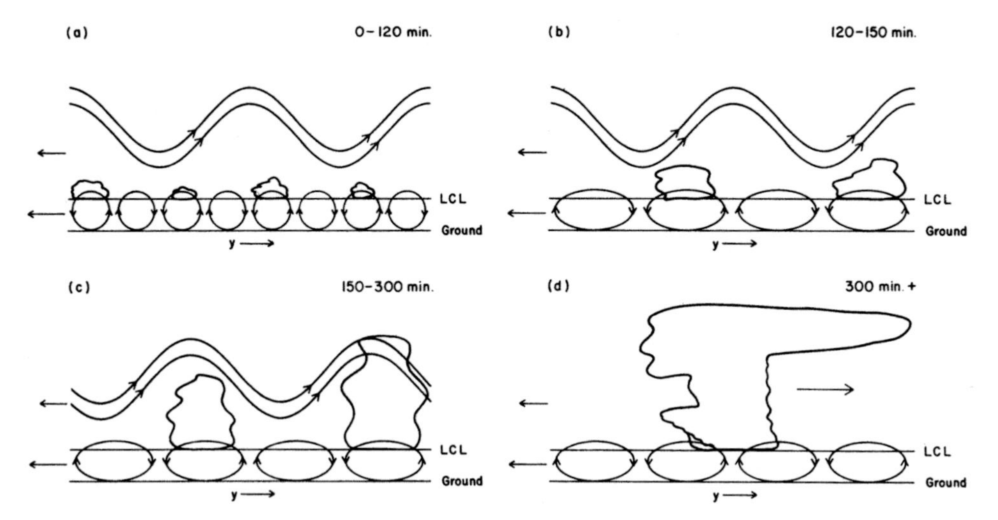

Figure 5.12 Illustration of an interaction between HCRs and internal gravity waves leading to the initiation of deep convective clouds. From Balaji and Clark (1988).

in roll broadening, to the extent that the longer-wavelength HCRs can be of sufficient scale to force deep convective growth. The favored selection of the few clouds for growth can be explained by the horizontal propagation of the gravity waves relative to the advection of the HCR field: only when the roll updrafts are in phase with the gravity wave ascent is a channel of deep lifting provided. The phase coupling depends on vertical variations in static stability, hence in the gravity-wave phase speed, and additionally on the vertical wind shear (see Chapter 2).

Although coupled HCRs and internal gravity waves have been observed in the real atmosphere, satisfactory observational evidence that this mechanism has initiated deep convective clouds is lacking.[33] This is in part because multiple mechanisms of mesoscale lift are often present on a given day, and the stronger, deeper lift will tend to mask other influences. Let us therefore turn our attention now to these stronger mesoscale forcings.

5.4 Mesoscale Baroclinic Circulations

5.4.1 Frontogenesis

The focus in this section is on the potential for cold fronts to initiate deep convection, but it should be understood that the following theoretical development applies to other frontal- and frontal-type systems. Typical along-cold-front lengths (several hundred kilometers) are of synoptic scale, but the across-cold-front distances of several tens of kilometers are well within the range of the mesoscale. Of specific interest are cross-frontal or transverse vertical circulations that respond to changes in the gradient of temperature. The quantitative characterization of such frontogenesis is given by the *frontogenesis function*, $\Im$. Traditionally, $\Im$ is defined as the rate of change of the magnitude of the horizontal potential temperature gradient following the motion of a parcel,

$$\Im \equiv \frac{D\left|\nabla_h \theta\right|}{Dt}. \tag{5.23}$$

This may be evaluated on an isobaric surface, but here we will assume evaluation at constant height. Various forms of $\Im$ exist, and depend on whether the material derivative is assumed to be with respect to 2D or 3D parcel motion, whether $\Im$ is evaluated at the ground (assumed to be a flat, horizontal surface), and whether along-front variability in potential temperature is considered. It is also possible to express $\Im$ in natural coordinates, and in terms of the wind based on quasi-geostrophic and semi-geostrophic principles. All of these can be reduced from a 3D vector form of (5.23), which we now pursue.

By definition, the vector frontogenesis function is

$$\vec{\Im} \equiv \frac{D\left(\nabla\theta\right)}{Dt}. \tag{5.24}$$

To obtain an equation governing $\vec{\Im}$, write the material derivative of the potential temperature gradient,

$$\frac{D}{Dt}\left(\nabla\theta\right) = \frac{\partial}{\partial t}\left(\nabla\theta\right) + \vec{V}\cdot\nabla\left(\nabla\theta\right), \tag{5.25}$$

and then the material derivative of the potential temperature operated on by the gradient operator,

$$\nabla\left(\frac{D\theta}{Dt}\right) = \nabla\left(\frac{\partial\theta}{\partial t}\right) + \nabla(\vec{V}\cdot\nabla\theta). \tag{5.26}$$

Because $\partial/\partial t$ and ∇ commute, we can eliminate the first terms on the right-hand side of (5.25) and (5.26) to yield

$$\frac{D\nabla\theta}{Dt} = \nabla\left(\frac{\partial\theta}{\partial t}\right) + (\vec{V}\cdot\nabla)\nabla\theta - \nabla(\vec{V}\cdot\nabla\theta). \tag{5.27}$$

A vector identity can be used to expand the third term on the right-hand side of (5.27):

$$\nabla(\vec{V}\cdot\nabla\theta) = \vec{V}\times(\nabla\times\nabla\theta) + \nabla\theta\times(\nabla\times\vec{V}) + (\vec{V}\cdot\nabla)\nabla\theta + (\nabla\theta\cdot\nabla)\vec{V}. \tag{5.28}$$

Using (5.28) in (5.27) and noting that the curl of the gradient of a scalar is identically zero, we are left with

$$\frac{D\left(\nabla\theta\right)}{Dt} = \nabla\dot{\theta} - \nabla\theta\times(\nabla\times\vec{V}) - (\nabla\theta\cdot\nabla)\vec{V}. \tag{5.29}$$

Equation (5.29) shows that the gradient of diabatic heating, the reorientation of the temperature gradient by the vorticity, and the compression and tilting of the temperature gradient, respectively, contribute to vector frontogenesis.[34]

As mentioned, (5.29) can be used to obtain specific, reduced forms of the frontogenesis equation. For example, frontogenesis in the x-direction, applied to a frontal surface in the y–z plane, is described by taking $\hat{i}\cdot$Eq. (5.29):

$$\frac{D}{Dt}\left(\frac{\partial\theta}{\partial x}\right) = \Im_x = \frac{\partial}{\partial x}\left(\dot{\theta}\right) - \frac{\partial u}{\partial x}\frac{\partial\theta}{\partial x} - \frac{\partial v}{\partial x}\frac{\partial\theta}{\partial y} - \frac{\partial w}{\partial x}\frac{\partial\theta}{\partial z}. \tag{5.30}$$

Further reductions of (5.30) follow from the neglect of along-frontal variations, or from the neglect of vertical motions (i.e., application near the ground).

These reduced forms suffice for our current purposes, because we are concerned not with the details of frontogenesis, but rather with the resultant circulation and its relative strength. Let us consider a front oriented in the y-direction and represent its vertical circulation through a streamfunction ψ,

$$u = \frac{\partial\psi}{\partial z}, \qquad w = -\frac{\partial\psi}{\partial x}, \tag{5.31}$$

so that

$$\nabla^2\psi = \xi, \tag{5.32}$$

where $\xi = \partial u/\partial z - \partial w/\partial x$ is the component of vorticity parallel to the front. Removing the explicit restriction of two-dimensionality imposed in the derivation of (5.21), a more complete equation governing time-dependent, inviscid changes in ξ is

$$\frac{D\xi}{Dt} = f\frac{\partial v}{\partial z} - \left(\frac{\partial v}{\partial z}\frac{\partial u}{\partial y} - \frac{\partial v}{\partial x}\frac{\partial w}{\partial y}\right) - \xi\left(\frac{\partial u}{\partial x} + \frac{\partial w}{\partial z}\right) - \frac{\partial B}{\partial x}, \tag{5.33}$$

where the first two right-hand terms represent the contribution from the tilting of planetary vorticity and x-component relative vorticity, the third is the contribution of horizontal stretching of y-component relative vorticity, and the fourth is the baroclinic generation or solenoidal term. The stretching term vanishes as a consequence of (5.31), and the tilting terms vanish if the flow is restricted to the x–z plane. This leaves

$$\frac{D}{Dt}\nabla^2\psi \simeq -\frac{g}{\overline{\theta}}\frac{\partial\theta}{\partial x}, \tag{5.34}$$

where we have approximated buoyancy as $B \simeq g(\theta - \overline{\theta})/\overline{\theta}$, and where $\overline{\theta} = \overline{\theta}(z)$ (see Chapter 2).

Equations (5.34) and (5.30) together show that a cross-frontal vertical circulation, with a rising branch on the warm side of the temperature gradient and a sinking branch on its cool side, will be strengthened by frontogenesis. Readers well versed in synoptic-scale dynamics will know that this physical connection is traditionally

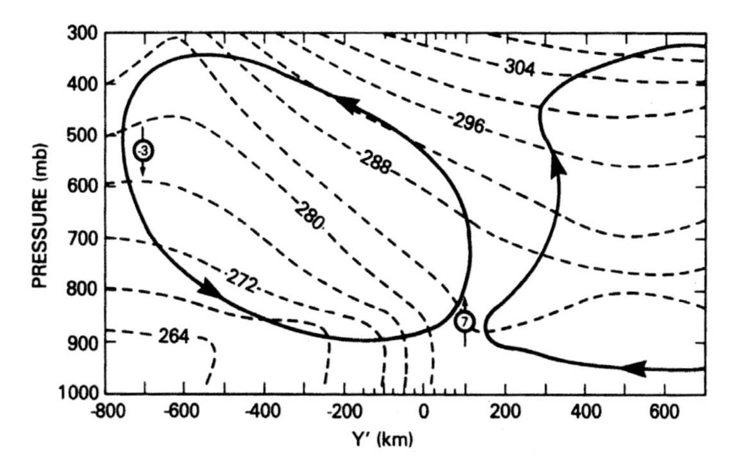

Figure 5.13 Thermally direct transverse circulation (bold arrowed lines) in a numerically simulated front. Contours are of potential temperature, and small circles indicate example magnitudes of vertical motion (cm s^{-1}) within the ascent and descent. From Koch (1984).

deduced through the Sawyer-Eliassen equation,[35] which is derived using semi-geostrophic principles and then solved to determine an analogous, cross-frontal streamfunction.

To more realistically assess the potential for cross-frontal circulations to initiate deep convection, we need to reintroduce effects implicitly or explicitly neglected in the preceding analysis, such as surface friction and turbulent mixing. Numerical models that include these effects simulate vertical wind speeds of $\geq$10 cm s^{-1} within the lowest ~2 km AGL in circulations (e.g., Figure 5.13). Layer lifting in the rising branch of the circulation will help destabilize the atmosphere, as described in Section 5.2. However, because parcels can be lifted through a ~2-km depth within ~2 to 3 hours, front-associated triggering of deep cumulus convection is also plausible.[36]

The specific lifting mechanism is the vertical circulation, given without explicit regard for the effect of the forward movement of the front. The cross-frontal winds at low levels in the cold air help maintain the temperature gradient and hence contribute to the frontogenetic circulation. It is possible for the frontal speed to exceed the low-level winds, suggesting frontal propagation and an unbalanced wind field, and moreover the idea that cold fronts – particularly shallow ones – sometimes behave as density currents (Chapter 2). The implication, as discussed further in Section 5.5.1, is that the nonhydrostratic high pressure that develops at the leading edge of such currents results in vertical pressure gradients capable of forcing strong vertical accelerations at low levels.[37]

In analogy with the coupled near-field mechanism described in Section 5.3, CI is known to occur when and where the rising branch of a frontal circulation interacts with other mesoscale circulations such as that associated with the *dryline*.[38] To appreciate this interaction more fully, an explanation of the dryline is now in order.

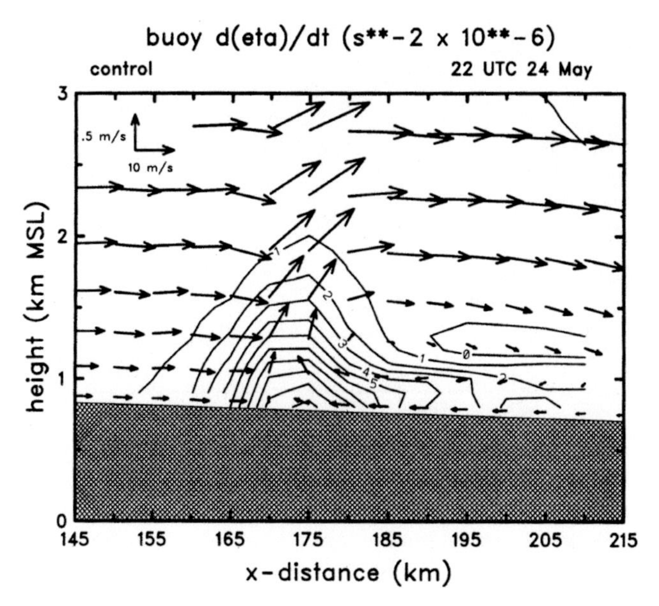

Figure 5.14 Evaluation of the solenoidal contribution to the along-dryline component of horizontal vorticity. Vectors show the transverse circulation of this simulated dryline. From Ziegler et al. (1995).

5.4.2 The Dryline

The dryline is an elongated (~500–1000 km), narrow (~1–20 km), and relatively shallow (~1–2 km) zone of concentrated moisture and temperature variation.[39] Its occurrence in the Great Plains region of the United States on ~30 percent of days during the months of April through June is due to (1) the warm, moist air that flows northward from the Gulf of Mexico, as on the west side of the semipermanent Bermuda High, and (2) the hot, dry air originating on the Mexican Plateau (see Section 5.1) and flowing eastward in the predominant midlatitude westerlies.[40,41] The dryline is generally oriented in the north–south direction, with a mean position near 101° W longitude.[42]

The dryline characteristic of primary relevance to the present discussion of CI is also a vertical transverse circulation. The Great Plains dryline circulation is composed of moist, easterly low-level flow, a substantial rising branch within the eastward-tilted zone of horizontal convergence, a westerly return flow aloft, and a compensating sinking branch east of the moisture gradient (see Figure 5.14). Evaluations of the equation governing along-line vorticity (e.g., (5.33) with $B = g(\theta_v - \overline{\theta}_v)/\overline{\theta}_v)$) reveal the importance of baroclinity in driving this thermally direct circulation (Figure 5.14).[43]

The moisture and temperature gradients that contribute to the baroclinity are intimately linked to the daily cycle of solar heating. Specifically, such gradients, and indeed the dryline formation itself, depend in part on the differential (west–east) sensible heating at the surface. The sensible heating is especially sensitive to the

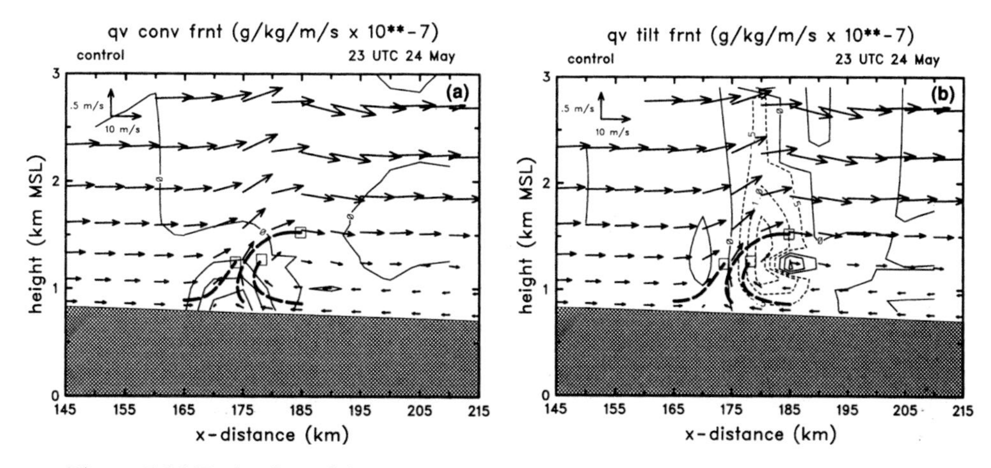

Figure 5.15 Evaluation of the water-vapor mixing ratio frontogenesis in a simulated dryline: (a) convergence term, and (b) tilting term (see text). Bold dashed lines show trajectories of air parcels originating in the surface layer. Vectors show the transverse circulation of this simulated dryline. From Ziegler et al. (1995).

west-to-east increase in soil moisture, as well as to horizontal variations in land-use and vegetation patterns.[44,45] There is an additional dependency on the westward increase in terrain height and thus on the westward decrease of moist-layer depth: as the shallow nocturnal inversion in the west is eroded by daytime heating, the correspondingly shallow moisture is mixed deeply to yield a relatively dry ABL. Concurrent mixing in the deeper moist layer in the east yields a relatively moist ABL; thus, the combined effect from the terrain is a west-to-east horizontal moisture increase.

The boundary-layer mixing also results in downward momentum transport, and the associated enhancement of low-level westerly winds and their horizontal gradient (i.e., $\partial u/\partial x$) aids in dryline formation. We see this effect in

$$\Im_{x,q_v} \simeq \frac{\partial}{\partial x}(\dot{q}_v) - \frac{\partial u}{\partial x}\frac{\partial q_v}{\partial x} - \frac{\partial w}{\partial x}\frac{\partial q_v}{\partial z}, \tag{5.35}$$

which is a version of (5.30) expressed in terms of water-vapor mixing ratio. The right-hand terms in (5.35) represent the respective contributions from moisture changes associated with diabatic processes and mixing, horizontal convergence of the horizontal moisture gradient, and vertical tilting of the vertical moisture gradient. Analysis of a numerically simulated dryline (Figure 5.15) shows that the convergence term contributes to $\Im_{x,qv} > 0$ and, hence, to frontogenesis at low levels.[46] The tilting term, conversely, contributes to $\Im_{x,qv} < 0$ and frontolysis above the ABL. Accounting for both terms, we find that the dryline intensity decreases with height, which places a limit on the depth of the circulation, and implicitly on dryline-initiated convection.

As alluded to, the basic wind and moisture distributions favoring dryline formation are linked to the regional physio-geography. However, synoptic-scale processes can supplement the moisture and wind gradients, setting the stage for a more intense

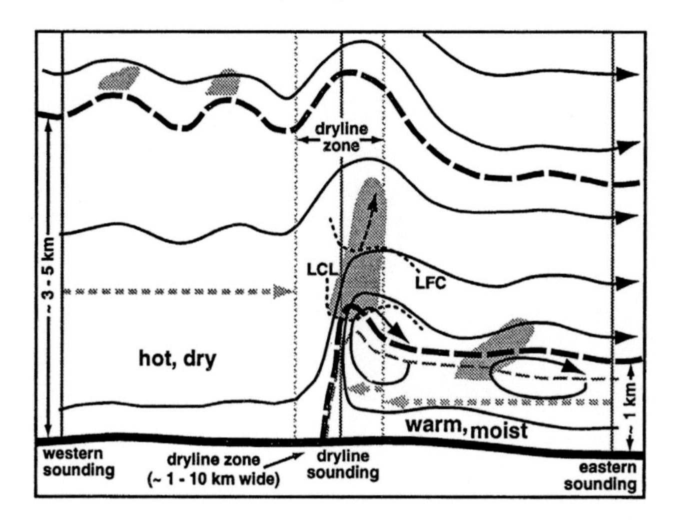

Figure 5.16 Conceptual model of convective initiation in a dryline. From Ziegler and Rasmussen (1998).

dryline.[47] For example, strong drylines in west Texas are promoted, on average, by an upper-level short-wave trough upstream of west Texas, with an associated ETC centered in the Texas Panhandle and eastern New Mexico. In contrast, weak drylines occur when an upper-level ridge axis is found upstream of west Texas, and broad low pressure at the surface exists over the southwestern United States and in the lee of the Rocky Mountains. One may be tempted to conclude that convection initiates readily in the presence of a strong dryline and associated synoptic-scale forcing, especially given vertical motions of several m s^{-1} in dryline circulations.[48] We know now from our discussions in the previous sections that the strength (and depth) of this external forcing must be examined relative to the local thermodynamic environment.

It appears necessary, in fact, that the depth of the mesoscale updraft be such that parcels following updraft-relative flow are lifted to their LCL and LFC before exiting the mesoscale updraft (Figure 5.16).[49] This requirement has a qualifier, in that deep layer lifting in the rising branch of the dryline circulation also helps destabilize the atmosphere, locally lowering the LCL and LFC, and reducing CIN.

Once initiated, deep convective clouds are most often observed between ~10 km west and ~40 km east of the leading edge of the moisture gradient.[50] Despite the implied two-dimensionality of the dryline circulation, deep convection seldom develops uniformly along the longitudinal extent of the dryline, but instead occurs with considerable along-line variability. It is apparent, therefore, that local enhancements in the mesoscale ascent (and/or in the thermodynamic environment) increase the local probability for CI. Candidate explanations for such include: (1) local horizontal bulges in the dryline, which cause local enhancements in horizontal convergence; (2) dryline–internal gravity wave interactions; (3) mesoscale areas of low pressure along the dryline, apparently due to localized differences in diabatic heating; (4) interactions between synoptic-scale fronts and/or other boundaries and the dryline, especially at so-called *triple points*; and (5) intersections/interactions between the dryline and HCRs.[51] Common to each of these is a constructive or positive phase relation between the supplementary mechanism's ascent and the mesoscale updraft of the dryline. Let us use (4) and (5) to illustrate this effect, bearing in mind that the basic concept applies to all five of the mechanisms.

Recall first that synoptic environments that favor stronger drylines may also possess ETCs that follow a climatological track through the southern Great Plains.[52] Depending on the precise movement of the cyclone, the orientation and extent of associated fronts, and the movement of the dryline, it is possible for the respective thermal and moisture boundaries to interact. The nature of the interaction depends on relative locations and orientations of the boundaries. Consider one specific example of a Pacific cold front that was roughly parallel to a dryline as it approached from the west.[53] The cold front had a transverse, thermally direct circulation that forced weak convective clouds as it approached the dryline (Figure 5.17a). Prefrontal subsidence aloft counteracted the dryline updraft and thus initially suppressed convection. Subsequent frontal movement relative to the dryline allowed for a merger of the rising branches of the two circulations, and CI ensued owing to the deep lifting of pre-dryline air.

Interactions between the dryline and a *line-perpendicular* thermal boundary (cold front, cold-air outflow boundary, etc.) are relatively common (Figure 5.17b). The point of interaction in these cases is referred to as a triple point, indicating a division of the warm and moist air east of the dryline, hot and dry air west of the dryline, and cool (and dry) air north of the thermal boundary. CI tends to occur near the triple point, where low-level horizontal convergence is maximized. It is possible for the dryline circulation to be displaced upward over a southward-progressing thermal boundary, thus locally deepening the dryline updraft and promoting CI.[54]

Elevated lifting is also known to occur in association with circulations such as HCRs. This is illustrated well by the case in Figure 5.17c, which depicts a field of HCRs with axes oriented at a nonzero angle to the dryline.[55] On this day, the dryline had a westward (or retrograde) motion during the midafternoon, and subsequently intersected the HCRs. Many of the localized intersections were accompanied by cloud formation east of the dryline, as shown in visible satellite imagery, and also in radar data via enhanced radar reflectivity values. The cloud formation itself was due to the local lifting of the HCRs by the dryline: the additive effect of the roll updrafts and the dryline updraft resulted in much deeper ascent than possible with the HCR or dryline alone. Hence, the along-line variability was equivalent to the spacing of the HCRs; subsequent variability resulted where the precipitating convection introduced cloud pools that interacted further with the dryline. Although the HCR–dryline interaction was enabled by a westward-moving dryline, such interaction and convection

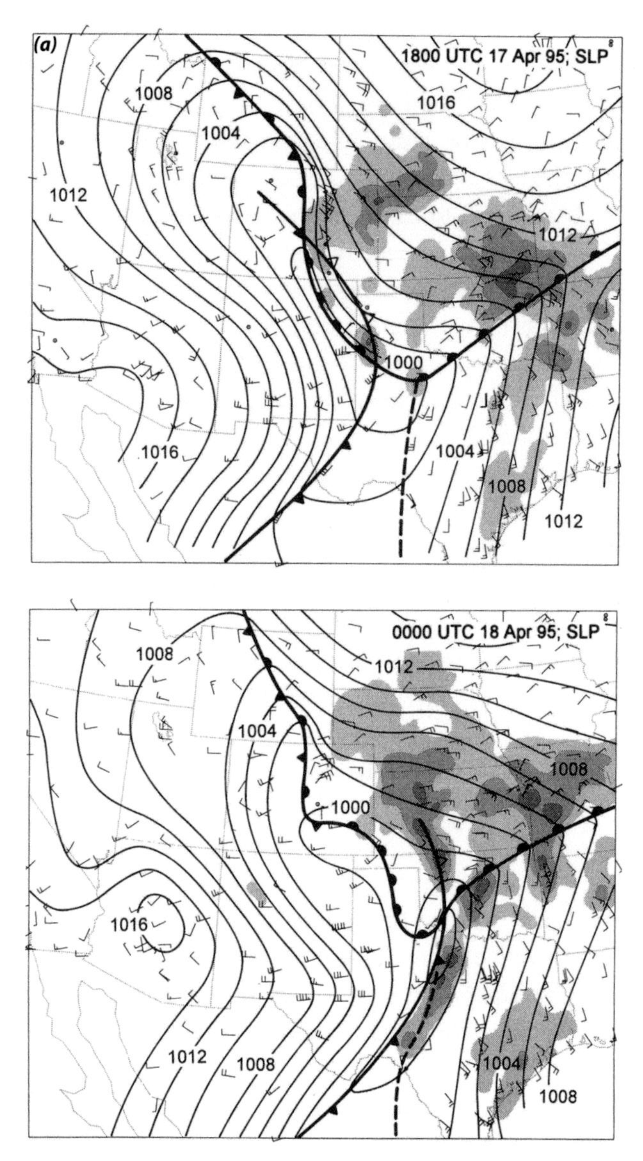

Figure 5.17 Three examples of dryline interactions that lead to convection initiation. (a) Parallel interaction between a cold front and a dryline, in which the rising branches of both circulations merged (Neiman and Wakimoto 1999). (b) Perpendicular interaction between a thermal boundary and dryline at a triple point (Weiss and Bluestein 2002). (c) HCR–dryline interaction (Atkins et al. 1998). In (a) and (b), contours are of sea level pressure and dewpoint temperature, respectively. In (a), shading is of radar reflectivity factor, whereas in (c), shading is of radar reflectivity factor (left) and spatial autocorrelation (right). Finally, in (b) and (c), the dryline is indicated by the scalloped line, and in (a), by a dashed line.

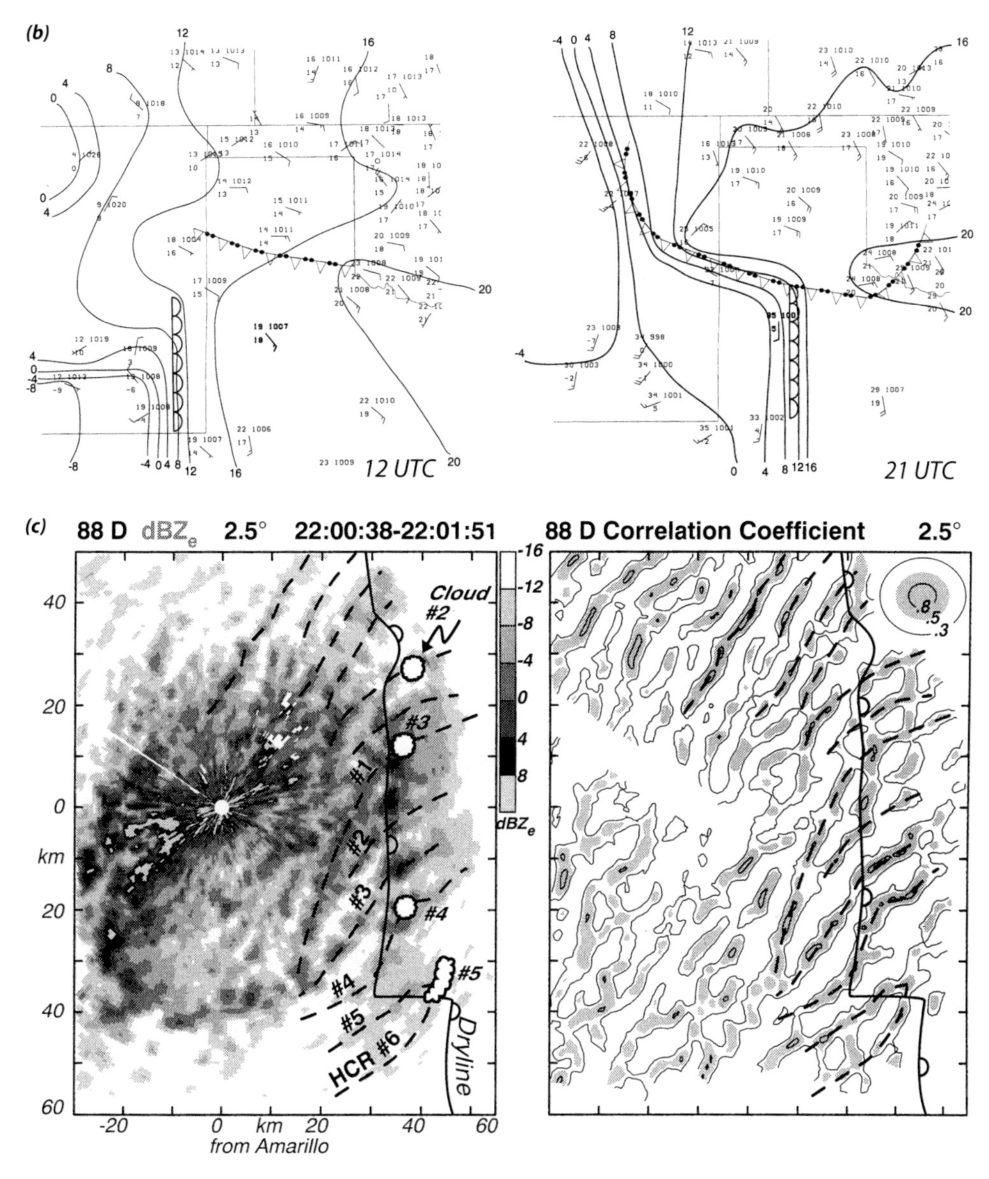

Figure 5.17 (*continued*).

initiation has been documented in eastward-moving drylines and HCRs east of the dryline.[56]

5.4.3 The Sea-Breeze Front

The idea of enhanced lifting from HCR–dryline interactions ostensibly has its origins in studies of Florida thunderstorms, in which convection initiation was attributed to an analogous interaction between HCRs and a sea-breeze front (SBF). Like the dryline,

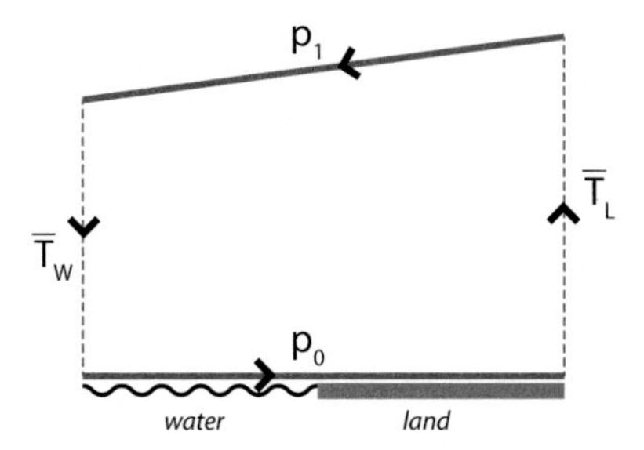

Figure 5.18 Schematic of sea-breeze circulation. Bold lines represent isobaric surfaces p_0 and p_1 (where $p_1 > p_0$). The vertically averaged temperature over land is $\overline{T}_L$, and the vertically averaged temperature over water is $\overline{T}_W$ (where $\overline{T}_W < \overline{T}_L$). Adapted from Holton (2004).

the sea breeze is associated with regional physio-geography, specifically, a large body of water (ocean, sea, lake) and the adjacent land.[57] Under initially cloud-free skies, the land surface heats more rapidly than does the water because of the higher specific heat of water. The hydrostatic pressure at the surface becomes lower over land, and in response to the resultant horizontal pressure gradient, an onshore surface wind develops. This is the sea breeze.

The sea breeze is the lower branch of a transverse circulation.

Although we can once again apply the horizontal vorticity equation to show the baroclinic origins of this circulation, let us alternatively make this argument using the circulation theorem

$$\frac{D\Gamma}{Dt} = -\oint R_d T \, d\ln p, \tag{5.36}$$

where Γ is circulation, defined as

$$\Gamma = \oint \vec{V} \cdot d\vec{l}. \tag{5.37}$$

The line integral is taken over a closed contour within a vertical plane perpendicular to the coastline, and $\vec{l}$ is a vector locally tangent to the closed contour.[58] If we let the top and bottom segments of the circuit be isobaric surfaces p_1 and p_0, respectively, (5.36) implies that only the vertical segments of the circuit contribute to circulation changes (see Figure 5.18). If we then approximate the temperature over these land and water segments with vertical averages $\overline{T}_L$ and $\overline{T}_W$, respectively, (5.36) can be integrated to yield

$$\frac{D\Gamma}{Dt} = R_d \ln\left(\frac{p_0}{p_1}\right)\left(\overline{T}_L - \overline{T}_W\right). \tag{5.38}$$

This neglects a number of factors that affect actual sea-breeze circulations, such as surface friction and details of the coastline, but it does provide the essential result that generation of circulation necessarily depends on a horizontal temperature contrast.

At the leading edge of the circulation is the SBF, which also marks the leading edge of the thermal contrast caused by the cooler marine air and the warmer inland air. The SBF behaves as a density current in its movement (see Chapter 2). The following empirical formula for this movement has been adapted from laboratory and theoretical studies:

$$V_{SBF} = b\sqrt{gd\frac{\Delta\rho}{\rho}} - 0.59V_s, \tag{5.39}$$

where b is an empirical constant ($= 0.62$), d is the depth of the sea-breeze circulation, and V_s is the cross-shore component of the lower-tropospheric synoptic-scale wind.[59] The convention implicit in (5.39) is that an onshore (offshore) wind is given by $V_s < 0$ (> 0), so that an onshore V_s would act to enhance the SBF movement. Equation (5.39) implicitly provides some initial information about convection initiation: an onshore synoptic-scale wind leads to a faster-moving (and inland-penetrating) SBF, but one that is often weaker and more diffuse.[60] An offshore synoptic-scale wind slows the SBF movement but results in enhanced horizontal convergence with the sea breeze, in frontogenesis, and ultimately in a stronger sea-breeze circulation.

Cloud bands have been observed to form parallel to the SBF, but as with drylines, deep cumulus convection is variable in location along the length of the SBF. As alluded to in Section 5.4.2, one explanation for this variability are HCRs normal to and in the environment of the advancing SBF; offshore synoptic-scale winds, and hence a deeper SBF, are assumed.[61] Roll circulations that are displaced upward by the SBF lead to deeper lifting – and thus to a higher potential for CI – where roll updrafts are in phase with the ascending branch of the sea-breeze circulation (Figure 5.19).[62] Convection initiation is also possible when HCRs are parallel to the SBF, when the rising branches of the respective circulations offer mutual enhancement.[63] Along-front variability is not strictly explained in this case, although variations in the local topography, coastline shape, and land surface would be contributing factors. In a similar vein, deep cumulus convection is observed to form when the SBF "collides" with gust fronts from convective storms originating elsewhere. This obviously is not an explanation for *initial* convection initiation, but does provide an appropriate segue to our discussion of "far-field" mechanisms, which prominently includes outflow boundaries.

5.5 Lifting Mechanisms from Nonlocal Sources

We now focus on lifting mechanisms that are themselves convectively generated, move from their respective source regions, and then eventually initiate convection at some

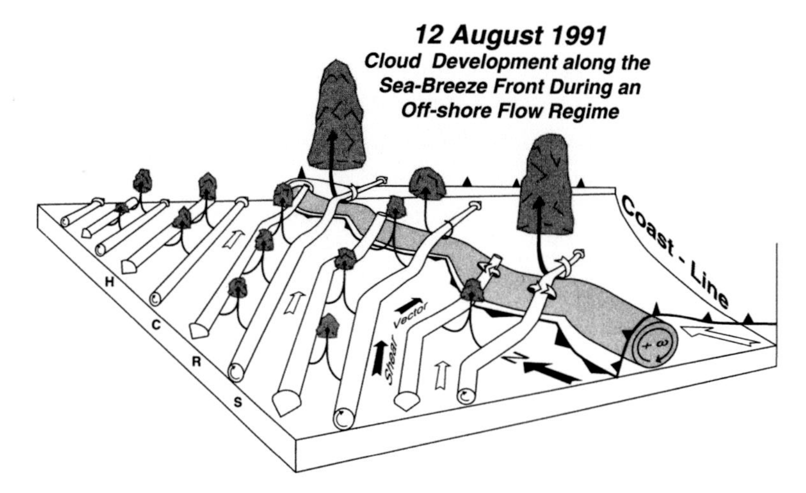

Figure 5.19 Schematic of HCR–sea-breeze interaction leading to convection initiation. From Atkins et al. (1995).

distance (and time) from this source. Thunderstorm outflow and gravity waves are well-known examples of such nonlocal processes. Each will be discussed separately, although multiple (local and nonlocal) mechanisms may be acting in concert to initiate convection. Indeed, this has been one of the recurring themes of this chapter.

5.5.1 Convective-Storm Gust Fronts

The archetypal precipitating convective storm generates an evaporatively chilled pool of air that spreads out laterally at the ground. At the leading edge of the cold pool and associated outflow is the gust front. In this section we explore a few of the characteristics of gust fronts and cold pools that relate to CI; a more general treatment of convective outflow dynamics is given in Chapter 6.

Gust fronts are usually included in the more general class of *boundary-layer convergence lines*. Colloquially referred to as *boundaries*, these are visualized in visible satellite images as cloud-arc lines, and in radar data by radar-reflectivity fine lines (and by convergence in low-altitude Doppler velocity) (see Chapter 2). The availability of satellite and radar data at high temporal resolution has facilitated quantification of the relative frequency of CI in association with boundaries. For example, in a study of eastern-Colorado convective storms, 79 percent of 418 convective-storm events were attributed to boundaries.[64] Of the 91 classifiable boundaries, 59 percent were in fact identified as gust fronts. Of these, 61 percent initiated storms: "storms" were objectively defined by the existence of a new radar echo with radar reflectivity factor ≥ 30 dBZ at $\sim$1 km AGL. Using a slightly different methodology and a different geographical domain, a separate study found that roughly 25 percent of (surface-based) initiation episodes were linked to a gust front, although gust fronts associated with

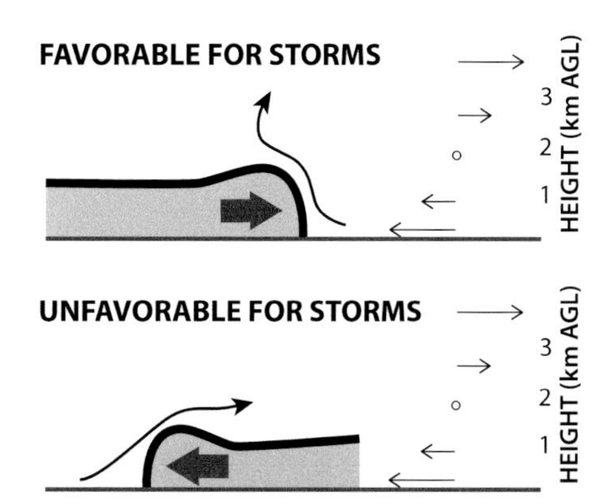

Figure 5.20 Schematic of favorable versus unfavorable relation between gust front and environmental wind shear. After Wilson et al. (1998).

ongoing, active convection were excluded in these statistics.[65] Nevertheless we can safely conclude from these and other studies that gust fronts represent a nonnegligible source of mesoscale lift and CI.[66]

Of the phenomena discussed so far in this chapter, gust fronts have the closest analogue in density currents, and thus it is appropriate to recall the theoretical density-current speed

$$V_{dc} = k\sqrt{gd\frac{\Delta\rho}{\rho}}, \tag{5.40}$$

where we are mindful of the many assumptions made in Chapter 2 in its derivation. As demonstrated by (5.40) and confirmed in laboratory and numerical modeling experiments, a deeper cold pool (d) is associated with a faster-moving gust front. It also results in more intense gust-frontal updrafts, with simulated and observed updraft values ranging from $\sim$1 to 10 m s^{-1}.[67]

The potential for CI depends not just on depth and speed, but also on how these relate to the environmental wind profile. Deep lifting is most favored when the environmental wind shear vector within the lowest few kilometers is oriented in the direction of gust-frontal movement, and hence generally perpendicular to the gust-front orientation. A particularly favorable wind profile has "steering level" ($\sim$3 km AGL) winds that are roughly equal to the gust-front speed, such that initial cells are vertically erect and also moving with the gust front.[68] In contrast, when the low-level wind shear vector is oriented opposite to the gust-front movement, parcels displaced upward over the gust front will have a shallow slope, rising slowly and progressively rearward of the gust front (Figure 5.20). This concept applies to sustenance of ongoing convective storms as well, and is cast in Chapter 8 in the relation between gust-front speed and environmental wind shear known as *RKW theory*.

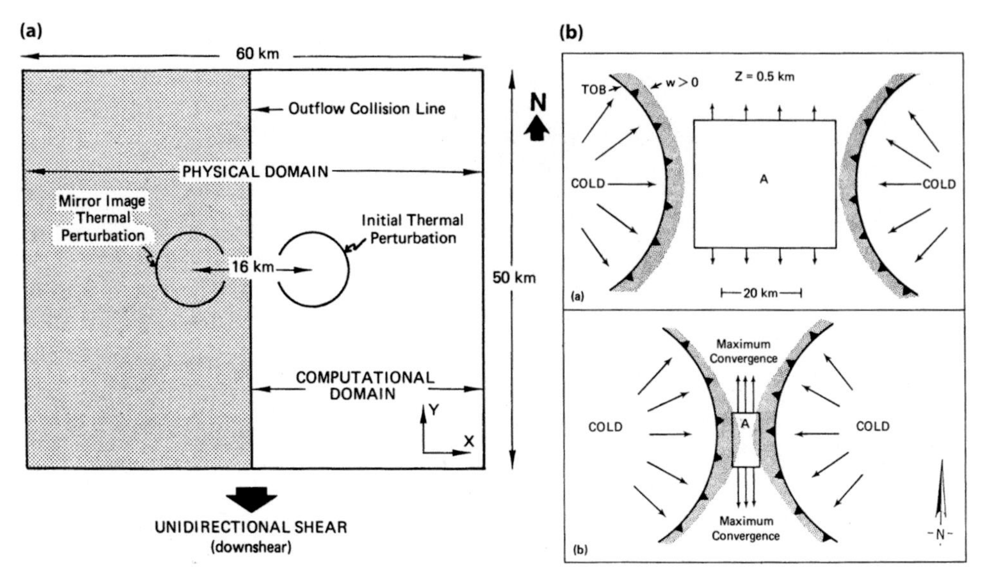

Figure 5.21 Numerical modeling of colliding gust fronts. (a) Modeling procedure. (b) Conceptual model of cold-pool interaction. From Droegemeier and Wilhelmson (1985a).

A compelling finding of the eastern-Colorado observational study is that 64 of the 327 boundary-initiated storms resulted from collisions between two or more boundaries.[69] The initiation itself was (presumably) at or near the point of collision, and the accompanying storms tended to be relatively more intense. "Head-on" collisions, or those involving boundaries nearly parallel on impact, were shown to be most productive in CI. Such collisions are readily modeled in an idealized framework (Figure 5.21a). Modeled convective clouds initiate first in the vertically diverging air between the colliding fronts, but subsequent clouds then form as the outflow spreads laterally along the collision plane (Figure 5.21b). The fate and intensity of the newly initiated clouds depend on low-level environmental moisture as well as the environmental wind profile.[70]

Much of the focus thus far on gust-front-aided CI has implicitly been on gust fronts with nonnegligible speeds and associated with an ongoing convective storm. Consider now the contrasting situation of CI aided by a remnant, quasi-stationary boundary, which is especially frequent during the summer months when synoptic-scale forcing is relatively weak.[71] It often begins with a nocturnal convective storm that decays by early morning but leaves behind its cold pool and laterally spreading outflow.[72] In absence of a convective source, the cold pool and thus density gradient gradually weaken in time, implying a slowing gust-frontal motion (e.g., (5.40)) and a weakening baroclinic circulation (e.g., (5.33)). Quasi-stationarity does not preclude CI, however, as proven in the sequence of satellite images in Figure 5.22. Besides parcel lifting, quasi-stationary boundaries are known to enhance the low-level horizontal convergence of moisture, which helps to pool moisture locally and hence aids in environmental

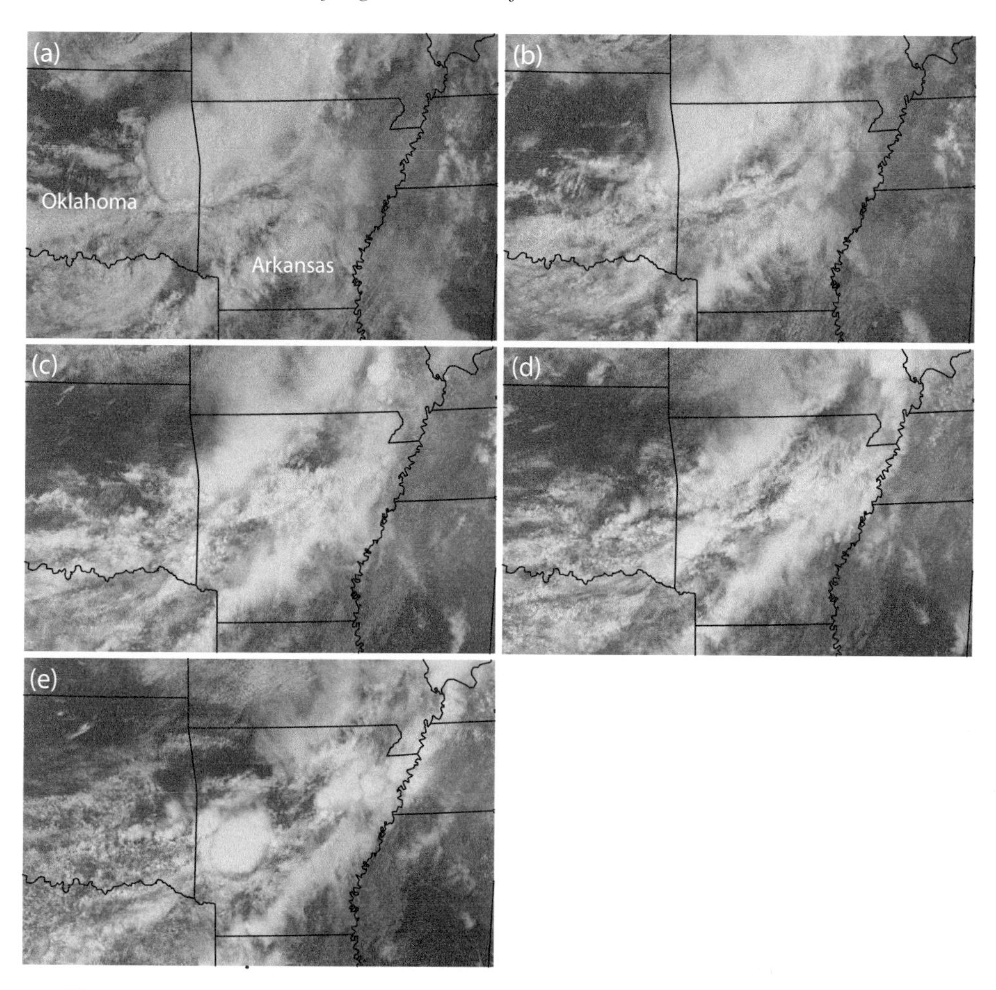

Figure 5.22 Sequence of satellite images on April 30, 2012, showing convection initiation involving a slowly moving gust front. (a)–(b) The gust front extending from north central Arkansas through eastern Oklahoma was associated with an convective storm from 1445 to 1545 UTC. (c) The storm had dissipated by 1645 UTC, with remnant outflow primarily in western Arkansas. (d)–(e) New convection initiated in southeastern Arkansas by 1745 UTC, just south of the outflow boundary (as also determined from surface winds; not shown).

destabilization.[73] As will be demonstrated in Chapter 9, such boundary processes can even play a role in the later evolution of the storm that it helped initiate.

CI in association with boundaries also exhibits spatial variability. As with drylines, there are many potential causes of this variability, but the specific effect of *misocylones* (see Chapter 1 and 8) is particularly compelling and worthy of mention here. Owing to a release of a horizontal shearing instability, which is a type of Kelvin-Helmholtz instability (see Chapter 2), a series of these vortices is known to form with roughly equal spacing along the boundary. Their role in CI is an indirect one, as exemplified in an idealized case of a north–south-oriented gust front. Individual misocyclonic circulations in the presence of a background shear flow result in zones of horizontal convergence between vortex centers.[74] Convergence zones north of the vortex center are also zones of enhanced moisture, owing to advection and convergence of environmental air: These zones are especially favored for CI. Because misocyclones tend to be surface-based, with an intensity that decreases with height, their centers coincide with downdrafts, owing to downward-directed pressure gradient forcing (see also Chapter 7): These locations are not favored for CI. A pattern of updraft formation north of the misocyclone centers therefore arises. A subsequent evolution from this pattern to one exhibiting non-supercell tornadoes is essentially the non-supercell tornadogenesis mechanism to be discussed in Chapter 8.

5.5.2 Gravity Waves

As mathematically developed in Chapter 2, gravity waves form as a result of the restoring force of gravity on vertical air displacements in a stably stratified fluid. A possible role of internal gravity waves in convection initiation has been discussed already in this chapter (Section 5.3.2). The context was a local or near-field lifting mechanism, wherein gravity waves were launched above the ABL by HCRs, and the waves and HCRs then became coupled to yield deep lifting of boundary-layer parcels. In this Section we consider internal gravity waves that are generated by an active convective cloud or similar means, propagate horizontally from this source, and subsequently initiate new cumulus convection. Nonlocal CI by gravity waves is constrained by the time (and hence distance) it takes for them to lose much of their amplitude and thus become attenuated. Gravity-wave aided CI additionally depends on where the wave amplitude is concentrated vertically, and thus on the vertical wavenumber.

Let us begin with some simplifications that will help illustrate the relevant concepts more clearly.[75] Assume for the moment that the undisturbed environment is in hydrostatic balance and at rest ($\overline{u} = U_0 = 0$), the wave motions are restricted to the x–z plane, and frequency σ is positive. The hydrostatic balance assumption implies that $m^2 \gg k^2$, and thus that the frequency-dispersion relation (2.97) can be written as

$$\sigma = Nk \Big/ \left[m^2 \left(1 + \frac{k^2}{m^2} \right) \right]^{1/2} \simeq Nk/m. \tag{5.41}$$

The phase speed is then

$$\begin{aligned} c_x &= \frac{\sigma}{k} = \frac{N}{m} \\ c_z &= \frac{\sigma}{m} = c_x \frac{k}{m} \end{aligned} \tag{5.42}$$

and correspondingly, the components of the group velocity are

$$\begin{aligned} c_{g,x} &= \frac{\partial \sigma}{\partial k} = \frac{N}{m} = c_x \\ c_{g,z} &= \frac{\partial \sigma}{\partial m} = -\frac{Nk}{m^2} = -c_x \frac{k}{m} \end{aligned}. \tag{5.43}$$

Equations (5.42) and (5.43) remind us that gravity waves are dispersive, and also reveal that waves with the longest vertical wavelengths (smallest wave numbers) have the fastest horizontal phase and group propagation. In the vertical, downward phase propagation implies upward group propagation. However, for "hydrostatic" gravity waves, wherein $c_x \gg c_z$, wave propagation is predominantly horizontal (in the troposphere; horizontal and vertical propagation both occur in the stratosphere).

A common practice, especially in idealized numerical modeling studies, is to assume that the tropopause acts as a rigid lid,[76] thus confining the waves to the troposphere and preventing vertical propagation of energy into the stratosphere. The vertical wavenumbers can then be expressed in terms of *vertical modes*, or integer fractions of tropopause height $z = Z_T$:

$$m = \frac{n\pi}{Z_T}, \tag{5.44}$$

where the nth vertical mode has wavelength $\lambda_z = 2Z_T/n$, the phase and group propagation become

$$c_x = \frac{NZ_T}{n\pi} = c_{g,x}, \tag{5.45}$$

and implicitly, $c_z \simeq 0$. Vertical modes $n = 1, 2, 3$ tend to contribute the greatest to the power spectrum, and also propagate away from the generating cloud sufficiently fast ($\sim$ several tens of m s^{-1}) and far ($\sim$ several tens of km) such that they can affect the cloud's environment.[77]

Figure 5.23 portrays the gravity wave response to a growing convective cloud in a 2D numerical model.[78] Away from the cloud and its associated diabatic processes, one can use perturbation potential temperature to identify the vertical displacements, with a negative perturbation indicating air that has been displaced upward (adiabatic cooling), and a positive perturbation indicating air that has been displaced downward (adiabatic warming). The dispersive nature of gravity waves is illustrated well in this sequence, with the $n = 1$ mode propagating rapidly to the eastern end of the domain ($x \sim$ 160–170 km) by $t = 40$ min, followed by the $n = 2$ mode by $t = 60$ min, and then by the $n = 3$ mode by $t = 80$ min. Notice that the $n = 1$ mode is represented in this tropospheric domain by a wave trough (downward displacement) with maximum amplitude at a height $\sim Z_T/2$. Thus, the $n = 1$ mode has a thermodynamically stabilizing effect, manifest primarily in a reduction of CAPE. The $n = 2$ mode similarly has a net stabilizing effect, specifically, with adiabatic warming at heights

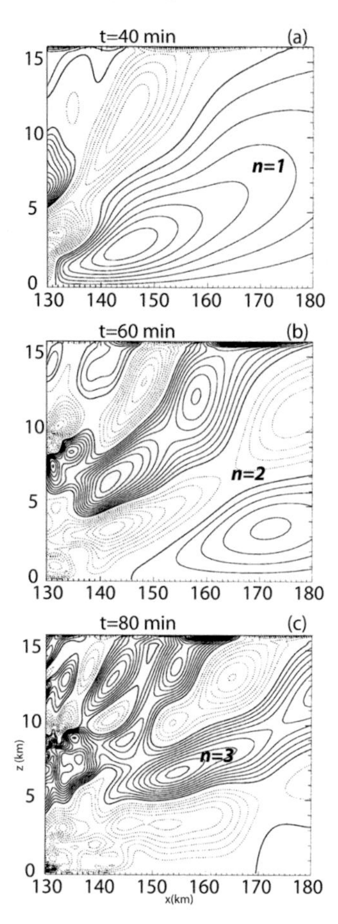

Figure 5.23 Time-sequence of perturbation potential temperature (contour interval of 0.2 K; negative value dashed), from a 2D model simulation of convectively generated gravity waves. (a) t = 40 min; (b) t = 60 min; and (c) t = 80 min. Original image cropped at x < 130 to emphasize the gravity wave response. From Lane and Reeder (2001).

below $Z_T/2$ and a consequential increase in CIN. The effect of the $n = 3$ mode, on the other hand, is that of thermodynamic destabilization owing to upward displacements at heights below $Z_T/3$, inclusive of the ABL. In this particular model realization, the $n = 3$ mode is associated with a mitigation of CIN and hence is most responsible for the initiation of convection.

The relationship between CI and specific gravity-wave modes depends in part on the respective depths of the troposphere (Z_T in the preceding example) and ABL, which govern where, in the vertical, the wave modes have their respective upward (and downward) displacements.[79] As alluded to earlier, the source-relative distances over which this mechanism can even apply depend on how far the modes can propagate before undergoing significant attenuation. One limit is imposed by the synoptic-scale dynamics via the Rossby radius of deformation, $\lambda_R = c_x/f$, regarded here as the

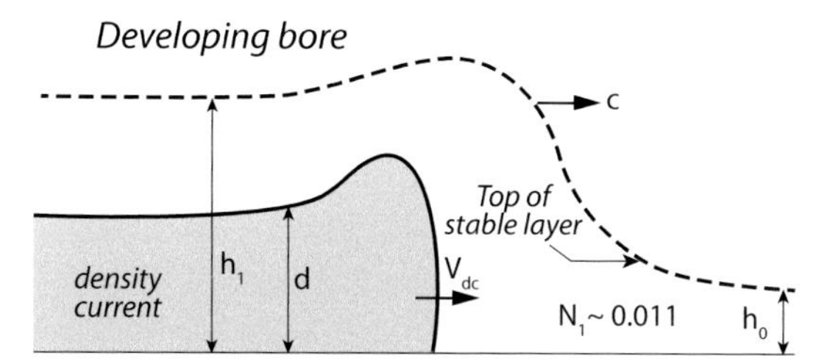

Figure 5.24 Schematic of the developing atmospheric bore shown in Figure 5.25. Indicated are: convective outflow depth d, NBL depth h_0, NBL static stability N_1, and bore depth h_1. After Knupp (2006).

distance over which waves are reduced to $1/e$ of their amplitudes.[80] For typical conditions in midlatitudes, wherein $N \sim 0.015$ s^{-1}, we find that $\lambda_R \sim 500$ km for the $n = 1$ mode. Now, recalling that c_x decreases in proportion to the order of the vertical mode (Equation (5.45)), we furthermore find that progressively slower-moving, higher-order modes are increasingly limited in the extent of their horizontal propagation, and thus in their ability to affect the cloud's far-field environment.

Horizontal gravity wave propagation is also affected by the presence of vertical wind through wave trapping. This is predicted using the *Scorer parameter*,

$$\ell^2 = \frac{N^2}{(U-c)^2} - \frac{d^2U/dz^2}{(U-c)}, \tag{5.46}$$

given here in terms of x-direction propagation (see Chapter 2). In vertical layers where $\ell^2 < k^2$, the corresponding wavenumbers decay exponentially with height; this criterion for wave trapping is met when the phase propagation is against the mean wind (U), and further, when such a mean wind is vertically distributed in the form of a jet. Putting this in the context of the current chapter, CI is limited when modes with upward displacements through the ABL are trapped within this layer. The preferential initiation (suppression) of convection in association with the downwind- (upwind-) propagating $n = 3$ mode has been offered as an example of the asymmetric effects of wave trapping.[81]

5.5.3 Atmospheric Bores

An atmospheric bore is a specific type of trapped gravity wave. It prevails in the highly stable nocturnal boundary layer (NBL), and often is generated when convective storm outflow (or similar process) moves into this nocturnal environment and displaces the air upward (Figure 5.24). The resultant (large-amplitude) internal gravity waves propagate on this layer of high static stability, with a horizontal speed that equals or even exceeds the speed of the density current.

Relevant variables in bore generation include convective outflow depth d, NBL depth h_0, NBL static stability N_1, and the static stability of the layer overlying the

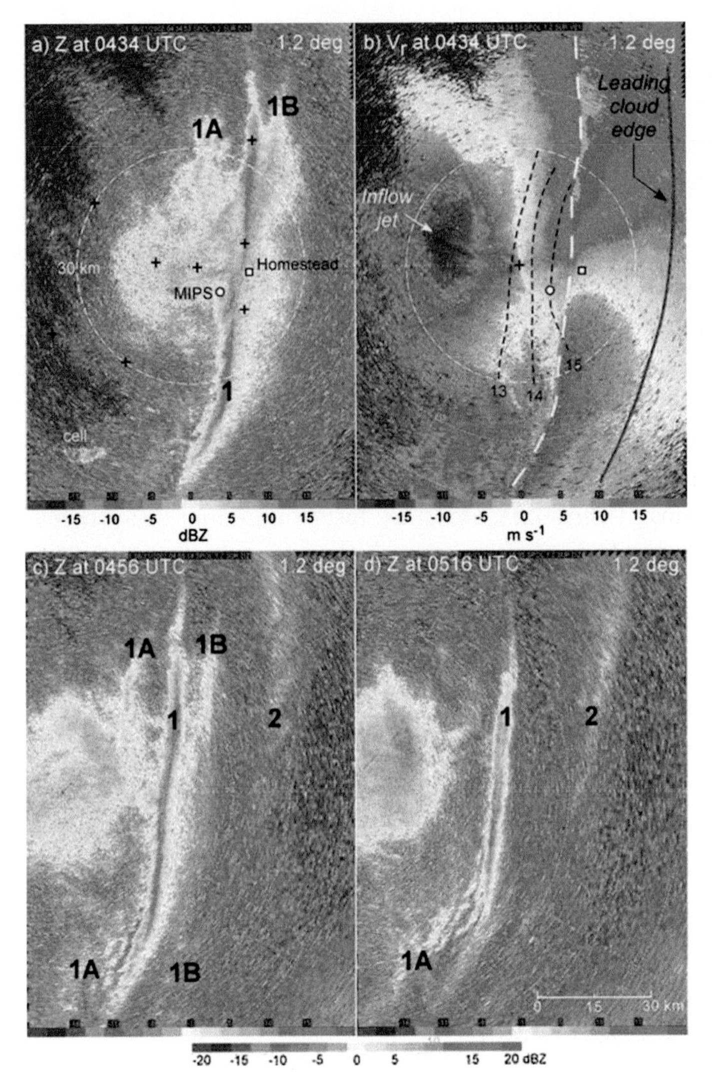

Figure 5.25 Example atmospheric bore evolution as observed by Doppler weather radar. Radar reflectivity factor (Z) is shown in (a), (c), and (d), and radial velocity in (b). From Knupp (2006). (For a color version of this figure, please see the color plate section.)

NBL N_2 (see Figure 5.25).[82] A typical NBL value associated with bores is $N_l \sim 0.015\ \mathrm{s}^{-1}$, with an overlying layer that is much smaller, such that $N_l/N_2 \sim 3$ to 4. The NBL depth is shallow compared with the outflow depth, with bore onset occurring in simulations and observations generally in the range $1.5 \leq d/h_0 \leq 2.5$.

Another relevant bore-generation variable, which is encapsulated in the Scorer parameter, is the environmental wind. A profile $U(z)$ with a jet that opposes the wave motion contributes negatively to the Scorer parameter above the jet level (Equation (5.46)). The wavenumbers that satisfy $\ell^2 < k^2$ in the layer above the jet will have their wave energy trapped below it. In the specific case of bores, the jet axis would be within or at least near the top of the NBL, as is observed in advance of some, though certainly not all, bores evaluated in the United States, Australia, and England.[83]

Once a bore has formed, its strength can be quantified by the (mean) bore depth h_1 relative to the NBL depth h_0; documented bores[84] have strengths that are clustered in the range $2 \lesssim h_1/h_0 \lesssim 3.5$. Notice that h_1 gives the height to which the NBL is vertically displaced (Figure 5.25), implying a plausible relation between bore strength and the potential for the initiation of secondary, nocturnal convection. The resultant convective storms initially (or even permanently) have inflow air that originates primarily above the NBL. Thus, bores are one candidate mechanism for the initiation of *elevated convection*.[85]

The frequency of atmospheric bores has not been well established in the scientific literature, although 24 bores were identified on 15 days during IHOP, suggesting their relative ubiquity.[86] Only three of these cases were linked to CI, suggesting that the frequency of bore-driven CI is relatively low. Of course, as with the other mechanisms discussed in this chapter, bores have been observed to work in tandem with various types of mesoscale boundaries to promote deep lifting of parcels to their LFCs.

Supplementary Information

For exercises, problem sets, and suggested case studies, please see www.cambridge.org/trapp/chapter5.

Notes

1 The treatment of Hess (1959) is followed here.
2 See also Schultz et al. (2000).
3 Emanuel (1994).
4 See also Rogers and Yau (1989).
5 Bryan and Fritsch (2000).
6 Sherwood (2000).
7 Ibid.
8 Emanuel (1994), Gilmore and Wicker (1998).
9 Doswell and Rasmussen (1994).
10 This statement depends on the precise computation of CAPE – specifically, whether the parcel originates from the surface (*SB*), based on the properties of a well-mixed layer (*ML*; typically the lowest 100 hPa of the sounding), or from a layer that yields the most-unstable parcel (*MU*; one with the maximum equivalent potential temperature).
11 Doswell and Markowski (2004).
12 Smith (1971).
13 As argued by Doswell (1987); see also Doswell and Bosart (2001).
14 Griffiths et al. (2000).
15 Roebber et al. (2002).
16 See the reviews by Doswell and Bosart (2001) and Stensrud (1996a).
17 Bluestein (1993).
18 Carlson et al. (1983); see also Stensrud (1993).
19 Carlson and Ludlum (1968).
20 Russell et al. (2008).

21 Banta and Schaaf (1987).
22 Kirshbaum (2011), Geerts et al. (2008), Banta and Schaaf (1987).
23 Kirshbaum (2011).
24 The simple form used here is from Emanuel (1994); other forms are found, for example, in Etling and Brown (1993) and Weckwerth et al. (1997).
25 There is some uncertainty to this range, with large-eddy simulation (LES) modeling studies revealing a slightly smaller range of values, and observational studies suggesting a much larger upper limit (e.g., Weckwerth et al. 1997 and Etling and Brown 1993).
26 Thompson and Edwards (2000).
27 Weckwerth et al. (1997).
28 Weckwerth et al. (1996).
29 This is the diameter of the updraft in its entirely. The archetypal updraft core, wherein the vertical windspeeds are strongest, is roughly half this diameter; see Chapter 6.
30 Balaji and Clark (1988).
31 Balaji and Clark (1988), Redelsperger and Clark (1990).
32 Bohme et al. (2007).
33 Weckwerth et al. (2008).
34 Davies (1994).
35 Bluestein (1993) and Holton (2004).
36 Koch (1984).
37 Shapiro et al. (1985).
38 Neiman and Wakimoto (1999), Wakimoto and Murphey (2010).
39 Ziegler et al. (1995), Ziegler et al. (2007).
40 Drylines have been identified in other parts of the world, such as in northern Australia; see Arnup and Reeder (2007).
41 Hoch and Markowski (2004).
42 Ibid.
43 Ziegler et al. (1995).
44 Ibid.
45 Pielke et al. (1997).
46 Ziegler et al. (1995).
47 Schultz et al. (2007).
48 Parsons et al. (1991).
49 Ziegler and Rasmussen (1998), Ziegler et al. (2007).
50 Ziegler and Rasmussen (1998).
51 Weckwerth and Parsons (2006).
52 Schultz et al. (2007).
53 Neiman and Wakimoto (1999).
54 Weiss and Bluestein (2002), Ziegler et al. (2007).
55 Atkins et al. (1998).
56 Peckham et al. (2004).
57 Pielke (1974).
58 Holton (2004).
59 Miller et al. (2003).
60 Arritt (1993).
61 HCRs can also exist in with on-shore flow, but the combined lifting will be relatively shallower because of the shallower SBF.
62 This has been observed by Atkins et al. (1995) and later simulated by Dailey and Fovell (1999).
63 Atkins et al. (1995).
64 Boundary identification required that the radar fine line/convergent flow be $>$ 10 km in length, over a minimum duration of 15 min; Wilson and Schreiber (1986).
65 Wilson and Roberts (2006).
66 Lima and Wilson (2008).
67 Droegemeier and Wilhelmson (1987).
68 Wilson et al. (1998).
69 Note again that the boundaries included, but were not limited to, gust fronts; Wilson and Schreiber (1986).

70 Droegemeier and Wilhelmson (1985a), Droegemeier and Wilhelmson (1985b).
71 Stensrud and Fritsch (1994).
72 Weaver et al. (2002).
73 Banacos and Schultz (2005).
74 Lee and Wilhelmson (1997), Marquis et al. (2007).
75 Much of the following draws from Lane and Reeder (2001).
76 In numerical models, a rid-top boundary condition is typically specified several kilometers above the nominal tropopause level, because convective motions are capable of vertically overshooting the stable tropopause. To prevent wave reflection, methods are used to damp and absorb wave motions in the layer between the tropopause and top boundary.
77 Lane and Reeder (2001), Mapes (1993), Marsham and Parker (2006).
78 Lane and Reeder (2001).
79 For example, Nicholls and Pielke (2000) show that weak uplift is provided by their $n = 2$ mode.
80 Mapes (1993).
81 Marsham and Parker (2006).
82 Knupp (2006).
83 Crook (1988).
84 Knupp (2006).
85 Marsham et al. (2011).
86 Wilson and Roberts (2006).

6

Elemental Convective Processes

Synopsis: Upon initiation, deep convective clouds may evolve into precipitating convective storms. The elemental storm processes are updrafts and downdrafts. Their dynamical structure is described in this chapter, as is that of the storm outflow, to which the updrafts and downdrafts are intimately linked. Consideration is then given to storm evolution, in the context of the single- and multicelled storms of lowest hierarchical rank.

6.1 Overview of the Convective Storm Spectrum

Once initiated, convective clouds may evolve into deep convective storms that produce precipitation at the ground, gusty surface winds, and sometimes hail, lightning, and tornadoes. The duration, intensity, and types of phenomena attendant to the storm are related largely to the storm morphology or *convective mode*. Observable structural characteristics can be used to classify storms as:

(1) Discrete, unicellular storms, including supercell storms
(2) Multicellular storms
(3) Mesoscale-convective systems (MCSs)

This classification – which is well supported by weather radar and satellite observations – borrows from the biological sciences, with *convective cells* regarded as the basic building blocks of convective storms. A convective cell has a definite, yet porous, boundary (*visible cloud edge*) like a biological cell wall. A convective cell may also divide and split into two cells (a *splitting supercell*), encounter and merge with other cells to become a larger cell (an MCS), have more than one "nucleus" (a *multicell*), and decay or grow depending on the availability of "nutrients" such as atmospheric moisture in its environment.

These classifications are not entirely distinct. For example, an MCS is inherently multicellular, but its horizontal length scale of $\sim$100 km is nominally – and perhaps artificially – used to distinguish it from a multicell (Chapter 8). The range of possible morpologies is, in reality, continuous across the *convective storm spectrum*. However, we still find it useful to discuss storms in terms of specific, recurring morphologies, if for no other reason than the fact that the probability of certain hazardous weather changes with mode: supercells are the more prolific tornado producers, and MCSs are regarded as the generators of long swaths of damaging straight-line surface winds. As we will see, the predominant dynamics also varies between modes, with supercells driven largely by rotational dynamics, and MCS intensity intimately linked to the dynamics associated with the convective-system cold pool. The convective mode is critically dependent on certain characteristics of the mesoscale and synoptic-scale environment, such as the amount of vertical wind shear and CAPE. Hence, under some circumstances, and of course with some degree of uncertainty, values of environmental parameters can serve as predictors of (and even proxies for; see Chapter 9) convective mode.

Chapter 7 is devoted to supercell storms, which are an important and special class of unicellular storms. In Chapter 8 we discuss MCSs, which often have a linear or at least quasilinear shape when viewed by radar. Short-lived unicellular storms, and less organized multicellular storms, are summarized in the following sections.

6.2 Components: Updrafts and Downdrafts

Composing convective cells are updrafts and downdrafts, which are distinct, localized air currents in the vertical plane. Both air currents are driven generally by buoyancy, although vertical gradients in pressure play a significant dynamical role in the currents of more organized convective storms, such as supercells. Both air currents are of fundamental importance to the cell.

6.2.1 Updrafts

Let us begin this section with an overview of the observed characteristics of updrafts. These have been determined in a number of ways, such as through in-situ measurements with an instrumented aircraft flying at constant altitude; radiosonde observations; direct, remotely sensed measurements using a wind profiler or other vertically pointing radar; and multiple Doppler radar observations and associated 3D wind retrievals (see Chapter 3). In the United States, these data have been collected predominantly in the Midwest, Great Plains, and Southeast regions, where major field campaigns have been conducted. Several field campaigns over tropical (oceanic and continental) regions have contributed additional convective cloud and storm data; these will be used later to explore global differences in updraft (and downdraft) properties.

In the midlatitudes, updraft speeds of $\gtrsim 10$ m s^{-1} are typical in cumulonimbi, with maximum updraft speeds in excess of 40 m s^{-1} found in mature severe

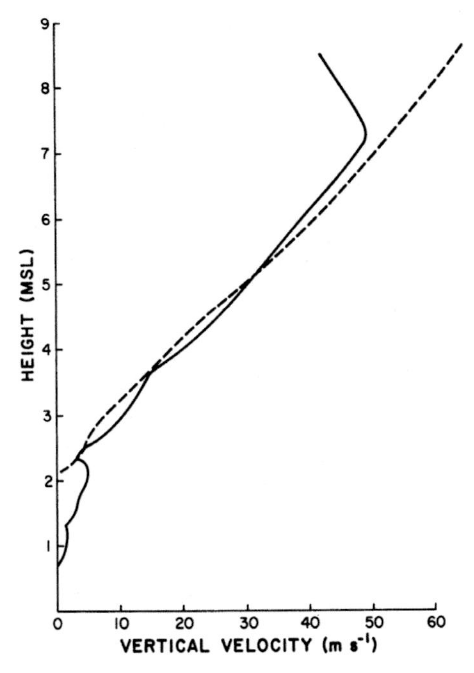

Figure 6.1 Vertical profile of vertical velocity computed from balloon ascent rate (solid line), and from parcel buoyancy $[2\mathrm{CAPE}(z)]^{1/2}$ (dashed line) computed from the same sounding. The sounding was launched within the updraft of a tornadic storm near Canadian, Texas. From Bluestein et al. (1988).

thunder-storms.[1,2] Figure 6.1 shows an updraft profile from a tornadic storm. It reveals a maximum updraft speed of nearly 50 m s^{-1} at 7 km (MSL), below which the speeds increase monotonically with height.[3]

This updraft profile is derived from the ascent rate of a radiosonde, but is compared in Figure 6.1 to a profile computed using a parcel's thermal buoyancy,

$$w(z) = [2\,\mathrm{CAPE}\,(z)]^{1/2} = \left[2\int_{z_{LFC}}^{z} B dz\right]^{1/2}, \tag{6.1}$$

where B is thermal buoyancy, and z_{LFC} indicates the height of the level of free convection (see Chapter 5). A surprising level of consistency with parcel theory is demonstrated through this comparison, especially considering the simplifications enumerated in Chapter 5. In particular, Figure 6.1 implies the ascent of updraft parcels that are relatively undiluted by environmental air.

Further evidence of relatively undiluted ascent is provided by flight-level data collected through the middle levels of a Montana supercell.[4] Within the ~6-km-wide core of this supercell updraft, updraft speeds exceed 30 m s^{-1}, and θ_e and liquid water content (LWC) have respective maxima (Figure 6.2). The turbulence intensity (eddy dissipation rate) is a minimum within the core, but has peaks just outside the core. Hence, despite this indication of considerable turbulent mixing at the lateral edges of the updraft, the core itself appears relatively insulated from the effects of *entrainment*.

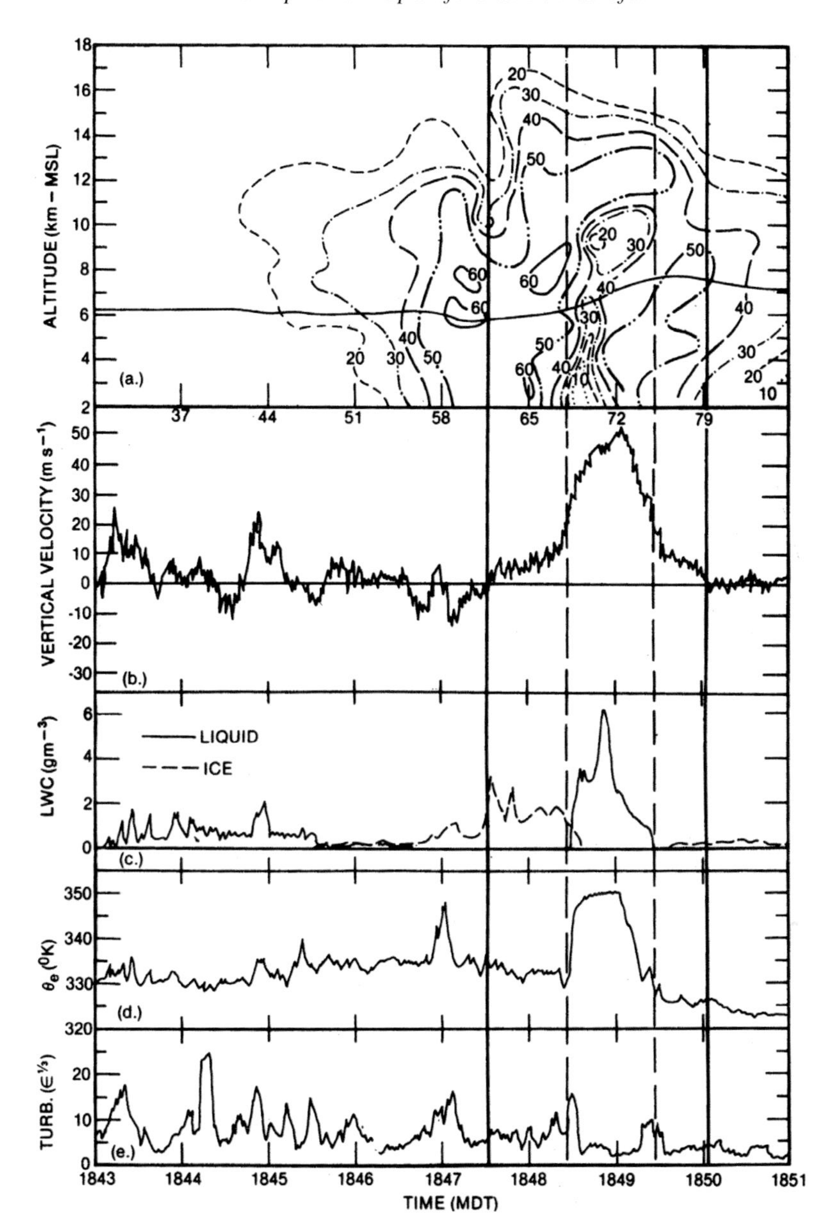

Figure 6.2 Time series of radar reflectivity, and corresponding flight-level data collected near eastern Montana through the midlevels of a supercell thunderstorm. Bold vertical line shows outer extent of the updraft, and dashed vertical line indicates the updraft core [and boundary of the weak echo region (WER)]; the horizontal extent of the core and WER is approximately 6 km. From Musil et al. (1986).

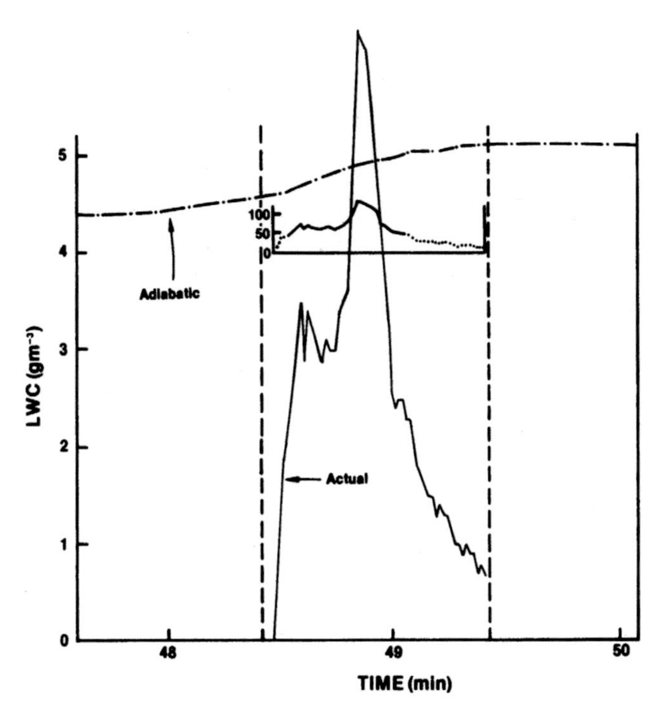

Figure 6.3 As in Figure 6.2, except only showing the measured LWC within the updraft core, and the corresponding ALWC (see text). The inset shows LWC/ALWC (%) over the same segment, with a solid line indicating the portion of the flight data where the ratio exceeds 50%. From Musil et al. (1986).

The degree of parcel dilution, and thus the effects of entrainment, can be quantified by comparing the measured LWC at some height to the adiabatic liquid water content (ALWC). The ALWC is the amount of water condensed assuming adiabatic ascent, in the absence of precipitation or evaporation (according to parcel theory).[5] Calculation of ALWC at some height is simply

$$ALWC(z) = \rho\left[\overline{q}_{v,s}\big|_{z_b} - q'_{v,s}(z)\right], \tag{6.2}$$

where $\overline{q}_{v,s}$ is the saturation mixing ratio determined from an environmental sounding at the height of the cloud base, and $q'_{v,s}$ is the saturation mixing ratio of an undiluted parcel lifted from the cloud base to the height z of evaluation. In the Montana supercell, the respective cloud-base temperature and pressure of 13°C and 730 hPa yield flight level ALWCs of 4.5 to 5 g m^{-3}. The sections of updraft wherein LWC/ALWC ≥ 0.5 are generally considered undiluted cores. As Figure 6.3 shows, such a relatively high LWC/ALWC ratio applies to almost half of the 6-km updraft.[6]

The existence of undilute, "adiabatic" cores in updrafts of less organized convective storms, and in particular, in tropical updrafts, has proven difficult to observe. The maximum LWC in tropical oceanic clouds rarely exceeds 0.4 $\times$ ALWC.[7] Moreover, the typical maximum updraft speeds (15 to 20 m s^{-1}), tend to be much less than that

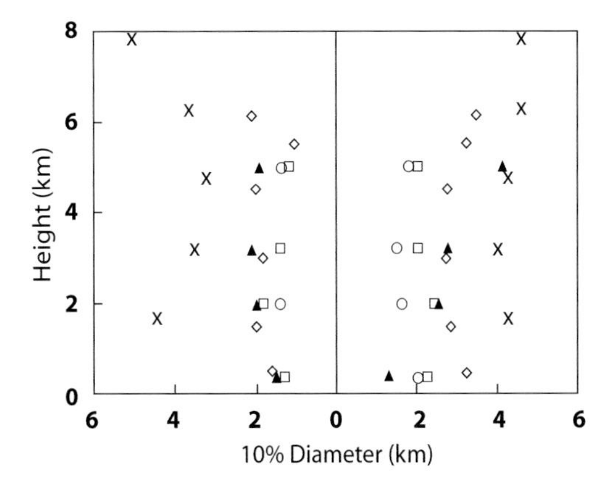

Figure 6.4 Comparison of core diameters of the strongest 10% of updrafts (right panel) and downdrafts (left panel) from samples collected in the tropical Atlantic (circles), the tropical Pacific off the coast of Australia (triangles), near Taiwan (squares), in hurricane rainbands (diamonds), and in Ohio and Florida (crosses). From Lucas et al. (1994).

predicted by parcel theory from typical CAPE (i.e., 1,500 J kg^{-1} to 2,000 J kg^{-1} which would yield maximum updraft speeds of 55 to 63 m s^{-1}).[8]

A relatively higher amount of entrainment in tropical convection may explain these differences. As will be explored later, entrainment rate is an inverse function of updraft diameter. Now consider Figure 6.4, which compares the diameters of the strongest 10 percent of updrafts sampled in regions within the tropical Pacific, the tropical Atlantic, and in Ohio and Florida during the Thunderstorm Project (see Chapter 1). Clearly, the updrafts in the tropical oceanic storms are narrower than those sampled in continental storms during the Thunderstorm Project.[9]

We pause here to ask why geography might relate to updraft diameter. One proposed explanation regards the depth of the atmospheric boundary layer (ABL), which is $\sim$500 m over the tropical oceans, $\sim$1 to 2 km over central Florida (subtropics), and $\gtrsim$2 km over the High Plains (continental midlatitudes).[10] In Chapter 5 we learned that ABL depth equates to the depth of horizontal convective rolls (HCRs), and that HCR depth is proportional to HCR width (aspect ratio is roughly 2–4). Thus, a deeper ABL begets wider HCRs that have the potential for initiating a broader area of moist convection, and, presumably, a wider updraft. Although compelling, it is unclear whether such a physical connection is this straightforward, especially in cases of convective systems, which can form as a result of the mergers of multiple cells. Moreover, as we have seen, convective initiation often results from a number of lifting mechanisms, which may or may not include HCRs.

The inverse relationship between updraft diameter and entrainment rate just mentioned depends fundamentally on the assumed behavior of the growing cumulus cloud, namely, as a *thermal*, *plume*, or *jet*.[11] Let us consider cumuli acting as thermals,

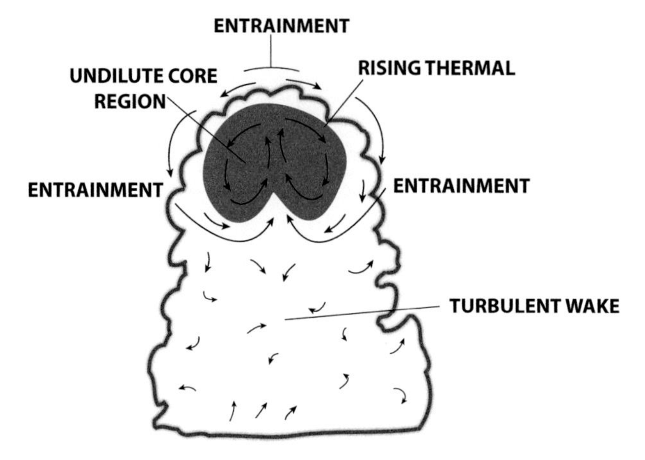

Figure 6.5 Conceptual model of growing cumulus cloud represented as a shedding thermal. Based on Blyth et al. (1988), with modifications courtesy of S. Lasher-Trapp (personal communication, 2011).

particularly those that "shed" cloudy material in their wake (Figure 6.5).[12] In essence, a thermal is a discrete buoyant entity that is released from some low-level reservoir, and then rises until its fluid has been diluted so that it is no longer buoyant. Thermals are modeled readily in the laboratory by introducing elements of a salt solution into a tank of water and then tracking them in time. Low-order numerical-model analogues have also been developed, such as the one we will use next to illustrate a simple treatment of entrainment in cumuli.[13]

Steady-state vertical excursions of a thermal are represented in a *one-dimensional Lagrangian model* through a reduced form of the vertical equation of motion,

$$\frac{Dw}{Dt} = w\frac{dw}{dz} = B - D - \Lambda w^2. \tag{6.3}$$

On the right-hand side of (6.3), B is thermal buoyancy ($= g(T_v - \overline{T}_v)/\overline{T}_v$, where T_v is virtual temperature, and the overbar represents an environmental value), and D equates to the vertical pressure gradient force (PGF) term. Because this model does not explicitly include pressure as a dependent variable, the PGF is often parameterized as $D = -0.33B$, which is interpreted as the amount of buoyancy needed to displace the air ahead of the thermal (see also the discussion of (6.18) later); this term is often neglected, as is the pressure effect in the buoyancy term (see Chapter 2). The third right-hand term is the reduction in vertical momentum owing to entrainment and mixing. In the 1D model, the entrainment rate is

$$\Lambda = \frac{1}{m}\frac{dm}{dz}, \tag{6.4}$$

representing the exchange of mass m between the environment and the rising (and/or sinking) entity. The general effect of entrainment is revealed for some fluid property A:

$$\frac{dA}{dz} = \left(\frac{dA}{dz}\right)_j + \Lambda(\overline{A} - A), \tag{6.5}$$

where subscript j indicates contributions to Lagrangian changes in A in absence of entrainment. The entrainment rate in the specific case of a thermal of core radius r_c is

$$\Lambda = \frac{3\alpha_\varepsilon}{r_c}, \tag{6.6}$$

which is based on laboratory experiments that reveal that rising thermals expand radially as

$$r_c = \alpha_\varepsilon z. \tag{6.7}$$

$\alpha_\varepsilon = 0.2$ is an empirical constant.

Coupled to the equation of motion is an equation that governs the Lagrangian change in the temperature T of the thermal core,

$$\frac{dT}{dz} = -\frac{g}{c_p} - \frac{L_v}{c_p}\frac{dq_v}{dz} + \Lambda\left[(\overline{T} - T) + \frac{L_v}{c_p}(\overline{q}_v - q_v)\right], \tag{6.8}$$

where the overbars on T and q_v denote environmental values based on an appropriate sounding. Equation (6.8) originates from a statement of conservation of *moist static energy*,

$$\hbar = c_p T + gz + L_v q_v, \tag{6.9}$$

and includes the environmental exchange term following (6.5). Other equations are included to govern the Lagrangian change in water vapor q_v,

$$\frac{dq_v}{dz} = -\frac{C}{w} + \Lambda(\overline{q}_v - q_v), \tag{6.10}$$

and in water substance (cloud condensate, rain, and ice species, depending on the desired complexity),

$$\frac{dq_i}{dz} = \frac{S_i}{w} + \Lambda(\overline{q}_i - q_i). \tag{6.11}$$

In (6.10), C is the net evaporation ($C < 0$) or condensation ($C > 0$) rate, and S_i in (6.11) indicates respective sources/sinks of the water substance class i; these would be represented as parameterizations, as described Chapter 4. The buoyancy term is the means through which (6.8), (6.10), and (6.11) are coupled to (6.3).

All dependent variables are assumed to have a top-hat radial profile within the core radius at all times. Accordingly, all variables undergo uniform – and

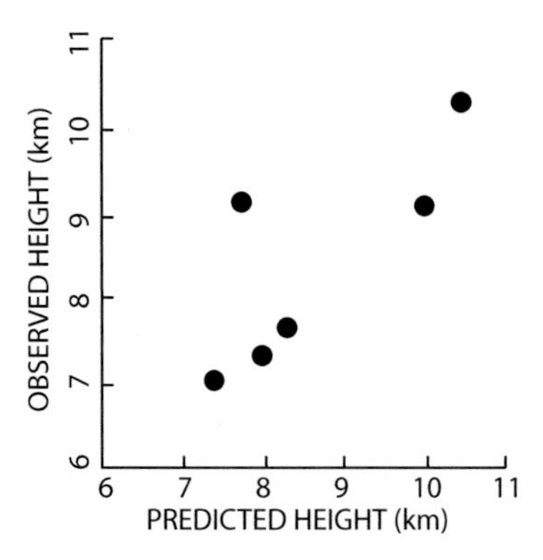

Figure 6.6 Comparison of cloud top measurements with predictions from a 1D model. From Warner (1970).

instantaneous – changes from entrainment and subsequent mixing; this is a major simplification of reality. From (6.3) we see that entrainment always reduces vertical momentum, because the quiescent air of the environment dilutes the rising air of the thermal. Modification of the thermodynamic variables by entrainment depends on their environmental values at a given height. For example, a relatively cooler, dryer environment will reduce T and q_v, and thus reduce buoyancy and, ultimately, vertical momentum. Put in the context of a cumulus cloud, entrainment affects the height to which the cloud can grow and the amount of liquid water it can contain.

According to (6.6), the parameterized entrainment rate decreases with core radius and, therefore, with height. One physical interpretation of this parameterization is that the torroidal circulation of the thermal – which, in large part, drives the entrainment (see Figure 6.5) – becomes progressively less vigorous as the core broadens; we will revisit this interpretation later.

With minor changes to the entrainment rate and to a few other terms in (6.8) and (6.10), the model formulation can be adjusted to treat a cumulus as a buoyant jet or plume.[14] Each of these treatments also posits an inverse relationship between entrainment and cloud core radius, and all three – thermal, jet, and plume – have underlying assumptions about *homogeneous mixing* of the entrained air, namely, that it:

- Occurs laterally
- Has a homogenous effect across the buoyant entity
- Is instantaneous in its dilution
- Operates continuously over all heights

Mixing in real cumuli can, at times, be rather inhomogeneous and otherwise violate the remaining assumptions.[15] It is not too surprising, therefore, that 1D models have difficulty reproducing observed cloud-top heights and LWCs (Figure 6.6).[16]

Thus, we consult two- and three-dimensional cloud-resolving models (CRMs), which are capable of a more realistic representation of entrainment and mixing. CRMs and complementary observations provide additional support for the idea that cumuli behave as thermals, and, moreover, of the attribution of entrainment to the toroidal circulation at the lateral interface between the environment and thermal.[17] Indeed, Figure 6.7 shows that (thermal-relative) air motion is downward along the thermal periphery, and upward into the thermal base and along its centerline. This primary vortex circulates environmental air from *above* the thermal top and then injects it into the thermal body from below. Such entrainment and mixing dilutes the buoyant air, which then reduces the buoyancy gradients and vortex forcing (see Equation (5.33)). With time, the ring vortex is stretched and deformed in association with the expanding thermal; in absence of a buoyancy gradient, the vortex is eventually dissipated by friction.

To reiterate a crucial point: air that is entrained into the thermal (and growing cumulus) core appears to originate at or above the thermal, rather than at the same altitude in the horizontal environment. A curious and perhaps nonintuitive outcome of this process is that the thermal core is susceptible to more dilution from entrainment than is the toroidal vortex just outside the core. Thus, a well-timed aircraft penetration or well-chosen model cross section would show a low LWC/ALWC ratio in the core, and relatively higher ratios in the flanks of the core (Figure 6.7).[18]

Most of the discussion thus far applies to growing but relatively shallow cumuli – that is, clouds with depths and widths of a few kilometers. But what about stronger, deeper updrafts? Is entrainment in these updrafts attributable to the ring vortices of a *sequence* of shedding thermals?

There is indeed evidence of multiple thermals in cumulus congestus, as provided by Doppler-radar scans showing temporally and spatially distinct reflectivity maxima.[19] There is also limited evidence to suggest the existence of a thermal sequence in organized storms such as supercells.[20] Consider Figure 6.8, which depicts a buoyancy (and wind) field retrieved from dual-Doppler observations of a tornadic supercell. Notice, in particular, the buoyancy maxima at different heights (i.e., $z = 1$ km, $z = 3$ km, and $z = 7$ km); these can be interpreted – cautiously – as multiple thermals at different stages of evolution.[21] It is noteworthy that the buoyancy maxima appear in the analysis only because the field is unsmoothed. Otherwise, the analysis would likely have shown a tilted, continuous channel of buoyancy, more likened to plume behavior.

In light of Figure 6.8, it is relevant here to consider how thermals and their associated updrafts evolve when the environmental winds are vertically sheared, a condition that favors highly organized storms such as supercells (Chapter 7). As a consequence of (differential) advection by the sheared environmental winds, an initial thermal will travel more (or at least equivalently) horizontally than vertically, losing its buoyancy via entrainment before reaching substantial heights. However, new thermals that

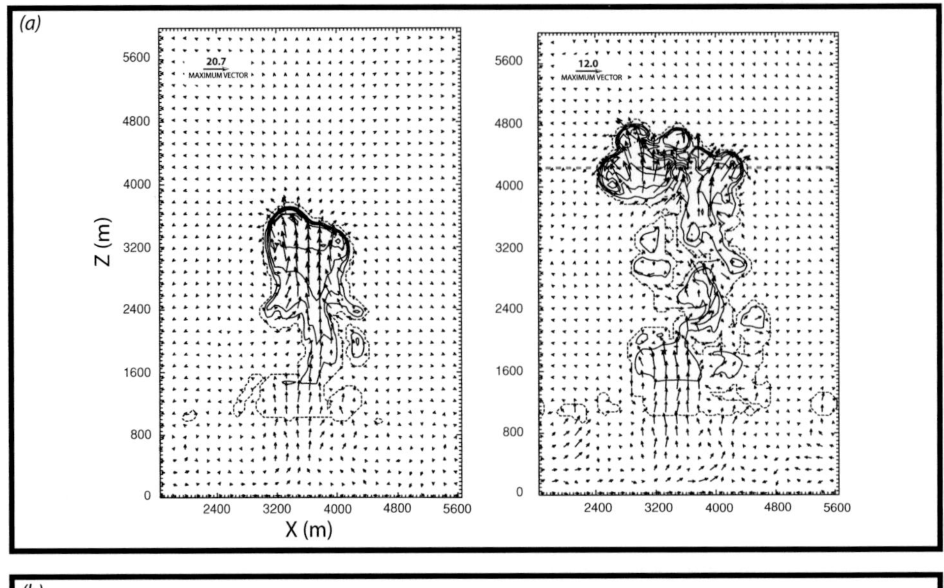

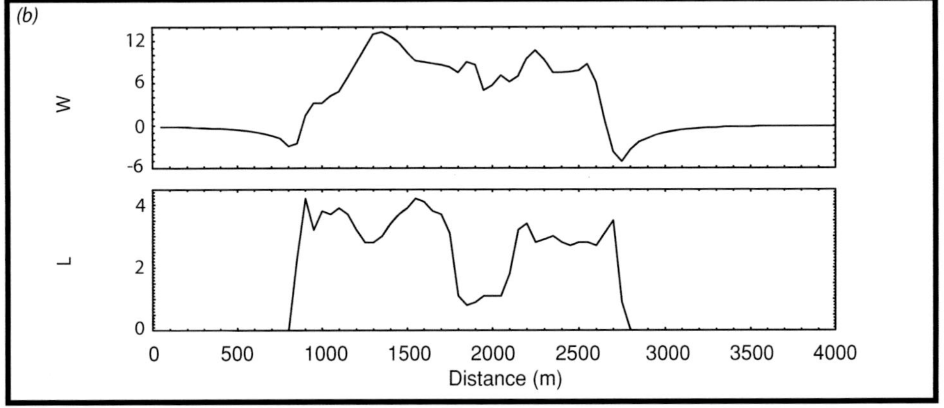

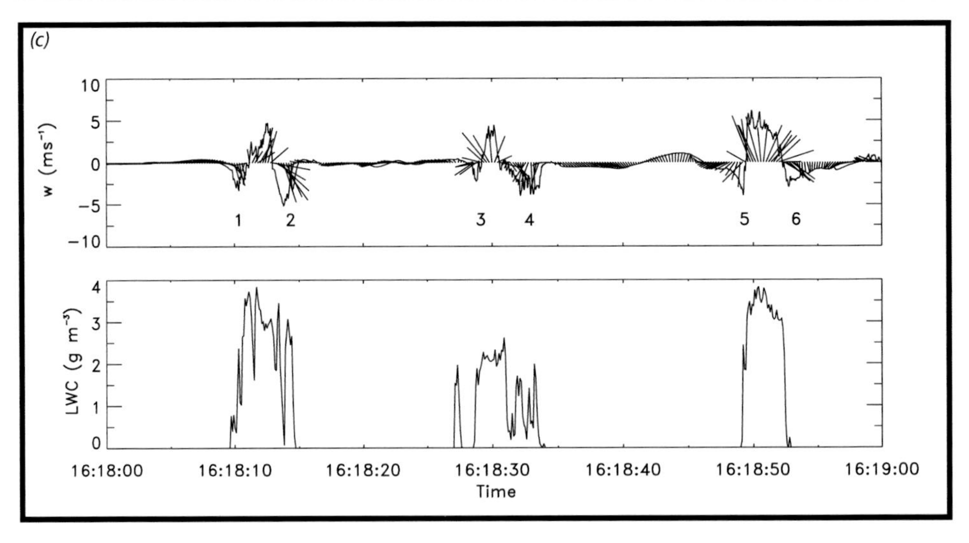

Figure 6.7 (a) Vertical cross sections of a 3D cloud-resolving model simulation of a Florida cumulus cloud at two times. Contours are of cloud LWC (contour interval 1 g m^{-3}), and vectors indicate flow in the plane. (b) Simulated properties of cloud updraft speed (m s^{-1}) and cloud liquid water content (g m^{-3}) along the dashed line in the cross section in the right panel in (a). (c) Observed properties of a Florida cumulus cloud from flight-track data, showing updraft speed and wind vectors, and LWC. After Blyth et al. (2005). Used with permission from John Wiley and Sons.

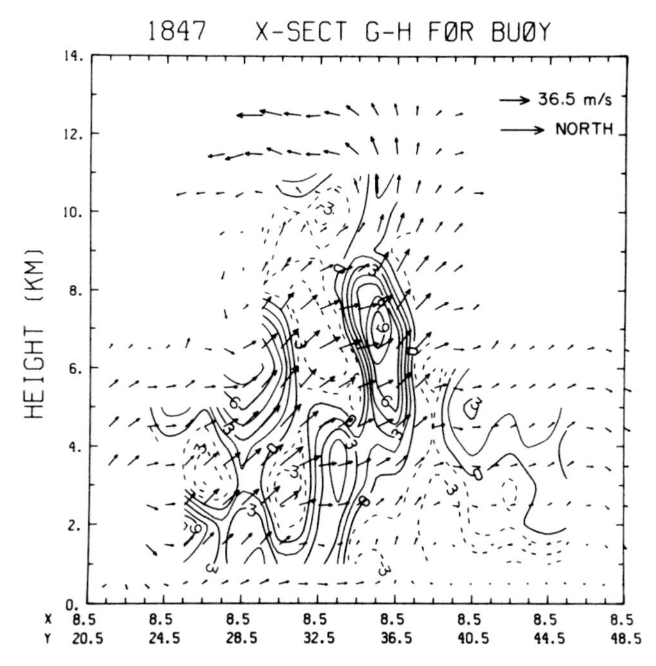

Figure 6.8 Vertical cross section of retrieved buoyancy in a supercell thunderstorm. This is a south–north section in the supercell inflow, beginning just at the rear-flank gust front, and terminating in the forward-flank outflow. The buoyancy field is unsmoothed, and not presented as a perturbation relative to an environmental sounding. From Hane and Ray (1985).

develop on the downshear side of this initial thermal will grow in, and entrain, the previous thermal's cloudy wake (Figure 6.9).[22] Although this and successive new thermals will also be differentially advected, they will also be able to rise to progressively greater heights because their buoyancy is less diluted than those of thermals growing in the presumably drier and cooler environmental air.

Additional, dynamical effects on updrafts are represented through the PGF term in the equation of motion,

$$\frac{\partial \vec{V}}{\partial t} = -\frac{1}{\overline{\rho}}\nabla p' + B\vec{k} - \vec{V} \cdot \nabla \vec{V}, \tag{6.12}$$

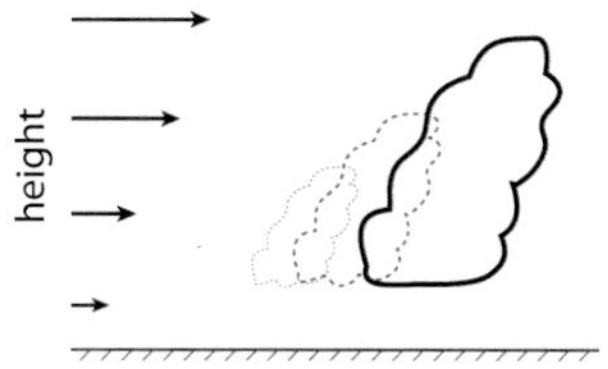

Figure 6.9 Growth of a cumulus cloud in vertical wind shear. Bold line shows the outline of a thermal and associated cloud that has developed through and downshear of the cloudy wakes (dashed lines) of earlier thermals.

where friction and Coriolis forcing are neglected, $B = -g\rho'/\overline{\rho}$ is now buoyancy in its most basic form, and p' is perturbation pressure. To expose these effects, we make use of the anelastic mass continuity equation,

$$\nabla \cdot \overline{\rho}\vec{V} = 0. \tag{6.13}$$

Upon taking $\nabla.$ of (6.12) multiplied through by $\overline{\rho}$, and then eliminating the left-hand side of the resulting equation with the use of $\partial/\partial t$ of (6.13), we obtain

$$\nabla^2 p' = \frac{\partial(\overline{\rho}B)}{\partial z} - \nabla \cdot \left(\overline{\rho}\vec{V} \cdot \nabla\vec{V}\right), \tag{6.14}$$

which alternatively can be written as

$$\nabla^2 p' = F_B + F_D. \tag{6.15}$$

Implicit in (6.15) is a decomposition of the perturbation pressure into contributions owing to the *buoyancy pressure forcing* (F_B), and the *dynamic pressure forcing* (F_D); that is,

$$p' = p'_B + p'_D, \tag{6.16}$$

where p'_B is the buoyancy pressure, and p'_D is the dynamic pressure.

The dynamic pressure forcing will be discussed in detail in Chapter 7, but let us consider here the buoyancy pressure forcing,

$$\nabla^2 p'_B = \frac{\partial\left(\overline{\rho}B\right)}{\partial z}. \tag{6.17}$$

If we make the approximation that $\nabla^2 p' \sim -p'$, which holds if p' is a sinusoidal function, we find that

$$p'_B \sim -\frac{\partial\overline{\rho}B}{\partial z}. \tag{6.18}$$

Let us then assume the existence of a spatially finite convective element, such as a thermal, with maximum (i.e., positive) buoyancy in its center. According to (6.18), buoyancy pressure will be high (low) at the top (bottom) of the element (Figure 6.10). The vertical gradient of the buoyancy pressure, or more specifically the vertical buoyancy pressure gradient force $\mathrm{VP_BGF} = -1/\overline{\rho}\,(\partial p'_B/\partial z)$, will therefore be negative within the element. Said another way, the vertical buoyancy pressure gradient results in a *deceleration that counteracts the acceleration due to buoyancy*. The deceleration accounts for the air that must be laterally displaced in advance of the rising thermal. It follows that as the size of the buoyant element (or updraft) increases, so does the amount of air that is laterally displaced and the corresponding deceleration.[23] In effect, this places an upper limit on updraft size, just as entrainment places a lower limit. Both arguments will be extended to downdrafts in the following section.

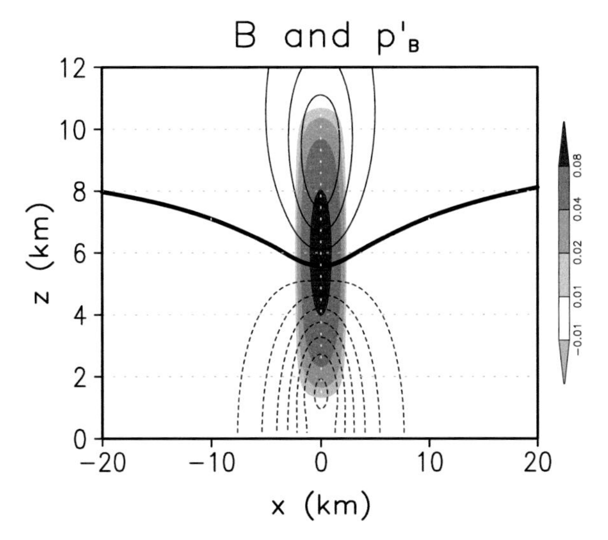

Figure 6.10 Vertical cross section of an idealized two-dimensional element of buoyancy (color-filled as shown; m s^{-2}) and the corresponding buoyancy pressure perturbation (contour interval 10 Pa, with $p'_B = 0$ indicated by the bold contour, and $p'_B < 0$ indicated by dashed contours). After Parker (2010). (For a color version of this figure, please see the color plate section.)

We close this section with a comment on how precipitation microphysics may affect the updraft properties. Tropical oceanic cumulonimbi illustrate this nicely. At levels where $T < -5$°C or at least $T < -15$°C, condensate within these updrafts freezes. The associated release of latent heat contributes to a gain in parcel buoyancy that, as can be deduced from (6.8) through (6.11), partially counteracts the loss in buoyancy to entrainment and mixing.[24] At lower heights, the fallout of precipitation within the updraft reduces the adverse effects of precipitation loading, and may also help to counteract entrainment. Thus, despite having diluted narrow cores that are not actually "hot," the resultant convective towers still grow deep enough to reach the tropical tropopause and attain significant intensity.

6.2.2 Downdrafts

In stark contrast to the preceding section on updrafts, we necessarily begin this section with a discussion of how downdrafts relate to precipitation microphysics, because they are largely inseparable.

In simplest terms, a *precipitating downdraft* commences when the contributing hydrometeors have grown sufficiently that their fallspeeds (V_f) exceed the speed of the ascending air in which they are embedded. The subsequent downdraft evolution requires explicit consideration of the: hydrometeor phase (liquid, frozen), species, and size distribution; thermodynamic environment through which the hydrometeors fall; and adiabatic and latent heating. Idealized models that bypass the details of cloud formation, as does the one presented next, can be used to isolate the influence of these factors.

The *Srivastava model* is a one-dimensional, time-varying model of a rainy downdraft.[25] It will be used here to quantitatively illustrate processes that affect downdraft formation and intensity. We begin the model description with an equation that governs the change in raindrop mass by diffusion of water vapor and capture of cloud condensate, respectively:

$$\frac{Dm}{Dt} = 4\pi r C_v D_v \left(\overline{\rho}_v - \rho_v\right) + \pi r^2 V_f \overline{\rho} q_c \, . \tag{6.19}$$

In (6.19), D_v is the diffusivity of water vapor in air, C_v is a ventilation coefficient, ρ_v is the water vapor density at the surface of the raindrop, $\overline{\rho}_v$ is the environmental water vapor density, $\overline{\rho}$ is environmental air density, and q_c is the cloud-water mixing ratio. The coefficients C_v and D_v are determined from empirical formulae, and fallspeed is approximated by

$$V_f = C_1 r^{1/2} \, , \tag{6.20}$$

where C_1 is also an empirical coefficient, and r is droplet radius.[26] Because a spherical water droplet of radius r has a mass $m = \rho_w (4/3) \pi r^3$, where ρ_w is the density of liquid water, (6.19) can be written in an alternate form:

$$\frac{Dr_i}{Dt} = \frac{\partial r_i}{\partial t} + w \frac{\partial r_i}{\partial z} = \frac{C_{v,i} D_v}{r_i \rho_w} (\overline{\rho}_v - \rho_v) + \frac{V_{f,i}}{4 \rho_w} \overline{\rho} q_c . \tag{6.21}$$

Here, index i refers to a raindrop size, which, more precisely, is a discrete *size bin* centered about r_i. Each bin has a number concentration n_i, or, the number of particles per unit mass of dry air. In absence of entrainment, drop breakup, and collection of other raindrops, the total number of raindrops is conserved. Thus,

$$\frac{\partial \left(n_i \overline{\rho}\right)}{\partial t} + \nabla \cdot \left[n_i \overline{\rho} (\vec{V} - \vec{V}_{f,i}) \right] = 0, \tag{6.22}$$

where this conservation statement is expressed in 3D for illustrative purposes. Upon expanding with the aid of the anelastic continuity equation (6.13), and then including the effects of entrainment (6.22) can be written in 1D as:

$$\frac{\partial n_i}{\partial t} + \left(w - V_{f,i}\right) \frac{\partial n_i}{\partial z} = -\left|w\right| \Lambda n_i + n_i \left(V_{f,i} \frac{d \ln \overline{\rho}}{dz} + \frac{\partial V_{f,i}}{\partial z} \right), \tag{6.23}$$

where Λ again is the entrainment rate, except now defined with slightly different coefficients, owing to an assumed jet-like behavior:

$$\Lambda = \frac{2\alpha_\varepsilon}{r_c} . \tag{6.24}$$

Laboratory experiments for jets suggest a value of $\alpha_\varepsilon = 0.1$. Given a lack of physical guidance on how the radius of the precipitating-downdraft column might change with height or time, the model formulation assumes a constant r_c. In the entrainment term of (6.23), drop-free environmental air is implicitly added instantaneously and uniformly across the downdraft column, diluting the number concentration; the use of an absolute value on w ensures a decrease regardless of the direction of vertical air motion.

Entrainment also affects the downdraft temperature, and is included in the equation based on the conservation of moist static energy,

$$\frac{\partial H}{\partial t} + w\frac{\partial H}{\partial z} = -w\frac{g}{c_p} + |w|\,\Lambda\left[(\overline{T} - T) + \frac{L_v}{c_p}\left(\overline{q}_v - q_v\right)\right], \tag{6.25}$$

where here we have

$$H = T + \frac{L_v}{c_p}q_v. \tag{6.26}$$

In (6.25), the contribution from dry adiabatic processes is represented separately in the first right-hand term. Thermodynamic changes in the downdraft are coupled to vertical velocity changes via

$$\frac{\partial w}{\partial t} + w\frac{\partial w}{\partial z} = B - \Lambda w^2. \tag{6.27}$$

Buoyancy is now defined as

$$B = g(\Delta T/\overline{T} + 0.61\Delta q_v - q_r - q_c), \tag{6.28}$$

where Δ represents the excess of the variable (T, q_v) within the downdraft over that of the environment. Notably, the PGF is neglected in (6.27), the validity of which we will explore later.

One model-implementation strategy is to specify a raindrop size distribution at the top boundary of the model domain, and then let this rain fall continuously into the domain. The specified rain core, which acts to initiate a downdraft, is assumed to have a horizontal dimension r_c that does not vary in time. The bulk properties of the rain are represented through the rainwater mixing ratio:

$$q_r(z,t) = \frac{4\pi\rho_w}{3}\sum r_i^3 n_i. \tag{6.29}$$

The rainwater mixing ratio is combined with the cloud-water and water-vapor mixing ratios to comprise *total water substance* $q_{TW} = q_r + q_c + q_v$, which is conserved in the absence of sources and sinks. When entrainment is included, the equation governing the total water substance is

$$\frac{\partial q_{TW}}{\partial t} + (w - V_*)\frac{\partial q_{TW}}{\partial z} = |w|\,\Lambda\left(\overline{q}_v - q_v - q_r\right) + q_r\left(V_*\frac{d\ln\rho}{dz} + \frac{\partial V_*}{\partial z}\right), \tag{6.30}$$

where V_* is the mass-weighted fallspeed of raindrops in the distribution; the cloud droplets are assumed to have a negligible fall velocity.

Before using the Srivastava model quantitatively, it is instructive to qualitatively examine the model processes and exchanges as they relate to downdraft formation and intensity. It is a straightforward deduction from (6.27) that the downward air accelerations represented in this model are due to negative buoyancy. Rainwater contributes to negative buoyancy through precipitation loading; the mixing ratios of ice hydrometeors similarly would have a precipitation loading effect. The largest loading is associated with the largest drops, by virtue of the cubic dependency of q_r on r in (6.29). Raindrop size, and hence precipitation loading, is reduced by evaporation, which acts more rapidly on smaller drops (see (6.21)), owing essentially to their large surface area relative to water mass. A reduction in q_r, the corresponding increase in q_v, and the uptake of latent heat of vaporization contributes to cooling of the downdraft. This is communicated to temperature via (6.25) and (6.26), and ultimately to vertical velocity through B in (6.27). The amount of evaporative cooling depends not only on drop size but also on the temperature and humidity of the environment relative to that in the downdraft column. An additional effect not yet considered is that due to adiabatic warming: from parcel theory, the temperature of (unsaturated) descending air increases at the dry adiabatic rate (i.e., the first right-hand term in (6.25)). The adiabatic warming opposes the evaporative cooling and precipitating loading, but evaporation and loading are themselves inversely related. Entrainment modifies all these processes.

The quantitative effects are revealed by Srivastava-model experiments in which the environmental conditions and specified precipitation are varied. In the experimental results presented in Figure 6.11, the environmental temperature lapse rate is 8°C km^{-1}, and the environmental relative humidity is a constant 70 percent.[27] Both conditions are applied over a relatively deep subcloud layer, with a cloud-base height of ~3.7 km AGL, and cloud-base temperature, relatively humidity, pressure, and vertical velocity of 0°C, RH = 100%, 550 hPa, and –1 m s^{-1}, respectively. Rain is specified at cloud base, and distributed according to

$$N(D) = N_o e^{-\lambda_r D}. \tag{6.31}$$

Recall that this is known as a Marshall-Palmer distribution, where $N(D)$ is the number concentration per unit volume of drops of diameter D, in diameter interval (or bin) $(D, D + dD)$, and N_0 is the intercept, set in this experiment to 0.08 cm^{-4}. The slope λ_r controls the numbers of small drops relative to larger drops. A slope of λ_r = 20 cm^{-1} is used, and is equivalent to q_r = 2.22 g kg^{-1} through

$$q_r = \frac{1}{\bar{\rho}} \int_0^\infty m(D) N_0 e^{-\lambda_r D} dD, \tag{6.32}$$

where m is mass as defined in (6.19).

As portrayed in Figure 6.11, the sharp peak in negative vertical velocity lags the thermal buoyancy peak as well as the advancing rain core. With time, evaporation

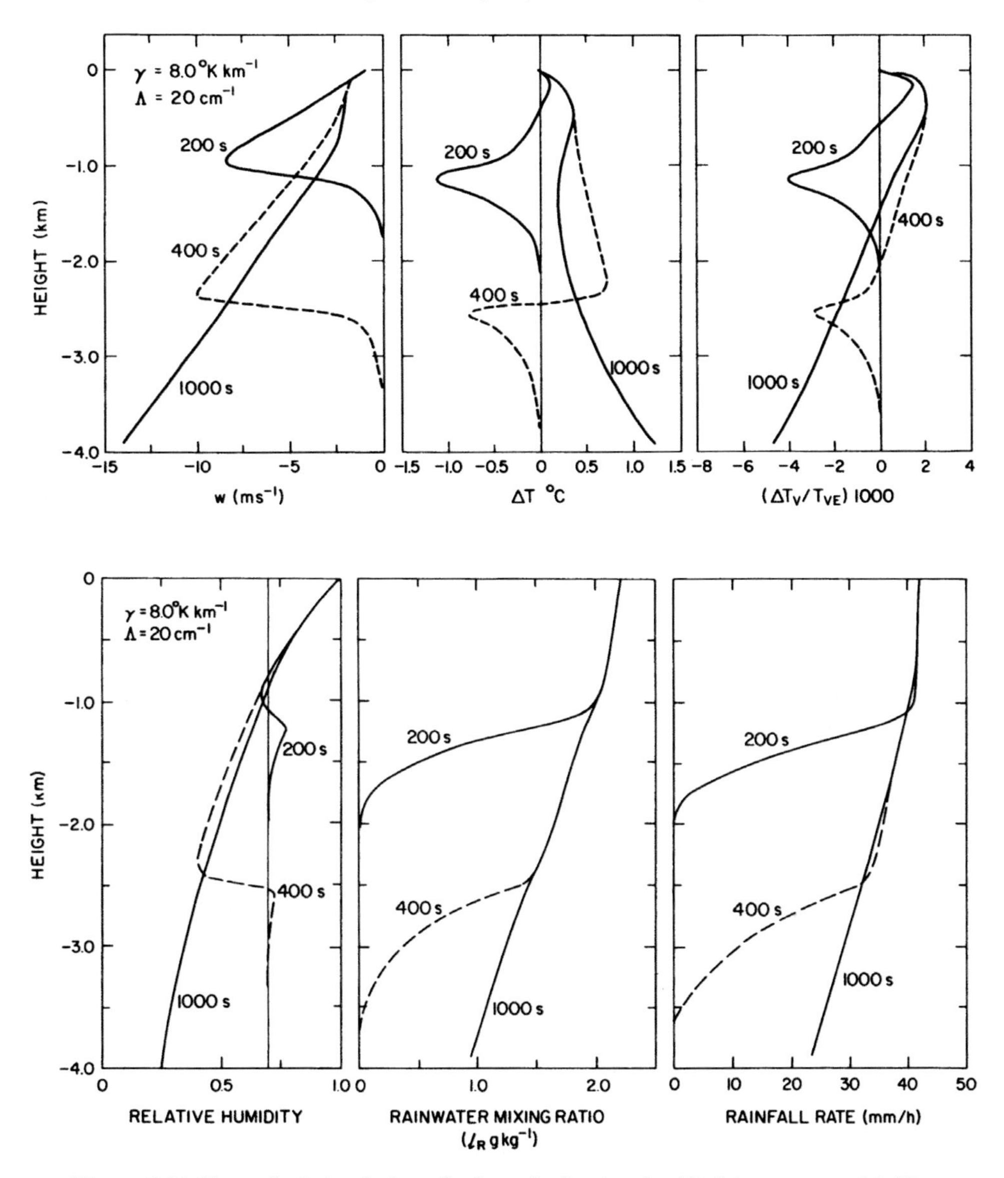

Figure 6.11 Numerical simulation of a downdraft using the 1D Srivastava model. The cloud-base/boundary conditions are $q_r = 2.22$ g kg^{-1}, $T = 0$°C, RH = 100% (from which q_v is determined), $p = 550$ hPa and $w = -1$ m s^{-1}. The results are from a simulation in which entrainment is neglected. From Srivastava (1985).

reduces the rain core to half of its cloud-base value, yet the peak in negative vertical velocity continues to be enhanced. This is explained in part because precipitation loading contributes explicitly to vertical *acceleration*, not vertical velocity. Thermal buoyancy also contributes to downward accelerating air, where and when evaporative cooling overcomes adiabatic heating. It is noteworthy that adiabatic heating and drying lead to downdraft air that is considerably warmer and drier than its environment by the time the downdraft core effectively reaches the ground (although the large, near-ground magnitudes are somewhat of an artifact of the open lower boundary in the 1D model; discussed later). Notwithstanding the continued precipitation loading, a temperature excess would seem to contradict the existence of a strong downdraft. Recall, however, that the buoyancy term accounts for humidity as well as temperature deviations, and thus the relative dryness of the downdraft causes it to be virtually cooler than its environment, as shown in the evolving profiles of the perturbation virtual temperature.

Considering evaporation of rain only (i.e., no ice), this and other experiments show that stronger downdrafts result when the (subcloud) environmental lapse rate is steepened, the specified cloud-base rainwater mixing ratio is increased, the size of drops comprising the rain is decreased, and the environmental humidity is increased (though still well below saturation). This latter result may seem counterintuitive, as one would expect lower humidity to increase evaporative cooling. Although this is true, a higher environmental humidity allows for the greater virtual cooling that arises as air dries during unsaturated adiabatic descent.

Downdrafts driven by ice particles have similar dependencies on the environment and precipitation specification, albeit with a couple of notable exceptions. First, in a relatively stable stratification (e.g., lapse rate is $\sim 6^{\circ}\mathrm{C\ km^{-1}}$), the strongest downdrafts are due to the melting of hail and subsequent evaporation; the extreme case is a *wet downburst* and corresponding microburst.[28] Second, in a deep layer with a nearly dry-adiabatic lapse rate, strong downdrafts result from the sublimation and also melting/evaporation of small, low-density snow particles; the extreme case is a *dry downburst* and corresponding microburst (Figure 6.12).[29]

Not yet addressed explicitly is *initiation* of precipitation-driven downdrafts. A downward acceleration will commence once positive thermal buoyancy is overwhelmed by precipitation loading, as promoted most effectively by large frozen hydrometeors, such as hail or graupel.[30] This often occurs in the middle-levels of the cloud rather than near the cloud top.[31] The hydrometeors may initially fall through cloudy air, but much of the cloud condensate is collected and added to the hydrometeor mass (see (6.20)). Consequently, the precipitating region and associated downdraft become visually distinct from the cloud.[32]

The size of this initial precipitating region, and hence the initial downdraft core, is a few kilometers in the particular case of downbursts (see Figure 6.12).[33] Because the hydrometeors that initiate a downdraft are grown in a related updraft, it is logical to expect a correspondence between downdraft and updraft core sizes. Indeed, statistics compiled from weather radar scans as well as in-situ sampling by aircraft bear this out (e.g., Figure 6.4).[34] Hence, tropical oceanic convection tends to have narrow downdrafts as well as narrow updrafts, relative to the larger drafts in midlatitude continental convection. It should be noted that many of these updraft/downdraft observations are from convective clouds in weakly sheared environments. In a strong

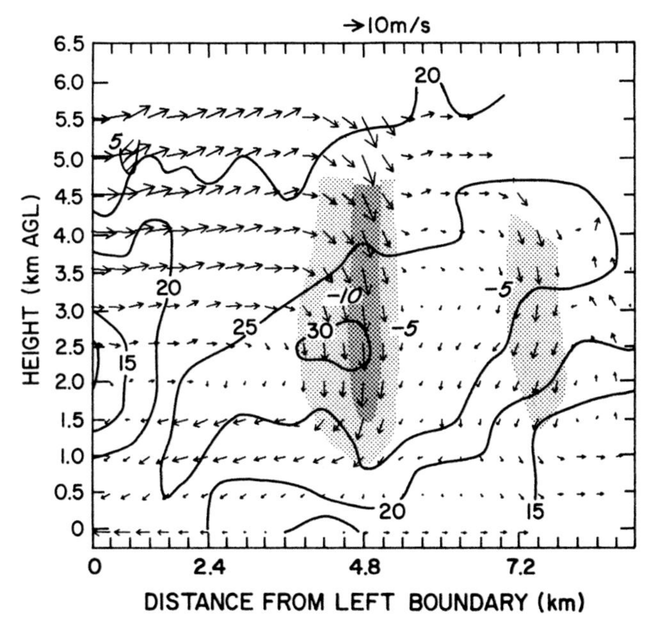

Figure 6.12 Vertical cross section of wind and radar reflectivity factor in a dry microburst. Derived from a dual-Doppler radar analyses. From Hjelmfelt et al. (1989).

environmental vertical wind shear, differential advection of hydrometeors should enlarge the precipitation region. However, even in supercell storms, the downdraft core sizes still correlate to the updraft cores sizes.[35]

As in updrafts, the entrainment rate is greater for narrower downdrafts, and entrainment of quiescent, precipitation-free air reduces downdraft intensity according to (6.23), (6.25), (6.27), and (6.30). In the 1D framework of the Srivastava model, "narrower" equates to core diameters <2 km. There is uncertainty in this particular threshold value, because it depends on the specified entrainment rate α_{ε}, and on the assumption of a height-invariant r_c.[36] Two-dimensional model simulations of downbursts, with parameterized turbulent mixing rather than a simple entrainment term proportional to $1/r_c$, suggest an optimum core diameter of approximately 725 m: smaller-diameter downdrafts are reduced in intensity by entrainment, and larger-diameter downdrafts are reduced in intensity by vertical gradients in buoyancy pressure. This larger-diameter reduction is explained by the fact that an increasing amount of air must be displaced laterally in advance of increasingly larger downdrafts. In other words, a large downdraft results in a positive $\mathrm{VP_BGF}$ that counteracts the negative or downward force from buoyancy, as analogously demonstrated in Section 6.2.1 for large updrafts.

It is appropriate here to recall the buoyancy pressure associated with updrafts, because one consequence of this pressure and its vertical gradient is actually a downdraft(s) near the updraft summit. We consider this to be an example of a *dynamically*

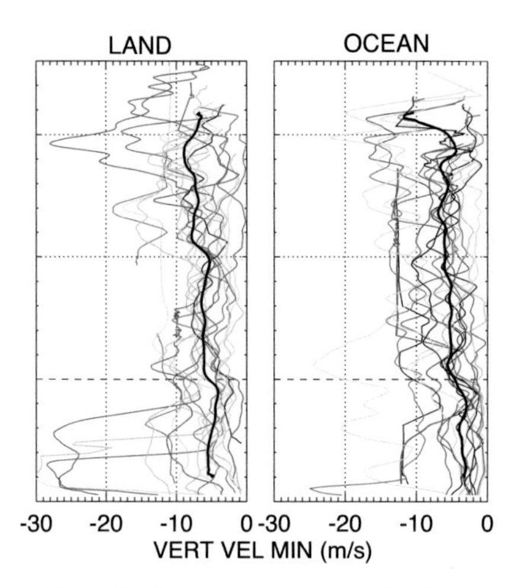

Figure 6.13 Vertical profiles of minimum vertical velocity in a variety of convective storms, over continental versus oceanic domains. The profiles are derived from a high-altitude airborne Doppler radar (EDOP). After Heymsfield et al. (2010). (For a color version of this figure, please see the color plate section.)

driven downdraft; others will be described in Chapters 7 and 8. When taken together with a precipitation-driven downdraft, the dynamically driven upper-level downdrafts result in a bimodal distribution in profiles of negative vertical velocity (Figure 6.13).[37] Figure 6.13 shows that some of these upper-level downdrafts are quite intense, with speeds exceeding –10 m s^{-1}, and comparable to the lower-level downdrafts. It shows moreover that upper-level downdrafts in land-based convection are relatively more intense than those based over the tropical oceans, which should not be too surprising in light of our previous discussions that established a geography–updraft size–buoyancy pressure gradient magnitude relationship.

To conclude this section, and to provide a transition to the next, we note that an additional effect of pressure results from the presence of the Earth's surface beneath the downdraft column. This rigid boundary, and the corresponding boundary condition of $w = 0$, requires a deceleration of air above the boundary. Air deceleration is achieved through an upward-directed pressure gradient force, manifest in 2D and 3D as a surface-based dome of high pressure (Figure 6.14).[38] The corresponding horizontal pressure gradient force accelerates air radially outward from the downdraft column, thus contributing to horizontal outflow. One implication of the vertical deceleration is an additional amount of time for evaporative cooling. Thus, in 2D and 3D models, temperature excesses do not tend to occur in the downdraft column at the ground, although they are still possible aloft.[39]

6.3 Convective Outflow

Upon reaching the ground, the air of the downdraft spreads laterally as outflow. The potential distance over which the outflow can then travel, or equivalently, the range

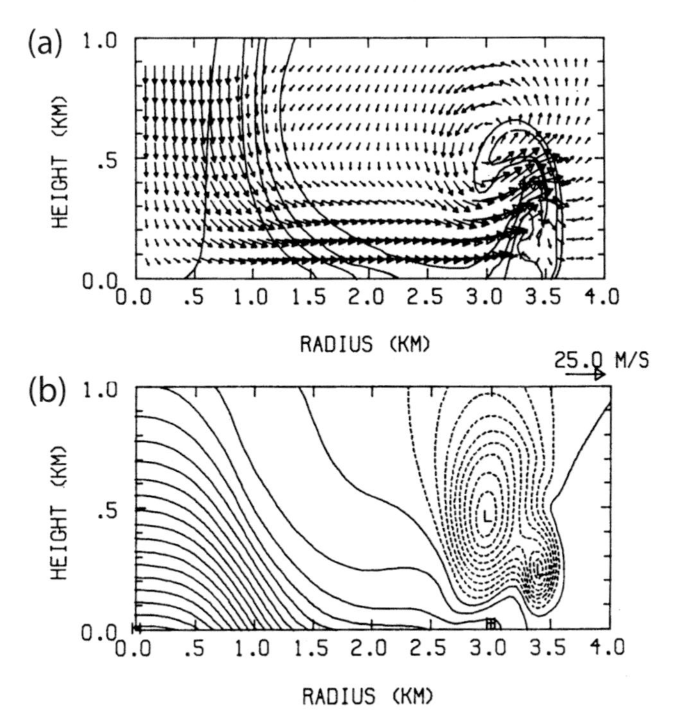

Figure 6.14 (a) Winds and contours of simulated radar reflectivity, and (b) perturbation pressure from a 2D numerical simulation of a wet microburst. From Proctor (1988).

over which characteristics of the storm are communicated back to the environment, depends in part on the amount of negative buoyancy in the downdraft air. The basis for this statement is the theoretical steady-state speed of a density current developed in Chapter 2:

$$V_{dc} = \mathscr{k}\sqrt{gd\frac{\rho'}{\rho_0}}, \tag{6.33}$$

and, in particular, the contribution to V_{dc} from ρ' , which is the difference between the respective densities of the current and environment; recall that (6.33) is valid under the strict assumptions given in Chapter 2, and that ρ_0 is the uniform density of the environment, d is a characteristic depth, and $\mathscr{k} = \sqrt{2}$ in the case of the theoretical upper limit on V_{dc} Convective outflows are considered to be dynamically similar to density currents; the maximum wind speed in real outflows does, however, tend to be faster than the equivalent steady speed of the current:

$$|\vec{V}|_{\max} \approx 1.5V_{dc}, \tag{6.34}$$

as has been determined empirically using observational data.[40]

In the archetypal convective outflow shown in Figure 6.15, vertical displacements of environmental air (gust-frontal lifting) occur relative to the head, which

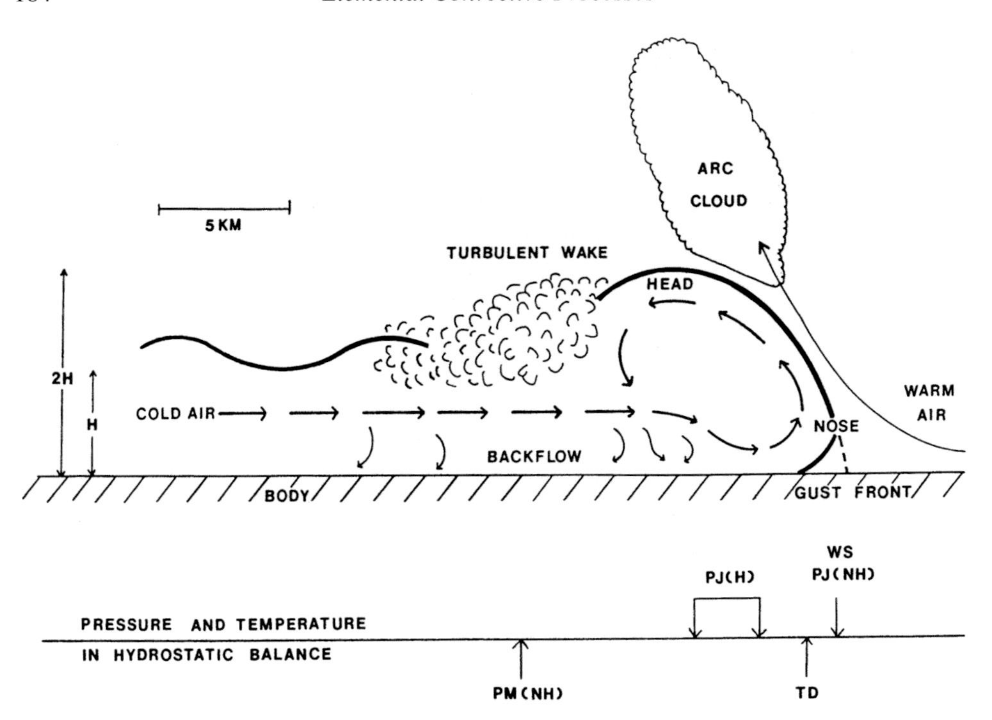

Figure 6.15 Schematic of a convective outflow in vertical cross section. From Droegemeier and Wilhelmson (1987).

is locally deeper than body depth d. Near the ground, the head has a noselike structure, present because friction retards the advance of the outflow at the ground. The head also contains a gust-front-relative, baroclinically generated circulation, whose vigor scales with the cross-frontal buoyancy deficit, as predicted by the horizontal vorticity equation.[41] Low perturbation pressure induced in the head circulation (see Figure 6.14) increases the near-ground horizontal pressure gradient and aids a horizontal acceleration well behind the gust front. This has been offered as one explanation of a secondary wind maximum observed after the initial, post-frontal jump in wind (direction and speed).[42] In association with the wind jump is a maximum in horizontal surface convergence. The convergence maximum, in turn, relates to locally high, nonhydrostatic (dynamic), perturbation pressure (Figure 6.15).

Rearward of the head is a region of turbulent eddies. The turbulent eddies derive from Kelvin-Helmholtz (KH) billows that form in the outflow–environment interface at the head, then propagate rearward or downstream, and then break. Recalling Chapter 2, KH instability (KHI) arises in the density-stratified shear layer between two overlying streams of (essentially) uniform velocity U. Instability onset is predicted by the Richardson number (Ri), which is the nondimensional ratio of buoyancy to inertia forces:[43]

$$\mathrm{Ri} = \frac{-(g/\overline{\rho})\partial\overline{\rho}/\partial z}{(\partial U/\partial z)^2}. \tag{6.35}$$

The necessary (though not sufficient) condition is Ri < 0.25, which explains well the existence of KH billows in simulated and observed outflows. The billows and related turbulent eddies are important because they entrain environmental air and act to dilute the density excess of the outflow.

Farther rearward of the head is the outflow body. Here, the wind speed is less than in the head, especially just above the ground, where surface friction induces an undercurrent or backflow. The perturbation pressure in the body attains a hydrostatic value dictated by the density stratification in the cool outflow air. The body depth is at or near the characteristic value d. From (6.33) we recall that d affects the speed, as does ρ'. Furthermore, as we learned in Chapter 5, the depth lends to the effectiveness by which the gust front can initiate new convective clouds or help sustain the outflow-generating storm. It is relevant to point out that in 2D model simulations, d is independent of the magnitude of imposed precipitation loading, but dependent on the downdraft radius r_c.[44] The latter can be reasoned quite simply using a mass continuity argument: assuming a symmetric downdraft and outflow, the greater the downward vertical mass flux $\rho w r_c$, the greater the outward horizontal mass flux $\rho U d$.

The archetype presented in Figure 6.15 is two-dimensional, but real outflows have along-frontal variability and thus strictly are 3D phenomena. One source of the variability is the release of *cleft and lobe instability* (CLI), aptly named for the series of clefts and lobes at the frontal head (Figure 6.16). The clefts originate as filaments with a roughly uniform spacing. Such spacing represents the most unstable wavelength that, along with the CLI growth rate, depends on the Reynolds number (Re) of the flow.[45] Inversely dependent on the Re is the height of the nose above the lower boundary. From Figure 6.15 we see that in the elevated nose, dense fluid overlays lighter, environmental fluid. Thus the individual filaments and clefts reveal the local occurrences of gravitational instability and resultant convective overturning. In the marginal cases of (1) Re $\to \infty$, the CLI is precluded because the lower boundary approaches free slip and hence the nose essentially is at the lower boundary; and (2) Re $\to 0$, the CLI is also precluded because the density current speed approaches zero.

Additional frontal variability may occur when environmental horizontal winds are significantly parallel to, but oriented in the opposite direction of the post-frontal winds. The resultant vortex sheet supports a horizontal shearing instability (HSI). Recall from Chapter 2 that this instability belongs to the same general class of hydrodynamic instability as KHI, save for the restoring force of buoyancy. As demonstrated in numerical simulations, the unstable shear zone (vertical-vortex sheet) can be perturbed by the clefts and lobes, which consequently give way to a series of equally spaced maxima in vertical vorticity.[46] The spacing, and hence the most unstable wavelength, is related to the thickness of the vorticity zone. The vorticity maxima have been shown to interact and merge to produce misocyclone-scale vortices that in turn can aid in convection initiation (Chapter 5) and perhaps intensify into tornadoes (Chapter 8).

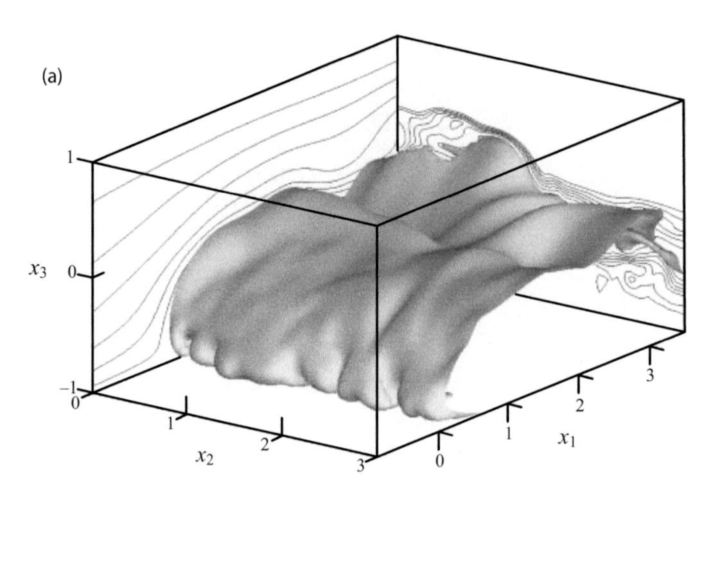

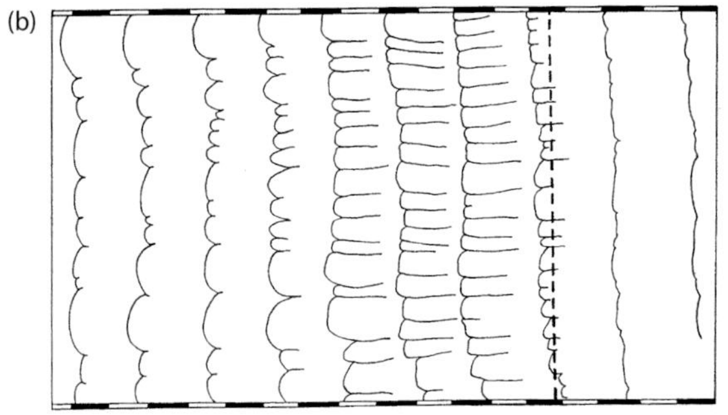

Figure 6.16 (a) Numerical and (b) laboratory simulations showing development of clefts and lobes at leading edge of density current. In (a), gray shading is a density isosurface ($\rho = 0.5$) and lines are instantaneous streamlines. In (b), lines show sketches of the cleft and lobe structure in plan view, at 1/3-s intervals. From Hartel et al. (2000). Used with permission from Cambridge University Press.

While creating spatial variability along the gust front, both mechanisms additionally serve to dilute the negative buoyancy of the outflow through horizontal (HSI) and vertical (CLI) mixing of environmental air. The KH billows play a similar role. Each of these mechanisms is related to the basic dynamics of the density current. Heterogeneity in the environment leads to further variability and dilution. Consider a heterogeneous surface whose virtual temperature is different than that of the approaching/overlying outflow. Fluxes of heat and moisture from the surface will alter the negative thermal buoyancy of the outflow.[47] The fluxes can be approximated by a bulk aerodynamic formula, with a resultant rate of dilution represented by $\sim (VC_d/d(\theta_{v,s} - \theta_v)$, where V is outflow speed, C_d is aerodynamic drag, and subscript s indicates a surface value. Hence, a virtually cool outflow that moves over a relatively warmer surface will be depleted of its negative buoyancy over time.

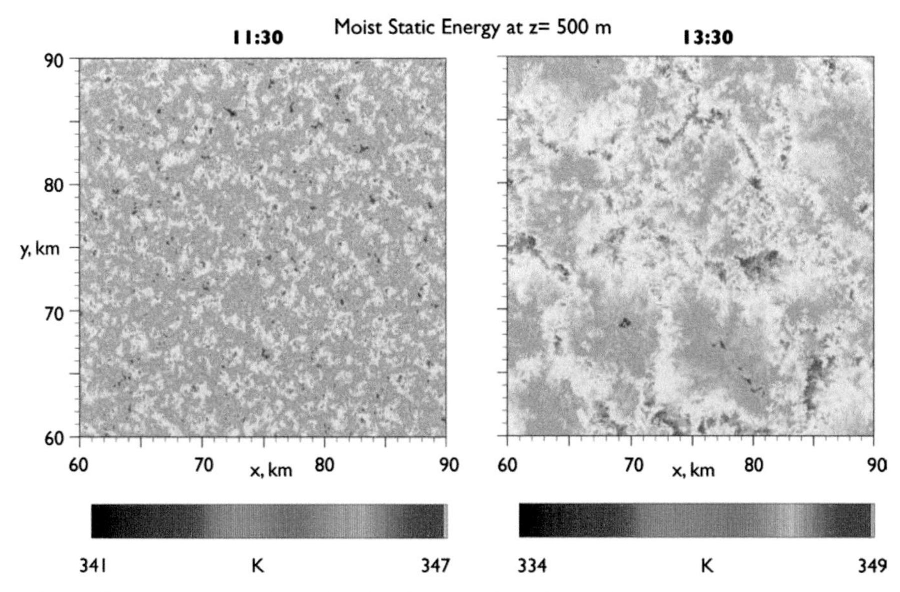

Figure 6.17 Example of a (simulated) transition from shallow to deep convective clouds. Shown are horizontal sections (at 500 m) of moist static energy. From Khairoutdinov and Randall (2006). (For a color version of this figure, please see the color plate section.)

Such buoyancy dilution from surface fluxes and entrainment also limits the distance over which the outflow can travel from its downdraft source.[48] As alluded to at the beginning of this section, the penetration range and associated outflow depth and speed all contribute to the idealized physical link of updraft→ downdraft→outflow→updraft. This linkage is revisited in Chapter 9, but it is worth discussing it here with reference to the tropics, because convective outflow has been proposed as a means to promote a transition from shallow convection to deep convection.[49] The basic premise is that the initial maritime clouds arising from boundary-layer thermals are too small in size to overcome the deleterious effects of entrainment (Section 6.2.1). Although these initial clouds do not grow to an appreciable depth, they are still capable of producing rain and hence outflow. New cloud formation is organized along the outflow boundaries, with an inherent scale that now permits deep growth. In the portrayal of this simulated transition in Figure 6.17, the outflow boundaries and newly developed deep convective clouds take the form of expanding cloud rings or cloud arcs.

Outside the tropics, one cycle of deep convective cloud formation and decay does not necessary require this separate, transitional stage involving shallow cumuli. It is indeed possible to identify distinct cumulus congestus/cumulonimbus that evolve through a single cycle within roughly one hour. In the convective-mode spectrum introduced earlier, these are the unicellular convective storms, which we now consider in more detail.

6.4 Unicellular Convective Storms

The continuous evolution of a typical "ordinary" or unicellular convective storm passes through three stages: a (*towering*) *cumulus stage*, a *mature (cumulonimbus) stage*, and a *dissipating stage* (Figure 6.18).[50] The towering-cumulus stage is present soon after convective initiation. Structurally, the growing cell has a deepening updraft, with convergent horizontal airflow below, and divergence horizontal airflow above. Precipitation particles – frozen and liquid – are growing in this updraft, but have not yet attained fallspeeds that exceed the updraft speed.

Larger, faster-falling particles are found during the mature-cumulonimbus stage. As just explained, these fall toward the ground in a negatively buoyant downdraft. A cold pool at the ground begins to form during this stage, leading to convective outflow that spreads laterally. Precipitation is eroding the lower part of the cloud (i.e., it is no longer visible as a cloud), but the upper cloud still exists and contains rising motion.

During the dissipating stage, the entire storm is engulfed in falling precipitation (and hence in downdraft), with an associated expanding surface outflow. The remainder of the cloud is eroded; remnants of the storm anvil, however, may continue to exist for some time before sublimating and/or evaporating.

In its entirety, a fully developed unicellular storm has a horizontal dimension of $\sim$10 km, but produces an outflow that may expand to more than twice this scale.[51] If we assume vertical motions of $W \sim 5$ to $10\ \mathrm{m\ s^{-1}}$, over a nominal atmospheric depth of $H \sim 10$ km, then the time scale for a parcel's rise through this atmosphere is

$$H/W \approx 16 \text{ to } 33\,\text{min}\,.$$

It follows then that parcel rise and subsequent descent (i.e., one deep convective overturning) has a time scale of $\approx$1/2 to 1 hr, which is consistent with observed unicell lifetimes.

As mentioned previously, convective mode depends on the local environment. The environment can be described quantitatively through a wide variety of parameters, although as discussed more in Chapters 7 and 8, two have particular relevance: CAPE and environmental wind shear over some vertical layer. A short-lived single cell can be supported by a broad range of CAPE, but is most apt to occur when the magnitude of deep-layer vertical wind shear is relatively low, for example:

$$|\vec{S}_{0-6}| = \mathrm{S}\,06 \leq 10\,\mathrm{m\,s^{-1}}$$

where $\vec{S}_{0-6} = \vec{V}_6 - \vec{V}_0$, and where $\vec{V}_6$ and $\vec{V}_0$ are representative mean values of the 6-km AGL and near-surface layers, respectively, (see Figure 6.19a).[52] Although not a strict threshold, the general condition on wind shear explains well the physical behavior: weak shear and storm-relative midlevel winds allow hydrometeors to fall through and quash the updraft, thus limiting the cell to one convective cycle.

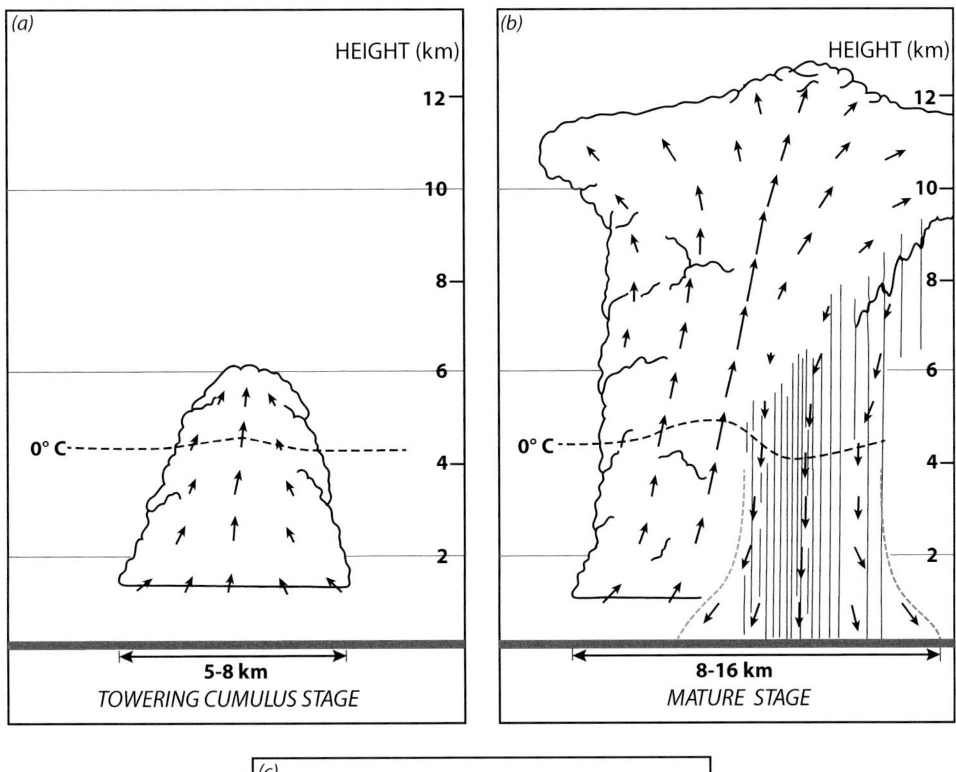

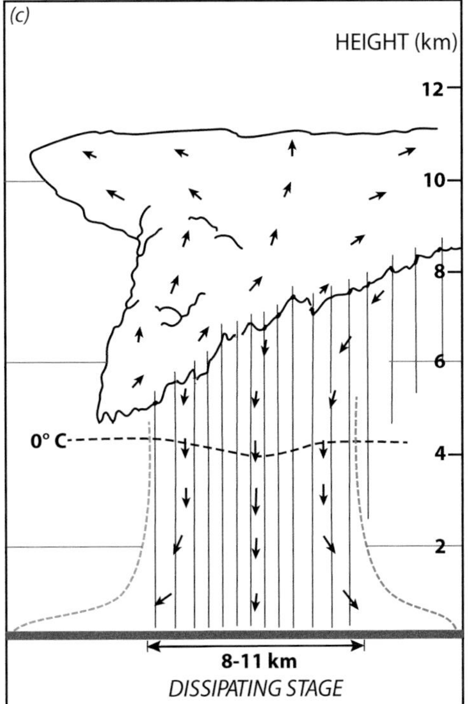

Figure 6.18 Schematic depiction of unicellular convective storm evolution from Byers and Braham (1949), as modified by Doswell (1985).

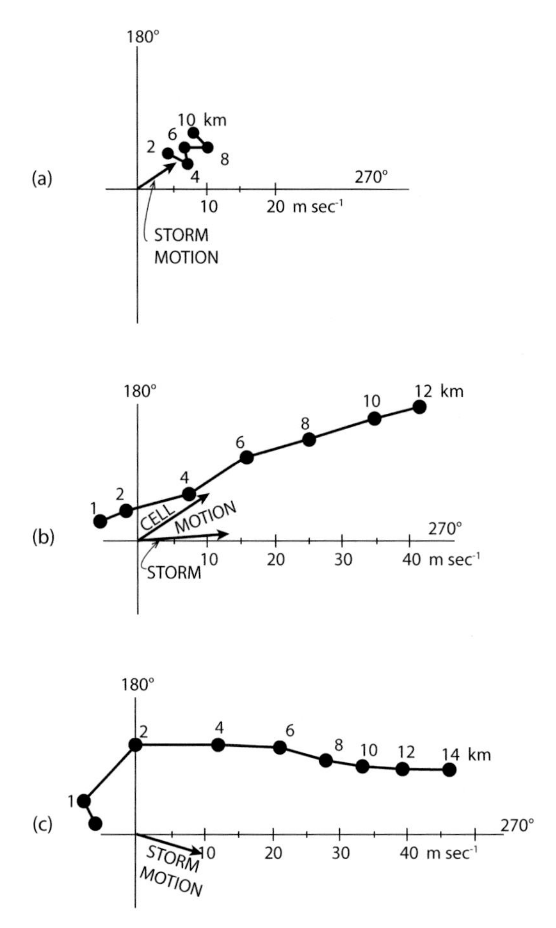

Figure 6.19 Composite hodographs constructed from observations of the environments of hailstorms near Alberta, Canada. (a) Single cell, (b) multicell, and (c) supercell storms. From Chisholm and Rennick (1972).

6.5 Multicellular Convective Storms

Organized multicellular storms literally are composed of a number of cells, each at different stages of development. In this regard, an MCS is a multicell storm, albeit one that tends to be large, long-lived, and associated with some unique dynamical properties. Hence, this section will serve as an introduction of sorts to Chapter 8.

The schematic representation of a multicell storm in Figure 6.20 shows a mature cell with a fully developed precipitating downdraft, decaying cells on its upshear flank, and growing cells on its downshear flank. Figure 6.20 also illustrates that this type of convection is self-sustaining, with downshear-cell initiation aided by lifting at the leading edge of the aggregate outflow of the multiple cells. Consequently, this mode of convection is longer lived than unicellular convection.

Key to storm sustenance is existence of intermediate environmental wind shear, $5 \lesssim \mathrm{S06} \lesssim 15\ \mathrm{m\ s^{-1}}$ (see also Figure 6.19b) and sufficiently large CAPE.[53] As with

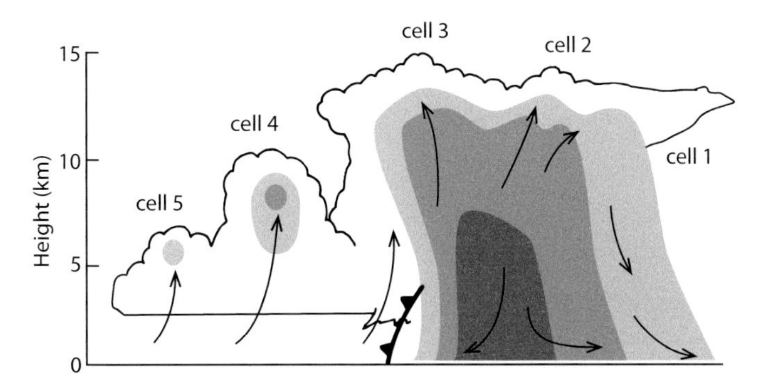

Figure 6.20 Schematic of a multicellular storm, adapted from Doswell (1985). Gray shading represents radar reflectivity factor. Environmental vertical wind shear vector points from right to left.

unicellular storms, these are not strict thresholds.[54] In fact, a *combination* of these two parameters relates better to the convective mode than either of them individually. Consider the *bulk Richardson number* (BRN),

$$\mathrm{BRN} = \frac{\mathrm{CAPE}}{(1/2)\left[(\Delta U)^2 + (\Delta V)^2\right]}. \tag{6.36}$$

Like Ri, BRN is a ratio of buoyancy to inertia forces, but alternatively can be viewed as a ratio of potential to kinetic energies of the environment. In the denominator, ΔU and ΔV are some quantification of the components of mean vertical wind shear over a deep layer.[55] BRN $\gtrsim$ 35 supports multicellular convection, although not unambiguously. Other factors, such as directional shear of the environmental winds, synoptic-scale forcing (QG vertical motions, frontal circulations), mesoscale boundaries, and the like also play a role in convective-storm organization.

Multicellular convective storms exhibit *discrete propagation*, or an apparent storm motion resulting from new cell growth on a storm flank.[56] This is separate from the motion of individual cells owing (initially) to advection, which tends to be in the direction of the mean environmental wind over the cloud-bearing layer of the troposphere (e.g., 850–300 hPa; see Chapter 8). Multicellular storm motion thus depends jointly on the individual cell motion and the location of new cell generation (see Figure 6.21). Such preferred generation is in the vicinity of the strongest (and deepest) lifting associated with the gust front, which depends in turn on the low-level, environmental winds. Two cases are portrayed in Figure 6.21: (a) new cell generation on the storm's right flank, with the low-level winds perpendicular to the cloud-bearing layer wind, and (b) new cell generation on the storm's forward flank, with the low-level winds parallel to but opposite the mean wind of the cloud-bearing layer.[57] Implicitly, the storm motion also depends on the gust-front motion itself: if the gust-front motion becomes much faster than the cell motion (as might be caused by a change in the thermodynamic environment), then the lifting and new-cell

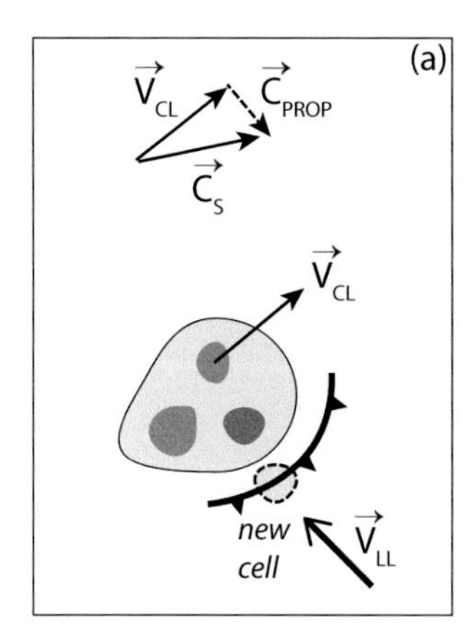

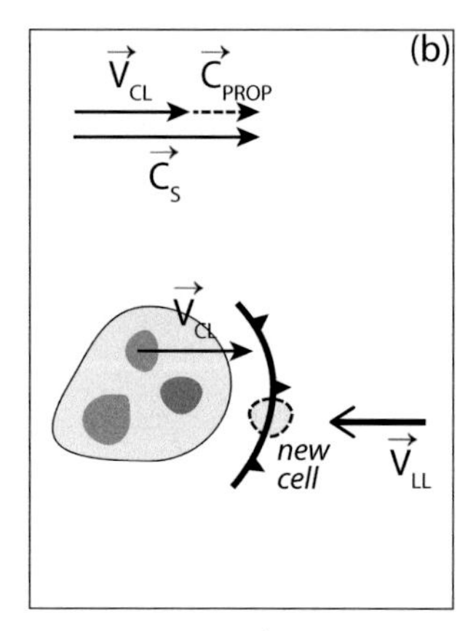

Figure 6.21 Motion of a multicellular convective storm $\vec{C}_S$ in the context of the motion and growth of the individual cells comprising the storm. Shadings indicate radar echoes. Here, $\vec{V}_{CL}$ is the cloud-bearing environmental wind vector, and affects the cell motion. The location of new cell generation affects the propagation $\vec{C}_{\text{PROP}}$, and depends in part on the low-level environmental winds $\vec{V}_{LL}$ relative to the gust-front motion. Storm motion $\vec{C}_S$ is the vector sum of $\vec{V}_{CL}$ and $\vec{C}_{\text{PROP}}$. Adapted from Houze (1993).

generation occurs progressively farther downshear from the existing cells, and discrete propagation eventually ceases.

A keen examination of a multicellular storm event would reveal an episodic growth and decay of cells. One explanation for this behavior is that as new cells become detached from the gust front and are swept rearward, the vertical circulation they induce effectively isolates them. A cell's circulation not only suppresses nearby convective motions but also disrupts the environmental inflow, thus hastening the cell's demise.[58] Idealized simulations show that this behavior is a function of environmental shear: it is realized in simple and more complex ways for moderate and strong shears, respectively, but tends to occur such that the intercell period is ~13 to 15 minutes.[59]

Supplementary Information

For exercises, problem sets, and suggested case studies, please see www.cambridge.org/trapp/chapter6.

Notes

1 For the purposes of this discussion, midlatitudes are those between 30° and 60° N (or S).
2 Lucas et al. (1994).
3 Bluestein et al. (1988), Davies-Jones (1974).
4 Musil et al. (1986).
5 Rogers and Yau (1989).
6 This statement comes with the caveat that the narrow spike in Figure 6.3, where the measured LWC exceeded the theoretical threshold of ALWC, is questionable. This exceedance could be attributed

to measurement errors in LWC, or to the use of unrepresentative cloud-base conditions for the ALWC calculations.

7 Zipser (2003).
8 May and Rajopadhyaya (1999), Lucas et al. (1994).
9 In this study (Lucas et al. 1994), a "core" is defined in flight-level data segments, when $w > 1$ m s^{-1} over at least 500 m distance.
10 Lucas et al. (1994).
11 A thorough discussion of these three flows can be found in Houze (1993), as can a review of the scientific debate regarding their application to atmospheric convection.
12 See Blyth et al. (1988); the idea of a thermal and associated toroidal circulation comes from Scorer and Ludlam (1953).
13 This follows from Houze (1993), but is based in part on the model developed by Malkus and Scorer (1955).
14 Houze (1993).
15 Ibid.
16 Warner (1970).
17 Blyth et al. (2005).
18 Blyth et al. (2005), Damiani et al. (2006).
19 Damiani et al. (2006).
20 As also has been conceptualized by Doswell (2001).
21 Caution is necessary in part because this interpretation is based on one cross section, at one time.
22 Hess (1959); see also Scorer and Ludlam (1953), who argue that new growth should be favored on the upshear side of an old thermal.
23 See also the discussion by Houze (1993), and Yuter and Houze (1995).
24 Zipser (2003).
25 This is based on the model developed by Srivastava (1985).
26 For example, for a radius interval 0.6 mm $< r <$ 2 mm, $C_1 = 2.01 \times 10^3$ cm$^{1/2}$ s^{-1}; see Rogers and Yau (1989).
27 Subsequent experiments by Srivastava (1985) using a height-invariant environmental mixing ratio, which better represents a well-mixed boundary layer, gave the same qualitative results as to be described next, but with reduced downdraft magnitudes.
28 Srivastava (1987), Proctor (1989).
29 Proctor (1989).
30 Hjelmfelt et al. (1989).
31 See Wakimoto (2001), particularly in reference to the Thunderstorm Project results.
32 See Doswell (2001), also in reference to the Thunderstorm Project results.
33 As with updrafts, a definition of a downdraft "core" is one with (negative) vertical motions exceeding -1 m s^{-1}.
34 Lucas et al. (1994).
35 Klemp and Wilhelmson (1978).
36 Doswell (2001) has suggested that precipitating downdrafts may behave as plumes, although thermal behavior is also apparent.
37 Yuter and Houze (1995), Heymsfield et al. (2010).
38 Note that the dynamic as well as buoyancy pressure effects were excluded from the 1D Srivastava model in part because of the exclusion of the PGF term in (6.27), but also because the lower boundary was considered to be open.
39 Proctor (1989).
40 Mahoney (1988), Goff (1976).
41 Droegemeier and Wilhelmson (1987).
42 Ibid.
43 Drazin (2002).
44 Proctor (1989).
45 Härtel et al. (2000).
46 Lee and Wilhelmson (1997).
47 Ross et al. (2004).
48 Tompkins (2001).
49 Khairoutdinov and Randall (2006), Tompkins (2001).

50 See Byers and Braham (1949), with parenthetical modifications proposed by Doswell (2001).
51 This is appropriate for the midlatitude convection; for tropical oceanic convection, the typical scales would be less.
52 Chisholm and Renick (1972), Weisman and Klemp (1982).
53 Chisholm and Renick (1972),Weisman and Klemp (1982).
54 Indeed, especially when based on numerical modeling studies, these values can be modulated by other model aspects, such as the treatment of the lower boundary condition, and the types of microphysical and turbulence parameterizations (e.g., Bryan et al. 2006).
55 In Weisman and Klemp (1982), for example, ΔU is the x-component difference between the mean, density-weighted, 0–6-km layer wind and surface-layer wind.
56 Marwitz (1972).
57 Houze (1993).
58 Fovell and Tan (1998).
59 Fovell and Dailey (1995).

7

Supercells

A Special Class of Long-Lived Rotating Convective Storms

Synopsis: This chapter is devoted to an in-depth discussion of the class of thunderstorms known as supercells. A hallmark of a supercell is a long-lived rotating updraft. Such mesocyclonic rotation influences the supercell dynamics, and in turn the supercell intensity, longevity, motion, and structure. Rotation generated near the ground and then become concentrated to form a tornado. A description of the sequence of processes leading to tornadogenesis is included in this chapter, as is a summary of parameters that help quantify the supercell environment. Some concluding remarks are made regarding tropical convective phenomena that share some of the same characteristics as supercells.

7.1 Characteristics of Supercell Thunderstorms: An Overview

Observations with weather radar from the 1950s through the 1970s revealed the existence of single large thunderstorms that persisted for periods of several hours, a time much longer than thought typical.[1] These intense storms also were observed to have an atypical movement – in directions to the right and/or left of, rather than parallel to, the cloud-bearing environmental winds. Finally, the storms were distinguished on radar by a persistent, internal circulation that coupled a primary updraft with downdrafts and rotation about a vertical axis.

This special class or mode of buoyant convection has become known as *supercellular* convection, borrowing again from the biological sciences: in addition to their characteristic longevity, deviant motion, and intensity, these cells undergo division (or *splitting*) and have an apparent autonomy. Such characteristics are largely attributable to the unique supercell dynamics, which are strongly controlled by the winds, moisture, and temperature in the supercell environment.

The archetypal supercell (Figure 7.1) is more intense and larger in scale than an ordinary convective storm, but has common components (see Chapter 6). Consider the updraft, which in the biological-sciences lexicon represents the nucleus of the

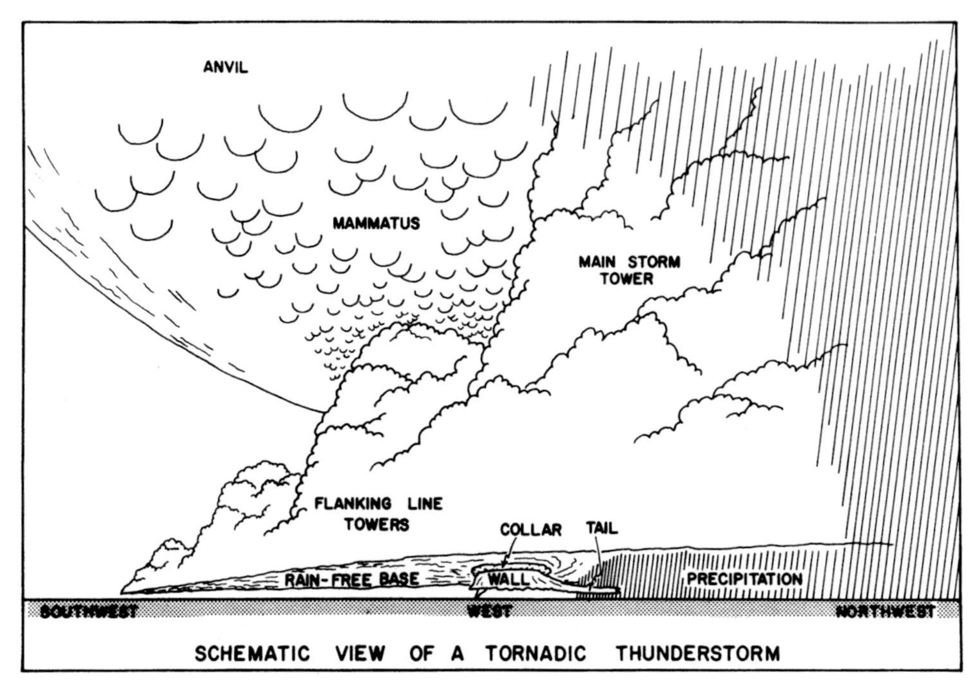

Figure 7.1 Schematic depiction of visual aspects of a tornadic supercell thunderstorm. From Doswell (1985).

supercell. The updraft extends through the troposphere but leans slightly in the direction opposite to that of the deep-layer vertical wind shear vector. At the statically stable tropopause, the supercell flow diverges horizontally, albeit asymmetrically, with a bias in the direction of the environmental upper-level winds. Particularly intense elements of the updraft may overshoot the tropopause, as shown rather strikingly in high-rate visible satellite images.[2]

As in ordinary convective storms, precipitation in supercells is generated in rising air and then lofted high into the storm. Strong storm-relative winds at middle and upper levels subsequently help evacuate the supercell updraft of lofted hydrometeors, so that they fall outside the updraft, rather than through it (Figure 7.2a). Hence, as visualized in horizontal and vertical radar scans, the updraft core is a region of relatively weak radar reflectivity, from which characterizations such as *weak echo region*, *bounded weak echo region*, and *weak echo vault* arise (Figure 7.2b). The falling precipitation contributes to downdrafts, especially forward and rearward of the updraft (Figure 7.3). The *forward-flank downdraft* (FFD) and *rear-flank downdraft* (RFD) benefit from potentially cool environmental air that enters the storm at middle levels and then aids evaporation and sublimation. Well-defined interfaces – or gust fronts – exist between the cool outflow air of these downdrafts and the warmer environmental air.

The supercell component not found in ordinary convective storms is the *mesocyclone*. The typical diameter of this cyclonically rotating vertical vortex is $\sim$5 to

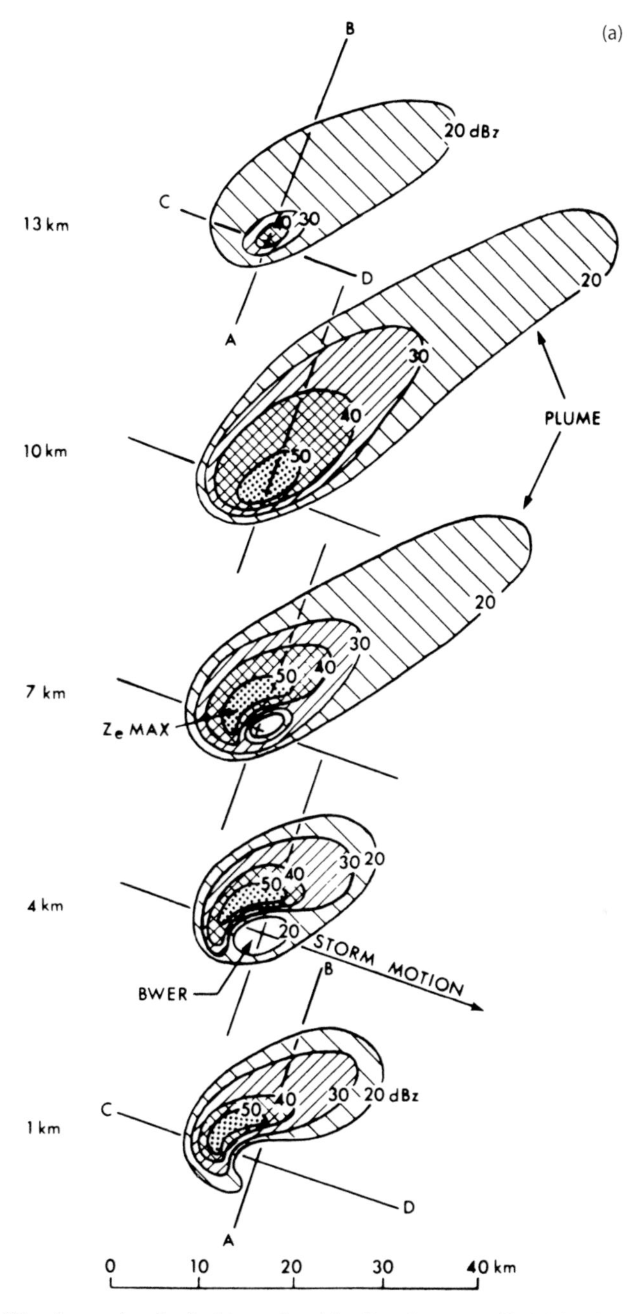

Figure 7.2 Weather radar depictions of an idealized supercell: (a) horizontal sections of radar reflectivity factor (hatched and stippled; values indicated) at various heights, and (b) vertical sections of reflectivity showing the storm updraft region and associated echo-free vault. Cross sections in (b) are indicated in (a). From Chisholm and Renick (1972).

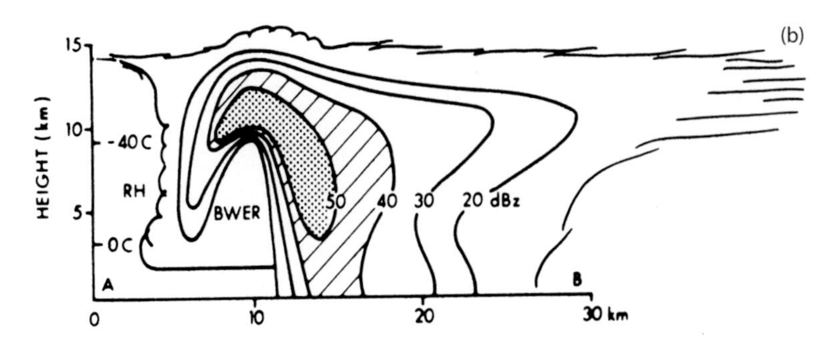

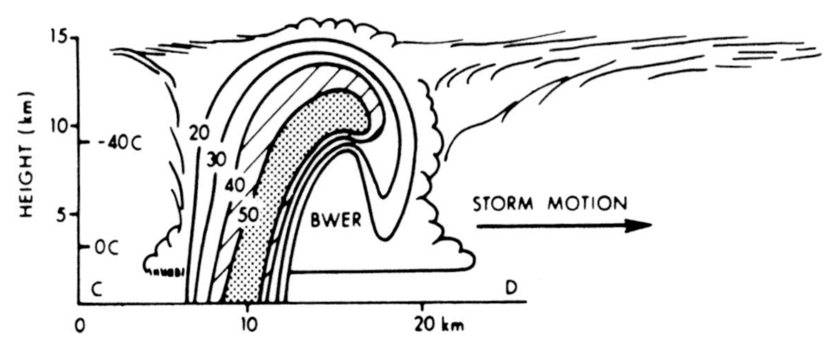

Figure 7.2 (*continued*).

6 km, although diameters much greater than 10 km are possible on rare occasions.[3,4] Tangential windspeeds of several tens of meters per second are typical, as is a vertical vorticity of $\sim 0.01\ s^{-1}$; the latter is often used as a nominal threshold for mesocyclone identification in numerical simulations. The mesocyclone does not exist in isolation, but rather is part of a coupled, 3D flow. For instance, in the middle levels of a mature supercell, the basic 3D flow is realized as a rotating updraft; at lower levels, the flow is more complicated, with rotating air that rises as well as sinks, thus forming a "divided" mesocyclone.[5] The formation and evolution of the mesocyclone are linked to the other supercell components. Figure 7.4 reveals some basic characteristics of this time evolution, and lends support to the distinction drawn hereinafter between (1) a *low-level mesocyclone*, or one extending from near the ground through ~1 km above the ground, and (2) a *midlevel mesocyclone*, or one that may span the remainder of the troposphere but tends to be centered about the middle troposphere (~5–6 km AGL). The distinction is primarily in terms of the formative mechanisms, as is demonstrated in the sections that follow.

The preceding text describes an archetype. However, as argued in Chapter 6, the physical forms assumed by real atmospheric convection span a continuous spectrum. It is common to find convective storms that possess one or more supercell-like features but that are not governed by the dynamics developed in Section 7.3. Indeed, from the approach taken herein, the litmus test of whether a storm is a supercell is the degree

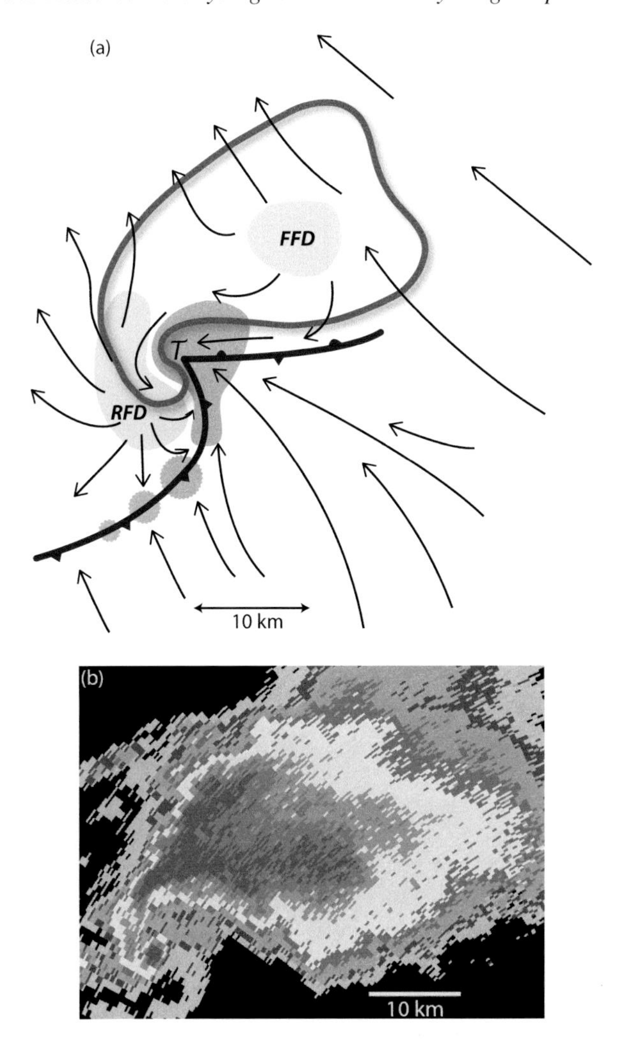

Figure 7.3 (a) Plan-view schematic of a supercell thunderstorm at low levels. Light (heavy) stippling shows regions of updrafts (downdrafts). Streamlines are of (storm-relative) near-surface flow. Bold contour shows ~40 dBZ contour. Adapted from Lemon and Doswell (1979). (b) Weather radar image of an actual (tornadic) supercell thunderstorm. (For a color version of this figure, please see the color plate section.)

to which its mesocyclonic rotation influences its dynamics, and in turn its intensity, longevity, and motion.

7.2 Midlevel Mesocyclogenesis in an Early-Stage Supercell

Our goal in this section is to explain the initial generation of vertical vorticity during the early stages of supercell development. Nominally, "early-stage" refers to the first ~30 min of evolution following the initiation of deep convection; *mesocyclogenesis* during this period occurs within the middle levels of the storm (Figure 7.4). As alluded to in Chapter 6, the rapid growth of a supercell and its components is enabled by an

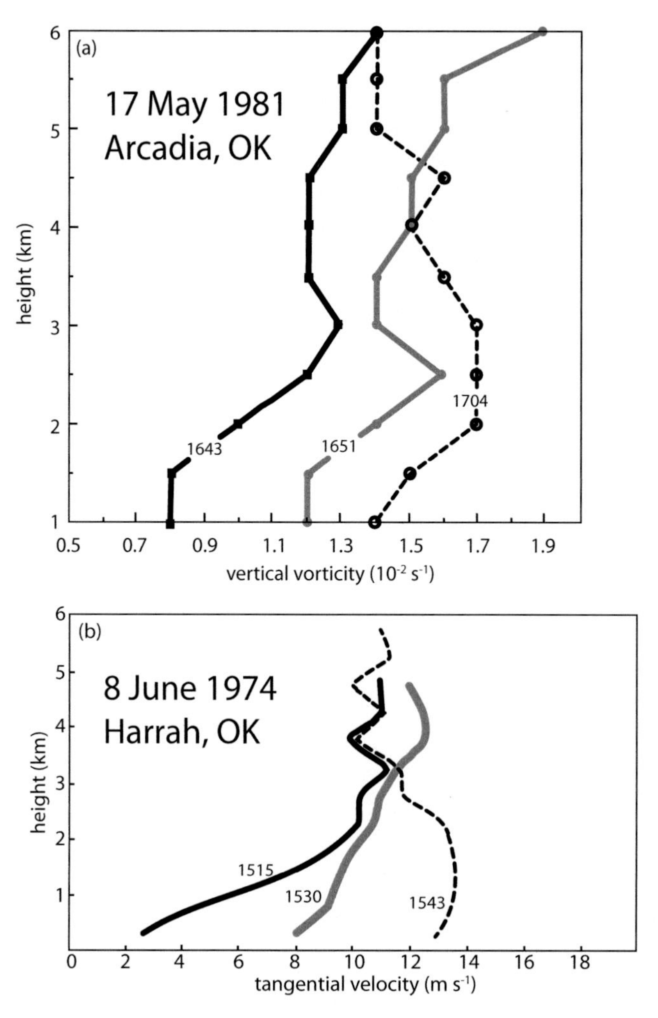

Figure 7.4 Vertical profiles of mesocyclone-associated (a) vertical vorticity, in the Arcadia, OK tornadic supercell of May 17, 1981 (from Dowell and Bluestein 1997), and (b) tangential velocity, in the Harrah, OK tornadic supercell of June 8, 1974 (from Bandes 1978). The Arcadia tornado began at approximately 1700 CST, and the Harrah tornado began at approximately 1546 CST.

environment characterized by large vertical wind shear over a deep tropospheric layer (see Section 7.7). Environmental vertical wind shear affects the supercell in a number of interrelated ways, but let us first consider its role in midlevel mesocyclogenesis.

We begin with a form of the vector equation of motion introduced in Chapter 2,

$$\frac{\partial \vec{V}}{\partial t} + \vec{V} \cdot \nabla \vec{V} = -\frac{1}{\overline{\rho}} \nabla p' + B\hat{k}, \tag{7.1}$$

where friction and Coriolis forcing are neglected, $\overline{\rho} = \overline{\rho}(z)$, p' is perturbation pressure relative to a hydrostatic base state, and B is buoyancy. An equation for the rate of change of the vertical component of vorticity ζ is obtained by taking the vertical component of the curl of (7.1),[6] yielding:

$$\frac{D\zeta}{Dt} = \vec{\omega}_H \cdot \nabla w - \zeta \nabla \cdot \vec{V}_H, \tag{7.2}$$

where $\vec{\omega}$ is the vorticity vector, and subscript H signifies a vector with horizontal components only; for instance,

$$\vec{\omega}_H = \left(\frac{\partial w}{\partial y} - \frac{\partial v}{\partial z}\right)\hat{i} + \left(\frac{\partial u}{\partial z} - \frac{\partial w}{\partial x}\right)\hat{j}. \tag{7.3}$$

The two right-hand terms of (7.2) are known as the "tilting" and "stretching" terms, respectively.

Equation (7.2) can be made more tractable by decomposing the velocity as the sum of a base state and an initially small perturbation relative to that base state:

$$\begin{aligned} u &= \overline{U}(z) + u' \\ v &= \overline{V}(z) + v' \\ w &= w'. \end{aligned} \tag{7.4}$$

Here, the perturbations represent the storm, and the base state is the horizontally homogeneous yet vertically varying environment. After substituting (7.4) into (7.2), and then assuming that products of perturbation quantities are negligibly small relative to other terms in the equation, a *linearized* version of the vertical vorticity equation can be written

$$\begin{aligned} \frac{\overline{D}\zeta'}{Dt} &= \frac{\partial \zeta'}{\partial t} + \overline{U}\frac{\partial \zeta'}{\partial x} + \overline{V}\frac{\partial \zeta'}{\partial y} = -\frac{d\overline{V}}{dz}\frac{\partial w'}{\partial x} + \frac{d\overline{U}}{dz}\frac{\partial w'}{\partial y} \\ &= \vec{\Omega}_H \cdot \nabla w', \end{aligned} \tag{7.5}$$

where $\zeta' = \partial v'/\partial x - \partial u'/\partial y$, and $\vec{\Omega}_H$ is the environmental horizontal vorticity vector. Our linearization procedure removed the stretching term, which governs the exponential and thus highly nonlinear process of vorticity amplification (see Section 7.6); this process is most relevant once vertical vorticity exists, and therefore its neglect in an examination of mesocyclone initiation is acceptable. What is retained in (7.5) is the process of tilting, expressed now as a function of the environmental vertical wind shear and the storm's vertical motion field.

Consider the canonical case of a westerly, unidirectional environmental wind profile, and assume that the perturbation vertical motion is in the form of a circular symmetrical updraft. Equation (7.5) becomes

$$\frac{\overline{D}\zeta'}{Dt} = \frac{d\overline{U}}{dz}\frac{\partial w'}{\partial y} \tag{7.6}$$

and shows, in this case, that the rate change of vertical vorticity is positive (negative) on the southern (northern) flank of the updraft core at some level. Looking downstream,

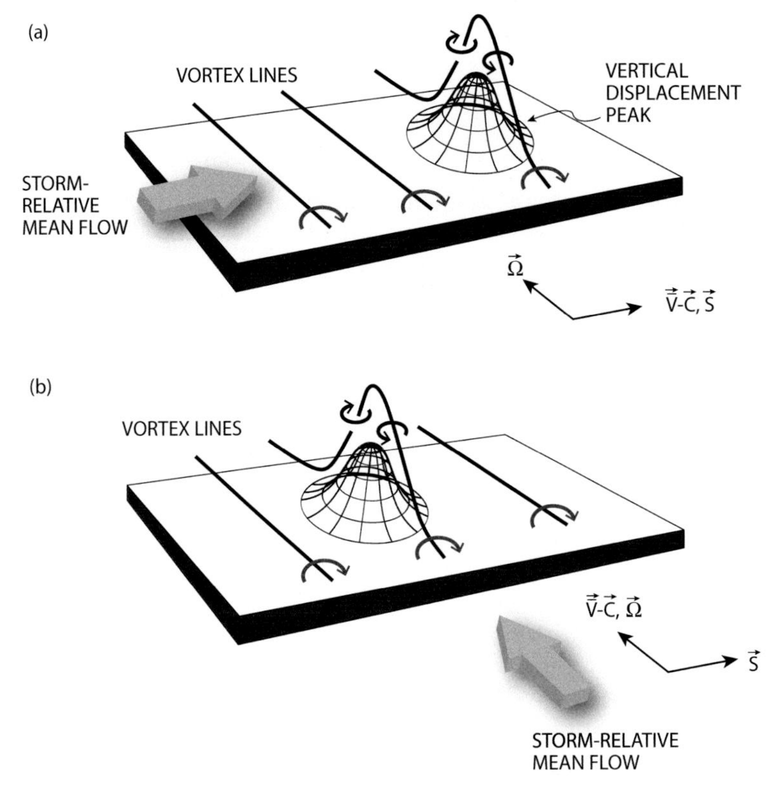

Figure 7.5 Characterization of two contrasting orientations of the storm-relative wind vector with respect to the environmental horizontal vorticity vector: (a) *crosswise* horizontal vorticity and (b) *streamwise* horizontal vorticity. $\vec{V}$ is the wind vector, $\vec{C}$ is the storm-motion vector, $\vec{\Omega}$ is the (horizontal) vorticity vector, and $\vec{S}$ is the wind shear vector. (1984).

this implies generation of cyclonic (anticyclonic) vertical vorticity on the right (left) flank of a Northern Hemispheric updraft (e.g., Figure 7.5a). The time-integrated contribution of such generation then leads to the existence of a symmetrical vertical-vortex couplet, as is routinely observed and numerically simulated.

During the early stage of supercell development, the average rotation (and net circulation) about the midlevel updraft is zero in the general case of unidirectional environmental vertical wind shear. Cyclonic updraft rotation awaits the influence of nonlinear supercell dynamics, as will be described in Section 7.3. Relevant to this description is the "streamwiseness" of the vorticity vector, or how the wind vector – and, more specifically, the *storm-relative* wind vector ($\vec{V} - \vec{C}$) – aligns with the environmental horizontal vorticity vector. Here we make use of the fact that the environmental shear vector,

$$\vec{S} = \frac{d\overline{U}}{dz}\hat{i} + \frac{d\overline{V}}{dz}\hat{j}, \tag{7.7}$$

is related to the environmental horizontal vorticity vector as

$$\vec{\boldsymbol{\Omega}}_H = \hat{k} \times \vec{S}. \tag{7.8}$$

In our current example of a westerly, unidirectional environmental wind profile, $\vec{\boldsymbol{\Omega}}_H = d\overline{U}/dz\hat{j}$. Because the initial storm-relative motion is westerly (owing to advection), it is straightforward to show that

$$(\vec{V} - \vec{C}) \cdot \vec{\boldsymbol{\Omega}}_H = 0. \tag{7.9}$$

In words, the storm-relative wind vector is initially *perpendicular* to the environmental horizontal vorticity vector, which thus defines *crosswise horizontal vorticity* (Figure 7.5a). The contrasting case is one in which the storm-relative wind vector is *parallel* to the environmental horizontal vorticity vector, thus defining *streamwise horizontal vorticity* (Figure 7.5b). The latter is exemplified by a storm with relative motion toward the south, but still in an environment of westerly wind shear. The development of a cyclonically rotating updraft in this case also depends critically on the supercell dynamics, as we now examine.

7.3 Supercell Dynamics

The essence of the supercell dynamics is revealed through the vertical equation of motion. Analysis of this equation addresses the following question: What induces changes to the supercell updraft in space and time? The answer must be contained in the buoyancy and vertical pressure gradient forces (VPGFs), assuming that the forces due to friction and Coriolis effects are comparatively small. The VPGF itself has a dual contribution because dynamical as well as (nonhydrostatic) thermodynamical effects can influence pressure. Indeed, recall the decomposition of perturbation pressure introduced in Chapter 6:

$$p' = p'_B + p'_D. \tag{7.10}$$

Using (7.10) in the vertical equation of motion gives

$$\frac{Dw}{Dt} = -\frac{1}{\overline{\rho}}\frac{\partial p'_D}{\partial z} - \left(\frac{1}{\overline{\rho}}\frac{\partial p'_B}{\partial z} - B\right), \tag{7.11}$$

where $-(1/\overline{\rho})(\partial p'_D/\partial z)$ is regarded as the dynamic pressure forcing, and $-(1/\overline{\rho})(\partial p'_B/\partial z - B)$, as the buoyancy pressure forcing.

Expressions for p'_B and p'_D follow from

$$\nabla^2 p' = \frac{\partial(\overline{\rho}B)}{\partial z} - \nabla \cdot (\overline{\rho}\vec{V} \cdot \nabla\vec{V}), \tag{7.12}$$

which, as discussed in Chapter 6, derives from the vector equation of motion and the anelastic mass continuity equation. We have shown previously that the first right-hand term of (7.12) contributes to the buoyancy pressure:

$$\nabla^2 p'_B = \frac{\partial(\overline{\rho}B)}{\partial z}. \tag{7.13}$$

On expanding the second right-hand term of (7.12), rearranging, and then applying the anelastic continuity equation, we can write a diagnostic equation for dynamic pressure as

$$\nabla^2 p'_D = -\overline{\rho}\left[\left(\frac{\partial u}{\partial x}\right)^2 + \left(\frac{\partial v}{\partial y}\right)^2 + \left(\frac{\partial 2}{\partial z}\right)^2 - \frac{d^2 \ln \overline{\rho}}{dz^2} w^2\right] - 2\overline{\rho}\left[\frac{\partial v}{\partial x}\frac{\partial u}{\partial y} + \frac{\partial w}{\partial x}\frac{\partial u}{\partial z} + \frac{\partial w}{\partial y}\frac{\partial v}{\partial z}\right], \tag{7.14}$$

or alternatively as

$$\nabla^2 p'_D = -\overline{\rho}\left(d_{i,j}d_{i,j} - \frac{d^2 \ln \overline{\rho}}{dz^2} w^2\right) + \overline{\rho}\left(\frac{|\vec{\omega}|^2}{2}\right), \tag{7.15}$$

where $d_{i,j} = \left(\partial u_i/\partial x_j + \partial u_j/\partial x_i\right)/2$ is the rate of deformation tensor expressed using summation notation.[7] In (7.14), the respective contributions to dynamic pressure are from fluid extension and fluid shear, which have equivalence to the "splat" and "spin" terms in (7.15).

Of particular interest in the supercell dynamics is the effect of rotation, as is represented in either the fluid shear term in (7.14) or the spin term in (7.15). To illustrate, let us assume a 2D, horizontal flow in solid-body rotation (e.g., $v = ax$, $u = -ay$, where a is some constant). Equation (7.14) or (7.15) can then be approximated as:

$$\nabla^2 p'_D \sim \zeta^2 \tag{7.16}$$

or by

$$-p'_D \sim \zeta^2, \tag{7.17}$$

where the additional approximation $\nabla^2 p' \sim -p'$ is used (see Chapter 6).[8] Equation (7.17) indicates that pressure is relatively low within a vertical vortex; because vertical vorticity is squared in (7.17), the pressure low occurs in association with cyclonic as well as anticyclonic rotation. Some vertical variation in the dynamic pressure is expected, because in a developing supercell, the vertical vorticity and hence pressure drop is initially largest in the middle troposphere (see Figure 7.4). It follows that the corresponding vertical pressure gradient and approximate dynamic pressure forcing

$$\frac{\partial p'_D}{\partial z} \sim \frac{\partial}{\partial z}(\zeta^2) \tag{7.18}$$

are positive in the lower troposphere, leading to localized vertical accelerations beneath midlevel vertical vorticity centers.

7.3.1 Case of a Straight Hodograph

Let us now apply (7.18) and attendant assumptions to the case of a unidirectional environmental wind profile. It is convenient to pause here and mention that, when plotted on a hodograph diagram, such a wind profile is represented as a straight line; following convention, we will refer to this profile and the associated curve as a *straight (line) hodograph*. Straight hodographs are not limited to unidirectional profiles, as Figure 7.6a demonstrates. A property of all straight hodographs is that the environmental *wind shear vector* $\vec{S}$ does not change direction with height. This follows from (7.7), which indicates that the wind shear vector is everywhere tangent to the hodograph curve. Because the environmental horizontal vorticity vector $\vec{\mathbf{\Omega}}_H$ is oriented normal and to the left of the shear vector (see (7.8)), it also follows that environmental horizontal vorticity vector is everywhere perpendicular to the hodograph curve. Hence, when the hodograph curve is a straight line, neither $\vec{S}$ nor $\vec{\mathbf{\Omega}}_H$ changes direction with height.

Recall from Section 7.2 that in an environment characterized by a straight hodograph, a midlevel cyclonic-anticyclonic vortex pair is initially generated in the updraft through the tilting process. Positive (rotational) dynamic forcing (Eq. (7.14); see also (7.18)) promotes vertical accelerations and subsequent new updraft growth beneath both members of the vortex pair (Figure 7.7; see also Figure 7.11). Low-level inflow is diverted to the new growth, effectively weakening the existing updraft and ultimately giving way to a precipitating downdraft. The result is a *storm split*, with development of new updrafts on the (north and south) flanks of the initial updraft. The two new updrafts are symmetrical–or mirror images–about the mean wind shear vector (Figures. 7.7, 7.8). Over time, through the tilting process described in Section 7.2, each new updraft develops its own midlevel vortex pair that induces a subsequent split. This cycle may repeat itself numerous times, with the cumulative tracks of the resultant storms resembling a tree with many branches.

Figures 7.7 and 7.8 indicate that storm splitting is but one of the effects of the rotational dynamic forcing. The other regards storm propagation: the locational shifts in the forcing and thus in new updraft growth cause left- and rightward deviations in motion relative to the motion of the initial updraft (and of the cloud-bearing environmental winds). In other words, the rotationally induced VPGFs lead to left-moving (LM) and right-moving (RM) storms.

7.3.2 Case of Curved Hodographs

The vertical wind profiles in the observed environments of cyclonically rotating (and tornadic) thunderstorms frequently have hodographs with clockwise curvature,

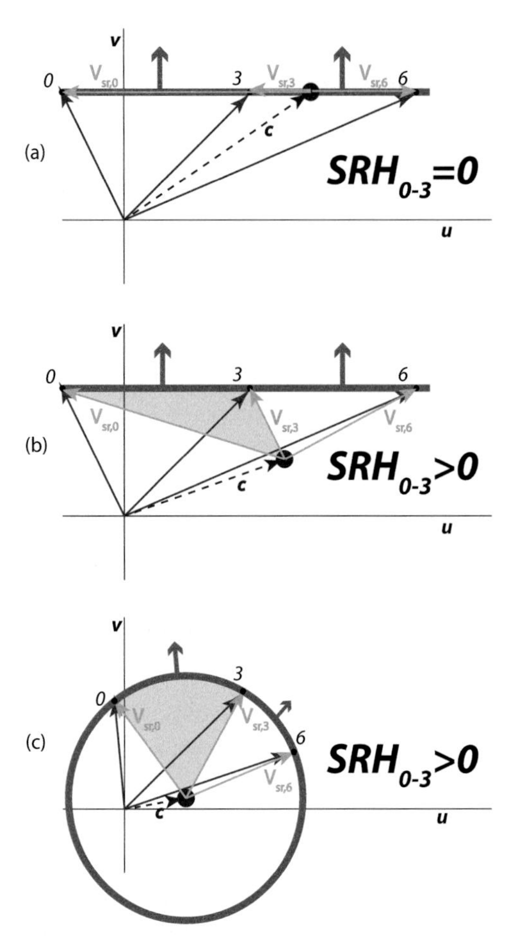

Figure 7.6 The hodograph (bold gray line) of an environmental wind (purple arrows). The environmental horizontal vorticity vector (direction indicated by red arrows) is everywhere normal to the shear vector, and hence to the hodograph curve. In (a) and (b), the hodograph shape is a straight line, and thus the environmental horizontal vorticity vector does not change with height. (a) The storm motion (blue dashed arrow) lies on the hodograph, and using this as a hodograph origin (blue circle), the storm-relative winds (orange vectors) are everywhere perpendicular to the horizontal vorticity vector. In this case, $SRH_{0\text{–}3} = 0$. (b) The storm motion lies off the hodograph, and the resultant storm-relative winds are not perpendicular to the horizontal vorticity vector. In this case, $SRH_{0\text{–}3} > 0$, with a magnitude equal to $-2 \times$ signed area (shaded region) swept out by the storm-relative wind vector between 0 and 3 km. (c) A circular hodograph with storm motion at the circle center. In this case, storm-relative winds (orange vectors) are everywhere parallel to the horizontal vorticity vector, thus maximizing $SRH_{0\text{–}3} > 0$. (For a color version of this figure, please see the color plate section.)

especially over the lower troposphere (see Figures 6.19c and 7.9).[9] Following the arguments presented in Section 7.3.1, this means that both the wind shear and horizontal vorticity vectors of these environments vary directionally with height over the levels of hodograph curvature. It is logical to ask whether there is a dynamical

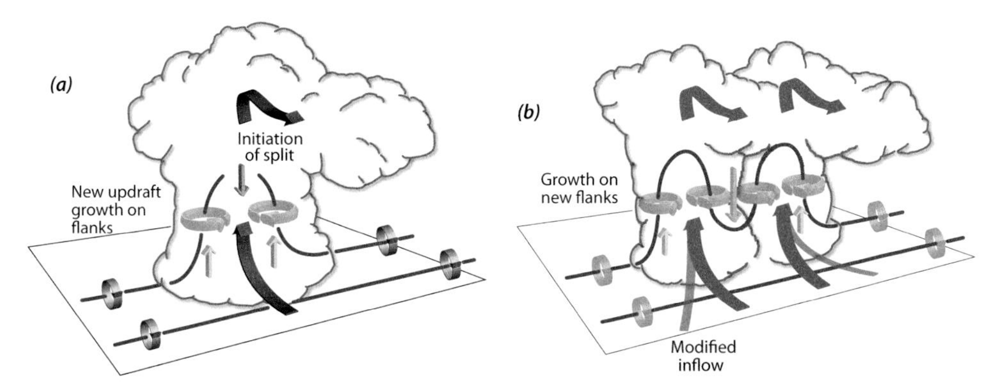

Figure 7.7 Schematic showing the location and effects of the dynamically (rotationally) induced vertical pressure gradient forcing on supercell evolution. In (a), the initial, midlevel vortex pair (gray ribbons) results in positive vertical pressure gradient forcing and subsequent vertical accelerations (yellow arrows) on the flanks of the initial updraft (red ribbons). As facilitated by a precipitation downdraft (blue arrow), a split of the initial cell ensues. In (b), the cell-splitting process bears two new cells, each of which generates a new midlevel vortex pair, and is accompanied by a modified low-level inflow (red ribbons). Positive vertical pressure gradient forcing is again found beneath each of the vortices; lifting with the inner vortices is impeded by precipitating downdrafts, and thus new updraft growth is favored beneath the outer vortices. Adapted from Klemp (1987). (For a color version of this figure, please see the color plate section.)

implication to this variation; the answer is revealed in part by the following set of idealized simulations.

Consider a storm that initiates in an environment characterized by a "half-circle" hodograph (wherein a semicircle is swept out, in a clockwise sense, by environmental winds over the 0–6 km layer, for example).[10] Notice from Figure 7.10 that this storm does not split into identical RM and LM storms as in the case of a straight hodograph, but rather splits into a dominant RM storm and a much weaker and shorter-lived LM storm (see also Figure 7.8).[11] Notice also that the RM updraft tends to acquire net cyclonic rotation much earlier in its life, as compared with the RM updraft of the straight-line hodograph case. This preferential enhancement of the cyclonically rotating RM storm is less pronounced in an environment characterized by a "quarter circle" hodograph, and more pronounced in an environment characterized by a "full circle" hodograph (a complete circle is swept out, in a clockwise sense, by environmental winds over the 0–12 km layer; Figure 7.10). Enhancement does not necessarily equate to storm longevity, however: in idealized simulations, a quarter-circle hodograph supports a supercell with a relatively stronger and longer-lasting updraft, and also one that most closely matches the archetype (see Figure 7.3). Not coincidentally, this is also the environment that most closely matches the composite hodograph from observations (see Figures 6.19c and 7.9).[12]

The dynamic pressure forcing in each of these cases is maximized on the right- to right-forward flank of the RM updraft, and results in updraft growth on this flank

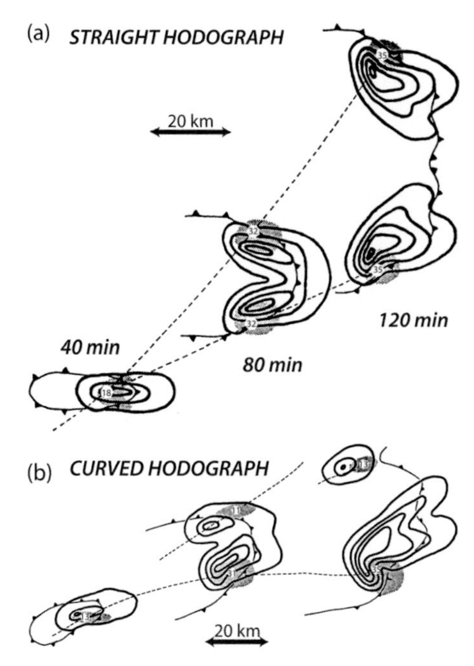

Figure 7.8 Time sequence of a convective cell that splits into (a) two mirror-image storms that evolve into cyclonically and anticyclonically rotating supercells and (b) two cells, with one evolving into a supercell. The environmental hodograph in (a) is straight and in (b) is curved. Contours are of rainwater mixing ratio at 1.8 km AGL, and the barbed line indicates the surface gust front. Shading shows regions where midlevel (4.6 km) updraft exceeds 5 m s^{-1}. From Weisman and Klemp (1986).

and in the attendant updraft propagation (Figure 7.11). Thus, as in the unidirectional shear case, the dynamic pressure forcing here is attributed largely to the rotationally induced VPGF. However, VPGFs associated with a *linear* interaction between the environmental wind shear and the updraft also contribute to the forcing. These effects, as described next in Section 7.3.3, explain well the preferential enhancement of the RM storm.

7.3.3 The Linear-Dynamic Pressure Forcing

To isolate the *linear* contribution to the decomposed pressure, we return to the vector equation of motion (7.1), linearize it about the base state velocity (7.4), and then take the divergence of the result. Omitting the buoyancy contribution yields a diagnostic pressure equation for the linear dynamics:

$$\nabla^2 p'_{LD} = -2\overline{\rho}\left[\frac{d\overline{U}}{dz}\frac{\partial w'}{dx} + \frac{d\overline{V}}{dz}\frac{\partial w'}{\partial y}\right]. \tag{7.19}$$

The linear-dynamic pressure forcing is formed by inversion of (7.19) and then by vertical differentiation of the result.[13] Physically, the forcing arises from an interaction

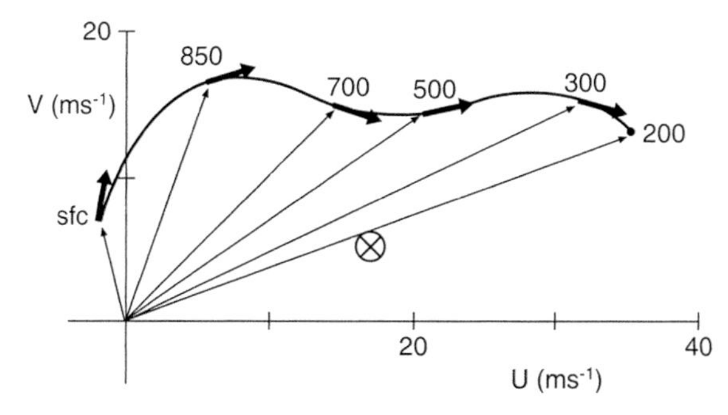

Figure 7.9 Composite hodograph based on radiosonde observations in the environments of 62 tornadic storm events. The composites are made by averaging the winds at each level relative to the estimated storm motion. Heavy arrows indicate the direction of the wind shear vector at each level (indicated in hPa). The (estimated) mean storm motion is denoted by X. From Weisman and Rotunno (2000), as adapted from Maddox (1976).

between the environmental wind shear and the updraft; the forcing and pressure are maximized on the updraft flanks. To show this more clearly, let us rewrite (7.19) as

$$\nabla^2 p'_{LD} = -2\bar{\rho}\vec{S} \cdot \nabla w' \tag{7.20}$$

or, approximately, as

$$p'_{LD} \sim \vec{S} \cdot \nabla w'. \tag{7.21}$$

If we again make the reasonable simplification that the vertical velocity distribution is circular-symmetrical about the updraft axis, (7.21) indicates that there will be a high-to-low pressure variation across the updraft, in the direction of the environmental wind-shear vector at that level. The vertical gradient of this pressure (and $\mathrm{VP_{LD}GF}$), therefore, will depend critically on the vertical change of the shear vector.

We can evaluate the $\mathrm{VP_{LD}GF}$ qualitatively by recalling that $\vec{S}$ is everywhere tangent to the hodograph curve. In the case of a straight hodograph, (7.21) requires that the linear-dynamic pressure be relatively high on the upshear flank of the updraft at all levels, and relatively low on the downshear flank at the corresponding levels (Figure 7.12a). If the magnitude of $\vec{S}$ increases with height, so also does the effect of the linear-dynamics pressure, such that descent (ascent) is forced on the upshear (downshear) flank owing to higher (lower) p'_{LD} aloft. The combined influence of upshear descent and downshear ascent aids in forward storm propagation, reinforces the storm inflow, and produces a net downshear tilt to developing updrafts, but does nothing to promote a preferential enhancement on the right or left updraft flanks.

Now consider the case represented in Figure 7.12b. The environmental wind-shear vector turns clockwise with height to form a semicircular hodograph. Such directional variation in $\vec{S}$ results in high-to-low pressure couplets that have a corresponding height variation (Figure 7.12b). The associated $\mathrm{VP_{LD}GF}$ over the lower-to-middle

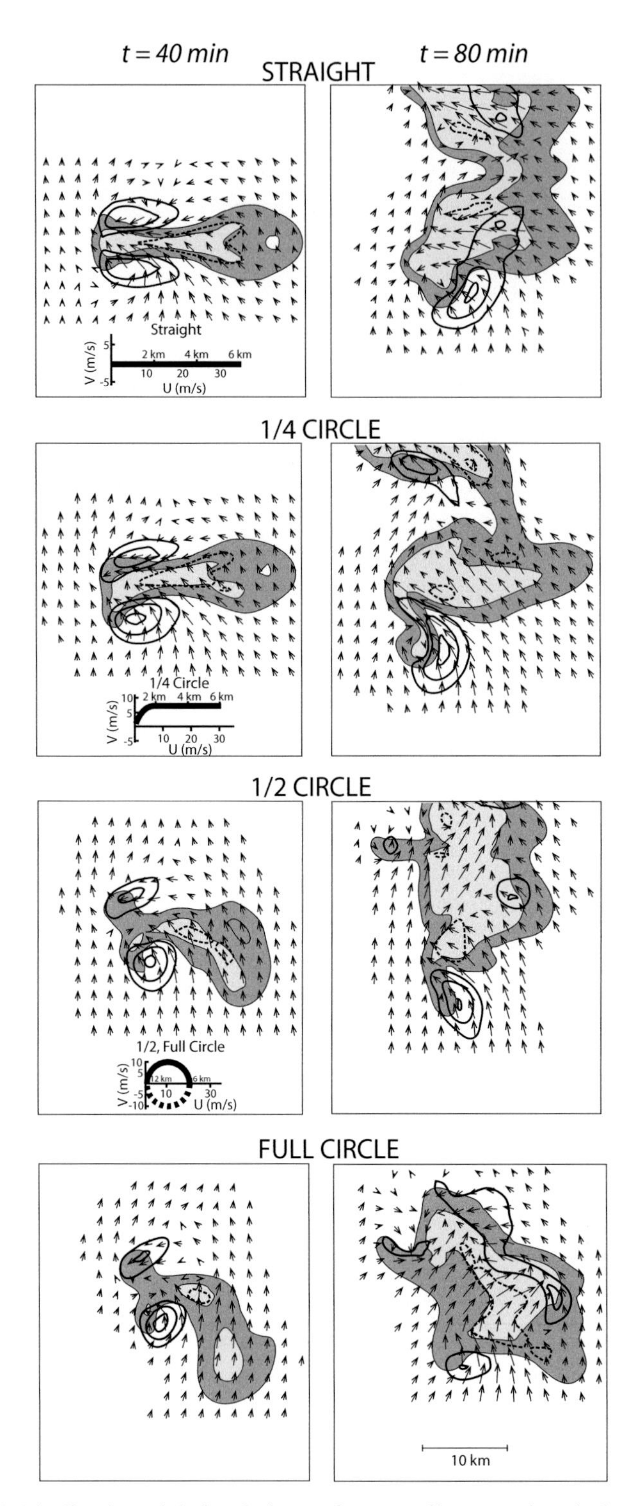

Figure 7.10 Idealized model simulations of supercell storms in wind environments characterized by straight, quarter circle, half circle, and full circle hodographs (see inset). Vertical velocity (contour interval 6 m s^{-1}, with dashed lines indicating downdrafts), rainwater mixing ratio (dark gray fill, 1–4 g kg^{-1}; light gray fill, >4 g kg^{-1}), and horizontal wind vectors show midlevel (z = 3 km) storm structure at t = 40 and 80 min. After Weisman and Rotunno (2000).

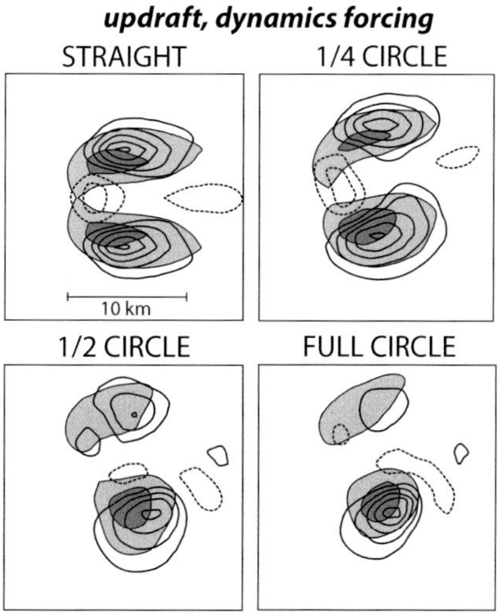

updraft, dynamics forcing

STRAIGHT 1/4 CIRCLE

10 km

1/2 CIRCLE FULL CIRCLE

Figure 7.11 Dynamic pressure forcing (see text) of vertical acceleration (contour interval 0.002 m s^{-2}), and vertical velocity (light shading, 4–14 m s^{-1}; dark shading, $>$ 14 m s^{-1}), of numerically simulated supercell storms at z = 3 km and t = 40 min. The supercells correspond to those shown in Figure 7.10, and thus are in environments characterized by straight, quarter circle, half circle, and full circle hodographs. After Weisman and Rotunno (2000).

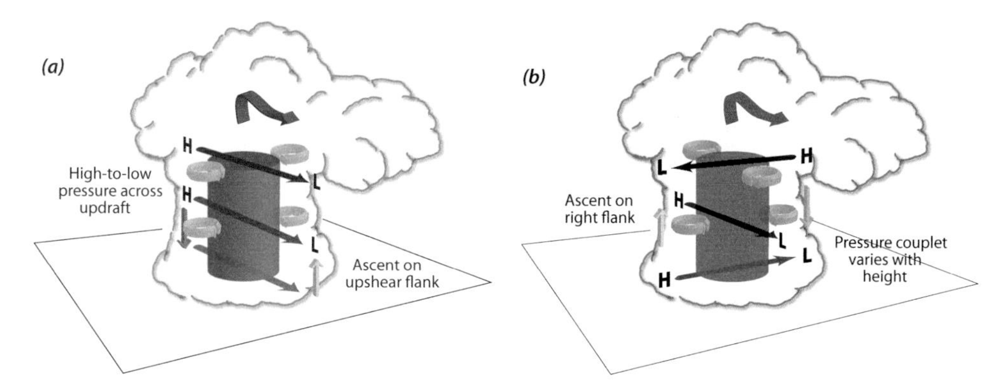

Figure 7.12 Qualitative depiction of the location and effects of the linear, dynamically induced pressure (H for high pressure, L for low pressure), as well as the accelerations responding to the associated vertical pressure gradient. Red cylinder represents the updraft, and the flat arrows show the shear vector orientation at the corresponding level. For reference, the location of vertical vortices at these levels is also shown (gray ribbons), which depict the nonlinear dynamic pressure effects. In (a), the supercell has formed in an environment with a straight hodograph. The linear-dynamics forcing results in upward acceleration and ascent (downward acceleration and descent) on the downshear (upshear) flank of the updraft. In (b), the supercell has formed in an environment with hodograph curvature (half-circle) over the lowest half of the troposphere. The linear dynamics forcing results in ascent (descent) on the right (left) flank of the updraft. Adapted from Klemp (1987). (For a color version of this figure, please see the color plate section.)

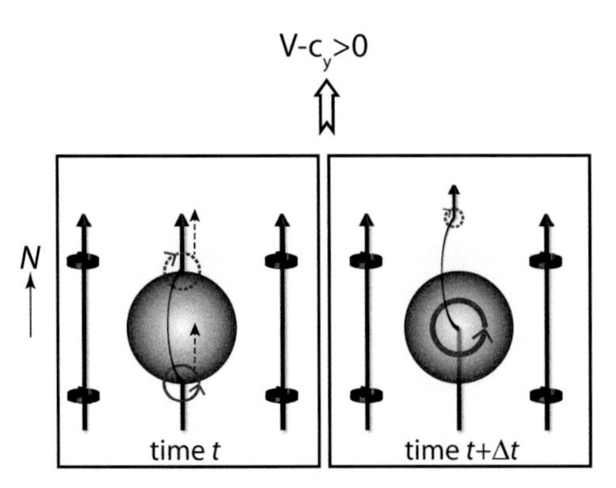

Figure 7.13 Schematic of tilting of streamwise horizontal vorticity (time t), and then subsequent advection by the storm-relative motion (time $t + \Delta t$). Bold lines are vortex lines, and gray shading represents an updraft. At $t + \Delta t$, vertical stretching is implied, such that the cyclonic vorticity is enhanced in the updraft, and the anticyclonic vorticity is diminished in the vertically motionless air. This process is assumed to apply at the height where $\overline{U} - c_x = 0$.

troposphere is positive (negative) on the right (left) flank of the updraft, thus forcing ascent (descent) on that flank.

In this curved hodograph case, the linear-dynamics pressure forcing also enhances (suppresses) the nonlinear-rotational dynamic forcing on the right (left) updraft flank. This is due to the fact that the axes of the shear-induced pressure couplets are perpendicular to the axes of tilting-generated vertical vorticity couplets. The dynamical consequence of this alignment of linear and nonlinear dynamic forcing is a preferential enhancement of cyclonically rotating RM supercells in environments possessing curved hodographs.

7.3.4 Development of Net Updraft Rotation

The eventual development of net updraft rotation, and ultimately of a mesocyclone, is common to all supercells, but the precise developmental pathway is a function of the characteristics of the environmental hodograph. Let us, once again, illustrate this with a straight hodograph, recalling that in the specific case of unidirectional westerly winds, the nonlinear dynamic pressure forcing induces a southward (northward) propagation in the RM (LM) storm (see Figure 7.7b). Over time, this induced propagation results in a component of storm-relative motion that is parallel to the environmental horizontal vorticity vector. This is the *streamwise environmental horizontal vorticity* (see Figures 7.5 and 7.13). Its importance in mesocyclone development is revealed by the linearized vertical vorticity equation (7.5) applied in a reference frame fixed to the moving storm[14]

$$(\overline{U} - c_x)\frac{\partial \zeta'}{\partial x} + (-c_y)\frac{\partial \zeta'}{\partial y} = \frac{d\overline{U}}{dz}\frac{\partial w'}{\partial y}, \tag{7.22}$$

where $\vec{C} = c_x\hat{i} + c_y\hat{j}$, and for this case, environmental wind vector $\vec{V}(z) = \overline{U}(z)\hat{i}$. For the sake of the current illustration, it is assumed in (7.22) that the (RM or LM) storm has reached a steady state and is moving at constant velocity $\vec{C}$. At the height in which $\overline{U} - c_x = 0$, which is regarded as a *critical level*,[15] (7.22) reduces to

$$-c_y \frac{\partial \zeta'}{\partial y} = \frac{d\overline{U}}{dz}\frac{\partial w'}{\partial y}. \tag{7.23}$$

Equation (7.23) represents a balance between *tilting of environmental horizontal vorticity*, and *advection by the storm-relative horizontal winds*. The outcome of the balance is clarified when (7.23) is integrated along a parcel trajectory

$$\zeta' = \frac{d\overline{U}}{dz}\frac{w'}{(-c_y)}. \tag{7.24}$$

Indeed, for the RM storm, in which $c_y < 0$, the anticyclonic vertical vorticity generated through the tilting of the streamwise horizontal vorticity is immediately advected into air with little vertical motion, while cyclonic vertical vorticity is advected into the updraft (Figure 7.13).[16] The updraft and cyclonic vorticity have become spatially correlated; the final intensification of this rotating updraft into a mesocyclone follows from vertical stretching, a nonlinear process excluded from (7.23).

We can generalize this connection between streamwise horizontal vorticity and net updraft rotation by expressing the steady form of (7.5), fixed again to the moving storm, in *natural coordinates*

$$U_s \frac{\partial \zeta'}{\partial s} = \Omega_s \frac{\partial w'}{\partial s} + \Omega_n \frac{\partial w'}{\partial n}, \tag{7.25}$$

where s denotes the direction ($\hat{s}$) everywhere tangential to the storm-relative flow, n denotes the direction ($\hat{n} = \hat{k} \times \hat{s}$) everywhere normal to the horizontal flow, $\vec{\mathbf{\Omega}}_H = \mathbf{\Omega}_s\hat{s} + \mathbf{\Omega}_n\hat{n}$, and $\vec{\mathbf{V}}(z) - \vec{C} = U_s\hat{s}$. If we assume that the environmental horizontal vorticity is purely streamwise, (7.25) can be reduced and similarly integrated to

$$\zeta' = \frac{\mathbf{\Omega}_s}{U_s} w', \tag{7.26}$$

where $\Omega_s = U_s d\overline{\Psi}/dz$, and $\overline{\Psi} = \tan^{-1}(\overline{V}/\overline{U})$ is the direction (angle) of the environmental wind, increasing counterclockwise.[17]

7.3.5 Storm-Relative Environmental Helicity

Equation (7.26) suggests a utility in being able to quantify the streamwiseness of the environmental horizontal vorticity. In this Section we introduce such a kinematic measure, and in doing so also introduce an alternative way of anticipating the development of rotation within convective updrafts.

Let us begin with *helicity*, defined[18] herein as the volume integral of the inner product between the 3D velocity and vorticity vectors:

$$\mathcal{H} = \int \hbar \, dV. \tag{7.27}$$

The quantity $\hbar$ is the *helicity density*, given by

$$\hbar = \vec{\omega} \cdot \vec{V}. \tag{7.28}$$

From (7.27) and (7.28) it should be clear that a fluid flow with large helicity is also one with large streamwise vorticity (recall Figure 7.5b). However, as implied previously, the quantitative characterization of streamwise vorticity depends on the reference frame of evaluation. For our purposes, the most physically relevant reference frame is the one fixed to a steady storm, moving at an assumed constant velocity $\vec{C}$.[19] If it is then assumed that the helical properties of storm-relative environmental winds can be treated as horizontal averages, and thus as horizontally integrated quantities over some (e.g., mesoscale) area, (7.27) reduces to a vertical integral

$$\mathrm{SRH} = \int_{z_B}^{z_T} \left[\vec{\boldsymbol{V}}(z) - \vec{C}\right] \cdot \vec{\boldsymbol{\Omega}}_H dz. \tag{7.29}$$

This is the *storm-relative environmental helicity* (SRH) (m^2 s^2), where $\vec{\boldsymbol{V}}(z) = \overline{U}\hat{\imath} + \overline{V}\hat{\jmath}$ is the environmental wind vector as before, and where integration limits z_B and z_T are, in practice, taken to be the surface level and ~3 km AGL, respectively; these limits nominally represent the layer of storm inflow.

Large (positive) SRH equates to large environmental streamwise vorticity in the storm-relative inflow, and in turn corresponds to the likelihood of a cyclonically rotating updraft (given that should an updraft occurs in this environment).[20] Tacitly assumed in an application of (7.29) to supercell prediction/diagnoses is that streamwiseness in the horizontal is converted to streamwiseness in the vertical through tilting. This follows from a conservation of helicity, although it is valid strictly in a barotropic, inviscid, and incompressible flow.[21] Notice that the basic vortex dynamics expressed through (7.2) are embodied in this use of helicity; the difference is in how the environmental vorticity is quantified.

To evaluate SRH, it is necessary to know the storm motion vector $\vec{C}$. *A posteriori*, this can be determined straightforwardly by tracking radar echoes of the observed storm over some time interval. *A priori*, storm motion can only be estimated, typically using empirical formulae based on the observed (or predicted) environmental winds. An example of a widely used formula is

$$\vec{C}_{RM} = \vec{\boldsymbol{V}}_{0-6} + \delta_s \left(\frac{\vec{S}_{0-6} \times \hat{k}}{|\vec{S}_{0-6}|} \right), \tag{7.30}$$

where, as before, RM indicates a right-moving supercell; $\vec{V}_{0-6}$ is the mean environmental wind over the 0–6-km layer; δ_s is the magnitude of the deviation of supercell motion from the mean wind, empirically determined to be 7.5 m s^{-1}; and $\vec{S}_{0,6}$ is the mean environmental shear vector over the 0–6-km layer, defined simply as $\vec{S}_{0-6} = \vec{V}_6 - \vec{V}_0$, where $\vec{V}_6$ and $\vec{V}_0$ are local mean values of the 6-km and near-surface layers, respectively.[22] The comparable formula for the motion of a left-moving supercell replaces the addition in (7.30) with subtraction. Note that because the effects of propagation (second right-hand term of (7.30)) on the estimated storm motion are based on vertical wind shear, the calculation of the *Bunkers motion*[23] is insensitive to the hodograph orientation. In other words, this method is *Galilean invariant*, as Figure 7.14 demonstrates.

SRH itself lacks Galilean invariance owing to its dependence on storm motion. To understand the significance of this property of SRH, we ask the following question: Can a straight hodograph have nonzero SRH? The answer is yes, but it is contingent on $\vec{C}$ and thus on the storm-relative environmental wind profile. We can prove this graphically by transforming the hodograph such that the terminus of $\vec{C}$ becomes the origin.[24] Notice in Figure 7.6 that the shape of the hodograph does not change with the transformation, nor does the local orientation of $\vec{\mathbf{\Omega}}_H$. In the case of storm motion *on* the straight hodograph (e.g., motion is due only to advection), $(\vec{V} - \vec{C}) \cdot \vec{\mathbf{\Omega}}_H = 0$ at all points on the hodograph, and thus SRH $= 0$. If storm motion lies *off* the straight hodograph (Figure 7.6b), which would generally be the case with supercells with a propagation component to their motion, then $(\vec{V} - \vec{C}) \cdot \vec{\mathbf{\Omega}}_H \neq 0$, and SRH $\neq 0$. The SRH magnitude increases as the storm-relative wind and horizontal vorticity vectors in the atmospheric layer of interest become more parallel. Equivalently, SRH magnitude increases with increases in area swept out on the hodograph plot by the storm-relative wind vectors in the layer (here, the integration limits; see Figure 7.6b).[25] As illustrated in Figure 7.6c, environmental wind profiles with curved hodographs have the potential for large values of SRH (see also Figure 7.9). An extreme example – one that maximizes environmental streamwise vorticity – is one of a circular hodograph with storm motion at the circle center.

The theoretical expectation from (7.29) is of a cyclonically rotating updraft in an environment of large, positive SRH. This is realized in numerically simulated convective storms, as confirmed by a strong statistical relationship between SRH and the maximum (midlevel) vertical vorticity generated in the resultant storms (Figure 7.15a), and an equally strong relationship between SRH and the degree of spatial agreement between vertical vorticity and vertical velocity (Figure 7.15b). Such spatial agreement itself is quantified as a *correlation coefficient*

$$r(w', \zeta') = \frac{\langle w'\zeta' \rangle}{(\langle w'^2 \rangle \langle \zeta'^2 \rangle)^{1/2}}, \tag{7.31}$$

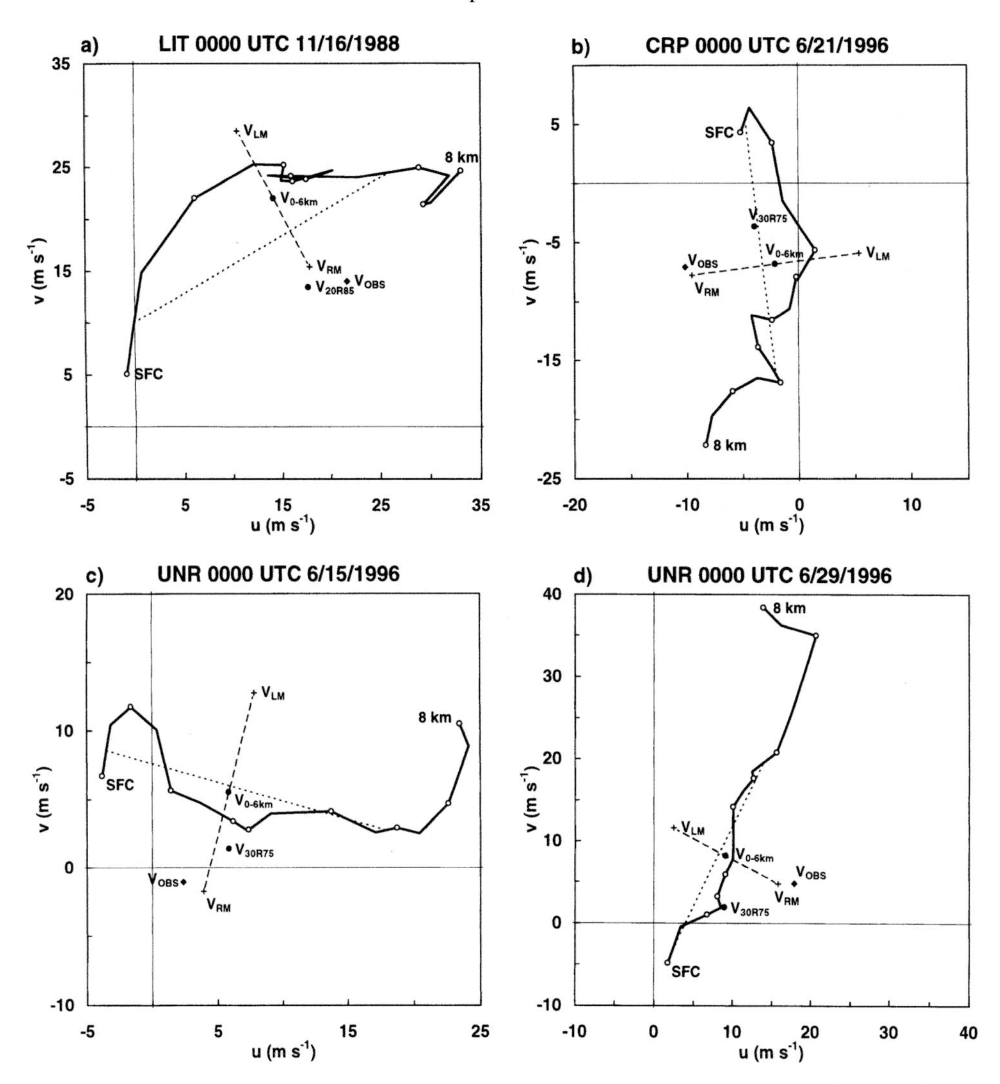

Figure 7.14 Example calculations of supercell motion according to the "Bunkers" formula. From Bunkers et al. (2000).

where $\langle\ \rangle$ represents a horizontal average and the value of r is averaged over some depth of the storm.[26] A vertically averaged correlation coefficient exceeding $\sim$0.75 is generally consistent with other characteristics used to define a storm as a supercell.[27] In fact, we will later use (7.31) as one means of discriminating mesocyclones from other mesoscale-convective vortices, such as those found in squall lines (Chapter 8).

The calculations in Figure 7.15 involve low- to mid- tropospheric winds and their effect on supercell dynamics and mesocyclogenesis. The disregard thus far of the environmental winds aloft does not imply their unimportance, however. As we will learn in the next section, the environmental winds in the mid- to upper troposphere play direct and indirect roles in the supercell morphology.

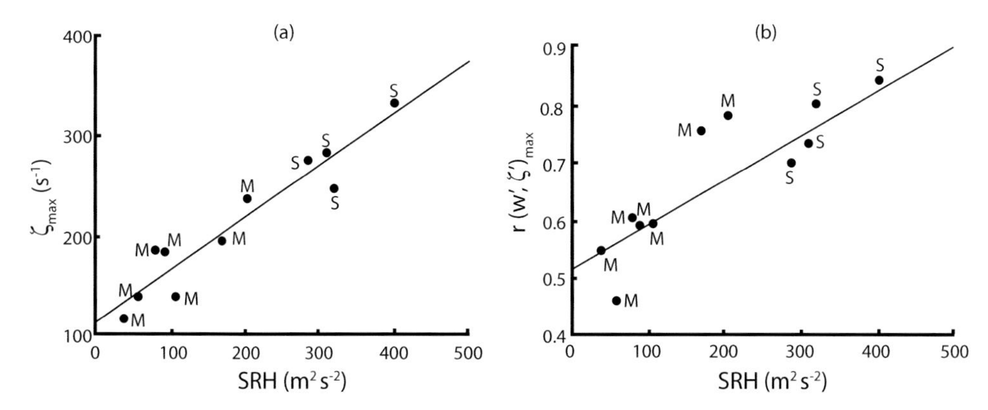

Figure 7.15 Quantification of the effects of SRH on convective storms that are numerically simulated in varying environments of SRH. (a) Scatterplot of SRH and the corresponding maximum (midlevel) vertical vorticity. (b) Scatterplot of SRH and the corresponding $r(w', \zeta')$, representing the degree of spatial agreement between vertical vorticity and vertical velocity. Symbols M and S indicate the convective mode (multicell or supercell) of the simulated storms. Straight line is linear regression fit to the data. After Droegemeier et al. (1993).

7.4 Interaction of Wind Field and Precipitation

Up to this point, the dynamical processes and characteristic structures have been described relative to an archetype, otherwise known as the "classic" supercell. Perusal of weather radar scans of mature supercells will quickly show that real storms deviate from this archetype to some extent, particularly in terms of the amount and distribution of precipitation relative to the updraft. As discussed here and then further in Section 7.5, this can be attributed largely to interactions between the precipitation and the internal/ambient airflow, and has implications on the supercell's ability to generate a sustained mesocyclone near the ground and even a tornado.

To characterize the nature of the precipitation distribution, the following supercell classes have been introduced: *low-precipitation* (LP), *classic* (CL), and *high-precipitation* (HP).[28] As with the spectrum of all convective storms, this supercell spectrum is continuous. LP supercells are associated with relatively little precipitation, although large hail can occur downwind of the updraft (Figure 7.16a). As supercells, they still exhibit storm-scale rotation, but have a relatively low probability of spawning a tornado. At the other end of this spectrum, HP supercells are associated with heavy precipitation that extends well into the rear-flank region of the updraft and occupies a significant area of the mesocyclone (Figure 7.16b). Tornadoes and other severe weather hazards are more probable with HP supercells, which additionally produce torrential rainfall that can lead to flash flooding. CL supercells have a precipitation amount and distribution midway between those of LP and HP supercells.

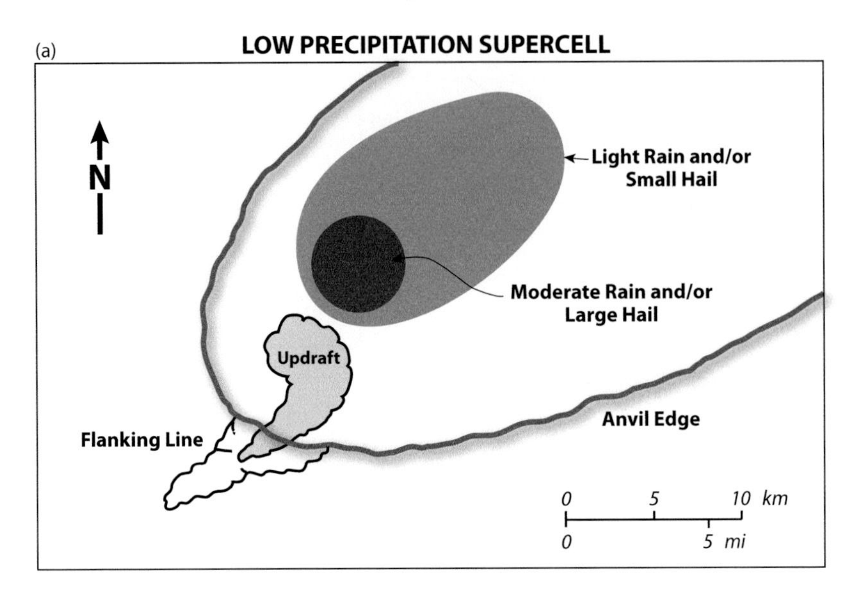

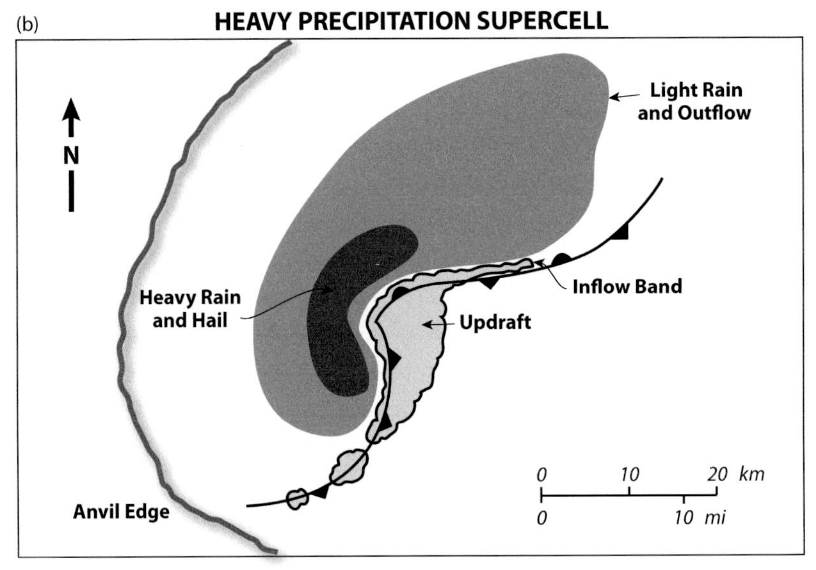

Figure 7.16 Schematic depictions of a (a) low-precipitation, and (b) high (or heavy) precipitation supercell, in plan view. After Moller et al. (1994).

The magnitude of the storm-relative environmental winds in the mid- to upper troposphere plays one possible role in determining the supercell precipitation characteristic.[29] For instance, it has been shown that LP supercells most likely form in an environment with relatively strong storm-relative winds at anvil level (e.g., 9–10 km AGL). These winds transport hydrometeors far downwind of the updraft, limiting recirculation of the hydrometeors through the updraft and hence limiting further growth. In contrast, HP supercells appear to be favored in an environment with relatively weak anvil-level, storm-relative winds that allow hydrometeors to experience subsequent growth. Such weak upper-tropospheric winds also allow these hydrometeors to fall and evaporate nearer to the updraft. The roles of cloud- and precipitation-scale processes are oversimplified here, particularly with regard to hydrometeor initiation and growth, which interact with the storm dynamics in ways not yet fully understood.

The environmental winds, and thus the proximity of the falling precipitation to the updraft, also influence the precipitation–mesocyclone interplay. To illustrate, imagine how some amount of precipitation that falls downstream, yet near the updraft, is transported by the midlevel mesocyclone to the left and then rear flank of the updraft. This spiraling precipitating downdraft results in cold air outflow that can aid in the generation of low-level vertical vorticity (Section 7.5), but also that can move beneath and through the base of the updraft, undercutting the updraft from its source of potentially unstable air;[30] the net effect depends in part on the low-level environmental winds relative to the strength of the outflow. The midlevel mesocyclone also acts to transport potentially cool environmental air from the rear of the storm toward the storm's forward and then left flank, consequently affecting the rate of evaporative cooling in these precipitating regions.[31] Both transports are a function of the size and strength of the midlevel mesocyclone, which relate back to the environmental vertical wind shear over the low- to mid-troposphere. Hence, the details of environmental winds over a large depth are important in supercell intensity and longevity, and, as just alluded to, have particular relevance to low-level mesocyclogenesis.[32]

7.5 Low-Level Mesocyclogenesis

We established in Sections 7.2 and 7.3 that mesocyclogenesis at midlevels is readily explained through the tilting of environmental horizontal vorticity by an updraft. In this section we learn that mesocyclogensis at and below the cloud base (lowest $\sim$1 km) also arises through a vorticity tilting process, albeit one that depends critically on the existence of a downdraft.

Recall from Figure 7.3 that the archetypal supercell has prominent downdrafts both forward and rearward of the updraft. The associated outflow in these FFD and RFD regions contribute to horizontal buoyancy gradients and, correspondingly, to horizontal vorticity oriented parallel to the respective gust fronts (e.g., Figure 7.17). Inflowing air parcels that pass through the buoyancy gradients en route to the updraft acquire streamwise horizontal vorticity in amounts that depend on the gradient magnitude and the storm-relative wind speed.

Baroclinic generation of streamwise horizontal vorticity can be estimated from the inviscid form of the equation for the streamwise component of horizontal vorticity:

$$\frac{D\omega_s}{Dt} = \omega_n \frac{D\Psi}{Dt} + \vec{\omega} \cdot \nabla u_s + \frac{\partial B}{\partial n}, \tag{7.32}$$

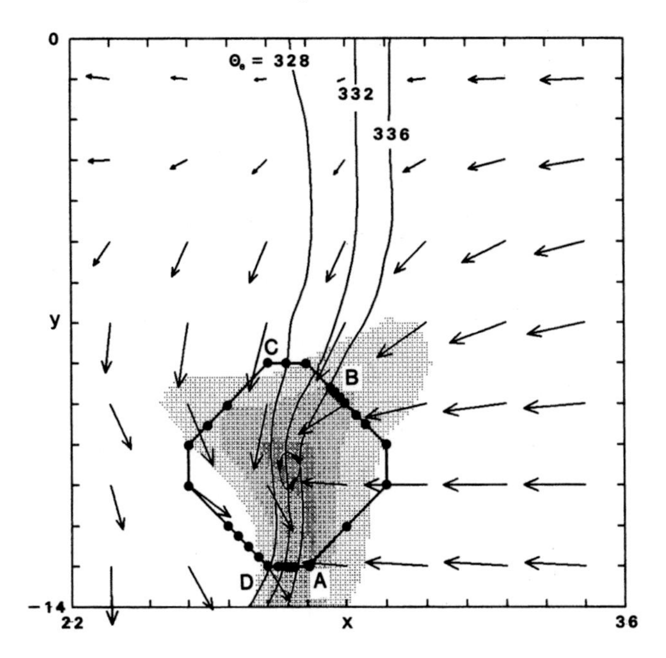

Figure 7.17 Horizontal vorticity (vectors), equivalent potential temperature (contours; K), and vertical velocity (shaded; m s^{-1}) at z = 0.25 km, in a numerical simulation of a supercell. The closed streamline shows the location of the maximum vertical vorticity, and thus of the developing low-level mesocyclone. The thick line and dots indicate the material curve, and accompanying parcel locations, used in an analysis of circulation. From Rotunno and Klemp (1985).

where $\vec{V} = u_s\hat{s} + w\hat{k}$ is the wind vector in (semi) natural coordinates, $\Psi = \tan^{-1}(v/u)$ is direction (angle) of the wind, increasing counterclockwise, and

$$\vec{\omega} = \left(-u_s\frac{\partial \Psi}{\partial z} + \frac{\partial w}{\partial n}\right)\hat{s} + \left(\frac{\partial u_s}{\partial z} - \frac{\partial w}{\partial s}\right)\hat{n} + \left(-\frac{\partial u_s}{\partial n}\right)\hat{k} \tag{7.33}$$

is the vorticity vector in (semi) natural coordinates.[33] The first right-hand term of (7.32) represents an exchange between crosswise and streamwise vorticity, and the second term embodies the effects of tilting as well as stretching. If we isolate the third right-hand term, which is baroclinic generation, and assume steady state on a horizontal plane, (7.32) can be approximated by:

$$\Delta\omega_s \approx \frac{\Delta B}{\Delta n}\frac{\Delta s}{u_s}. \tag{7.34}$$

where Δs is some distance increment tangential to the storm-relative flow, $\Delta B/\Delta n$ is the buoyancy gradient normal to the flow, and u_s is the speed of the storm-relative flow. [34] Now, assume in (7.34) that the buoyancy gradient in the FFD or RFD outflow is due mostly to temperature changes and hence can be approximated as $g/\theta_0(\Delta\theta/\Delta n)$, and, that the storm-relative flow is parallel to the isentropes. Then, for example, let the potential temperature gradient be $\Delta\theta/\Delta n = 1°\mathrm{C\ km}^{-1}$, and the (storm-relative) speed of low-level parcels moving parallel to the isentropes be $u_s = 15$ m s^{-1}. We

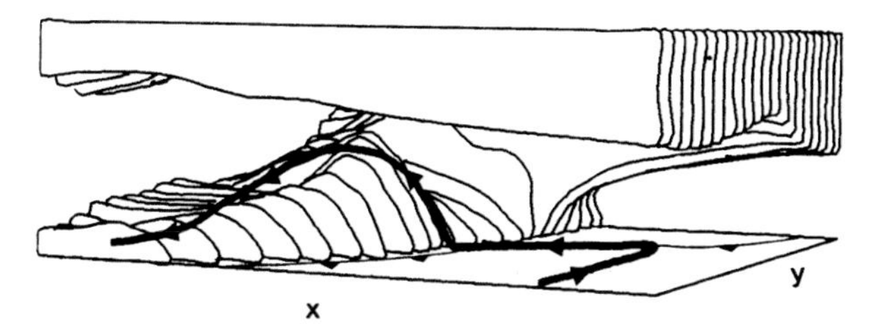

Figure 7.18 Three-dimensional perspective of a simulated supercell storm and a vortex line. The vortex line (bold) passes through the forward-flank baroclinic zone as well as the location of maximum vertical vorticity. The storm itself is represented as a constant surface of equivalent potential temperature ($\theta_e = 331$ K). From Rotunno and Klemp (1985).

find from (7.34) that over a distance of $\Delta s = 5$ km, these parcels acquire horizontal vorticity of $\Delta\omega_s = 10^{-2}$ s^{-1}, which is an amount comparable to or even exceeding typical environmental horizontal vorticity.

Once the parcels encounter a horizontal gradient of vertical velocity (updraft or downdraft), such streamwise baroclinic vorticity is tilted into the vertical (e.g., Figure 7.18). The resultant vertical vorticity is then amplified through vortex stretching to result in a mesocyclone at low levels.

Support for the connection between horizontal baroclinic vorticity and low-level mesocyclogenesis is provided by idealized numerical modeling experiments in which evaporation of precipitation is disallowed.[35] A rainy FFD (and RFD) is still produced in these experiments, but the lack of evaporation precludes existence of chilled outflow and, hence, surface baroclinity. Consistent with theory, significant low-level vertical vorticity is absent in the resultant storm. Additional support is provided by observational datasets from field campaigns.[36] Indeed, evaluations of (7.34) using surface data collected in FFD outflow give values of $\Delta\omega_s$ that range from 0.2×10^{-2} to 1.5×10^{-2} s^{-1}. This baroclinic horizontal vorticity is observed in storms with significant vertical vorticity at low levels, though not necessarily storms that subsequently spawn tornadoes; possible reasons why a low-level mesocyclone alone is a insufficient condition for tornadogenesis will be explored in Section 7.6.

For a different perspective of the mesocyclogenesis process, we consult the circulation theorem:

$$\frac{D\Gamma}{Dt} = \oint B\hat{k} \cdot d\vec{l}, \tag{7.35}$$

recalling that $\vec{l}$ is a vector locally tangent to the closed contour of integration.[37] Equation (7.35) is valid in absence of Coriolis effects and for an inviscid, incompressible fluid; it deviates slightly from the expression used in Chapter 5 to explain the sea-breeze circulation, but both expressions include only a baroclinic generation term. The circulation

$$\Gamma(t) = \oint \vec{V} \cdot d\vec{l} \tag{7.36}$$

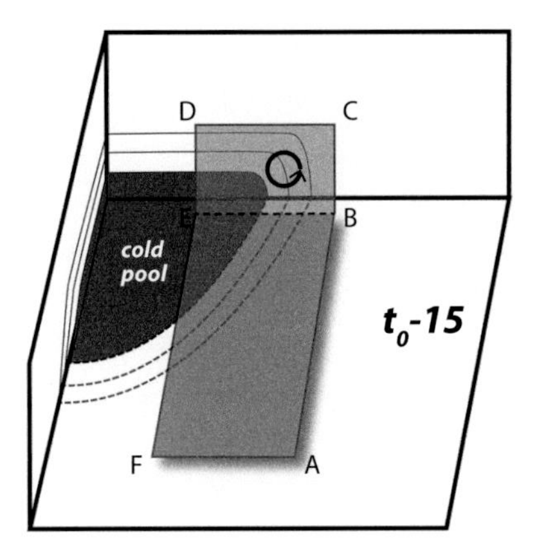

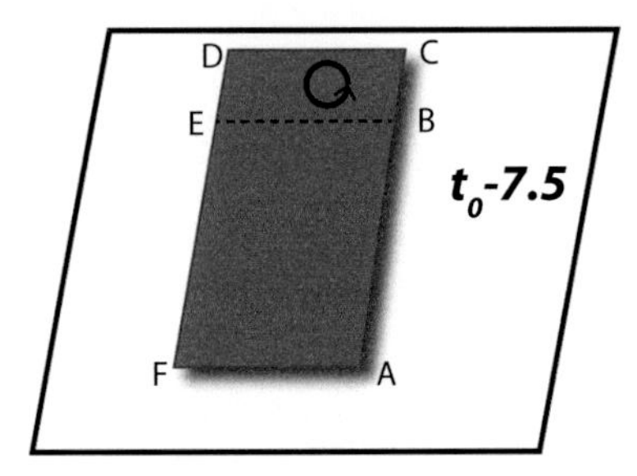

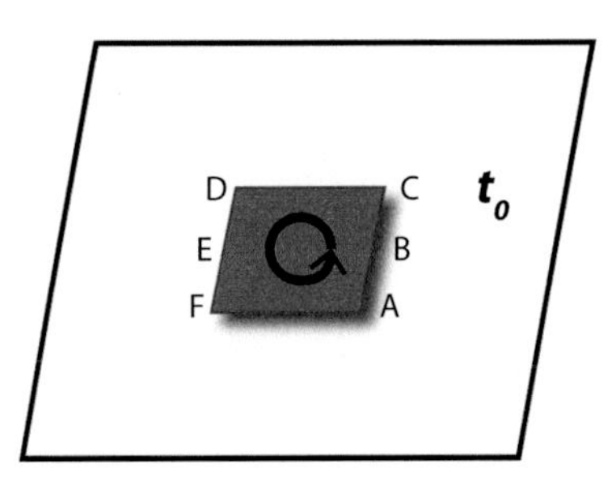

Figure 7.19 Three-dimensional evolution of a material curve that initially (t_0) enclosed the vertical vorticity maximum at low levels (black curve). Letters indicate individual parcels, and accordingly, segments of the material curve. In (a), gray shading and contours are of buoyancy, and indicate the cold pool.

is integrated around a *material curve* that by definition is composed, at all times, of the same fluid parcels.[38] Thus, the strategy often employed in analyzing (7.35) is to surround the low-level mesocyclone by a curve in the horizontal plane (see Figure 7.17), and then evaluate the (backward) time history of the curve, the associated circulation, and its generation.

Consider the idealized material curve shown in Figure 7.19, which is based on actual curves from analyses of mesocyclogenesis in simulated supercells.[39] At the early time (t_0 – 15 min), notice that the area bounded by the curve is considerably larger than at the initial time (t_0; Figure 7.19c). Notice also that a section of the curve is in the vertical plane, and cuts through a surface-based pool of negatively buoyant air (Figure 7.19a). From (7.35), circulation is generated only through the vertical segments of integration; in the case shown in Figure 7.19a, segments DE and BC contribute to a net positive circulation. At the intermediate time (t_0 – 7.5 min; Figure 7.19b), baroclinic generation diminishes because the curve is primarily in the horizontal plane. This geometric change is brought about by the descent of the parcels comprising the vertical section of curve. Said another way, the curve is flattened by a downdraft.[40] The subsequent decrease in the area within the curve (Figure 7.19c) suggests an increase in the average vertical vorticity of fluid enclosed by the curve. It also implies a horizontal convergence of the parcels, such that the curve is beneath an updraft at this final time.

The circulation perspective of low-level mesocyclogenesis is equivalent to the vorticity perspective, as shown by an application of Stokes' theorem to (7.36):

$$\Gamma = \oiint \nabla \times \vec{V} \cdot d\vec{A}, \tag{7.37}$$

where A is the area enclosed by the material curve. Circulation is the vorticity normal to and integrated over area A. In Figure 7.19a (Figures 7.19b,c) positive circulation relates to the horizontal (vertical) vorticity vectors normal to the vertical (horizontal) plane. The geometric change from Figure 7.19a to 7.19b corresponds to tilting; the change from Figure 7.19b to 7.19c, in which area shrinks, corresponds to stretching.

These perspectives provide information only on how mesocyclone-scale rotation is generated near the ground. We now know that this is a necessary, but not sufficient, condition for tornadogenesis.[41] Any successful explanations for tornadogenesis must address this insufficiency issue.

7.6 Tornadogenesis

Tornadogenesis is regarded herein as the culmination of a multistep process, in which deep-layer mesocyclonic vertical vorticity ($\zeta \sim 10^{-2}$ s^{-1}) is concentrated into a tornadic vortex with $\zeta \sim 10^{0}$ s^{-1}. This 10^2 order-of-magnitude increase in vertical vorticity is attributed primarily to vortex stretching, as demonstrated schematically in Figure 7.20. A quantification of vortex stretching and its role in tornadogenesis follows from an approximation of the vertical vorticity equation (7.2):

$$\frac{1}{\zeta}\frac{D\zeta}{Dt} \approx -\nabla \cdot \vec{V}_H. \tag{7.38}$$

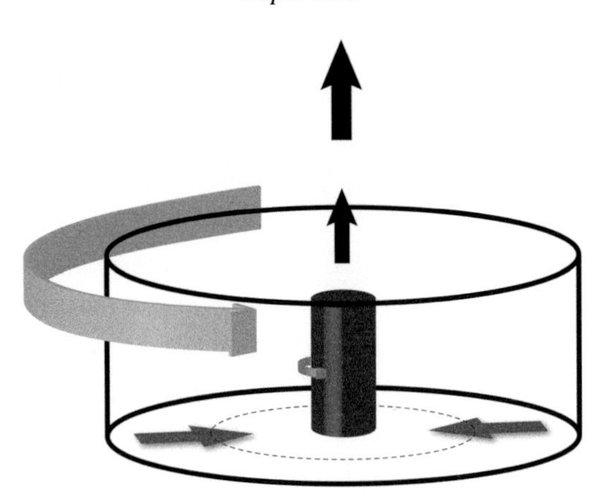

Figure 7.20 Illustration of how vortex stretching leads to tornado formation. The outer cylinder and associated gray arrow represent the mesocyclone. Black arrows represent the horizontally converging, vertically diverging flow associated with the convective circulation. The inner cylinder represents the nascent tornado.

where for mathematical tractability we have isolated the vortex stretching term, with the recognition that the omitted tilting term is relatively small yet still nonzero even within a mature tornado. Integrating (7.38) along a parcel trajectory gives,

$$\ln\left[\frac{\zeta(t)}{\zeta_0}\right] = -\int_0^t \nabla \cdot \vec{V}_H dt \tag{7.39}$$

and then upon assuming constant horizontal convergence $\mathcal{d} = -\nabla \cdot \vec{V}_H > 0$ over the interval of integration, we have

$$\zeta(t) = \zeta_0 \exp\left(\mathcal{d}t\right). \tag{7.40}$$

Let the convergence at low levels be $\mathcal{d} = 5 \times 10^{-3}\ \mathrm{s}^{-1}$, which is a reasonable value in observed storms, and then let initial vertical vorticity be the nominal mesocyclonic value of $\zeta_0 = 10^{-2}\ \mathrm{s}^{-1}$. From (7.40), the corresponding timescale for tornado formation is 15 min; a doubling of convergence reduces this timescale to 7.5 min. The timescales based on this simple analysis are in line with observed tornadogenesis from Doppler weather radar. However, lacking from this analysis is an explanation of the relative infrequency of tornadoes: it is estimated that only ~40 percent (~15 percent) of all low-level (midlevel) mesocyclones are tornadic.[42]

Imagine again the archetypal supercell, wherein air parcels gain horizontal and then vertical vorticity from the processes described in Section 7.5, and then exit the RFD (or FFD) at low levels before converging with inflow air beneath the base of the updraft. The rotating updraft region during this pretornadic stage can be modeled in a 2D axisymmetric framework that lends itself to simple experimentation on tornadogenesis (Figure 7.21).[43] Ascent of low-level parcels is driven by a fixed, buoyancy-like *body*

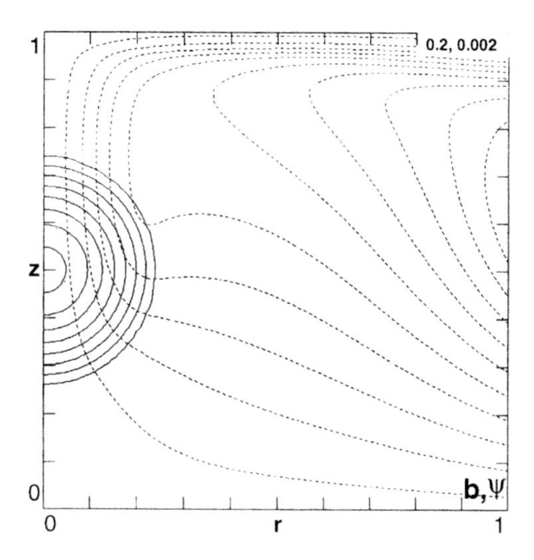

Figure 7.21 Example configuration of a 2D (r, z) axisymmetric model with convection forced by a fixed body force (solid contours). The steady-state streamfunction associated with the body force is indicated by the dashed contours. From Trapp and Davies-Jones (1997).

force in the vertical equation of motion. The thermodynamic environment in the model is specified so that it includes properties of the RFD air. Finally, mesocyclone-scale rotation is prescribed at the outer boundary of the model.

An increase of the mesocyclonic rotation into a concentrated, surface-based vortex (i.e., tornadogenesis; see Figure 7.22a) depends to a large degree on the strength and depth of the convective circulation, and in turn on the fixed body force and buoyancy. Experiments in which the potential temperature lapse rate is systematically increased to 6 K km^{-1} increasingly preclude extension of the convective circulation down to the lowest levels. As a consequence, the vortices attempting to form in the presence of increasingly stable environments become suspended aloft and fail to make contact with the surface (Figure 7.22b).[44] This is one characterization of *tornadogenesis failure*.

Consistent with the 2D model experiments, the observed thermodynamic properties of such "failed" (nontornadic) supercells are indeed different from those of tornadic storms. Specifically, the RFD and associated outflow in tornadic supercells tend to be ~3 to 5 K warmer (as quantified by θ_e), and have ~300 J kg^{-1} more surface-based CAPE than nontornadic storms.[45] The disparity in these properties can be a result of differences in the vertical path of parcels in the RFD: a shallower downdraft means a shorter parcel path with less time for evaporative cooling. Alternatively, it can result from differences in the negative buoyant forcing of the RFD: the implied smaller role of thermal buoyancy in a shallower downdraft means a relatively larger role by precipitation loading. Both these plausible explanations demonstrate that subtle

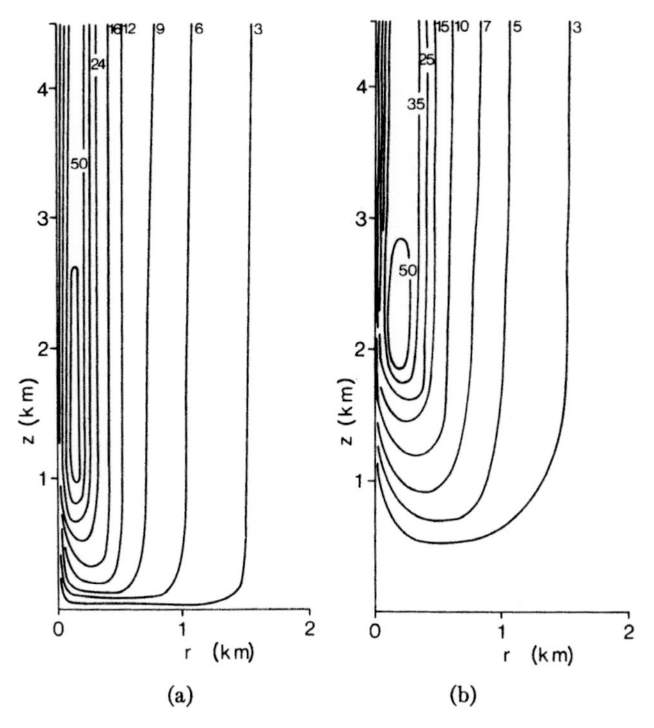

Figure 7.22 Tangential velocity from 2D axisymmetric modeling experiments in which (a) the base state temperature stratification is neutral (constant potential temperature), and (b) the base state temperature stratification is stable (potential temperature increase of 6 K km^{-1} over the lowest 1 km). From Leslie and Smith (1978).

details control whether or not a tornado forms, and hence begin to explain the rarity of tornadoes.

It has been difficult to identify an unambiguous *kinematic* distinction between nontornadic and tornadic storms, but certain kinematic properties do influence the tornadogenetical behavior. Returning to our 2D, axisymmetric framework, this behavior is revealed through an application of the conservation of angular momentum:

$$V(r, z)r = A_m = \text{const}, \tag{7.41}$$

where V is tangential velocity, r is radial distance from the central vertical axis, and A_m is angular momentum. Conservation requires that V increases when r decreases, and the height at which V first becomes large is the height at which parcels penetrate closest to the axis. This axial penetration is controlled primarily by the convective circulation.

Let us consider two cases: (1) the convective circulation is deep and extends through the lowest levels, and (2) the circulation is elevated (Figure 7.23). Assume that the vertical vorticity, and hence tangential velocity, is uniform along the outer boundary of our axisymmetric domain, and that the hydrostatic stability is relatively

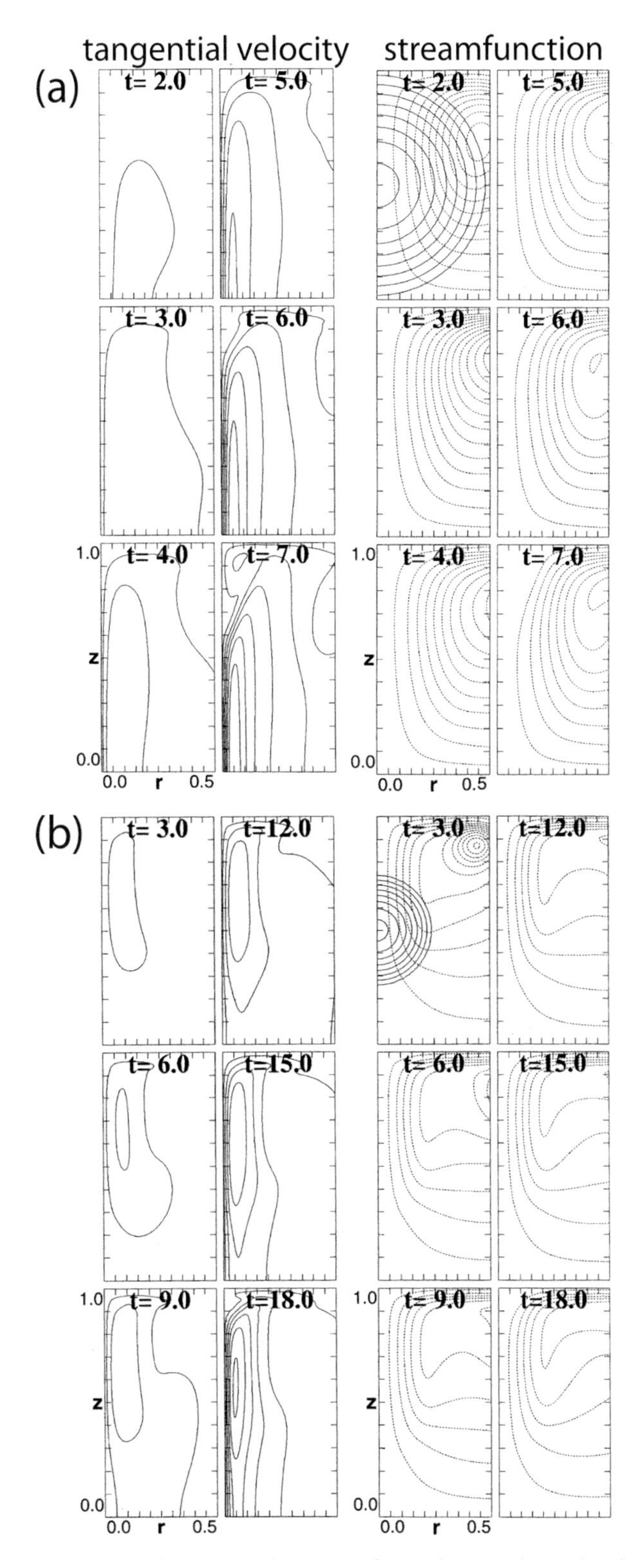

Figure 7.23 Two modes of tornado-like vortex formation, as shown by time sequence of tangential velocity and streamfunction from 2D axisymmetric modeling experiments. (a) Fixed body force is broad, as is the corresponding convective circulation. (b) Fixed body force is more isolated, and the corresponding circulation does not extend initially to low levels. From Trapp and Davies-Jones (1997).

low. In case 1, parcels penetrate to the axis first at low levels. Vortex formation – tornadogenesis – thus commences in the lowest levels and rapidly builds upward (Figure 7.23a). Because the horizontal (radial) convergence in case 1 is strongest in the lowest levels, this mode of tornadogenesis can also be explained in terms of vortex stretching and subsequent vertical advection.[46] In case 2, parcels penetrate to the axis first at middle levels, where corresponding vortex stretching is strongest, and vortex formation begins aloft. The subsequent descent of the nascent tornado owes to the *dynamic pipe effect*:[47] rotation-induced low dynamic pressure p_D leads to a vertical acceleration beneath (and above) the vortex core. Horizontal accelerations respond, allowing a new level of parcel penetration and vortex formation. The process repeats itself at progressively lower levels until the vortex has bootstrapped itself to the ground (Figure 7.23b).

In this idealization of tornado development, the supercell and its environment are only crudely represented. In the next section, we focus explicitly on the environment, and the parameters used for its quantification.

7.7 Supercell Environment Quantification

Having established the theory behind the development of supercells and attendant characteristics, it is worthwhile to examine the degree with which specific environmental parameters from soundings can be used to anticipate such development.

Refer back to Figure 7.15, and notice that a $SRH_{0\text{–}3}$ threshold of $\sim$250 $m^2\ s^{-2}$ is suggested for supercell occurrence. This is based on a *parameter space* study, in which input parameters are varied through a series of numerical modeling experiments. Observed supercells, particularly those that have produced significant tornadoes (damage ratings $\geq$ F2/EF2), are found in environments with equivalently high $SRH_{0\text{–}3}$ (see Table 7.1), whereas non-supercellular convective storms occur, on average, when the observed $SRH_{0\text{–}3}$ is $\sim$ 50 $m^2\ s^{-2}$.[48] For reference, the magnitude of the vector difference between the surface and 6-km environmental winds $|\vec{S}_{0-6}| =$ S06 is >20 m s^{-1}, on average, in the environments of supercells, but <10 m s^{-1} in environments of non-supercells. Evaluation of *bulk shear* over shallower layers such as 0–1 km also reveal differences between supercell and non-supercell environments, as does SRH over a 0- to 1-km layer (Table 7.1). Because the inflow (and overall) depths will vary considerably in a sample of actual storms, alternative forms of these vertical-shear parameters with objectively determined inflow layers, rather than the predetermined layers (0–3 km, etc.), are also used.[49]

Thermodynamic parameters from soundings also have some utility in discriminating significant tornadic supercells from non-supercells, although, as with the vertical-shear parameters, the precise values vary by dataset and analysis methodology. For example, in a study that used soundings derived from numerical weather prediction model analyses, the mean-layer (ML) CAPE in environments of significant tornadic

Table 7.1 *Parameter values from soundings derived from Rapid Update Cycle model analyses, and from radiosonde observations, in close proximity to occurrences of significant tornadic supercells (associated with tornadoes that produced damage ≥F2/EF2) and of non-supercellular convective storms. See text for description of parameters. The values should not be treated as strict thresholds. Compiled using data from Thompson et al. (2003) and Rasmussen and Blanchard (1998).*

	Significant tornadic supercell	Non-supercell
$SRH_{0\text{–}3}$	250 $m^2\ s^{-2}$	50 $m^2\ s^{-2}$
S06	20 $m\ s^{-1}$	10 $m\ s^{-1}$
$SRH_{0\text{–}1}$	185 $m^2\ s^{-2}$	15 $m^2\ s^{-2}$
S01	10 $m\ s^{-1}$	4 $m\ s^{-1}$
MLCAPE	2300 $J\ kg^{-1}$	1300 $J\ kg^{-1}$
LCL	1029 m (AGL)	1919 m (AGL)
BRN	40	300

supercells is ~2300 J kg^{-1}, whereas MLCAPE is ~1300 J kg^{-1} on average in environments of non-supercells.[50] Findings of relatively lower lifting condensation level (LCL) heights in significant tornadic supercells are consistent with theories that a more humid subcloud environment is more favorable for tornadogenesis.[51] A 0–1-km mean relative humidity >65 percent combined with $SRH_{0\text{–}1}$ >75 $m^2\ s^{-2}$ has been shown to characterize well the environments of significant tornadic supercells.[52]

This RH-SRH pair represents a parameter combination. Another is BRN, which we saw in Chapter 6. In environments of significant tornadic supercells (non-supercells), the mean BRN is ~ 40 (~300). Other composite parameters are too numerous to list here. Most involve some combination of bulk shear and/or SRH with CAPE, and are based on a statistical regression or similar technique. A challenge in the development of these parameters is to find values with broad applicability, even in regions not known for high occurrences of supercells.

7.8 Analogues in Tropical Convection: Vortical Hot Towers

It is well established that supercells exist in rainbands of landfallen tropical cyclones. Structurally, these tend to be miniature versions of the supercells observed in the southern Great Plains of the United States, for example.[53] There are, however, other tropical convective phenomena, namely *vortical hot towers* (VHTs), that share some

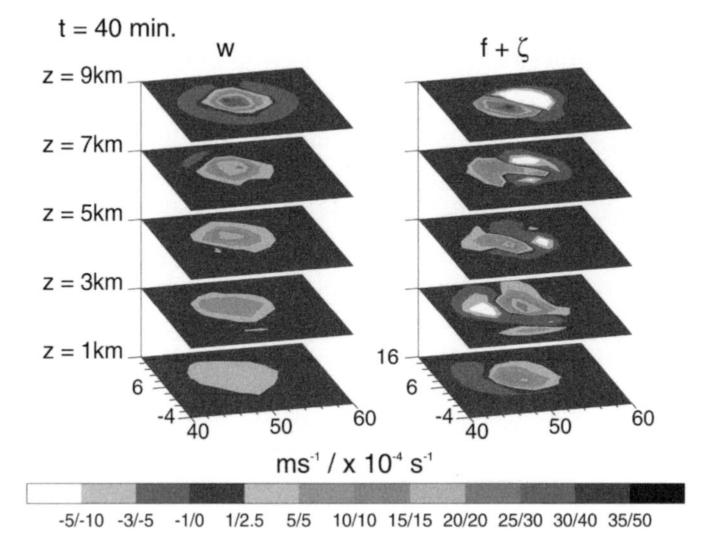

Figure 7.24 Vertical velocity and absolute vertical vorticity associated with a numerically simulated vortical hot tower. From Montgomery et al. (2006). (For a color version of this figure, please see the color plate section.)

of the same characteristics as supercells. Although the degree to which VHTs are governed by the dynamics developed in Section 7.3 is currently unclear, it is most convenient to introduce VHTs in this chapter, especially to allow a continuation of our previous efforts to compare and contrast tropical and midlatitude convection.

VHTs are rotating tropical cumulonimbi that occur in the vorticity-rich mesoscale environment of developing tropical cyclones.[54] They have horizontal scales ~10 km, and comparable vertical scales, implying aspect ratios that are near unity. Within the tower cores, vertical vorticity is of a magnitude at or near 10^{-2} s^{-1} (Figure 7.24). Such vertical vorticity is generated by tilting of environmental horizontal vorticity and by subsequent stretching, just as it is for supercell mesocyclones.[55] One significant difference here is that the "environment" is that of the weak cyclonic circulation of a tropical disturbance; hence, the environmental horizontal vorticity arises from the vertically sheared tangential wind of this disturbance. Tilting results in a vortex couplet (see Figure 7.24); the cyclonic member is then preferentially enhanced through vortex stretching within this environment of positive absolute vertical vorticity. We will see a similar preferential enhancement of cyclonic mesovortices in the low levels of quasilinear mesoscale convective systems (Chapter 8).

In contrast to nonvortical tropical convection (Chapter 6), the high helicity inherent in VHTs renders them more resistant to lateral entrainment, thus allowing for more efficient convective processes; an analogous conclusion has been made about the effects of helicity on supercells, especially in the context of a suppression of turbulent dissipation and thereby enhancement of supercell longevity.[56,57] Diabatic heating by an ensemble of these relatively efficient VHTs drives a thermally direct toroidal circulation, with low-to-mid tropospheric inflow and upper-tropospheric outflow. The toroidal circulation then converges the cyclonic vertical vorticity of the VHTs, and that of the environment, to yield a warm-core tropical vortex.[58]

The hypothesized VHT pathway to tropical cyclogenesis has similarities to the tornadogenesis model presented in Section 7.6. As we will see in Chapter 8, there are also similarities to vortical processes in midlatitude mesoscale convective systems.

Supplementary Information

For exercises, problem sets, and suggested case studies, please see www.cambridge.org/trapp/chapter7.

Notes

1 Browning and Ludlum (1962), Browning and Donaldson (1963), and Browning (1964).
2 An animation can be found at http://www.umanitoba.ca/environment/envirogeog/weather/overshootanim.html.
3 A comparable anticyclonically rotating vortex is a *mesoanticyclone*.
4 Wakimoto et al. (2004).
5 Lemon and Doswell (1979).
6 Alternatively, begin with the horizontal component equations of (7.1), and then differentiate, following $\zeta = \partial v/\partial x - \partial u/\partial y$, to form (7.2).
7 Thus, for example, $d_{1,1} = \partial u/\partial x$, $d_{1,2} = d_{2,1} = (\partial v/\partial x + \partial u/\partial y)/2$, etc. See Rotunno and Klemp (1985), and Davies-Jones (2002), for more discussion of both forms of the diagnostic equation for dynamic pressure.
8 This approximation – and qualitative arguments based on it – applies only in the interior of the flow, well away from flow boundaries. A full inversion of the Laplacian would require properly posed boundary conditions; see Davies-Jones (2002).
9 Maddox (1976), Davies-Jones (1984).
10 Much of this is based on Weisman and Rotunno (2000) and accompanying discussions in the literature.
11 An identical environmental wind profile except for winds turning *counterclockwise* with height would favor the LM storm, and the anticyclonic rotation of its updraft would be preferentially enhanced.
12 Weisman and Rotunno (2000).
13 It is assumed again that properly posed boundary conditions are used in the inversion of the Laplacian; see Davies-Jones (2002).
14 This analysis approach follows Lilly (1979) and Lilly (1986a).
15 In Lilly (1979), this level is $\sim$1.75 km.
16 The LM storm, which has a propagation $c_y > 0$, has the reverse outcome.
17 Lilly (1982).
18 Droegemeier et al. (1993).
19 Davies-Jones (1984).
20 Davies-Jones (1984), Droegemeier et al. (1993).
21 Strict conservation also requires that the domain of consideration have rigid boundaries at which the normal component of the vorticity vector vanishes; see Droegemeier et al. (1993).
22 Bunkers et al. (2000).
23 Ibid.
24 This draws from Doswell (1991).
25 SRH is determined graphically as minus twice the signed area swept out by the storm-relative wind vectors in the layer of interest; see, e.g., Davies-Jones (1984) and Droegemeier et al. (1993).
26 The average is typically performed over a subdomain centered on the storm, where w exceeds some threshold, such as 1 m s^{-1}; see, e.g., Droegemeier et al. (1993).

27 This is the threshold proposed by Droegemeier et al. (1993), but other values may be appropriate depending on the depth and layers used in the vertical averaging.
28 Doswell and Burgess (1993), Moller et al. (1994).
29 Rasmussen and Straka (1998).
30 Brooks et al. (1994).
31 Rotunno and Klemp (1985).
32 Brooks et al. (1994).
33 Adlerman et al. (1999).
34 Klemp and Rotunno (1983).
35 Rotunno and Klemp (1985).
36 Shabbott et al. (2006).
37 As expressed in Rotunno and Klemp (1985).
38 We assume an integration in a counterclockwise sense, such that a curve enclosing a cyclonic vortex has positive circulation.
39 Rotunno and Klemp (1985), Davies-Jones and Brooks (1993), Trapp and Fiedler (1995).
40 Davies-Jones and Brooks (1993).
41 Trapp (1999).
42 Trapp et al. (2005a).
43 Smith and Leslie (1978).
44 Leslie and Smith (1978).
45 Markowski et al. (2002).
46 Trapp and Davies-Jones (1997).
47 Smith and Leslie (1978).
48 Thompson et al. (2003).
49 Thompson et al. (2007).
50 Thompson et al. (2003).
51 Markowski et al. (2002).
52 Thompson et al. (2003).
53 McCaul (1987), McCaul and Weisman (1996).
54 Hendricks et al. (2004).
55 Montgomery et al. (2006).
56 Henricks et al. (2004).
57 Lilly (1986b).
58 Montgomery et al. (2006).

8

Mesoscale Convective Systems

Synopsis: A mesoscale convective system (MCS) is composed of precipitating convective clouds that interact to produce a nearly contiguous, extensive area of precipitation. Chapter 8 describes MCS structure and organization, and then explains the dynamical links among structure, longevity, and intensity. The quasilinear MCSs, especially those that have leading edges that "bow" outward, can produce swaths of damaging "straight-line" surface winds. Proposed mechanisms for this wind production are described, including one that involves vertical vortices at low levels. The chapter also includes discussions of mesoscale convective complexes, another common organizational mode, as well as of the remnant vortices that these and other MCSs often generate.

8.1 Overview of MCS Characteristics and Morphology

A mesoscale convective system (MCS) is an organized collection of two or more cumulonimubus clouds that interact to form an extensive region of precipitation.[1] As observed in weather radar scans at low elevation angles, the precipitation is nearly contiguous, especially at the leading edge of the system (Figure 8.1a). Indeed, this characteristic is one that allows for an observational distinction between an MCS and a group (or line) of discrete cells; it additionally has consequences on the MCS dynamics, as will be explained shortly. Another characteristic is the time scale, which typically is much longer than the ~1 h life cycle of the individual cumulonimbi comprising the system (see Chapter 6). As was demonstrated using scaling arguments in Chapter 2, such longevity suggests a relative importance of the Coriolis force as well as of the pressure gradient and buoyancy forces in the equations of motion.[2] A convenient MCS time scale is f^{-1}, which at midlatitudes is ~3 h, although MCSs are known to have lifetimes thrice this amount. A final characteristic is a length scale L of ~100 km, often defined by the contiguous leading-edge precipitation. These length and time scales are consistent through $L \sim U/f$, assuming a wind speed $U \sim 10$ m s^{-1}.

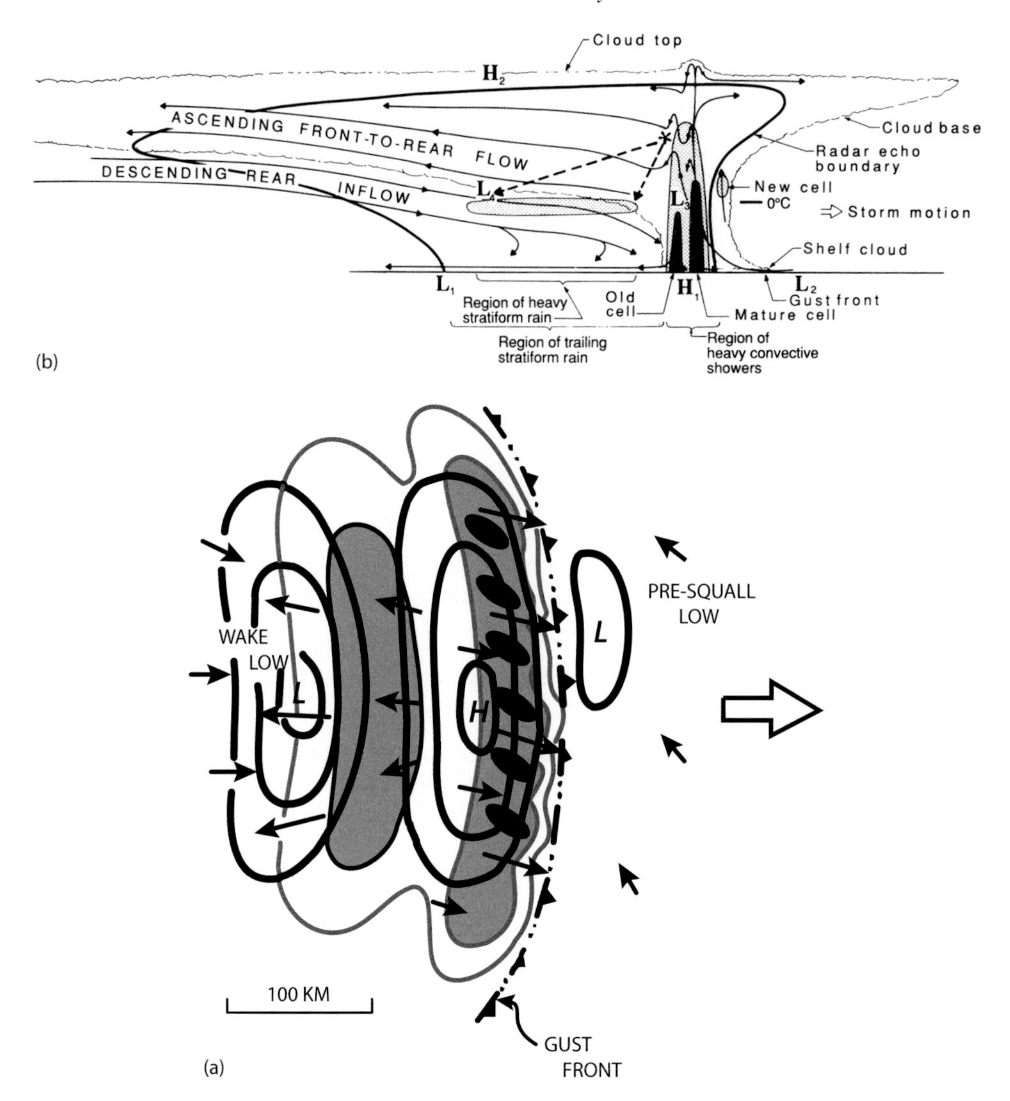

Figure 8.1 Conceptual model of a squall line with trailing stratiform precipitation. (a) Plan view, with low-level radar echoes denoted by thin gray contours and shading, and pressure (1-mb interval) indicated by thick contours, and arrows depicting surface flow. After Loehrer and Johnson (1995). (b) Vertical cross section, showing approximate 2D structure. From Houze et al. (1989).

MCSs are often conceptualized as squall lines, which are linear or at least quasi-linear systems (Figure 8.1a) and thus treated essentially as 2D. In a representative vertical plane, the salient features of a squall line include a narrow (~10 km) zone of deep convective updrafts at the leading or downstream edge, a mesoscale updraft within a sloping current that flows from this front edge toward the rear of the system, and a mesoscale downdraft within a sloping current that descends from the rear of the system toward the front (Figure 8.1b). Convective showers occur at the leading edge.

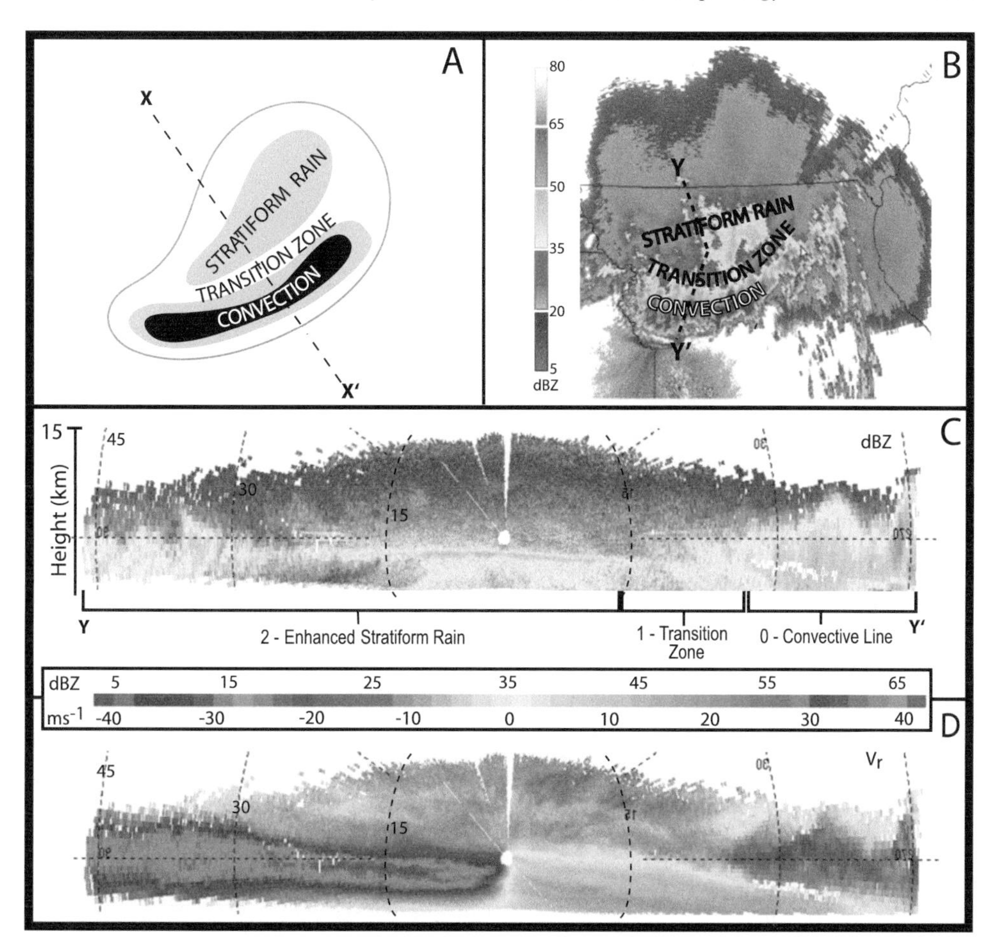

Figure 8.2 Schematic depiction (A), and corresponding Doppler radar observations (B)–(D), of a mature squall line with a trailing stratiform region. Radar reflectivity in (B) is from a WSR-88D scan at 0.5° elevation. Radar reflectivity and Doppler velocity in (C) and (D), respectively, are from quasi-vertical scans from an airborne Doppler radar (NOAA P-3). After Smith et al. (2009). (For a color version of this figure, please see the color plate section.)

A transition zone often follows the convection zone, in the form of a narrow channel of lower reflectivity (see Figure 8.2), which is then followed or trailed by extensive, stratiform precipitation.[3]

Such *trailing stratiform* (TS) precipitation is a prominent feature in this conceptual model. Analyses of radar data from a large set of MCSs do, in fact, reveal that the TS precipitation (e.g., Figure 8.3a) is prevalent in MCSs.[4] However, the radar data also show the frequent occurrence of *leading stratiform* (LS) and *parallel stratiform* (PS) precipitation modes, in which the stratiform precipitation falls, respectively, in advance of, or along the direction of orientation of, the convective line (Figures 8.3b,c).

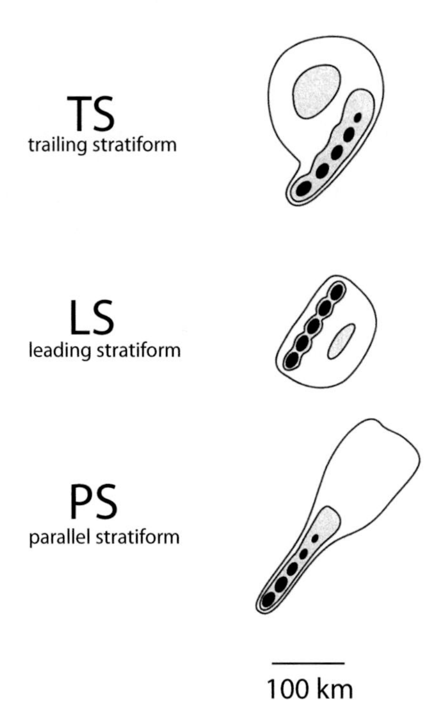

Figure 8.3 Idealized weather-radar presentations of quasilinear MCSs with (a) trailing stratiform (TS), (b) leading stratiform (LS), and (c) parallel stratiform (PS) precipitation regions. After Parker and Johnson (2000).

The mode of MCS precipitation is influenced significantly by the (line-perpendicular component of the) mid- and upper-tropospheric storm-relative winds and their role in hydrometeor transport and growth. The TS precipitation, for example, originates from particles that form in the convective region and then are ejected rearward. Subsequent growth – particularly of ice-phase hydrometeors – occurs in the ascending, storm-relative front-to-rear flow at mid to upper levels (Figures 8.1 and 8.2). Upon falling relative to the mesoscale updraft, the hydrometeors grow additionally through aggregation but then begin to melt and evaporate near and below the 0°C level.[5] The changeover to mixed-phase precipitation at this level is revealed on radar by the characteristic "bright band" signature (Figure 8.2c; see also Chapter 3).

The melting, evaporation, and sublimation of the stratiform precipitation contribute to a mesoscale downdraft en route to a surface-based pool of relatively cold air. The downdraft is part of the descending rear-to-front flow (also termed *rear inflow*) that resides below the ascending front-to-rear flow. As will be described further in the following sections, both airstreams are coupled back to the cold pool and convective updrafts through circulations internal to the convective system, and their dynamical interactions with the near-storm environment.

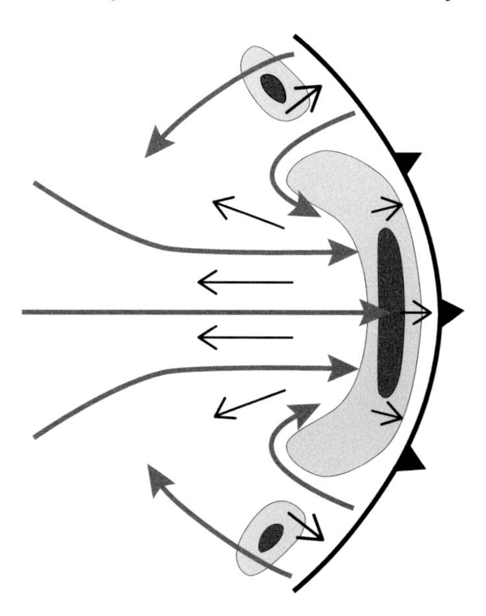

Figure 8.4 Schematic of a mature bow echo. Gray shading represents precipitation intensity, bold gray (thin black) arrows show system-relative flow at mid (low) levels, and barbed line indicates the position of the gust front. Adapted from Weisman (1993).

One particular realization of this coupling results in a convex shape to the convective line. This shape is a salient feature of a special class of quasilinear MCSs known as *bow echoes* (Figure 8.4). Named after their presentation in fields of radar reflectivity factor (see Figure 8.5), bow echoes owe their characteristic structure – and association with damaging surface winds – in part to the development of a *rear inflow jet* (see Section 8.2.2). Multiple bowing segments can occur in an extensive squall line, which would then be subclassified as a *squall-line bow echo*[6] (Figure 8.5b); the radar-reflectivity presentation of a squall-line bow echo is often described by a *line-echo wave pattern* (LEWP). Less commonly, bow echoes are also known to result from the upscale growth of HP supercells (Figure 8.6).[7] Although it is debatable whether the resultant *cell bow echoes* merit classification as MCSs, their occurrence does highlight the broad range of sizes, lifetimes, and intensities of this convective mode.

Bowing and nonbowing MCSs alike typically have, at some point during their lifetime, a pressure distribution analogous to that of the archetype (Figure 8.1). At midlevels, low pressure (a *mesolow*) is induced hydrostatically within the convection zone and rearward, and contributes to a horizontal pressure gradient that helps force the horizontal component of the rear inflow (see Section 8.2).[8] At the surface, high pressure (a *mesohigh*) forms in association with the precipitation-cooled downdraft air. A *wake low* is found near the rear edge of the stratiform precipitation, due to adiabatic compressional warming. This particular characteristic can be modified by the presence of a *mesoscale-convective vortex* (MCV; see Section 8.4 and Chapter 9).[9]

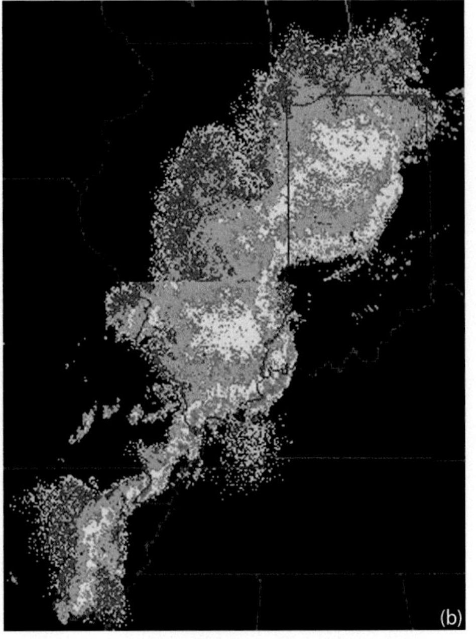

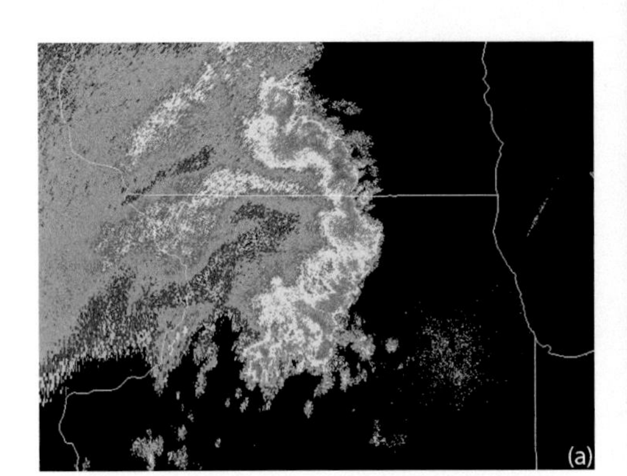

Figure 8.5 Radar reflectivity imagery of example bow-echo events: (a) classic bow echo, July 11, 2011, and (b) squall-line bow echo, or LEWP, April 19, 2011. Both are from scans at 0.5° elevation angle, and (b) is a composite of data from two radars. (For a color version of this figure, please see the color plate section.)

Vortices besides MCVs also locally modify the pressure distribution and lend to the three-dimensionality of the system. Of particular relevance to the morphology of bow echoes are midlevel ($z \sim 3 - 5$ km AGL) "bookend" or "line-end" vortices (Figure 8.4). Literally, these vortices occur at the lateral ends of the convective line, with cyclonic (anticyclonic) rotation on the system-relative left (right) end, looking downstream in the Northern Hemisphere. The line-end vortices arise from the tilting of horizontal vorticity generated in the leading gradient of the surface cold pool. Thus, the cold pool and system-scale airstreams are also coupled to vertical vortices

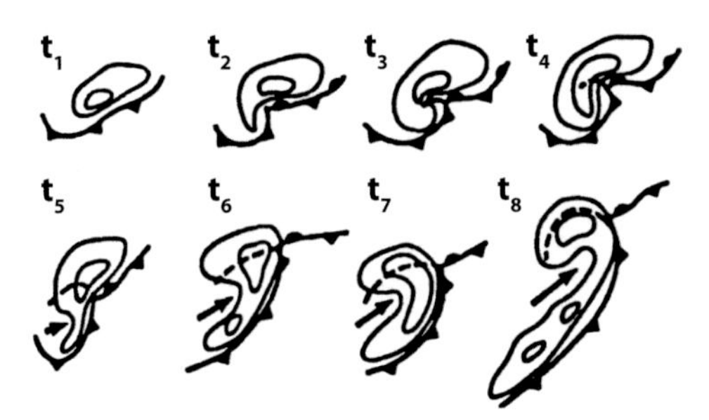

Figure 8.6 Schematic depiction of a transition from an HP supercell to a bow echo. From Moller et al. (1994).

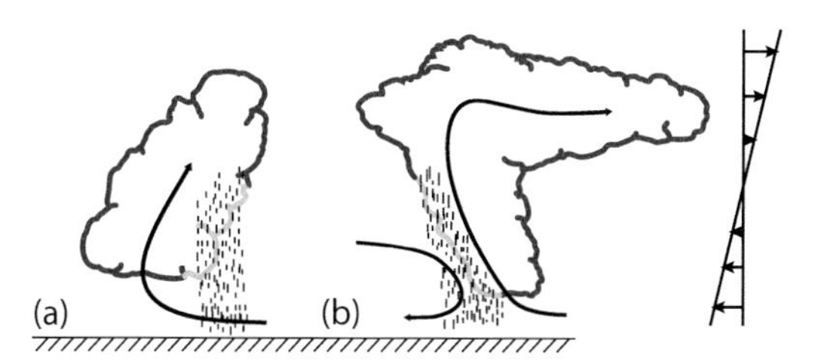

Figure 8.7 Contrasting cases of idealized precipitating updrafts that are tilted (a) in the direction of the environmental vertical wind shear vector, and (b) in the direction opposite to that of environmental shear. Adapted from Rotunno et al. (1988).

(see Sections 8.2–8.3), implying the need to treat each of these components when forming an explanation of MCS evolution.

8.2 Dynamical Explanations for MCS Evolution

In this Section we learn how the internal circulations of an MCS are influenced by the near-storm environment. Of particular interest here are cases of relatively intense and long-lived quasilinear MCSs; this interest is pursued further in Section 8.3, which emphasizes bow-echo systems that produce very strong surface winds. No consideration is given to lines of supercells because, by virtue of discussions in Chapter 7, Section 8.1, and later in Section 8.3, such discrete convective storms are not considered herein to be MCSs.[10]

8.2.1 The Rotunno-Klemp-Weisman Theory

The effects of environmental vertical wind shear on updraft tilt and cell evolution were introduced in Chapters 6 and 7, but we revisit this topic here because it is at the heart of the MCS dynamics. Let us impose an assumption of two-dimensionality, so we can restrict our attention to a vertical slab. Consider first the case of a 2D updraft that is tilted in the direction of the environmental shear vector; that is, *downshear-tilted*: because of differential advection, precipitation that is generated in the updraft is ejected downshear, and deposited into the environment ahead of this storm. The falling precipitation then diabatically cools the environmental inflow (Figure 8.7a). If the inflow layer is deep, the sublimation/melting/evaporation of precipitation actually helps destabilize the environment because such diabatic cooling increases with height within the inflow layer.[11] But, for the current line of discussion, assume that the inflow layer is relatively shallow, so that the potential buoyancy of the inflow parcels is significantly reduced by the diabatic cooling. Once ingested into the updraft, these stabilized parcels will lead to the updraft's eventual decay.

It is possible for a 2D updraft to be tilted in the direction opposite that of the environmental shear, with precipitation falling upshear rather than into the inflow

(Figure 8.7b). An *upshear-tilted* updraft can arise if the cold pool moves rapidly out ahead of the updraft, with a horizontal speed that exceeds the speed of the vertically lifted air at the gust front; the resulting convective updraft is shallow and tilted over the cold pool.

The essence of the so-called *Rotunno-Klemp-Weisman* (*RKW*) *theory* is an "optimal" state in which these competing effects of vertical shear and the cold pool (speed) balance each other.[12] An outcome of this balance are vertically erect updrafts, which suffer less from decelerations due to the vertical gradient of buoyancy pressure (p'_B; see Chapter 6) than do sloped updrafts, and thus are able to realize more of the positive buoyancy.[13] Another outcome is restraint of the convective outflow by the low-level winds (and wind shear), and hence a state in which the outflow does not move out too far ahead of the updraft. Deep lifting at the gust front is therefore possible, with new cells generated immediately downshear of the initial cell. Continuous generation of successive cells downstream/downshear helps propagate the system, and lends to the system longevity.

The RKW theory is expressed in terms of a horizontal vorticity balance. It can be derived by assuming a 2D squall line that is oriented in the y-direction, and thus with a consideration of the vorticity component in the y-direction, $\xi = \partial u/\partial z - \partial w/\partial z$. By neglecting friction and Coriolis effects, and with this 2D assumption, the equation governing y-component vorticity (e.g., (5.33)) can be simplified and written as

$$\frac{\partial \xi}{\partial t} = -\frac{\partial}{\partial x}(u\xi) - \frac{\partial}{\partial z}(w\xi) - \frac{\partial B}{\partial x}, \tag{8.1}$$

where the flux form of the advection terms follows from application of the chain rule

$$\frac{\partial}{\partial x}(u\xi) + \frac{\partial}{\partial z}(w\xi) = u\frac{\partial \xi}{\partial x} + w\frac{\partial \xi}{\partial z} + \xi\left(\frac{\partial u}{\partial x} + \frac{\partial w}{\partial z}\right)$$

and then use of the incompressibility assumption to eliminate the last right-hand term. We seek a local balance between horizontal vorticity generated in buoyancy gradients and horizontal vorticity associated with the environmental shear. A symmetric, idealized buoyant updraft supports positive and negative vorticity generation on its right and left flanks, respectively (Figure 8.8a). The buoyancy gradient of the cold pool introduces a negative-vorticity bias to the left flank (Figure 8.8b), but this can be balanced by import of equivalent amounts of positive vorticity on the right flank from the environment (Figures 8.8c,d). The horizontal buoyancy gradient, and hence vorticity generation due to the cold pool, span the depth of the cold pool, which typically is a few kilometers deep; this is the depth over which the dynamics underlying the RKW theory are most relevant.

A means to quantify the vorticity balance is obtained by integrating (8.1) over a control volume fixed to the moving cold pool, as done in Chapter 2 (see Figure 2.7).

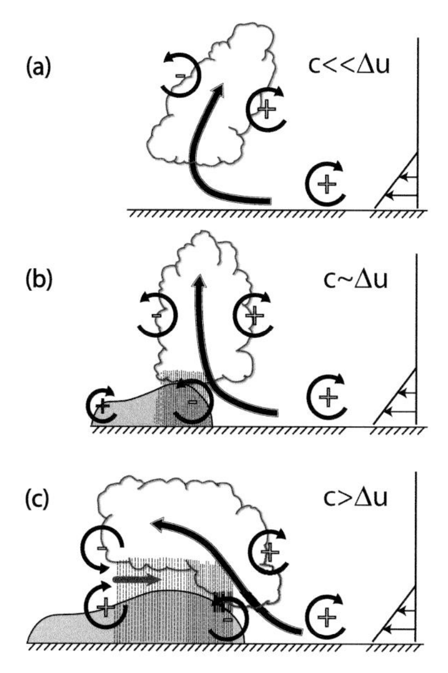

Figure 8.8 Horizontal vorticity balance, leading to vertically erect updrafts in a 2D squall line. Positive and negative vorticity is indicated by a "+" and "–", respectively, and shading represents the cold pool. (a) The downshear bias of a buoyant updraft in a sheared environment ($c/\Delta u \ll 1$), and (b) a balance between the horizontal vorticity in the cold pool, the environment, and buoyant updraft ($c/\Delta u \sim 1$). In (c), the upshear bias owing to the expansion and strengthening introduction of a cold pool is characterized as $c/\Delta u > 1$. Based on Rotunno et al. (1988) and Weisman (1992).

With respect to the top ($z = h$), bottom ($z = 0$), right ($x = R$), and left ($x = L$) edges of the control volume, the integral is:

$$\int_L^R \int_0^h \frac{\partial \xi}{\partial t} \, dz dx = -\int_L^R \int_0^h \frac{\partial}{\partial x}(u\xi) \, dz dx - \int_L^R \int_0^h \frac{\partial}{\partial z}(w\xi) \, dz dx - \int_L^R \int_0^h \frac{\partial B}{\partial x} \, dz dx. \tag{8.2}$$

Let: depth d of the cold pool be such that $d < h$; the boundaries of the volume relative to the cold pool be fixed; and the lower boundary be free slip and impermeable ($\partial u/\partial z = 0$ and $w = 0$ at $z = 0$). Equation (8.2) then reduces to

$$\frac{\partial}{\partial t} \int_L^R \int_0^h \xi \, dz dx = \underbrace{\int_0^h (u\xi)_L \, dz}_{(a)} - \underbrace{\int_0^h (u\xi)_R \, dz}_{(b)} - \underbrace{\int_L^R (w\xi)_h \, dx}_{(c)} + \underbrace{\int_0^h (B_L - B_R) \, dz}_{(d)}, \tag{8.3}$$

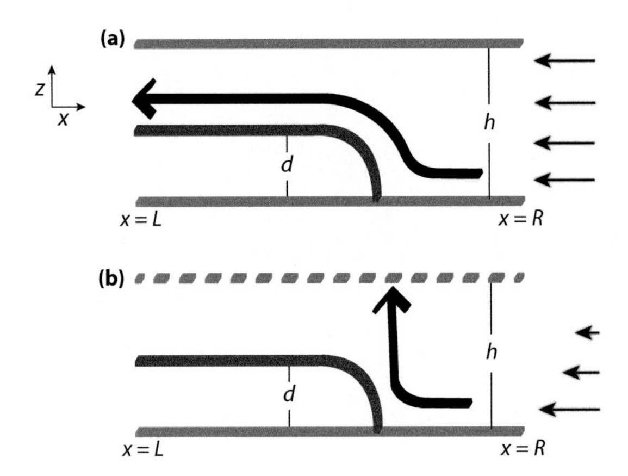

Figure 8.9 Control volume domain. (a) Case of no vertical wind shear, with flow (black ribbon) exiting the left boundary. (b) Case of vertical wind shear, with flow exiting the top boundary. Cold pool boundary has a dark gray outlined.

where

(a) is the horizontal flux of ξ at the left boundary;
(b) is the horizontal flux of ξ at the right boundary;
(c) is the vertical flux of ξ at the top boundary; and
(d) is the net ξ generation due to buoyancy.

We seek a solution to (8.3) in which the respective contributions from the right-hand terms sum to zero, and thus in which the left-hand side vanishes, which we now assume it to do.

Notice that the control volume is specified so that the leading edge of the cold pool is well away from $x = R$, and, in general, so that the flow is horizontal (or quiescent) at the lateral boundaries (Figures 2.7 and 8.9). Accordingly, $w \approx 0$ and thus $\xi \approx \partial u/\partial z$ at $x = L$ and $x = R$, and terms (a) and (b) take the form

$$\int (u\xi)\, dz \approx \int \frac{\partial}{\partial z}\left(\frac{u^2}{2}\right) dz = \frac{u^2}{2}. \tag{8.4}$$

Notice also that, in a relative sense, the buoyancy approaching the cold pool (B_R) is negligible compared with the buoyancy in the cold pool (B_L), which is restricted to $z \leq d$. With the additional simplification that the left, lower boundary is in a region of stagnant flow relative to the moving gust front (see Figure 6.15), and hence that $u\,(x = L, z = 0) = 0$, (8.4) reduces to

$$0 = \frac{u_{L,h}^2}{2} - \left(\frac{u_{R,h}^2}{2} - \frac{u_{R,0}^2}{2}\right) - \int_L^R (w\xi)_h\, dx + \int_0^d B_L dz. \tag{8.5}$$

In (8.5), the second right-hand term represents the low-level environmental wind shear, and the fourth right-hand term can be evaluated as will be shown soon. A clue on how to contend with the first and third right-hand terms comes from recalling that the optimal state of the RKW theory is one characterized by vertically erect updrafts over the outflow. Thus, instead of allowing the flow to exit the left boundary horizontally as it would in the simple case of a density current (Figure 8.9a), we require that the flow exits the domain top ($z = h$) as a vertical jet that arises from a reorientation of the low-level ambient flow by the cold pool (Figure 8.9b). With this requirement, as well as the facts that $d < h$ and the control volume is fixed to the moving cold pool, it can be surmised that the third right-hand term becomes

$$\int_L^R (w\xi)_h \, dx \approx \int_L^R \frac{\partial}{\partial x}\left(-\frac{w_h^2}{2}\right) dx = \frac{w_{h,L}^2}{2} - \frac{w_{h,R}^2}{2} = 0 \tag{8.6}$$

and that $u^2{}_{L,h}/2 = 0$. Equation (8.5) then reduces to the balance condition:

$$\Delta u = c, \tag{8.7}$$

where

$$\Delta u = u_{R,h} - u_{R,0} \tag{8.8}$$

and

$$c^2 = 2\int_0^d B_L \, dz. \tag{8.9}$$

Equation (8.9) is the speed of a density current, having a more familiar form if we let $B_L = g\Delta\theta/\theta_0$, where $\Delta\theta$ is some constant temperature deficit within the cold pool, and θ_0 is the base-state potential temperature.

Equation (8.7) represents the theoretical criterion for an "optimum" squall line. There has been some debate about how well this applies to squall-line intensity and longevity as originally proposed, because squall lines are indeed known to be long-lived, and be associated with multiple reports of severe winds, even when not optimal in terms of (8.7).[14] However, when intensity is quantified through average and maximum surface winds, and total surface rainfall, the most intense squall lines are shown to be the ones in which $\Delta u \approx c$.[15]

The strict physical interpretation of (8.7) is that the export of negative horizontal vorticity generated by the cold pool balances the import of positive horizontal vorticity associated with the low-level vertical wind shear (Figure 8.8b). Greater than optimum low-level shear ($c < \Delta u$) corresponds physically to a state in which cells are triggered at the gust front but then become downshear-tilted. Squall lines tend to be characterized by this state in their early stages of development, especially when the cold pool has not yet fully formed (wherein $c \ll \Delta u$; see Figure 8.8a). Less than optimum low-level shear ($c > \Delta u$) corresponds physically to a state in which cells triggered at the gust front are quickly swept rearward over the advancing cold pool (Figure 8.8c). Squall lines tend to evolve naturally to this stage, owing to a strengthening (and expansion) of the cold pool over time, and with the associated effect of rear inflow. In the real atmosphere, inhomogeneity of the environmental conditions can also upset the balance condition.

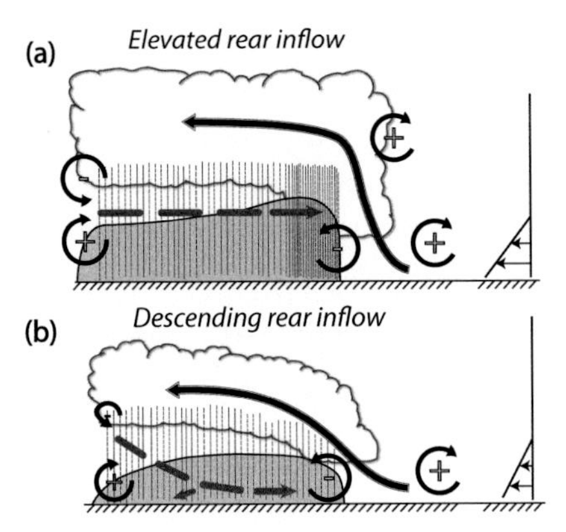

Figure 8.10 As in Figure 8.8, except for a mature MCS. In (a), the rear inflow remains elevated, whereas in (b) the rear inflow descends well behind the leading edge. Based on Weisman (1992).

8.2.2 The Rear Inflow

Besides thermodynamically enhancing the mesoscale descent and cold pool, the rear inflow also plays a dynamical role, especially in mature MCSs. One description of this role is offered through a simple extension of RKW theory: positive horizontal vorticity generation within the rear edge of the cold pool, combined with the negative vorticity generation on the left flank of upshear-tilted updrafts, helps rebalance the system (Figure 8.10a); inclusion of the u_L terms in the condition (8.7) accounts for this effect.[16] From this perspective, the *rear-inflow jet* (RIJ) is a manifestation of horizontal vorticity generation of opposite signs. Implicitly, the RIJ reinforces the vertical lifting of inflow parcels, and thus aids system maintenance.

An alternative description is given in terms of the vertical equation of motion, and involves a more explicit role in parcel lifting. The essence is in how low-level horizontal convergence in association with the outflow is enhanced. As can be deduced from Figure 8.10b, rear inflow that descends to the surface and then spreads laterally well rearward of the gust front contributes only to shallow and/or weak convergence (Figure 8.10b). In contrast, rear inflow that remains elevated until reaching the leading edge contributes to a deep layer of convergence (Figure 8.10a). Recall from (7.15) that dynamic pressure p'_D is high where such "splat" is large, as it is at the gust front (Figures 8.11 and 8.11f). The resultant upward directed vertical pressure gradient force (VP_DGF) extends over a deep layer in the elevated rear-inflow case, leading to deep vertical accelerations at the gust front; in the descended rear-inflow case, vertical accelerations are forced over a relatively shallow layer.

Information about the forcing of the rear inflow itself, and its concentration into a jet, is also given in the 3D pressure field. Apparent in Figures 8.11d,e is the hydrostatically induced mesolow mentioned in Section 8.1. The following analysis indeed shows a significant contribution of buoyancy pressure to the rear inflow. Also apparent in

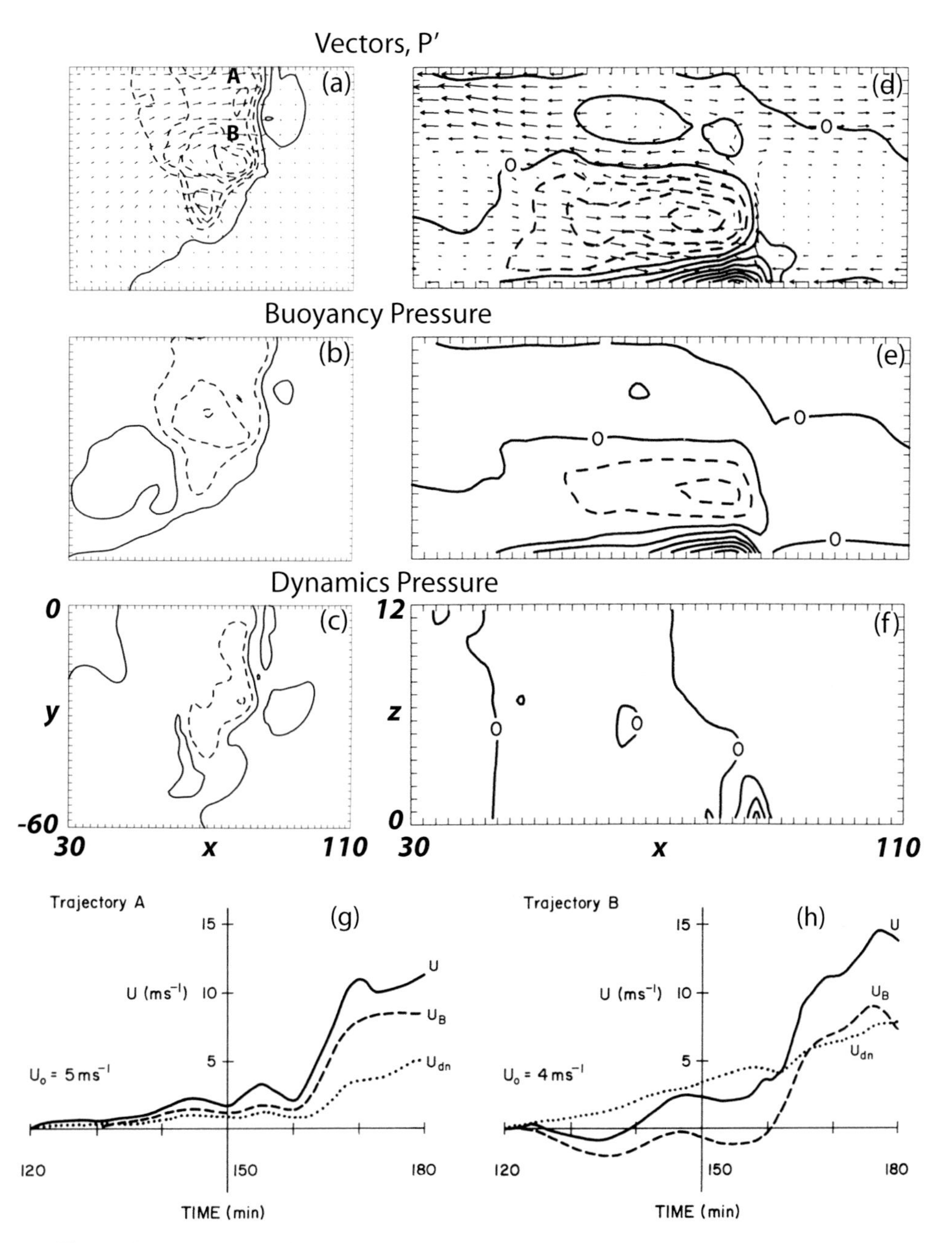

Figure 8.11 Analysis of a numerically simulated bow echo. Midlevel ($z = 2.5$ km) horizontal cross section of (a) perturbation pressure and perturbation horizontal winds, (b) buoyancy pressure, and (c) dynamic pressure, during a mature stage. Corresponding vertical cross section (through $y = 0$ plane) of (d) perturbation pressure and perturbation winds in the plane, (e) buoyancy pressure, and (f) dynamic pressure. Panels (g) and (h) show the time-integrated analysis of the decomposed pressure gradient force for two parcels (indicated in (a)), over a 60-min interval terminating at the time of analysis in (a)–(c). From Weisman (1993).

Figure 8.11, however, is low dynamic pressure and its correspondence to the midlevel line-end vortices. The line-end vortices straddle the RIJ, suggesting an additional connection worthy of our exploration.

Thus, consider the following simple representation of the processes governing a westerly RIJ:

$$\frac{Du}{Dt} = -\frac{1}{\rho_0}\frac{\partial p'}{\partial x}. \tag{8.10}$$

which is an approximate form of the horizontal equation of motion.[17] As done in Chapters 6 and 7 with the vertical equation of motion, the PGF term in this equation is decomposed into contributions from p'_D and p'_B:

$$\frac{Du}{Dt} = -\frac{1}{\rho_0}\frac{\partial p'_D}{\partial x} - \frac{1}{\rho_0}\frac{\partial p'_B}{\partial x} \tag{8.11}$$

(cf. Eqs. (7.13) and (7.14)). The right-hand terms of (8.11) can be evaluated at a specific time using numerical model output or suitable gridded observations, with RIJ generation deduced from the instantaneous fields (Figures 8.11a–c). This approach implicitly assumes that the instantaneous forcing relates directly to the time-integrated effect, which is the case here but may not always be so in other instances (see Section 8.3.2). To avoid this assumption, an alternative approach is to integrate (8.11) along relevant parcel trajectories:

$$u(\vec{x}, t) = u_o(\vec{x}_o, t_o) + u_D(\vec{x}, t) + u_B(\vec{x}, t), \tag{8.12}$$

where the subscript o indicates the initial position and time of the trajectory, and

$$\begin{aligned} u - u_o &= \int_{t_o}^{t} \frac{Du}{Dt}\, dt \\ u_D &= -\int_{t_o}^{t} \frac{1}{\rho_0}\frac{\partial p'_D}{\partial x}\, dt. \\ u_B &= -\int_{t_o}^{t} \frac{1}{\rho_0}\frac{\partial p'_B}{\partial x}\, dt \end{aligned} \tag{8.13}$$

Let us consider an evaluation of (8.12) with output from a numerical simulation of a bow echo. One of the judiciously chosen parcel trajectories in Figure 8.11 originates (at t = 180 min in the simulation) near the apex of the bow echo and then passes, backward in time, through the core of the RIJ (parcel A in Figure 8.11a). Consistent with the discussion above, the integration of (8.12) along this trajectory shows that the buoyancy pressure gradient contributes nearly two-thirds of the time-integrated wind speed of the parcel (Figure 8.11g). The other judiciously chosen trajectory similarly terminates in the RIJ core, but originates near a line-end vortex (parcel B in Figure 8.11a). For this parcel, the time-integrated contributions from the buoyancy

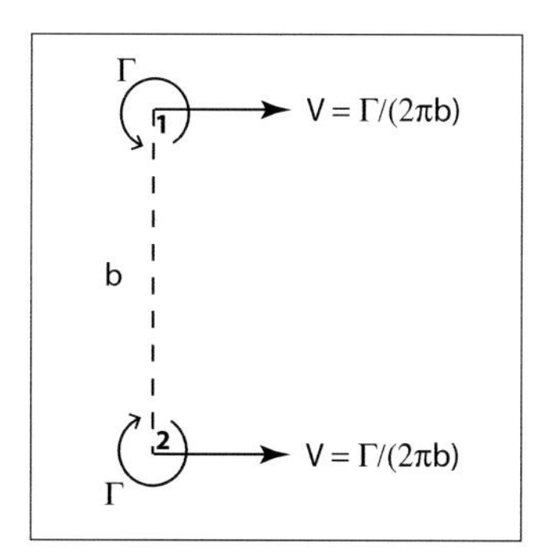

Figure 8.12 Interaction of two line vortices of equal strength. Based on Kundu (1990).

and dynamic pressure gradients are nearly equal (Figure 8.11h), thus quantifying the dynamical connection alluded to previously.

Physically, the dynamical effect of the line-end vortices is analogous to that of the vortex pair that forms at the ends of a knife blade drawn through water. Imagine how the vortex pair continues to move forward relative to the fluid, even after the blade is removed from the water.[18] The forward translation is a result of the mutual interaction of vortices with equal but oppositely signed circulation (Γ). The effect of vortex 1 induced at the location of vortex 2, and, vice versa, the effect of vortex 2 at the location of vortex 1, is an identically oriented velocity with speed $V = \Gamma/(2\pi b)$ (Figure 8.12). The vortex-system translation depends on b, the separation distance between the two vortices, and on the vortex strength.[19] In the context of an MCS, this translation depends on the length scale of the leading-edge updraft and on the strength of the line-end vortices, and equates to an enhancement of the rear inflow.

To reiterate a point made in Section 8.1, these dependencies are actually part of a more comprehensive interaction between the line-end vortices, rear inflow, cold pool, and environment: the vortices develop as horizontal vorticity, generated in the buoyancy gradients of the cold pool, is tilted in updrafts, either near the leading edge or in the more gradual front-to-rear ascent behind the leading edge (see Figure 8.13).[20] The updrafts are enhanced by the RIJ, as described earlier. The cold pool is also enhanced by the rear inflow, but in turn contributes to the rear-inflow forcing. And of course, external to this coupling is the environment, which strongly controls system-scale components and structure (see Section 8.5).

8.2.3 The Presence of a Nocturnal Stable Atmospheric Boundary Layer

A typical assumption in theoretical and numerical modeling studies of MCS dynamics is of a thermodynamic environment with a well-mixed atmospheric boundary layer (ABL). MCSs are, however, frequently sustained (and even form) well after local sunset, in the presence of a nocturnal, stable ABL. The sustenance of an MCS under

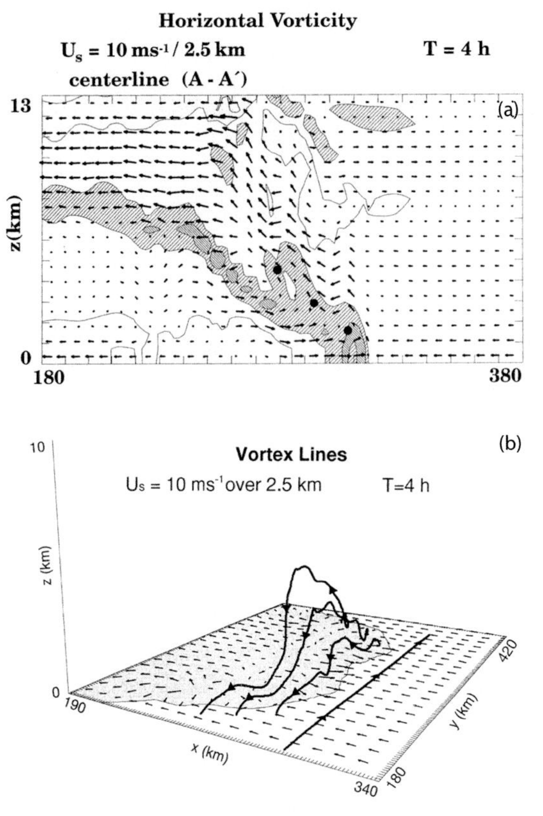

Figure 8.13 Processes associated with line-end vortexgenesis in a simulated bow echo. (a) Vertical cross section, through the bow-echo centerline, of winds and the *y*-component of horizontal vorticity. (b) 3D perspective of vortex lines, showing how the vortex lines are tilted by the bow-echo updraft. The dots in (a) indicate the location where the vortex lines intersect that vertical plane. From Weisman and Davis (1998).

these conditions depends on the thermodynamics of the cold pool relative to that of the environment: when air parcels originating near the ground in a nocturnal ABL are potentially cooler than the air in the cold pool, they flow under ("undercut") rather than being lifted over the cold pool.[21] On the other hand, air parcels from higher levels which are potentially warmer than the cold pool can still be lifted to free convection and hence contribute to convective-system sustenance. Such lifting is said to result in "elevated" convection, with elevated convective cells composing the system.

The occurrence of nocturnal MCSs and elevated convection is enabled largely by synoptic- and (nonconvective) mesoscale processes such as *nocturnal* low-level jets (LLJs). These LLJs are linked dynamically to the diurnal cycle of solar heating, and are known for their ability to rapidly transport warm, moist air poleward during the overnight hours. The vertical component of the nocturnal LLJ additionally aids in parcel destabilization and, in some circumstances, assists in lifting parcels to free convection (see also Chapter 5).[22] Given a jet core level of ~1 to 2 km AGL, the vertical wind profile associated with a LLJ is inherently one of pronounced low-level vertical shear below (and above) the core. Hence, in addition to moisture and instability, the nocturnal LLJ provides environmental shear that benefits storm-scale organization, in the manner described in Section 8.2.1. This benefit applies to parcels that comprise elevated convection, as well as to surface parcels, when/if the MCS has its convective roots in the ABL.[23]

An intriguing aspect of nocturnal MCSs is that some have stages in which the system-generated outflow appears not to play a direct role in the parcel lifting.[24] Rather, the outflow initiates gravity waves and/or bores (see Chapters 2 and 5) that serve to displace air upward. These parcel displacements are modulated by the environmental vertical shear, through its influence on the amplitude of the gravity wave/bore within the low-level stable layer.[25]

Nocturnal MCSs are also intriguing in their ability to generate strong surface winds: the nocturnal ABL should limit the downward penetration of mesoscale downdrafts in the stratiform region, thus limiting transports of rear inflow to the ground (and the generation of low-level mesovortices; see Section 8.3.1); nevertheless, strong and even damaging surface winds are still observed with some nocturnal MCSs. As discussed next, the former also has implications on MCS movement.

8.2.4 Movement

As with the movement of other convective-storm modes, MCS movement is the combined effect of individual convective-cell motion plus the system propagation.[26] The cell motion is primarily an advective effect, and corresponds well to the mean environmental wind of the cloud-bearing layer. Using routinely available observations, a simple estimate of the mean cloud-bearing layer winds is

$$\vec{V}_{CL} = \left(\vec{V}_{850} + \vec{V}_{700} + \vec{V}_{500} + \vec{V}_{300}\right)/4, \tag{8.14}$$

where subscripts indicate the isobaric levels (hPa) of evaluation. It should be understood that the values are from a vertical profile representative of the environment of the MCS.[27]

System propagation $\vec{C}_{PROP}$ encompasses the rate and (system-relative) location of new cell formation, and hence depends jointly on the speed and direction of the gust front as well as that of the low-level environmental inflow. Equation (8.9) accounts for the density current, but other processes contribute to the gust front speed, namely, *convective momentum transport* (CMT). CMT includes horizontal as well as vertical fluxes, but the downward flux of horizontal momentum is most relevant here. Principally, downdrafts in the MCS transport both the environmental flow and the storm-generated horizontal winds (the RIJ) into the surface cold pool.[28] We will see

(a)

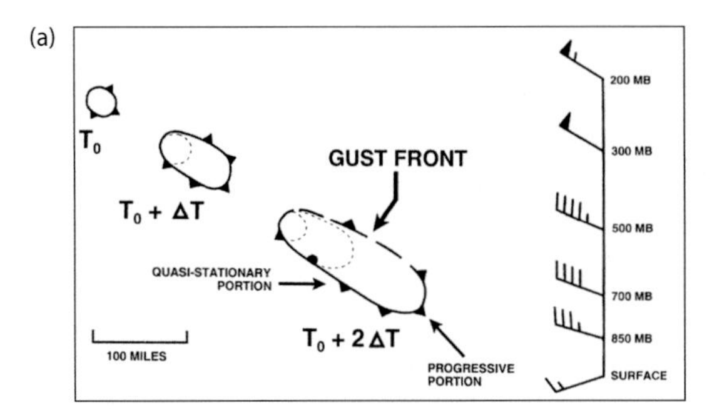

(b)

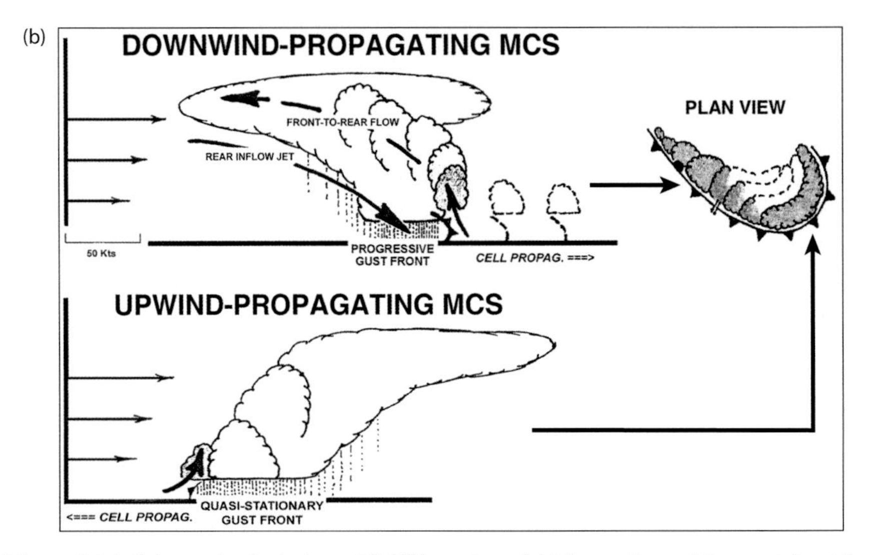

Figure 8.14 Schematic depiction of MCS motion. (a) Elongation of the gust front in the direction of the mean wind. (b) The associated cells can propagate "upwind," along the quasi-stationary portion of the gust front, or "downwind," along the progressive portion of the gust front. From Corfidi (2003).

in Section 8.3.1 that such transport also contributes to the straight-line wind hazard in MCSs.

Combining the cell motion and propagation components gives the MCS movement:

$$\vec{C}_{MCS} = \vec{V}_{CL} + \vec{C}_{PROP}. \tag{8.15}$$

The orientation of motion vector $\vec{C}_{MCS}$ depends on the orientation of the gust front relative to the mean environmental wind. Owing to the downshear transport of hydrometeors, the cold pool will, in time, tend to elongate in the direction of the mean wind, further resulting in gust-front segments that both are parallel (and quasi-stationary) as well as perpendicular (and progressive) to the mean wind (Figure 8.14).[29] Two types of cell propagation are then possible, depending on the availability of environmental moisture and on the static stability relative to the segments. The first is upwind propagation, in which new cells develop in regions of strong low-level convergence

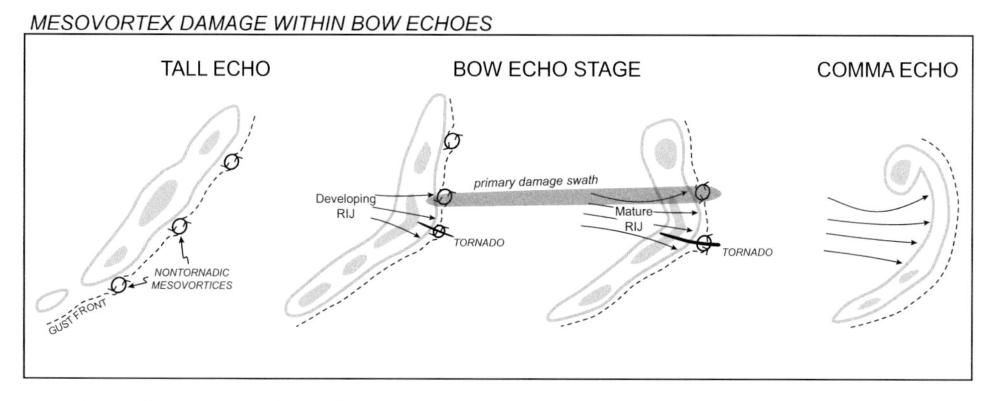

Figure 8.15 The Fujita (1979) conceptual model of a bow echo, as revised by Atkins et al. (2005) to include a mesovortex-induced wind swath.

near the rear or upwind end of the parallel segment, and subsequently move along this gust-front segment, creating a succession or "train" of cells. As discussed in the next section, such propagation often leads to flash flooding. The second type is downwind propagation. Cell development occurs in regions of strong low-level convergence downwind of the perpendicular gust-front segment, thus leading to downwind propagation and a progressive gust front. This explains the tendency of MCSs to become oriented perpendicular to the shear vector.

8.3 Weather Hazards Associated with MCSs

B 8.3.1 Straight-Line Surface Winds from the RIJ

MCSs, and bow echoes in particular, are well known for their propensity to generate damaging "straight-line" winds. In the common conceptual model of a bow echo, intense surface winds occur at the apex of the bow, in association with the RIJ core (Figure 8.15).[30] The RIJ originates at altitudes of few kilometers AGL, extends from tens of kilometers rearward of the convection zone toward the bowed leading edge, and then descends to the ground.

The convex echo shape itself is a result of RIJ-enhanced propagation of convective cells, and often signifies that intense surface winds are present.[31] Bow-echo genesis, then, is literally attributed to RIJ formation.[32]

Recall that the RIJ is forced primarily by horizontal gradients in pressure. Midlevel line-end vortices contribute dynamically to the pressure forcing, with the overall magnitude depending on the coupled influence of other system components, and ultimately on the environment (Section 8.2.2). The relationship between the RIJ and surface winds is also a function of the environment, as can be inferred from arguments made in Section 8.2.2 (see also Figure 8.10). For example, when the low-level shear (e.g., 0–3 km layer) is moderate to strong[33] (e.g., $|\vec{S}_{0-3}| \sim 20$ m s^{-1}), the rear inflow

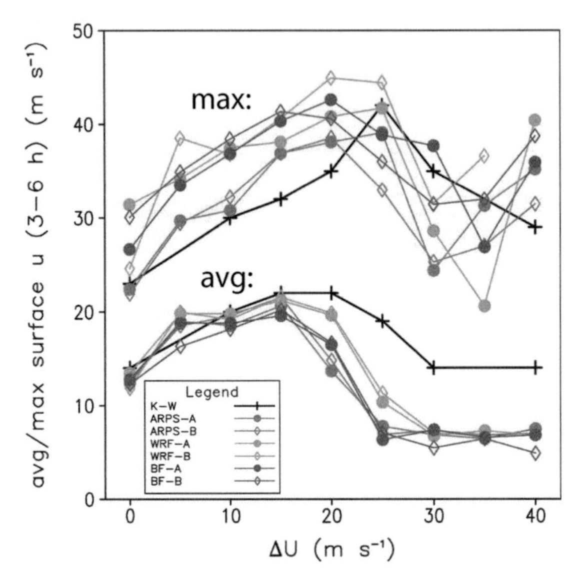

Figure 8.16 Maximum surface winds generated by numerically simulated squall lines as a function of unidirectional environmental shear over the 0–5-km layer. All other parameters, including an environmental CAPE ~2200 J kg^{-1}, are held constant. The simulations are performed using four different numerical models (see Bryan et al. (2006) for details). The maximum surface winds are relatively weaker when the environmental shear exceeds ~25 m s^{-1} because the convective cells become downshear tilted and also organized into more 3D, supercell-like entities rather than MCSs. From Bryan et al. (2006). (For a color version of this figure, please see the color plate section.)

remains elevated until reaching the leading edge (see Figure 8.10a), whereupon it descends (is transported downward) to result in a thin strip of intense, post-frontal surface winds.[34] In contrast, an environment of weak-to-moderate low-level shear (e.g., $|\vec{S}_{0-3}| \sim 10$ m s^{-1}) results in an MCS with rear inflow that descends to the surface and then spreads laterally well rearward of the gust front (see Figure 8.10b), producing strong winds over a relatively larger area, but with a relatively weaker magnitude.

The parameters that distinguish environments of severe MCSs from nonsevere MCSs are further quantified in Section 8.5, but it is convenient here to summarize the specific influence of environmental shear on the elevation of the RIJ, and hence on the generation of severe surface winds. First, the convective dynamics discussed in Section 8.2.1 are used to support the argument that, in the presence of sufficiently large CAPE (e.g., ~2000 J kg^{-1}), updraft intensity increases in step with increases in vertical shear. More intense updrafts beget higher precipitation rates, which in turn contribute to potentially colder outflow air. Finally, colder surface outflow relates to relatively lower pressure in the mesolow and thus, following (8.12), stronger pressure gradient forcing of the rear inflow at midlevels. Figure 8.16 encapsulates the end result, showing an increase in maximum surface windspeeds in model-simulated squall lines as the environmental shear is increased through a range of values up to ~ 25 m s^{-1}.[35]

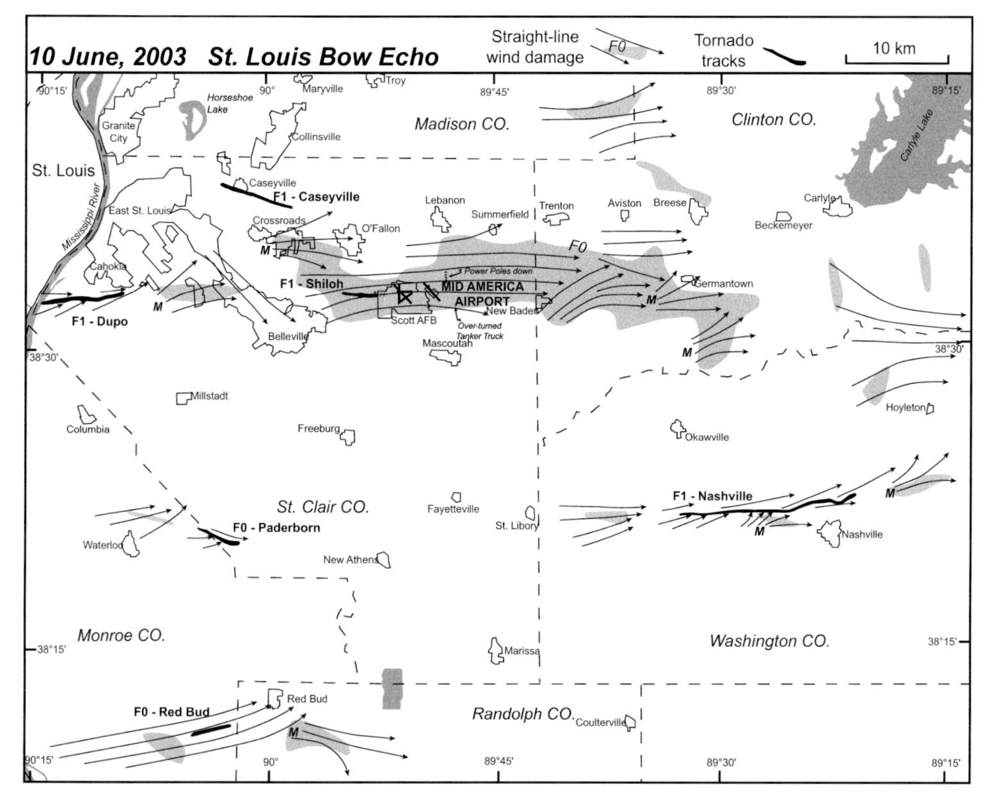

Figure 8.17 Analysis of wind damage from the June 10, 2003 bow echo event that occurred in the vicinity of St. Louis, Missouri. *M*s indicate location of microburst damage, and shading shows regions of F0 or greater straight-line wind damage. From Atkins et al. (2005).

The RIJ's contribution to strong surface winds is enhanced by the downward transport of environmental momentum. This effect is perhaps best exemplified in "cool-season" MCSs (e.g., MCSs occurring during boreal autumn). Despite the typically low CAPE found in cool-season environments, MCSs during this time of year generate damaging winds by exploiting very strong ($\sim$50 m s^{-1}) mid-tropospheric environmental winds. In the relatively higher CAPE and weaker-wind environment of "warm-season" MCSs, on the other hand, the broad swath of RIJ-associated winds may be punctuated with isolated pockets of much stronger winds associated with microbursts (Figure 8.17).[36] Indeed, warm-season environments do allow the formation of strong, cellular updrafts, with corresponding strong downdrafts – even downbursts – within convective systems.[37] All these mechanisms are known to help promote *derechos*, which are particularly long-lived convective events associated with widespread wind damage (Figure 8.18). As originally defined,[38] a derecho is not a unique convective mode, but rather is an MCS that often, though not always, is organized as a squall line/bow echo at some time during its extensive and hazardous lifetime (e.g., Figure 8.18).

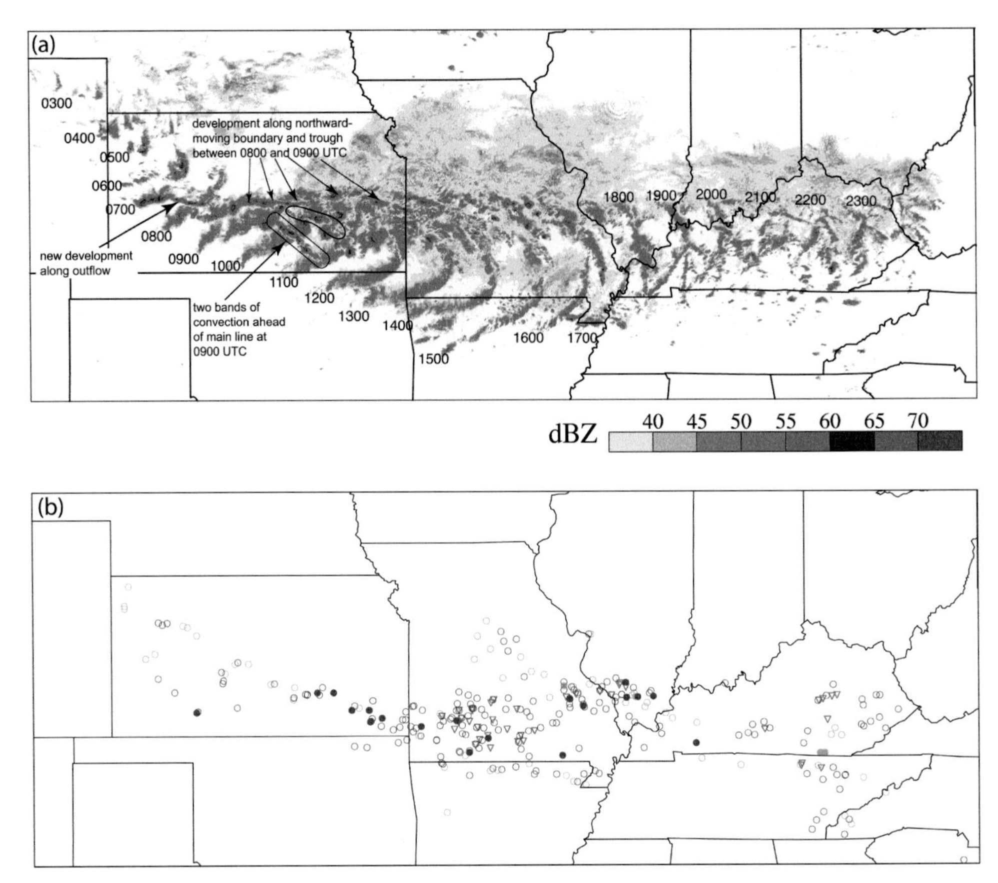

Figure 8.18 Chronology of the derecho that occurred from 03 UTC to 23 UTC, May 8, 2009. (a) Hourly composite radar reflectivity (dBZ). (b) Location of severe weather reports: open (closed) green circles indicate hail $\geq$ 0.75 in ($\geq$ 2.0 in), open blue (closed) circles indicate wind damage or wind gusts $\geq$ 26 m s^{-1} (wind gusts measured or estimated $\geq$ 33.5 m s^{-1}), and red triangles indicate tornado reports. From Coniglio et al. (2011). (For a color version of this figure, please see the color plate section.)

As we will learn next, subsystem–scale (length $\sim$ 5–10 km, time $\sim$ 1 h) vertical vortices that occur at *low levels* ($z \lesssim 1$ km AGL) are additional promoters of hazardous winds in quasilinear MCSs. Although similar in many ways to the midlevel line-end vortices discussed in Section 8.2.2, the predominance of cyclonically rotating vortices suggests a few key differences both in their formation and their dynamical implications.

8.3.2 Mesovortices (and Related Vortical Topics)

The low-level vortices often found in multiple numbers along the leading edge of squall lines and bow echoes are a wind hazard and potential host to tornadoes (see Figure 8.19). Such vortices have been termed *mesovortices* to distinguish them

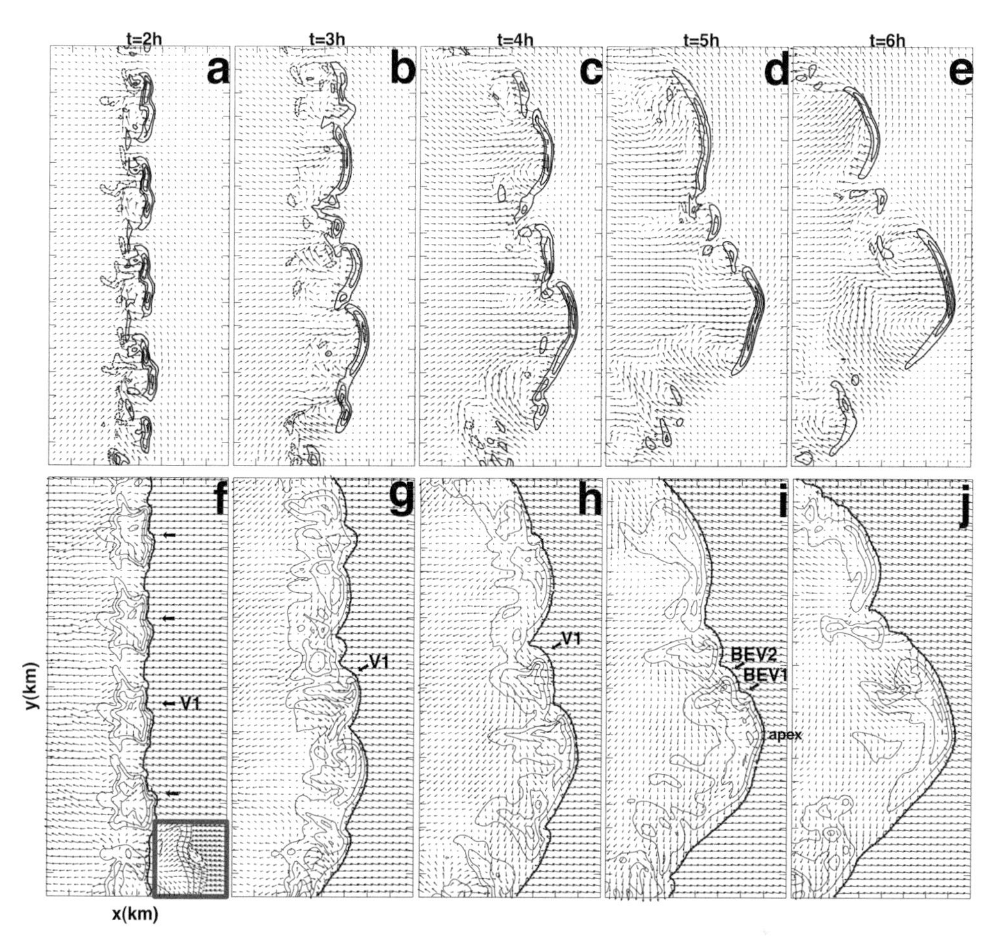

Figure 8.19 Evolution of the horizontal structure of a simulated squall-line bow echo. (a)–(e) Horizontal wind vectors, and contours of vertical velocity (5 m s^{-1} interval) at $z = 3$ km, at $t = 2$, 3, 4, 5, and 6 h. (f)–(j) Horizontal wind vectors, rainwater mixing ratio (solid contours at 1, 3, and 5 g kg^{-1}), and perturbation potential temperature (dashed contour, at –1K), at z = 0.25 km, at $t = 2$, 3, 4, 5, and 6 h. In (f), the inset shows vortex V1 in a 15 × 15 km subdomain. From Trapp and Weisman (2003).

from supercell mesocyclones, which have similar scales. Their genesis has been attributed – although not always justifiably – to the release of a horizontal shearing instability (HSI) (see Chapters 2, 5, and 6). Support for this attribution draws from observations and simulations of MCSs exhibiting cyclonic-only mesovortices rather than cyclonic-anticyclonic couplets. Low-level mesovortexgenesis can also be explained through the tilting of horizontal vorticity and subsequent vortex stretching, with nuanced differences in the possible means by which these processes are brought about.[39] This latter mechanism is shown schematically in Figure 8.20, in the context of an *early-stage* MCS. Here we see tilting, by a rainy downdraft, of crosswise horizontal vorticity generated within the low-level horizontal buoyancy gradient of the surface cold pool; in a mature MCS, convective downdrafts just rearward of the

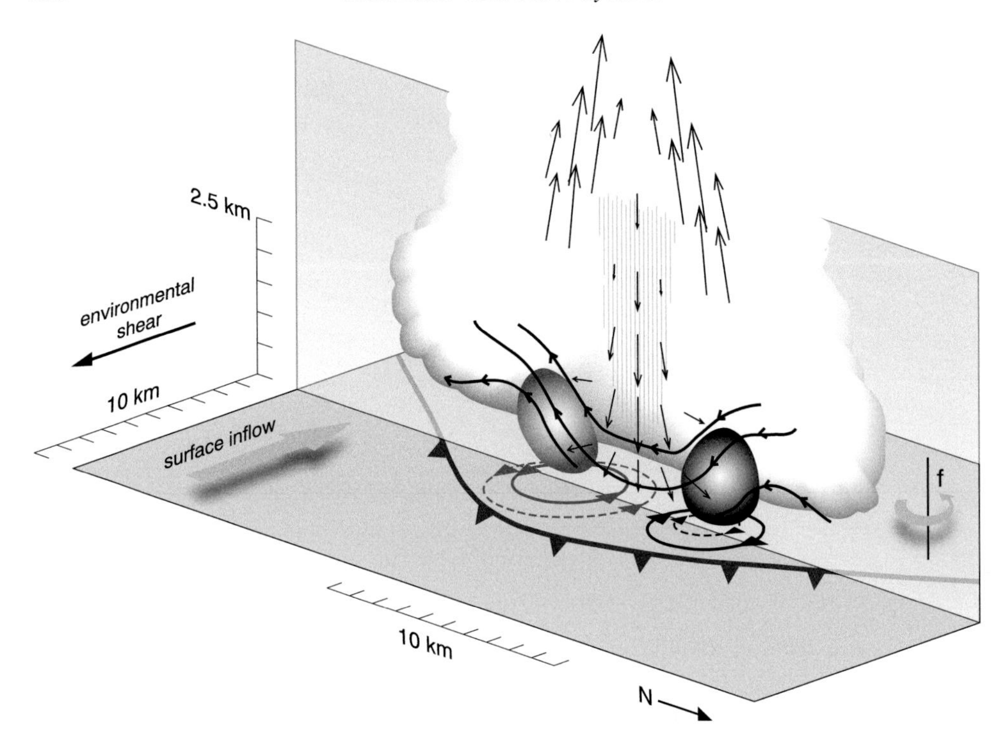

Figure 8.20 Schematic of mesovortexgenesis in the early stages of a quasilinear MCS. Vortex lines (black) are tilted vertically by the downdraft (vectors, and blue hatching), to result in a surface vortex couplet (cyclonic vertical vorticity is red; anticyclonic is purple). The dashed red and purple circles represent the future state of the vortex couplet, which is due in part to the stretching of planetary vorticity (f) as shown. During the mature stage, relevant vortex lines would have opposite orientation, and thus the resultant vortex-couplet orientation would be reversed. From Trapp and Weisman (2003). (For a color version of this figure, please see the color plate section.)

leading updrafts tilt the crosswise horizontal vorticity associated with, but below, the RIJ core.[40] Although crosswise-vorticity tilting results in a low-level vortex couplet, the cyclonic member is preferentially, and rapidly, intensified through stretching of the newly generated positive relative vorticity as well as through stretching of planetary vorticity. The anticyclonic member, on the other hand, is weakened and eventually diminished by stretching of planetary vorticity. Hence, planetary vorticity breaks the symmetry initially set through tilting of crosswise horizontal vorticity.

Let us pause here to note that an inherently time-dependent process such as vortexgenesis is revealed most convincingly through a time-dependent analysis of the appropriate equation. This was the motivation behind the procedure used to diagnose RIJ generation in Section 8.2.2 (see (8.12) and (8.13)), and now is used to diagnose low-level vortexgenesis. Thus, consider the inviscid form of the absolute vertical vorticity equation,

$$\frac{D\zeta_a}{Dt} = \vec{\omega}_H \cdot \nabla w - \zeta_a \nabla \cdot \vec{V}_H, \tag{8.16}$$

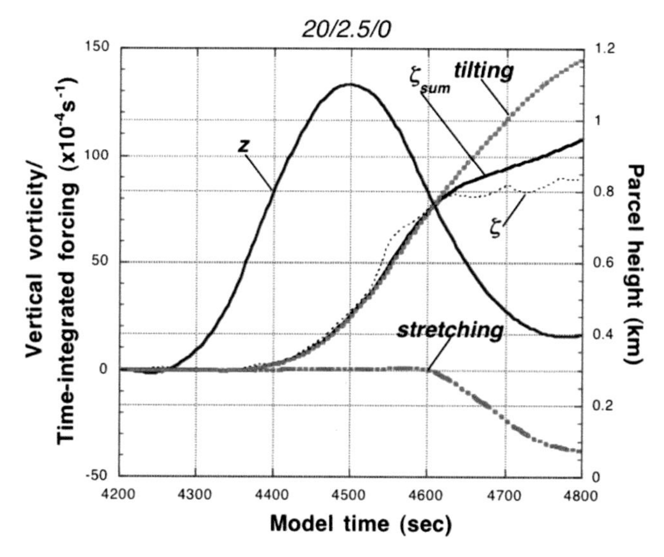

Figure 8.21 Time-integrated analysis of terms of the vertical vorticity equation (Eq. (8.17)), along a parcel trajectory that terminates in a simulated bow-echo mesovortex: z is the parcel height, ζ_{sum} is the sum of the time-integrated contributions from vortex stretching and tilting, and ζ is the instantaneous vertical vorticity at the parcel location. From Trapp and Weisman (2003).

where $\zeta_a = \zeta + f$ is the absolute vertical vorticity. Equation (8.16) is integrated along a parcel trajectory as

$$\zeta_a(\vec{x}, t) = \zeta_a(\vec{x}_o, t_o) + \int_{t_o}^{t} \vec{\omega}_H \cdot \nabla w \, dt - \int_{t_o}^{t} \zeta_a \nabla \cdot \vec{V}_H \, dt, \tag{8.17}$$

where the subscript o again indicates the initial position and time of the trajectory. The basic procedure – for the current problem as well as for supercell-mesocyclone investigations – is to populate a region of interest (here, the vortex or vorticity maximum) with parcels, compute the trajectories of these parcels backward in time over a period of tens of minutes, and then evaluate (8.17) over this interval. Parcels that terminate in a mature vortex will inevitably have a *vorticity budget analysis* that is dominated almost exclusively by the effects of vortex stretching. Hence, to identify the origin of rotation, one would choose an integration period that ends at a time when the vortex is just beginning to become significant,[41] which can be quantified by some threshold value of vertical vorticity (such as $\zeta = 10^{-2}\ \mathrm{s}^{-1}$). An evaluation of (8.17) along a trajectory that terminates in such a *nascent* mesovortex is given in Figure 8.21. Noting that the largest gains in ζ correspond to parcel descent, this analysis confirms the generation of cyclonic vertical vorticity through tilting in a downdraft.[42] Parcels with forward trajectories that originate in this nascent mesovortex show subsequent rapid gains in vertical vorticity through stretching of relative as well as planetary vorticity.

The emphasis on the role of planetary vorticity in the genesis of a meso-γ-scale vortex may be surprising to the reader, especially given traditional scale analyses (e.g., Chapter 2) that often argue the neglect of Coriolis force terms because of their comparably small magnitudes at this scale. However, recall that the traditional scale analysis quantifies only the instantaneous, order-of-magnitude contribution of terms in the governing, time-dependent equations; the scale analysis does not, on the other hand, account for the time-integrated effect of these terms. To illustrate such effect, let us approximate (8.16) as we similarly did in Chapter 7:

$$\frac{1}{\zeta_a}\frac{D\zeta_a}{Dt} \approx \mathcal{C}, \tag{8.18}$$

where $\mathcal{C} = -\nabla \cdot \vec{V}_H$ is horizontal convergence. Integrating (8.18) under the assumption that $\mathcal{C}$ is constant over the time interval $0 \leq t \leq \tau$, we can solve for τ:

$$\tau = \frac{1}{\mathcal{C}} \ln\left(\frac{\zeta_a}{\zeta_{a_o}}\right). \tag{8.19}$$

Let $\mathcal{C} = 10^{-3}\ \mathrm{s}^{-1}$ over the interval, and let initial relative vertical vorticity be zero, with the implication that $\zeta_{a_0} = f \sim 10^{-4}\ \mathrm{s}^{-1}$ (at midlatitudes). The time τ required to amplify, through vortex stretching, this initially small vertical vorticity to a value $\zeta_a \sim 10^{-2}\ \mathrm{s}^{-1}$ is 1.2 h. If the horizontal convergence is doubled, then $\tau = 0.26$ h. Historically, such calculations were used to argue the role of planetary vorticity in the development of supercell mesocyclones and tornadoes. However, idealized cloud models routinely produce strong, low-level mesocyclones in supercells when Coriolis forcing is absent from the model. The same idealized cloud models fail to simulate low-level mesovortices in MCSs without inclusion of Coriolis forcing.[43]

It is prudent now to return to the topic of Section 8.3, because modeling and observational studies have associated low-level mesovortices with intense, *nontornadic* surface winds. The revised conceptual model in Figure 8.15 depicts mesovortex-induced wind swaths north of the bow-echo apex, in addition to a RIJ-induced swath behind the apex (and localized microbursts). To be clear, the total ground-relative wind is a sum of the MCS motion plus the mesovortex-induced winds, which in particular is why the mesovortex winds are characterized as "straight-line" rather than "rotating." The strongest ground-relative winds are along the right-front quadrant of the vortex with respect to the direction of motion; for an eastward-moving system, this will be on the southern flank of the vortex.

In addition to vortex signatures in Doppler velocity (see Chapter 3), the Doppler-radar presentation of mesovortices often includes hooklike appendages and weak-echo regions in radar reflectivity factor. It might seem logical, then, to interpret these as observations of supercells "embedded" within the mesoscale-precipitating system. However, further thought should lead to a realization that this interpretation is

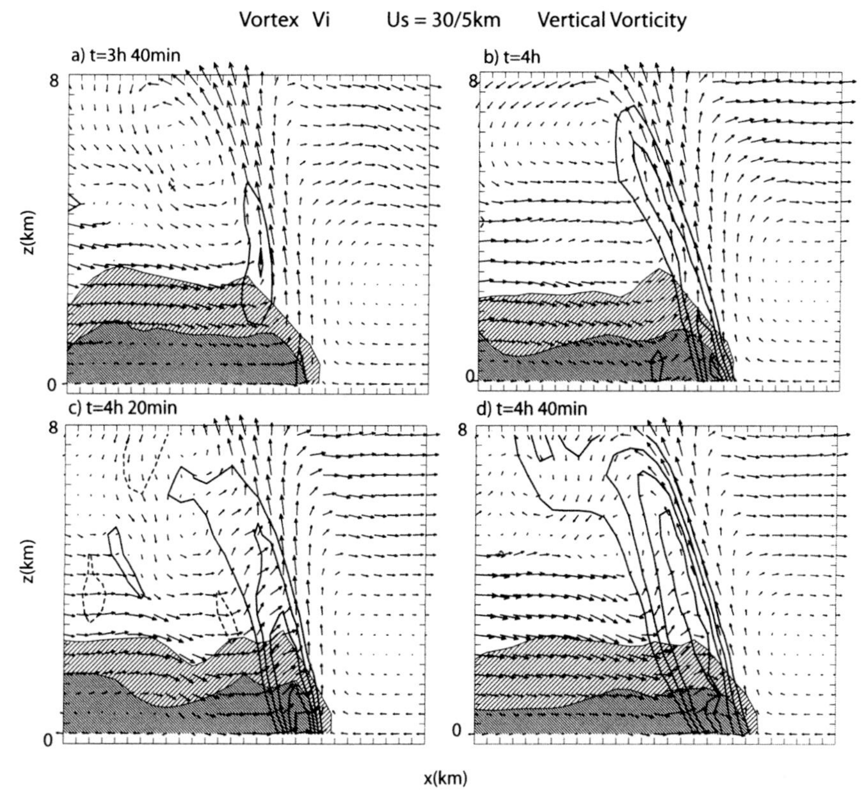

Figure 8.22 Vertical cross section of wind (vectors), vertical vorticity (contours; 0.0004 s^{-1} interval), and perturbation potential temperature (dark hatching, < –4 K; light hatching, –1 to –4 K), through a mesovortex within a numerically simulated bow echo. From Weisman and Trapp (2003).

inconsistent with supercell structure and dynamics. For example, high spatial correlation between MCS updraft and vortex tends to be limited to the lowest few kilometers of the system, especially in mature MCSs, which have at least some upshear tilt (Figure 8.22). Indeed, calculations of the linear correlation coefficient $r(w', \zeta')$ over the depth of numerically simulated mesovortices yield values of $\sim$0.1 to 0.2, which are small compared with typical values in isolated supercells.[44] Consider also that the forcing of MCS updrafts is typically maximized within a surface-based, several-kilometer layer. The primary contributor to this forcing is the cold pool (Figure 8.23), with the important implication that mesovortices do not play a significant role in the system propagation. As discussed in Chapter 7, this is in contrast to the typical case of a supercell, in which the rotationally induced dynamic pressure forcing (as well as the buoyancy pressure forcing) is maximized at middle levels, and plays a critical role in supercell propagation. Thus, vortices in squall lines and bow echoes do not exhibit the dynamical hallmarks of supercells.

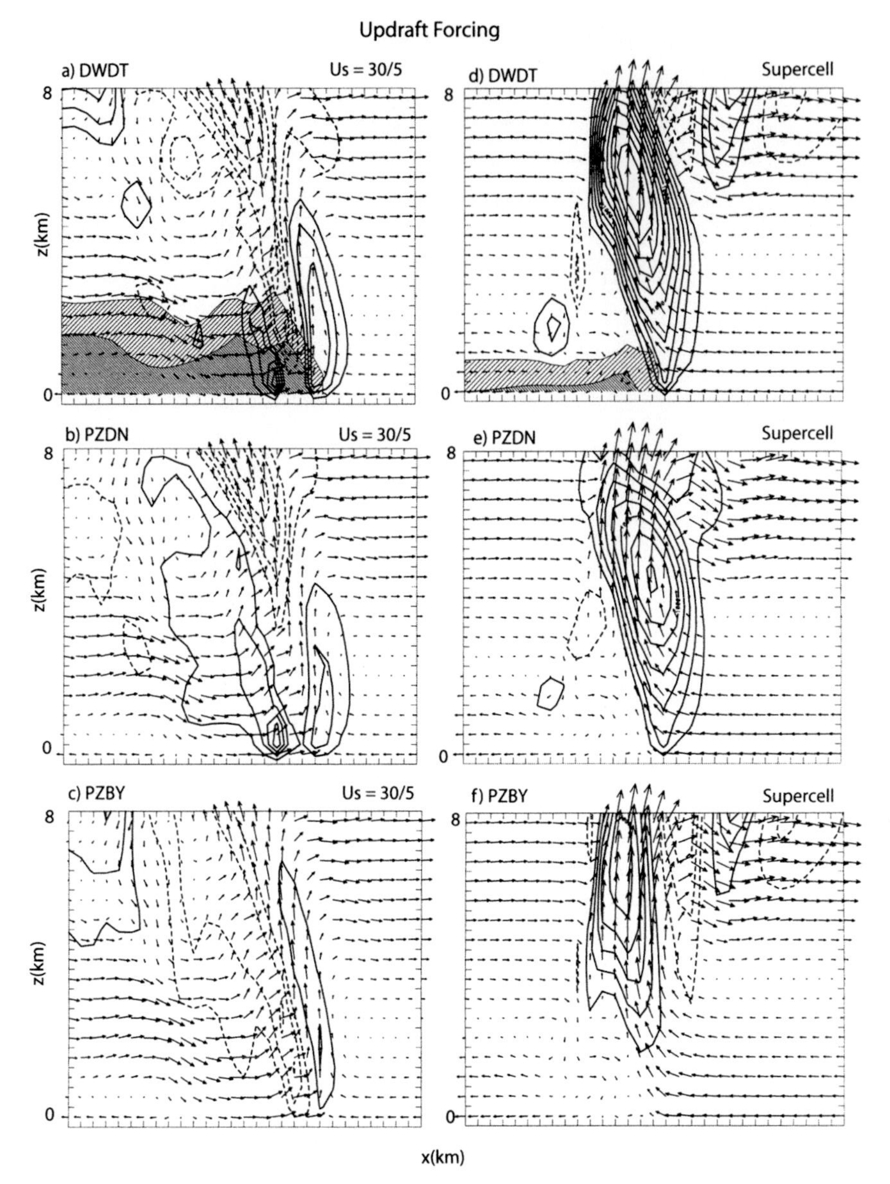

Figure 8.23 Comparison of updraft forcing, evaluated using decomposed pressure in the vertical equation of motion, between a simulated bow echo (a)–(c), and simulated supercell (d)–(f). DWDT is the total forcing, PZDN is the dynamic pressure forcing, and PZBY is the buoyancy pressure forcing (see text). Contour interval is 0.002 m s^{-2}, with the zero contour omitted. From Weisman and Trapp (2003).

8.3.3 Tornadoes

Mesovortices north of the bow-echo apex are known to host tornadoes, as are LEWPs and the rotating comma head (or northern line-end vortex) of a single bow echo.[45] More than one mechanism has been offered to explain the associated tornadogenesis,

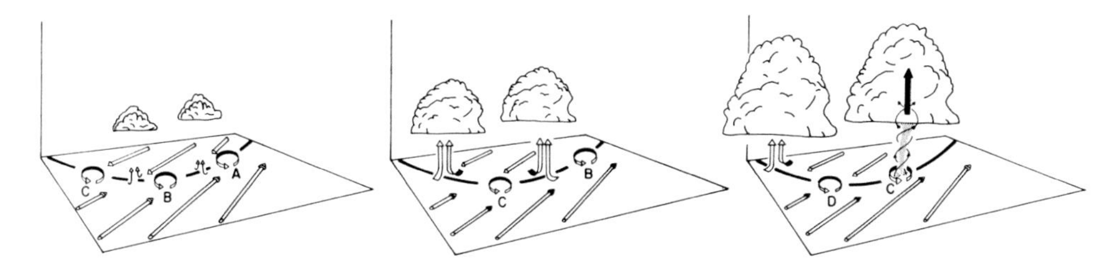

Figure 8.24 Conceptual model of nonsupercellular tornadogenesis. From Wakimoto and Wilson (1989).

including (1) the concentration of vertical vorticity following low-level mesovortexgenesis as just described and (2) interactions between an external "boundary" and the MCS (see Chapter 9). In both, the tornadogenesis is characterized as "nondescending," because it begins first at low levels, and then progresses upward (e.g., Figure 8.22; see also Chapter 7).

A third mechanism, which encapsulates *nonsupercellular tornadogenesis*, begins with the release of HSI, and then is followed by vortex merging and growth.[46] As originally conceived, this mechanism involves the positioning of cumulus congestus or cumulonimbi over a low-level horizontal shear zone (vertical-vortex sheet) that becomes unstable.[47] Recall from Chapter 5 that the deep cumuli themselves can be linked to misocyclones that evolve out of the HSI release. The misocyclones are subsequently stretched into tornado-scale vortices by the convective updrafts (Figure 8.24). In the case of quasilinear MCSs, the horizontal shear zone is associated with the extensive gust front, and the MCS updrafts are the agents of stretching. The resultant tornadoes are often weak, especially when the MCS updrafts are especially sloped or vertically limited. Strong and violent tornadoes in MCSs have been observed, however.

8.3.4 Heavy Rainfall and Flash Flooding

Flooding is another recognized MCS hazard. The particular flooding that can follow heavy rainfall over a time scale of $\sim$6 h or less is, by definition, *flash flooding*; flooding over longer time scales (periods of days or weeks) is usually referred to as *river flooding*. Surface hydrological processes are critical in both types of events, but the discussion herein is limited to meteorological processes, particularly of flash floods.

Although a range of precipitating-system types can generate large to even extreme amounts of rainfall and hence contribute to flooding, convective systems having "training line/adjoining stratiform" (TL/AS) precipitation, and "backbuilding/quasi-stationary" (BB) precipitation (Figure 8.25), appear to be the most prolific.[48,49] These systems share a common trait of being able to generate large rainfall rates R, over a several-hour duration D, at some point in space.

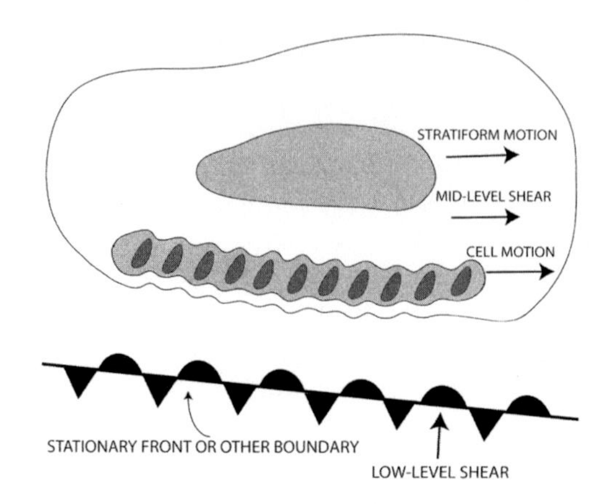

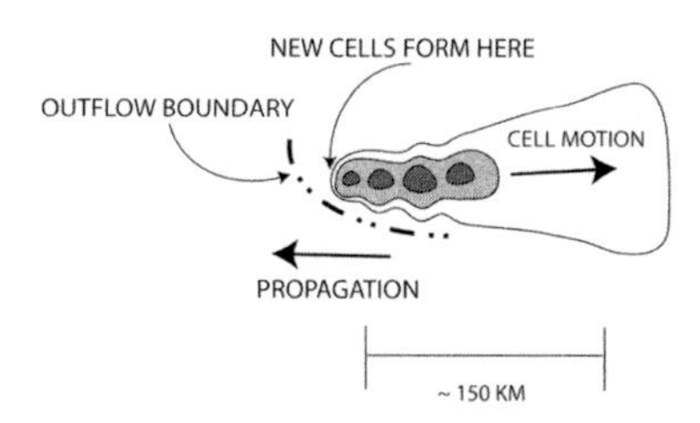

Figure 8.25 Schematic weather-radar depiction of (a) training line/adjoining stratiform and (b) backbuilding/quasi-stationary MCS organizations. From Schumacher and Johnson (2005).

Consider R, expressed here as

$$R = \langle w\, q \rangle \; P_e, \tag{8.20}$$

where $\langle w\ q \rangle$ is a measure of the upward flux of moisture, and P_e is the precipitation efficiency.[50] Fundamentally, rainfall rate depends on the amount of water vapor that is condensed into a cloud, and on the fraction of this condensate that returns to the ground as rain. The precipitation efficiency can be related to the time- and volume-integrated mass of water precipitated to the ground, divided by the integrated input water mass, where the volume is that of (and in the reference frame following) the convective precipitating system. Implicitly, P_e must account for the amount of the condensate that: is transported downwind of the precipitating cloud (only later to evaporate); is evaporated in-cloud, owing to entrainment of relatively dry ambient air; and reaches the ground as precipitation. The ambient humidity and winds, the microphysical details such as drop size distribution, and the rate of entrainment all influence P_e. In this regard, it is relevant to note that the deleterious effects of entrainment are reduced in MCSs, as embedded cells are more insulated from the undisturbed environment than are isolated cells.

Duration D depends on the precipitating system's size and movement, especially that relative to the orientation of the system's major axis. The effective system movement includes that of individual cells (see Chapter 6): cell motion in the direction parallel to the MCS major axis gives the impression of a "train" of cells traveling along the same hypothetical "track" (Figure 8.25a); a quasi-stationary frontal boundary facilitates such "training." Movement also includes propagation, and hence the manner in which the cells develop and dissipate. Cell "backbuilding" and the BB organization applies here, with new cells initiating in advance of a gust front that moves in a direction opposite to that of the deep-layer winds and, hence, to that of the cells (Figure 8.25b). Finally, large D results simply from MCSs with large regions of active convection. These anomalously large systems merit further discussion, as is provided in the next section.

8.4 Mesoscale Convective Complex

In the spectrum of organized convective storms, there is a class of especially large, long-lived, multi-cellular convective systems known as mesoscale convective complexes (MCCs). MCCs are observed worldwide, including at tropical latitudes.[51] MCC anvil shields and the associated region of coldest cloud tops appear nearly circular in infrared (IR) satellite imagery (Figure 8.26). This characteristic allows for an objective classification,[52] namely, that:

(1) The anvil cloud shield has a continuous IR brightness temperature $\leq -32°$C over an area $\geq 100{,}000$ km^2.
(2) This outer shield contains an interior cloud shield with brightness temperature $\leq -52°$C over an area $\geq 50{,}000$ km^2.
(3) The eccentricity of the IR-defined cloud top is ≥ 0.7 when the interior cloud shield reaches its maximum size.

An additional requirement is that these conditions persist for at least 6 h.

Given this time scale and a nominal anvil-shield diameter of $\sim$350 km, it should be apparent from the discussion in Section 8.3.4 that flash flooding is one hazard often attributed to mature MCCs. MCCs can also spawn tornadoes, hail, and damaging winds, especially during the MCC genesis stage, which is composed of individual convective cells.

Implicitly, the MCC classification scheme accounts for the total area and intensity of convective ascent. The morphological pathway leading to MCC organization is not, however, a consideration in the satellite-based criteria listed previously. For this, we consult weather radar observations collected simultaneously with the satellite observations. Two basic MCC types are suggested.[53] The canonical, "type-1" MCC is formed from cells that initiate above the cold air of a synoptic-scale front; initiation is added by the lift/destabilization associated with lower-tropospheric winds oriented

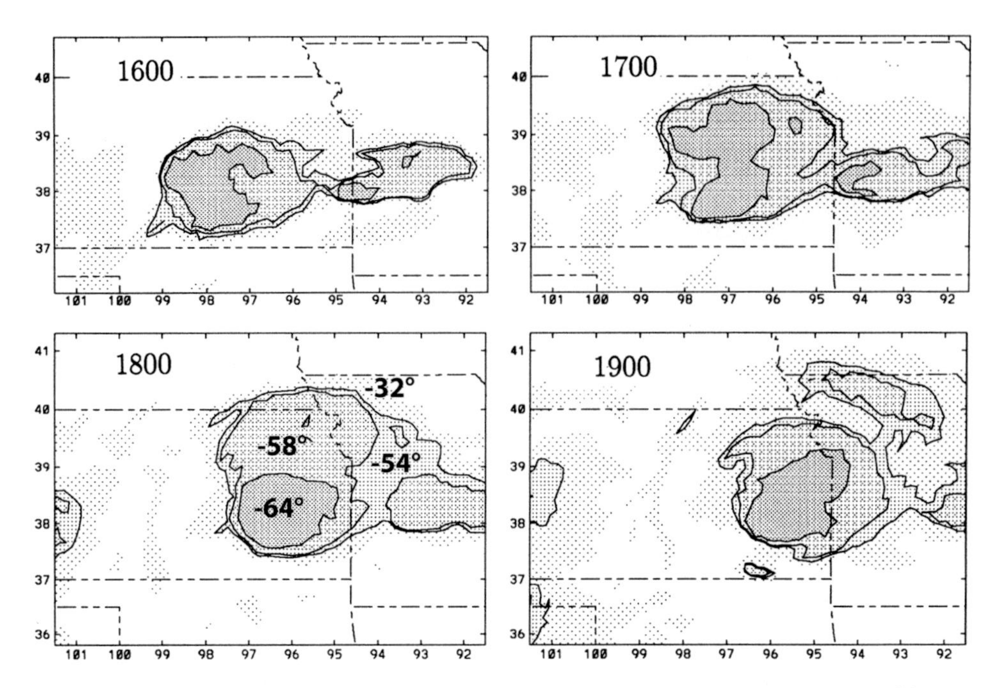

Figure 8.26 Example of a mesoscale convective complex, as shown in IR satellite imagery at four different times (UTC). Brightness temperatures and corresponding shading are indicated. From Nachamkin et al. (1994).

perpendicular to the frontal zone. The downdrafts and attendant rain-chilled air of these cells typically have insufficient negative buoyancy to penetrate the vertical extent of the frontal air, and consequently are able to form only weak mesohigh pressure (perturbations of a few hPa) and a weak cold pool at the surface. Subsequent convective growth and development arises from a diabatic coupling with the larger-scale flow. Within the region of convection, release of latent heat of condensation elicits a response in the form of convergence of mid-tropospheric air and ascent. The mid-tropospheric convergence acts to amplify planetary vorticity, leading to a mid-tropospheric MCV (see Chapter 9) and mesolow pressure. This 3D mesoscale circulation helps sustain the convective complex in the presence of sufficient environmental CAPE and water vapor.

"Type-2" MCC events are borne out of quasilinear MCSs, and hence out of boundary-layer-based convective processes. These events are most apt to form in the warm sector of midlatitude cyclones, where the environmental CAPE and low-level vertical wind shear are substantial. The upscale growth of the MCS to MCC size occurs in concert with the development and strengthening of the surface cold pool, and of midlevel, system-scale vortices. Ultimately, type-2 MCC events can also be associated with MCVs.

MCCs are mostly nocturnal events, and linked often to the presence of a nocturnal, physio-geographically forced LLJ; this is offered as one explanation why MCCs occur in specific geographical regions such as the central United States, even though the general meteorological conditions supporting them, such as a mobile short-wave

Table 8.1 *Average characteristics of the environmental wind profiles of intense, long-lived MCSs. Compiled from Thorpe et al. (1982), Weisman et al. (1988), Weisman and Rotunno (2004), and Cohen et al. (2007).*

	Magnitude	Layer, Depth
Vertical wind shear	15 m s^{-1}	0–2.5 km
	30 m s^{-1}	0–5 km
	30 m s^{-1}	0–10 km
System-relative inflow	20 m s^{-1}	0–1 km
Ground-relative mean wind	20 m s^{-1}	6–10 km

trough and warm-advection at the 850-hPa level, are fairly common.[54] Accordingly, the demise of MCCs follows the daytime cessation of the nocturnal LLJ and its isentropic transport of high θ_e air. If an MCV has formed, it can persist well past the MCC demise; in fact, the MCV dynamics can help initiate new convective cells that in turn can evolve into a new MCC (or MCS; see Chapter 9).

8.5 Environments

MCSs are known to exist within vastly different thermodynamic environments, including those nearly devoid of *surface-based* static instability. However, the most intense and long-lived MCSs appear to be favored when mean-layer (ML) CAPE is moderate to large ($\gtrsim$2000 J kg^{-1}) and the corresponding temperature lapse rate is $\gtrsim$7° km^{-1} over mid-tropospheric layers.[55] Other parameters have been proposed as discriminators of such intense MCS environments from weak MCS environments. For example, a vertical difference in θ_e has been used to quantify the potential dryness of the downdraft-bearing layers, and the potential moistness of the low levels. Sounding data analyses suggest that $\Delta\theta_e \lesssim -25$ K over a 0–5-km layer provides a good indication of an environment supportive of strong downdrafts and hence of the possibility of a derecho event.[56]

Because strong vertical drafts exist in MCSs and supercells alike, there is overlap in the thermodynamic environments ascribed to these convective modes. Their respective environmental wind profiles, on the other hand, have noticeable differences on average: relative to a composite hodograph from environments of tornadic supercells, the composite squall line hodograph exhibits less curvature in the low levels, and in general, the orientation of the shear vector has less variation with height (see Figure 6.19). Environments of (severe) squall lines are also characterized often by strong vertical shear in the low levels (e.g., 0 to ~3-km layer), and weaker shear aloft[57] (Figure 8.27). There has been considerable debate over whether other characteristics of the hodograph might also be most important in squall-line development and maintenance.[58] Table 8.1 summarizes some of these characteristics, which include approximate values of mid-tropospheric shear and mean wind.

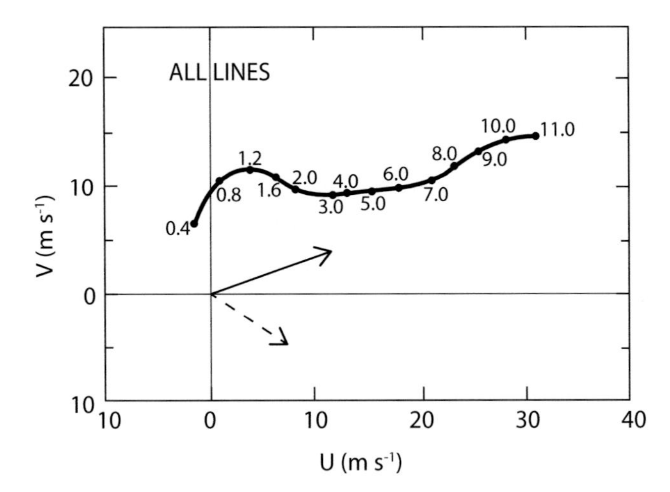

Figure 8.27 Composite hodograph constructed from radiosonde observations of the environments of 40 severe squall lines in Oklahoma. Numbers on hodograph are heights (km MSL). Solid arrow indicates mean cell motion, and dashed arrow indicates mean line motion. From Bluestein and Jain (1985).

These antecedent conditions arise from relatively distinct synoptic-scale patterns, the most common of which consists of a 500-hPa trough upstream of the (eventual) severe MCS occurrence.[59] The associated midtropospheric flow east of the trough axis is often coupled with a similarly oriented LLJ (Figure 8.28), which provides for the nearly unidirectional shear profile; the LLJ also supplies significant low-level heat and moisture, and hence contributes to the static instability. This is sometimes termed the *dynamic pattern*, which is observed in conjunction with severe (and derecho-associated) MCSs at all times during the year.[60] A recurring pattern during the summer months consists of a 500-hPa ridge, coupled to an anticyclonically curved LLJ and a quasi-stationary surface frontal boundary (Figure 8.28). This *warm-season pattern* is often characterized by a northwesterly flow aloft and thus with a midtropospheric ridge axis upstream of the MCS location.[61] A ridge with axis either downstream or proximate to the MCS location also qualifies as a warm-season pattern, as does a situation with midtropospheric zonal flow, when coupled to an 850-hPa flow and surface thermal boundary that are nearly parallel to the flow aloft.[62] Although transverse circulations induced by jet streaks may be present in some of these situations, the general implication is that the synoptic-scale (e.g., quasi-geostrophic) forcing in warm-season patterns is weak relative to that found in association with the dynamic pattern. These environmental characteristics have seasonal (and regional) tendencies, with warm-season environments having relatively weaker shear, higher CAPE, and less synoptic-scale forcing, and cool-season environments with stronger shear, less CAPE, and more synoptic-scale forcing.[63]

Although originally identified for midlatitude MCSs, the warm-season conditions have a tropical analogue.[64] Not coincidentally, MCS organization as a squall line is

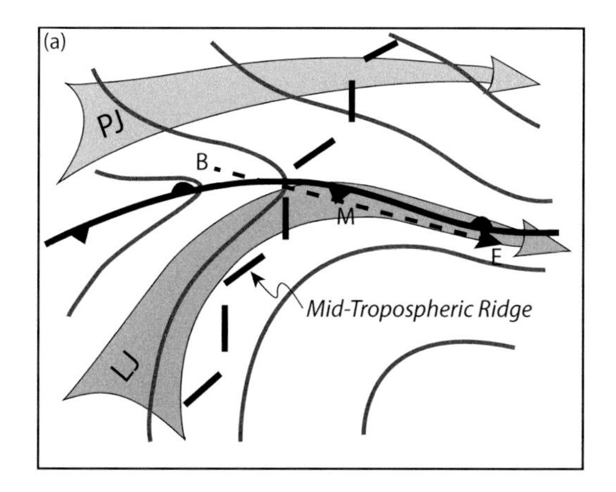

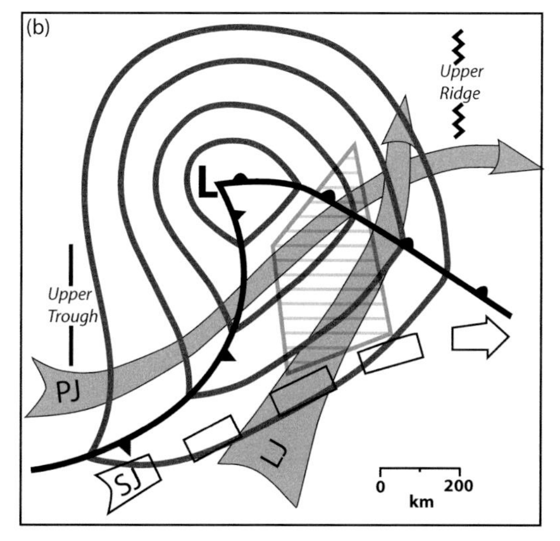

Figure 8.28 Idealized synoptic patterns for long-lived, damaging MCSs (derechos). (a) "Warm-season" pattern. (b) "Dynamic" pattern. PJ, LJ, and SJ denote polar jet, low-level jet, and subtropical jet, respectively. In (a), the line B-M-E represents the likely track of the convective system. In (b), the hatched region indicates the likely region of MCS development. From Johns (1993).

common in the tropics, especially within oceanic regions. Tropical environmental CAPE and wind shear support squall lines with the same dynamical behavior as at midlatitudes, and also with some of the same basic attendant weather hazards, especially heavy rain. This organized convection also plays a critical role in the heat and momentum transports within the larger-scale tropical atmosphere. These will be discussed more in Chapter 9.

Supplementary Information

For exercises, problem sets, and suggested case studies, please see www.cambridge.org/trapp/chapter8.

Notes

1 Zipser (1977).
2 See also the argument made by Parker and Johnson (2000) regarding the consistency between the time and space scales of MCSs.
3 Houze et al. (1990).
4 Parker and Johnson (2000).
5 See Smith et al. (2009) for a more in-depth description of microphysical observations in squall lines.
6 Klimowski et al. (2004).
7 Klimowski et al. (2004), Moller et al. (1994).
8 e.g., Loehrer and Johnson (1995), Fritsch and Forbes (2001).
9 James and Johnson (2010).
10 This also excludes from consideration here what have been termed "cellular" convective lines (James et al. 2005).
11 Parker and Johnson (2004).
12 Rotunno et al. (1988).
13 Parker (2010).
14 Evans and Doswell (2001).
15 Bryan et al. (2006).
16 Weisman (1992).
17 As does Weisman (1993).
18 Kundu (1990).
19 Weisman (1993).
20 Weisman and Davis (1998).
21 Atkins and Cunningham (2006), Parker (2008).
22 Augustine and Caracena (1994).
23 French and Parker (2010).
24 Parker (2008), Marsham et al. (2011).
25 French and Parker (2010).
26 Newton and Newton (1959).
27 Corfidi (2003).
28 Mahoney et al. (2009).
29 Corfidi (2003).
30 Fujita (1979).
31 The exception is when a strong nocturnal inversion is in place, which limits downward penetration of the MCS downdrafts. See Atkins and Cunningham (2006) and Parker (2008).
32 Weisman (2001).
33 Here and elsewhere, environmental shear values are from published simulation results using the Klemp-Wilhelmson (KW) model. The intermodel comparison by Bryan et al. (2006) suggests that these threshold values are high relative to those from other models, and thus should be regarded only as approximates.
34 Trapp and Weisman (2003).
35 The maximum surface winds become relatively weaker once the environmental shear exceeds $\sim$25 m s^{-1} because the convective cells become downshear tilted and also organized into more 3D, supercell-like entities rather than MCSs.
36 Atkins et al. (2005).
37 Weisman and Trapp (2003).
38 Johns and Hirt (1987). *Derecho* is a Spanish adjective for "straight" or "upright."
39 Trapp and Weisman (2003), Atkins and St. Laurent (2009), Wakimoto et al. (2006).
40 In the mature MCS, this tilting process results in a vortex couplet with an orientation opposite to that shown in Figure 8.20 for the early-stage MCS; see Trapp and Weisman (2003).
41 Davies-Jones and Brooks (1993).
42 Trapp and Weisman (2003).
43 Ibid.
44 Weisman and Trapp (2003).
45 Trapp et al. (2005b).
46 Carbone (1983), Wheatley and Trapp (2008).

47 Brady and Szoke (1989), Wakimoto and Wilson (1989).
48 The qualifier "extreme" can have many definitions. For example, Schumacher and Johnson (2005) required the 24-h precipitation to exceed the 50-yr recurrence interval amount at one or more rain gauges. However, the choice of recurrence interval is subjective, and the amount can be determined via a number of different statistical methods. See also Hitchens et al. (2010).
49 Schumacher and Johnson (2005).
50 Much of the following discussion draws from Doswell et al. (1996).
51 Laing and Fritsch (1997).
52 Maddox (1980a).
53 Much of the following discussion draws from Fritsch and Forbes (2001).
54 Fritsch and Forbes (2001).
55 Cohen et al. (2007).
56 Ibid.
57 Bluestein and Jain (1985), Thorpe et al. (1982), Rotunno et al. (1988), Weisman et al. (1988).
58 Weisman and Rotunno (2004), Stensrud et al. (2005).
59 Coniglio et al. (2004).
60 Johns (1993).
61 Ibid.
62 Coniglio et al. (2004).
63 Ibid.
64 Barnes and Sieckman (1984).

9

Interactions and Feedbacks

Synopsis: This chapter addresses ways in which convective storms affect and are affected by external processes. Perhaps the most familiar of such interactions involve convective-storm "remnants" like outflow boundaries and mesoscale-convective vortices. Chapter 9 also explores convective influences on the synoptic-scale dynamics, especially through the diabatic heating due to the convective storms. Finally, the roles of mesoscale-convective processes over longer time scales are considered. This includes feedbacks involving the land surface type, and global radiative forcing, and the formation of precipitating convective clouds.

9.1 Introduction

Taken in order of increasing length scale, time scale, and complexity across these scales, some of the ways in which convective storms affect and are affected by processes external to the storms are explored in this chapter. Frequently implicated in such interactions on the mesoscale are storm "remnants" introduced in Chapters 6 and 8, namely, outflow boundaries and mesoscale convective vortices (MCVs). Both are convectively generated, persist long after the demise of the generating storms, and thereafter help initiate new convective storms. The diabatic heating due to convective storms influences the synoptic-scale dynamics, particularly in terms of surface cyclone development and intensification. This interaction is two-way, because an effect of a deepening cyclone is to enhance transports of heat and water vapor. Locally, this helps to sustain ongoing, and support subsequent, convective activity.

External processes also promote multiscale interactions with slower responses. Consider, for example, feedbacks between the land surface and convective precipitation. These are realized over time scales of months or even seasons, and contribute to droughts and floods. The hydrological cycle is affected by naturally varying as well as anthropogenic forcing on even longer time and larger space scales. Indeed, the global radiative forcing owing to elevated greenhouse gas concentrations has been proposed to encourage decadal trends in the frequency and intensity of convective precipitating storms.

9.2 Boundaries

In the lexicon of mesoscale meteorology, a "boundary" usually refers to the leading edge of the cool-air outflow of a convective storm or system, although in principle it could be equated to other air mass separators, including a synoptic-scale cold front and a dryline. A horizontal gradient in potential temperature (or in equivalent or virtual potential temperature) is one demarcation. Another is a horizontal gradient in horizontal winds.

The role of boundaries in convection initiation was described at length in Chapter 5. Thus, this section will be brief and qualitative, but with a subtly different focus, as exemplified by the following hypothetical scenario: an MCS develops during the late afternoon and subsequently produces an extensive, southward-moving cold pool and associated outflow boundary. The MCS dissipates at local sunrise, and dilution of the cold pool by solar heating and eddy mixing effectively weakens the boundary's movement, but does not completely eliminate the cold pool itself. By early afternoon, the boundary is now quasi-stationary (and now also a storm remnant), and helps to initiate new convective storms. These storms organize into a new MCS by early evening, which produces a new cold pool and boundary, thus aiding the initiation of yet another round of convective storms, and the cycle continues.

Although this particular feedback explicitly spans the mesoscale, the synoptic- and larger scales still play a critical role, by providing the environmental conditions necessary for convective-storm formation. Stated in a slightly different way, the larger-scale atmosphere also controls the prevalence of boundary feedbacks. Consider, in particular, how mesoscale forcing in the midlatitudes becomes increasingly more prominent during the warm season as the mean jet stream shifts poleward and the midlatitude synoptic-scale activity is reduced. Indeed, some seasonal dependence on the relative frequency of feedbacks and interactions involving mesoscale boundaries is expected.

Not all interactions necessarily involve remnant boundaries, and not all represent feedbacks that extend over more than one diurnal cycle.[1] Some simply result in storm morphology changes, as in cases of interactions leading to updraft rotation and even tornadogenesis.[2] For example, boundaries rich in vertical vorticity are known to provide a local source of low-level rotation that becomes amplified upon an encounter with a cumulonimbus cloud.[3] In essence, this is the nonsupercellular tornadogenesis mechanism described in Chapter 8, yet it can also involve supercells originating at a distance from the boundary. A related interaction is conceptualized in Figure 9.1,

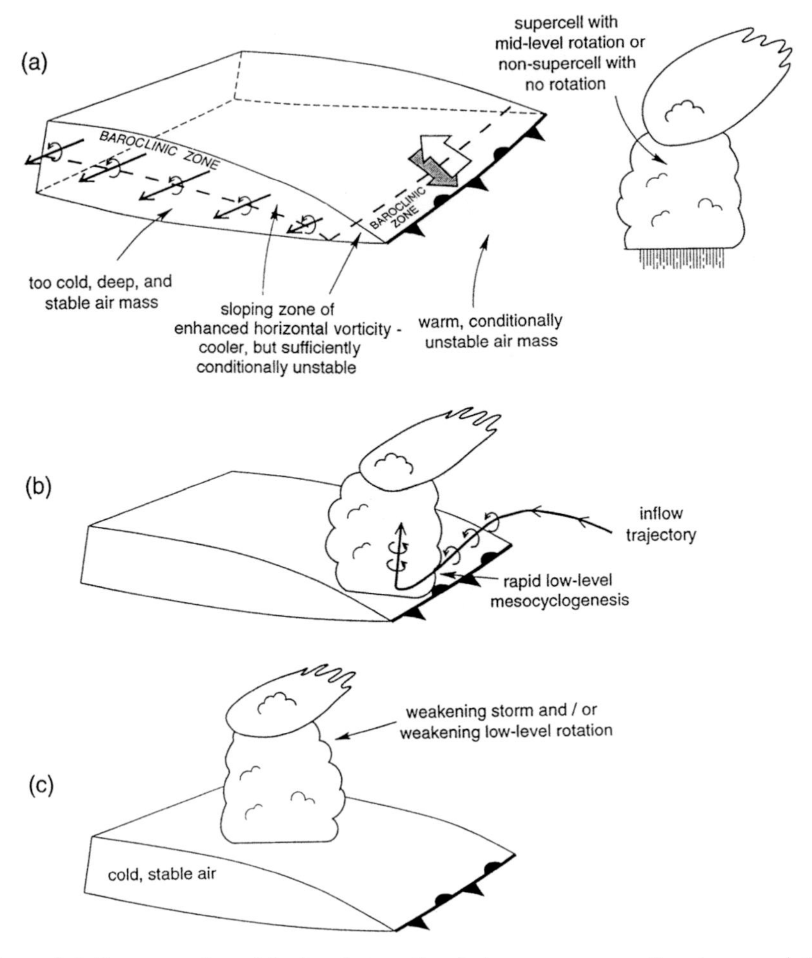

Figure 9.1 Conceptual model of an interaction between a supercell and a preexisting thermal boundary. Horizontal vorticity associated with the boundary is tilted into the vertical by the supercell updraft as the supercell crosses the boundary. From Markowski et al. (1998).

wherein an external boundary provides a source of horizontal baroclinic vorticity that is vertically tilted by the storm;[4] in actuality, the result of this type of interaction depends in part on the motion of the storm with respect to the orientation of the boundary, and in part on the boundary's thermal and kinematic characteristics.[5] A final morphological-altering interaction depicted in Figure 9.2 resulted in a low-level mesovortex that later was associated with wind damage and a tornado.[6] Bow-echo–boundary interactions such as this one have been implicated in numerous tornado and damaging-wind events although the nature of the interaction is still unclear.[7] In fact, because of the range of the processes involved, an open question common to all of these interactions is to the degree to which the boundary is necessary and sufficient for the morphological change.

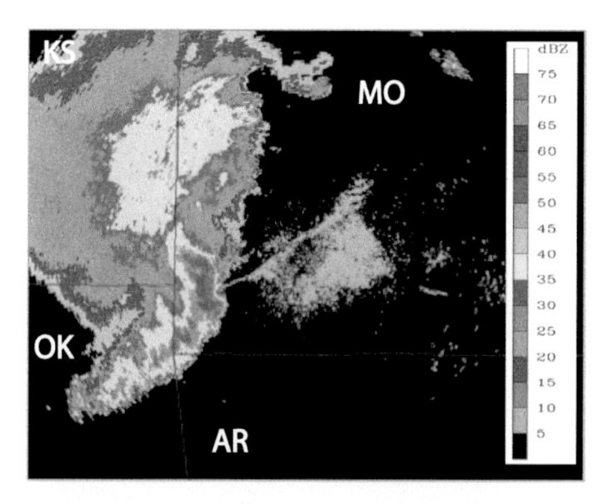

Figure 9.2 Radar reflectivity factor from the 0.5° scan of the Springfield, Missouri WSR-88D, at 1230 UTC on July 4, 2003. The SW-NE–oriented thin line shows the boundary. The point of interaction with the asymmetric bow echo corresponds to a low-level mesovortex that was associated with wind damage and a tornado. State boundaries (and state abbreviations) provide scale. (For a color version of this figure, please see the color plate section.)

9.3 Mesoscale-Convective Vortices

The mesovortex that resulted from the bow-echo–boundary interaction shown in Figure 9.2 is small in size relative to other vertical vortices generated by MCSs. The largest in this vortex spectrum is the MCV, which has a length scale proportional to the ~100-km scale of the convective system. The time scale of MCVs ranges from several hours to several days, thus often exceeding the time scale of their host MCS.[8] Figure 9.3 reveals both characteristics, but, more importantly, demonstrates the reason why MCVs are treated in this chapter: this convective remnant has the profound ability to induce motions that can initiate new or "secondary" convective storms, which in turn can regenerate an MCV.

MCVs are found often in the middle troposphere, within the stratiform precipitation rearward of the leading convective updrafts. Their initial vorticity originates partly from midlevel, line-end vortices generated through tilting of horizontal vorticity (Chapter 8). Ultimately, however, a contribution from planetary vorticity appears necessary to produce a dominant cyclonic vortex (in the Northern Hemisphere).[9] The relative plus planetary vorticity is increased by the persistent midlevel horizontal convergence fostered by the combined effect of elevated rear inflow and ascending front-to-rear flow; such horizontal convergence and associated stretching is also known to result from a mesoscale downdraft.[10]

An examination of the circulation theorem[11] supports this perspective of MCV formation:

$$\frac{D\Gamma}{Dt} = \int B\hat{k} \cdot d\vec{l} - f\frac{DA}{Dt}, \tag{9.1}$$

where the contribution to circulation from planetary vorticity is enabled through inclusion of the Coriolis parameter f, and where A is the area within the closed

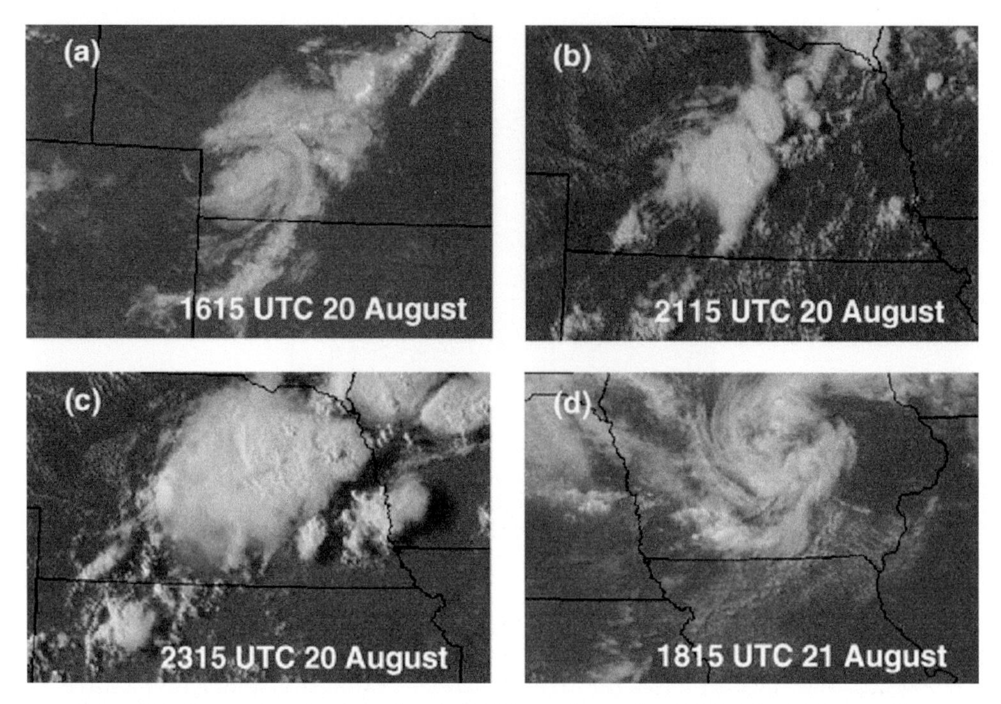

Figure 9.3 Example of a long-lived MCV as shown by GOES-8 visible satellite imagery. In (a), the cloudiness is the remnant of a nocturnal MCS and an associated MCV (as indicated by the cloud swirl). Early afternoon convection (b) evolves into an MCS (c) by early evening. The MCS decays by early morning, but not before regenerating the MCV revealed in the cloud swirl in (d). From Trier et al. (2000a).

material circuit projected into the horizontal plane, $\vec{l}$ is the local tangent to the closed contour of integration, B is buoyancy, and Γ is the relative circulation

$$\Gamma(t) = \int \vec{V} \cdot d\vec{l}.$$

As discussed in Chapter 7, the first right-hand term in (9.1) contributes to the generation of circulation from baroclinity. To understand the second right-hand term, imagine a material circuit that initially has zero relative circulation but lies completely in a horizontal plane on Earth. If the circuit shrinks in time ($DA/Dt < 0$), as would occur in the presence of horizontal convergence, positive circulation is gained.

Potential vorticity provides another perspective of MCV formation. Recall from Chapter 5 that

$$\mathrm{P} = g\,(\zeta_\theta + f)(-\partial\theta/\partial p)\,, \tag{9.2}$$

where the subscript θ indicates evaluation on an isentropic surface. As a maximum in P, an MCV is dually characterized by large absolute vertical vorticity and large static stability. The later is realized as a local compression of isentropes within the MCV

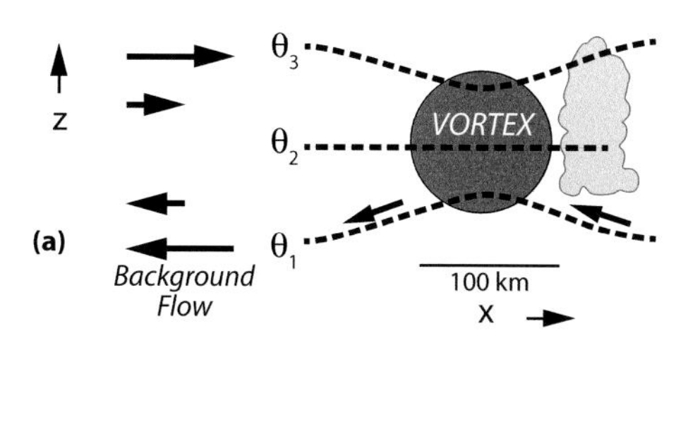

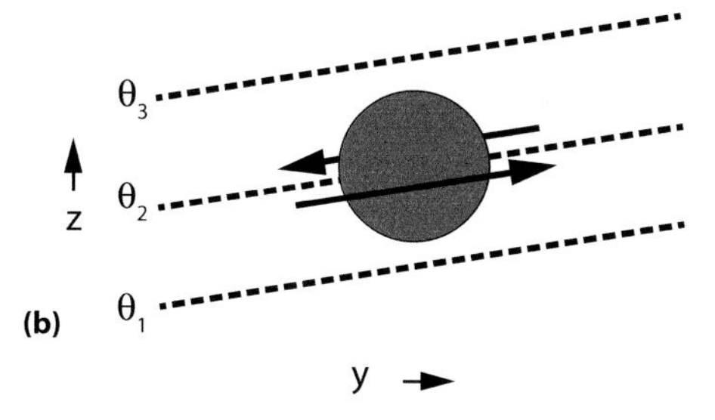

Figure 9.4 Vertical cross sections through an idealized MCV, as represented as an anomaly in potential vorticity. (a) Perspective from the south. (b) Perspective from the east. The dashed lines are isentropes, which, from the perspective in (a), are vertically compressed within the MCV core. In (a) isentropic uplift on the downshear (east) side of the MCV favors the initiation of new convective cells. From Trier et al. (2000a), as adapted from Raymond and Jiang (1990).

core (see Figure 9.4), which, as we will see shortly, is paramount to the MCV-induced motion.

The isentrope compression in the MCV core corresponds to a thermal perturbation of a few degrees Celsius, and is suggestive of a critical process in MCV development – namely, diabatic heating.[12] This is revealed in the equation[13] that governs potential-vorticity generation

$$\frac{D_\theta \mathrm{P}}{Dt} = -g\frac{\partial \theta}{\partial p}(\nabla \times \vec{V} + f\hat{k}) \cdot \nabla \left(\dot{Q}\right) - g\frac{\partial \theta}{\partial p}\hat{k} \cdot \nabla \times \vec{F}, \tag{9.3}$$

where the two right-hand terms involve diabatic heating and friction, respectively, and where in isentropic coordinates, $D_\theta \mathrm{P}/Dt = \partial \mathrm{P}/\partial t + u(\partial \mathrm{P}/\partial x)_\theta + v\,(\partial \mathrm{P}/\partial y)_\theta$, and $\nabla = \hat{i}\,(\partial/\partial x)_\theta + \hat{j}\,(\partial/\partial y)_\theta + \hat{k}\,(\partial/\partial \theta)$. Neglecting the contribution from friction as well as that from horizontal gradient of diabatic heating, (9.3) reduces to

$$\frac{D_\theta \mathrm{P}}{Dt} \simeq \mathrm{P}\frac{\partial}{\partial \theta}\left(\dot{Q}\right). \tag{9.4}$$

Equation (9.4) indicates that generation of P is large where the vertical gradient of diabatic heating is large. In MCSs, particularly in the stratiform region, this corresponds to the middle troposphere, according to observations of a sharp maximum

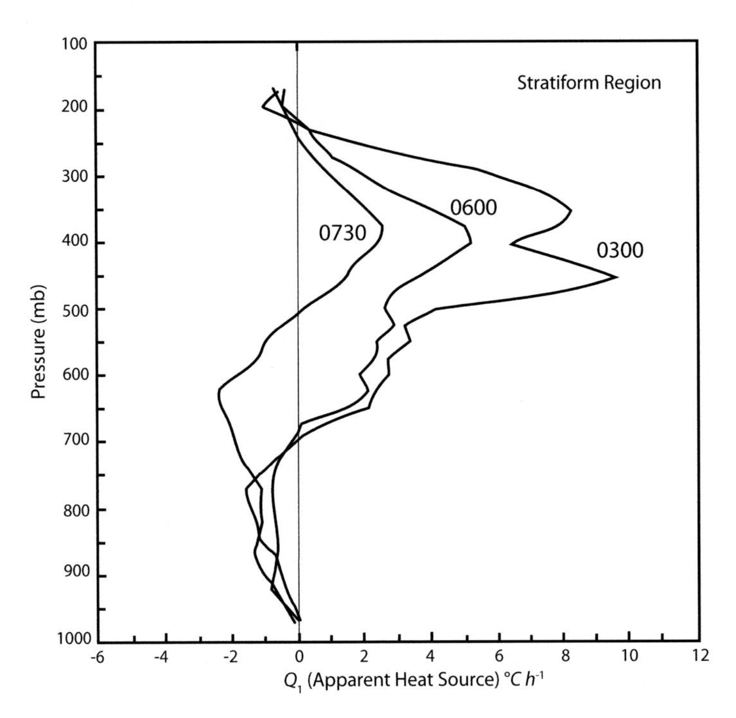

Figure 9.5 Vertical profiles of diabatic heating within the stratiform region of a squall line. These estimates, at three different times within the squall line, are a combined measure of radiative heating, latent heating from condensation, melting, and sublimation, and vertical and horizontal transports of sensible heat. From Gallus and Johnson (1991).

in diabatic heating in the upper half of the troposphere (see Figure 9.5). Thus, convective heating in an MCS, in the presence of planetary vorticity and, hence, absolute vertical vorticity can generate a potential vorticity anomaly that we recognize as an MCV.

The basic interaction between an MCV and its baroclinic environment during one diurnal cycle is illustrated in Figure 9.4. In this particular example, the deep westerly environmental shear is comprised of low-level easterly, and upper-level westerly, winds. Isentropic ascent (descent) owing to this wind profile and the MCV-modified isentropes occurs on the downshear (upshear) side of the vortex, within the low to middle troposphere (Figure 9.4a).[14] Additional ascent and descent are induced by the tangential winds of the vortex itself, via the flow along the isentropic surfaces of the environment in which the vortex is embedded (Figure 9.4b).

The significance of this interaction, and the induced ascent/descent in particular, depends on the effects of vortex strength, vortex aspect ratio (L/H, where L is a horizontal length scale, and H is a vertical depth scale), and the environmental vertical wind shear over the vortex depth ($\Delta U/H$, where ΔU is the vertical change in environmental winds).[15] As can be deduced from Figure 9.4b, the first-order effect of stronger tangential winds is stronger isentropic ascent (and descent). The first-order effect of environmental wind shear is a downshear tilt to the vortex, owing to differential advection. The joint effects of environmental shear and vortex strength modulate the orientation and extent of the vortex tilt, and hence the vortex-relative locations of induced ascent and descent. Finally, the joint effects of environmental shear and vortex aspect ratio lead to a time scale equal to $L/\Delta U$, which in this context means that a large vortex in weak environmental shear favors long-duration induced motion.

To generalize:

- MCV-induced parcel displacements are maximized beneath the midtropospheric potential vorticity anomaly, near the radius of the strongest vortex winds.[16]
- The associated lifting and environmental destabilization aids the initiation of new convective clouds.
- This secondary convection reinvigorates the MCV through vortex stretching (or diabatic generation of potential vorticity), and the cycle of convective decay → MCV continuance → convection initiation begins anew.

Not all MCVs persist long past the demise of the generating MCS, and not all secondary convection evolves into a new MCS. On the other hand, multiple cycles are possible, culminatating in a *serial* MCS. The number of cycles, and hence this feedback itself, depends critically on the thermodynamics and dynamics of the evolving environment.

9.4 Interaction between Deep Moist Convection and Synoptic-Scale Dynamics

Like MCVs, synoptic-scale cyclones are also known to interact beneficially with convective-scale processes. Such interactions are realized in multiple ways, including through changes in potential vorticity. It is convenient, then, for us to return immediately to (9.4), but let us do so with an initial neglect of diabatic heating, and accordingly, of the convective-scale processes. Although seemingly imprudent, this allows for a reduction of (9.4) to a statement of potential-vorticity conservation that we will exploit to examine relevant synoptic-scale processes. Facilitating our examination is a construct known as *isentropic potential vorticity (IPV) thinking*.[17]

To apply IPV thinking, let us assume the initial existence of a mid-to-upper-tropospheric positive anomaly in P. As depicted in Figure 9.6a, the anomaly is a local maximum in potential vorticity. Figure 9.6a also indicates that the anomaly penetrates downward and induces a horizontal circulation in the lower troposphere. The penetration depth is effectively the vertical scale of the anomaly, and depends directly on the length scale of the anomaly and inversely on the static stability of the environment; it follows that relatively large anomalies induce relatively strong wind fields.[18] The induced circulation locally advects P (or potential temperature, near the surface) to form an anomaly at this new level (Figure 9.6b). However, the new

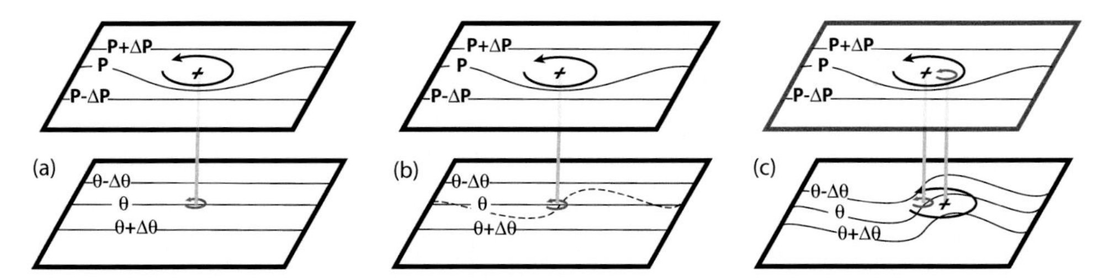

Figure 9.6 Application of IPV thinking to explain surface cyclogenesis. (a) Initial positive potential-vorticity anomaly aloft penetrates downward to induce a circulation at the surface. (b) The circulation locally advects potential temperature to create a surface anomaly. (c) The new surface anomaly penetrates upward to induce a circulation aloft, which then will locally alter the original potential vorticity anomaly. Adapted from Bluestein (1993).

anomaly – here the surface cyclone – in turn penetrates upward and induces its own circulation at upper levels (Figure 9.6c). Local advection by this induced circulation reinforces the upper-level anomaly, which then reinforces the low-level anomaly through the process just described. Such a mutually beneficial interaction between the upper and lower anomalies will persist as long as the anomalies remain locked in this spatial arrangement.

Diabatic heating associated with deep moist convection is capable of modifying this arrangement, depending on where the convective heating occurs relative to the anomalies. Recall that the vertical gradient of diabatic heating typically is large and positive (negative) in the lower (upper) half of the troposphere (Figure 9.5), and thus that diabatic generation of potential vorticity is positive below (negative above) the level of maximum heating (Equation (9.4)). In the case of deep moist convection downstream of the upper-level anomaly (and in proximity to the low-level anomaly, as usually observed), the low-level anomaly is directly enhanced through generation of positive potential vorticity. The convective heating also indirectly enhances the upper-level anomaly, with generation of negative potential vorticity aloft effectively causing the positive anomaly upstream to be even more "anomalous" (Figure 9.7).[19] In this relatively common case, then, the respective upper and lower anomalies both are intensified, and both remain in the mutually beneficial phase-locked state.

The basic dynamics of these interactions are also embodied in quasi-geostrophic (QG) theory. Consistent (as they must be) with IPV thinking, arguments based on QG theory also indicate a sensitivity to the vertical distribution of the diabatic heating, as well as to the horizontal distribution and magnitude of the heating.[20] A feedback loop envisioned through QG theory is as follows:

- Synoptic-scale advections and ascent condition the environment and then help initiate convection.
- The subsequent diabatic heating enhances the development of a surface cyclone.
- A deepening cyclone elicits a response in the form of a lower-tropospheric wind increase.

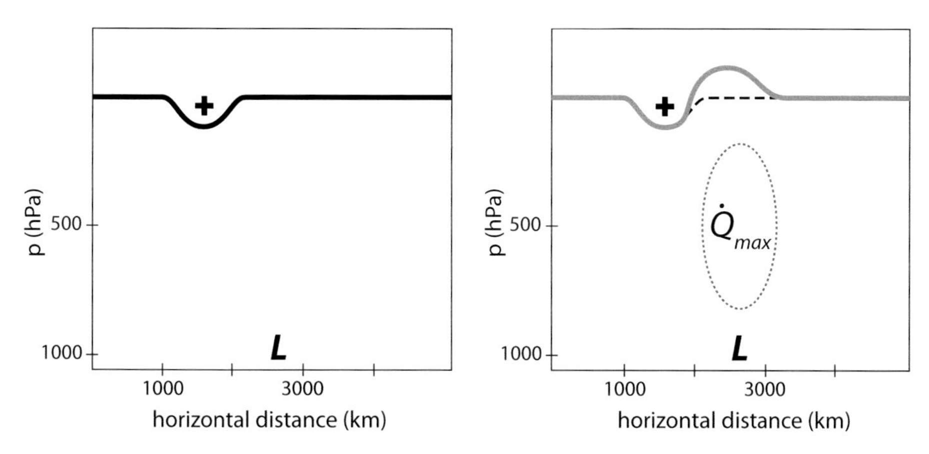

Figure 9.7 Demonstration of an effect of diabatic heating on upper- and lower-tropospheric potential vorticity anomalies. Adapted from Martin (2006).

- The enhanced winds locally increase the water vapor (and temperature) through horizontal advection, which acts to replace the moisture processed by the convection, and thereby helps to sustain ongoing, and support subsequent, convective activity.
- The associated diabatic heating enhances the surface cyclone.[21]

Because deep moist convection is treated here in a bulk sense, an open question regards whether or not the details of individual storms matter. For example, in terms of their effect on the synoptic scale, do a few widely spaced but long-lived supercells equate to a cluster of weaker convective storms over an equivalent area? It is relevant to mention that convectively generated cold pools comprise yet another effect on the synoptic scale. Especially when generated from a series of large MCSs, the cold pools can enhance the synoptic-scale baroclinity, which in turn aids cyclogenesis.[22]

The influence of deep convective storms is not limited to midlatitude surface cyclones. For example, in tropical latitudes over Africa, MCS-associated diabatic heating and potential vorticity generation near the entrance region of an African easterly jet can excite *African easterly waves* (AEWs).[23] The AEWs thereafter propagate toward the west, with the potential to develop into tropical cyclones over the Atlantic Ocean. MCVs that are generated by easterly-wave–associated MCSs are theorized to contribute to such tropical cyclogenesis, perhaps in a manner involving the vortical hot towers described in Chapter 7.[24]

Large areas of persistent convection can also excite and/or develop a coupling with a variety of other wave modes, including *Rossby waves*, which literally have global-scale influence. (Barotropic) Rossby waves are a consequence of absolute-vorticity conservation following horizontal (denoted by the subscript H) motion:

$$\frac{D_H(\zeta + f)}{Dt} = 0, \tag{9.5}$$

which holds in an atmosphere assumed to be inviscid and barotropic. As we learned in Chapter 2, Rossby waves have a frequency-dispersion relation

$$\sigma = \overline{u}k - \frac{k\beta}{K^2} \tag{9.6}$$

and zonal phase speed

$$c = \overline{u} - \frac{\beta}{K^2}, \tag{9.7}$$

where σ is frequency, c is phase speed, $K = \sqrt{k^2 + l^2}$ is horizontal wavenumber magnitude, $\beta \equiv df/dy$, and $\overline{u}$ is the base-state (westerly) wind. We also learned from Chapter 2 that Rossby waves are dispersive, or in other words, that long-wavelength (small-wavenumber) Rossby waves generally are retrogressive with respect to the base state westerly wind, and short-wavelength (large-wavenumber) Rossby waves are progressive. In the more general – and herein, more relevant – case of a baroclinic atmosphere, Rossby waves are a consequence of potential-vorticity conservation:

$$\mathrm{P} = g(\zeta_\theta + f)/(-\partial p/\partial \theta) = \text{const}, \tag{9.8}$$

as based on the adiabatic, inviscid form of (9.3). P is rewritten slightly in (9.8) to emphasize the denominator, which we can regard as the effective depth of a fluid column. Variations in column depth will excite Rossby waves, as occurs when potential-vorticity–conserving fluid columns interact with topography. In the case of a north-south topographic barrier,[25] the result is a leeside series of alternating troughs and ridges – that is, a Rossby wave train (see Figure 9.8).

Nearly stationary diabatic heating has an analogous effect.[26] Rossby waves can be excited at upper levels by convective generation of potential vorticity and subsequent potential vorticity conservation within a base-state flow. The wave propagation is in climatologically favored zones known as *Rossby waveguides*. The relevant waveguides for waves excited in the central United States – a region of frequent MCS occurrence and, thus, a likely Rossby wave source region – extend northeastward to the North Atlantic, and southwestward from the west coast of North America to the south-central Pacific.[27] In the example in Figure 9.9, convectively excited Rossby waves are apparent in both waveguides during the month of July 1992; this was a period in which anomalously large rainfall and persistent convection occurred in the Great Plains and Upper Mississippi Valley regions. Note that the southwestward-oriented waveguide pertains most to convective feedbacks because it indicates upstream wave propagation. Such upstream-propagating Rossby waves are thought to enhance the westerly flow over the southern Rocky Mountains. This fosters the development of a persistent lee trough east of the mountains, which then provides a favorable environment for subsequent convection in the central United States.

It should now be evident that convective heating can have nonnegligible effects on – and nonnegligible interactions with – the synoptic and global-scale circulations.

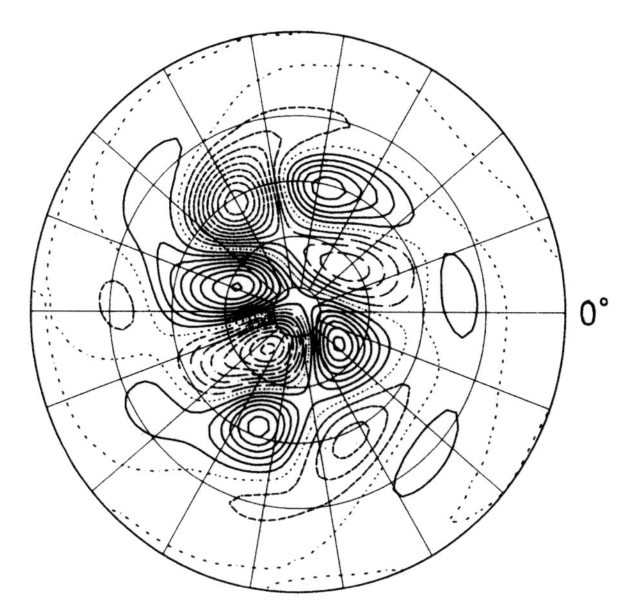

Figure 9.8 Rossby wave train, as depicted in 300-hPa geopotential height perturbations (contour interval of 2 dam) in the Northern Hemisphere. The Rossby waves are orographically forced. From Hoskins and Karoly (1981).

Indeed, if improperly represented in NWP models, for example, this particular upscale feedback can be the source of significant errors in weather-model forecasts (e.g., Chapters 4 and 10). As we will learn in the next section, precipitation itself has a nonneglible feedback with the underlying land surface. The effects of feedbacks involving the land surface and its properties can be far reaching in space and time, and, if similarly misrepresented, can be the cause for additional error in forecasts of weather and climate.

9.5 Land–Atmosphere Feedbacks

Mesoscale-convective processes can participate in feedback mechanisms that have time scales of a few hours to days, as just discussed, but also of months and even seasons. A positive feedback with a relatively slow response time involves convective precipitation and soil moisture.[28] The *soil moisture–rainfall feedback* described next is known for its role in fostering persistent rainfall, and indeed has been invoked to help explain the record flooding in the U.S. Midwest during the summer of 1993.

9.5.1 Soil Moisture–Rainfall Feedback

The soil moisture–rainfall feedback has a number of interrelated theoretical components (Figure 9.10).[29] The first involves radiative transfer (see Chapter 4): absorption

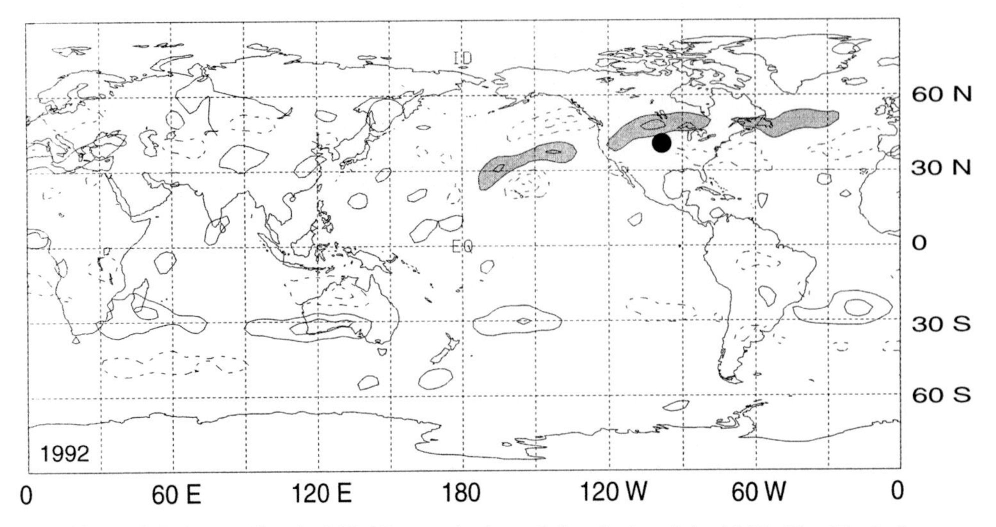

Figure 9.9 Anomalies in 200-hPa vertical vorticity during July 1992. The black dot represents the Rossby-wave source region. Shaded anomalies are those that lie within the Rossby waveguides. Contour interval is 1×10^{-5} s^{-1}. From Stensrud and Anderson (2001).

of incoming solar or shortwave (↓SW) radiation is enhanced in the presence of wet soils, because wet soils tend to be darker in color and also support denser, greener vegetation; both aspects yield a relatively lower surface albedo (A_{sfc}). This effect competes with low-level cloudiness, which is enhanced with soil moisture. Low-level clouds reduce surface temperature, thereby reducing emission of surface longwave (↑LW) radiation, by virtue of the Stefan-Boltzmann law. The clouds, however, re-emit outgoing longwave radiation, and the downward longwave radiation (↓LW) is further enhanced by the increase in low-level atmospheric water vapor brought about by a larger surface latent heat flux. All effects contribute to the net surface radiative flux:

$$F_{rad}^{sfc} = F_{SW}^{\downarrow}(1 - A_{sfc}) - F_{LW}^{\uparrow} + F_{LW}^{\downarrow}, \tag{9.9}$$

and are quantified in Figure 9.11a–c using regional climate model experiments focused on the 1993 Midwest flood event. In these experiments, net shortwave (longwave) radiation decreases (increases) with increasing soil moisture. When combined in (9.9), the net (all-wave) surface radiative flux is shown to increase with increasing soil moisture.[30]

Recall now from Chapter 4 that any increases in the net surface radiative flux must be balanced by the energy lost to the atmosphere as sensible heat (↑SH), the energy lost as latent heat from the evaporation of water at the surface (↑LH), and the heat flux into the ground (↓G):

$$F_{rad}^{sfc} = F_{SH}^{\uparrow} + F_{LH}^{\uparrow} + F_{G}^{\downarrow}. \tag{9.10}$$

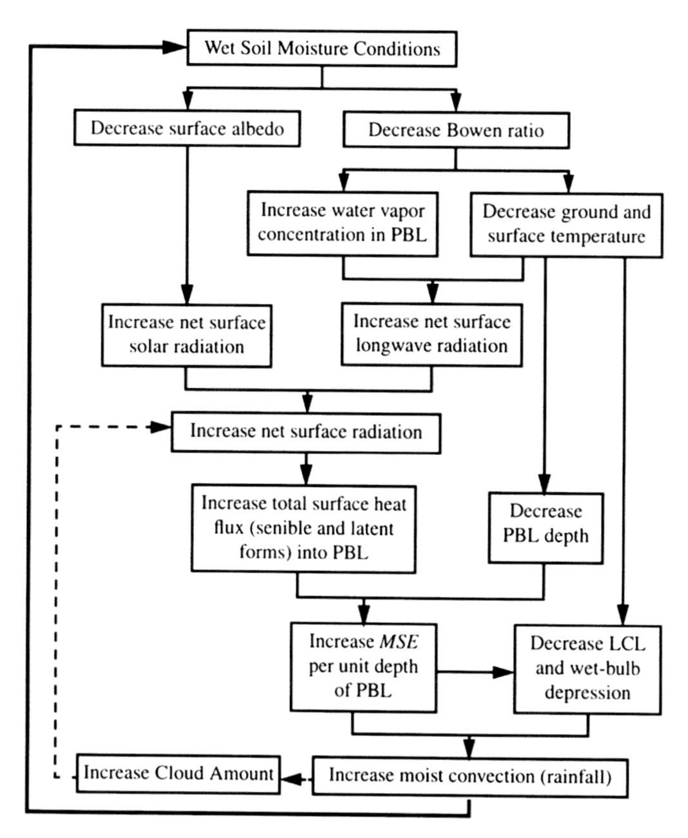

Figure 9.10 The soil moisture-precipitation feedback deconstructed: The pathways that lead from anomalously wet soil, to enhanced precipitation, back to enhanced soil moisture. From Pal and Eltahir (2001).

In the surface energy balance, surface energy storage is neglected, as is the heat flux associated with the melting (freezing) of snow or ice (water), because this is not particularly relevant for the current discussion; over long time scales, the ground heat flux is also negligible. The effects of the surface sensible and latent fluxes can be quantified through the moist static energy (see also Chapter 6) evaluated within the atmospheric boundary layer (ABL):

$$\hbar = c_p T + gz + L_v q_v. \tag{9.11}$$

Consulting the regional climate model simulations again, we find that increases in soil moisture are realized as increases in moist static energy (Figure 9.11). The moist static energy increases are due to water-vapor mixing ratio increases within the ABL, as given by increases in the latent heat flux; temperature contributions to moist static energy, in contrast, decrease with increasing soil moisture, in accord with decreases in sensible heat flux.

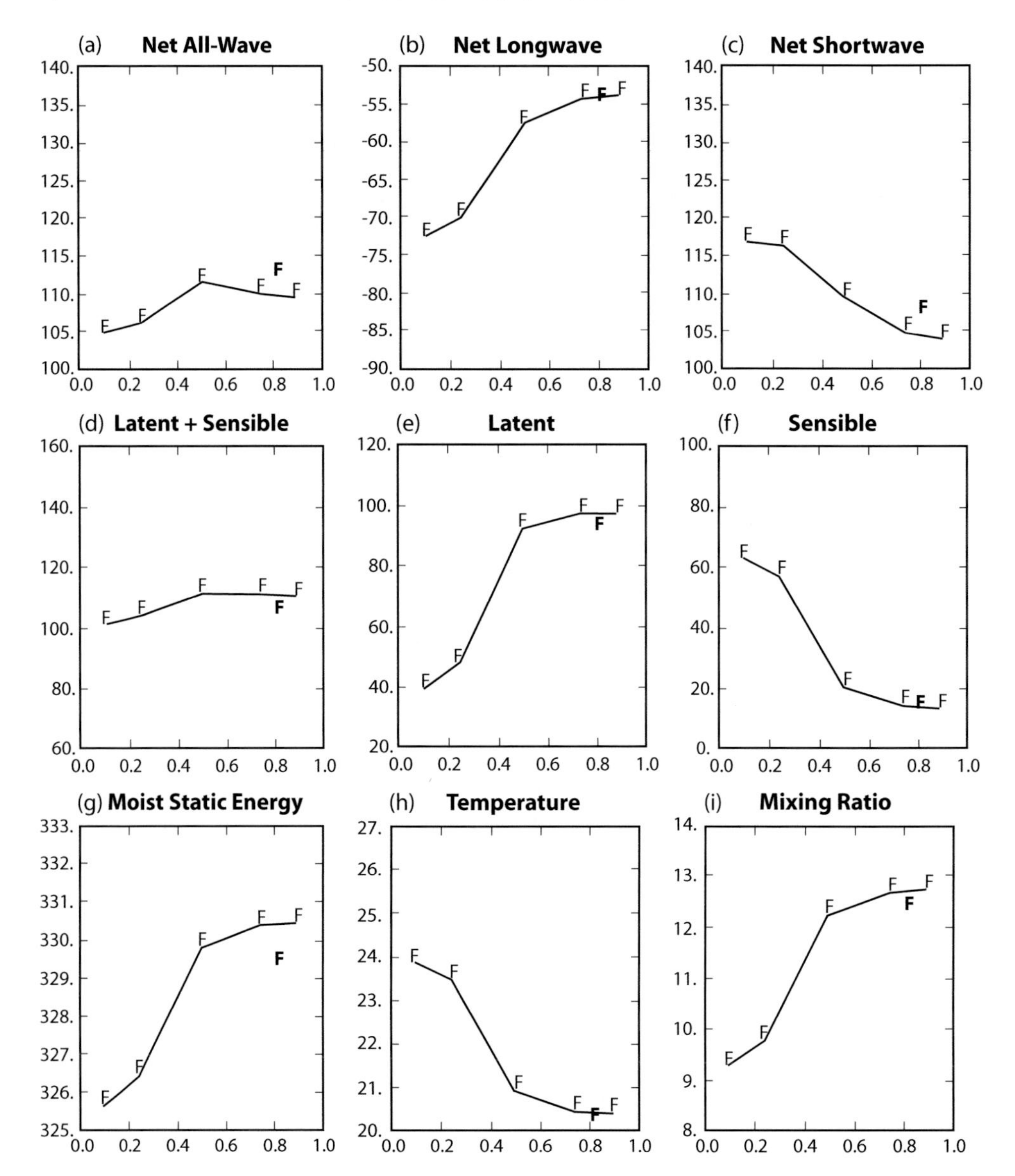

Figure 9.11 Quantifications of soil moisture-precipitation feedback components (see Fig. 9.10), as a function of initial soil saturation in regional climate model experiments. The simulations are based on the summer 1993 U.S. flooding event. (a) Net all-wave surface radiative flux (W m^{-2}), (b) net longwave surface radiative flux (W m^{-2}), (c) net shortwave surface radiative flux (W m^{-2}), (d) sum of surface latent and sensible heat fluxes (W m^{-2}), (e) surface latent heat flux (W m^{-2}), (f) surface sensible heat flux (W m^{-2}), (g) moist static energy (kJ kg^{-1}), (h) temperature (C), and (i) mixing ratio (g kg^{-1}). Bold F denotes the control experiment. All values represent seasonal averages, over a Midwestern U.S. subdomain. After Pal and Eltahir (2001).

This response is manifest as a reduction of the *Bowen ratio* (ratio of sensible heat flux to latent heat flux). A reduced Bowen ratio has ramifications on ABL depth, because a lowered sensible heat flux implies less turbulent mixing in the vertical, a decreased ABL growth rate, and, ultimately, a relatively shallower ABL. Above the ABL, where soil moisture has little direct impact, moist static energy generally decreases with height.[31] Hence, the vertical profile of moist static energy resulting from anomalously wet soils is associated with decreased static stability and increased convective available potential energy (CAPE). Convective clouds that are realized in this enhanced environment will have more intense vertical motions (Chapter 6) and, therefore, the potential to generate relatively heavier precipitation, thus closing the feedback loop.

This linkage between soil moisture and rainfall is in situ. However, the local effects can grow upscale and consequently induce long-term remote effects as well. For example, when the midwestern United States was experiencing record flooding in 1993, drought prevailed in the southwestern United States. The persistence of anomalous dry conditions in the Southwest can be attributed in part to a soil-moisture–rainfall feedback, albeit a negative feedback in this case. Additionally, a lack of latent heating and corresponding enhanced sensible heating promoted an increase in hydrostatic pressure, and, ultimately, anomalous anticyclonic flow in the lower troposphere over the drought region.[32] The attendant synoptic-scale subsidence (or at least reduced upward motion) further anchored the drought. The anomalous anticyclone also affected the atmosphere circulation elsewhere, such as in the Midwest,[33] in the form of a southward-displaced jet stream and storm track. Hence, local convective feedbacks, and their local and remote upscale effects, add layers of complexity to large-scale atmospheric variability. The challenge is to separate out the relative importance of each.

9.5.2 Heterogeneities in the Land Surface

A variation on the theme of soil-moisture feedbacks occurs, literally, when soil moisture varies spatially. Applying the surface energy balance arguments introduced in Section 9.5.1, such surface heterogeneity leads to a horizontal variation in surface sensible heat fluxes.[34] More generally, variations in F_{SH} can be random or nonrandom, and arise between forested and adjacent unforested regions, urban and adjacent agricultural landscapes, areas of irrigated and adjacent nonirrigated land, and so forth. If the individual regions or areas are sufficiently large ($\gtrsim$10 to 100 km in length scale), the horizontal gradient in F_{SH} effectively leads to a thermally direct, vertical circulation (see Chapter 5).

As with the sea-breeze circulation, *landscape-induced mesoscale circulations*[35] have a lower branch that flows nearly horizontally from areas of low to high F_{SH}, and then a rising branch above the high F_{SH} region. Vertical motions within the rising

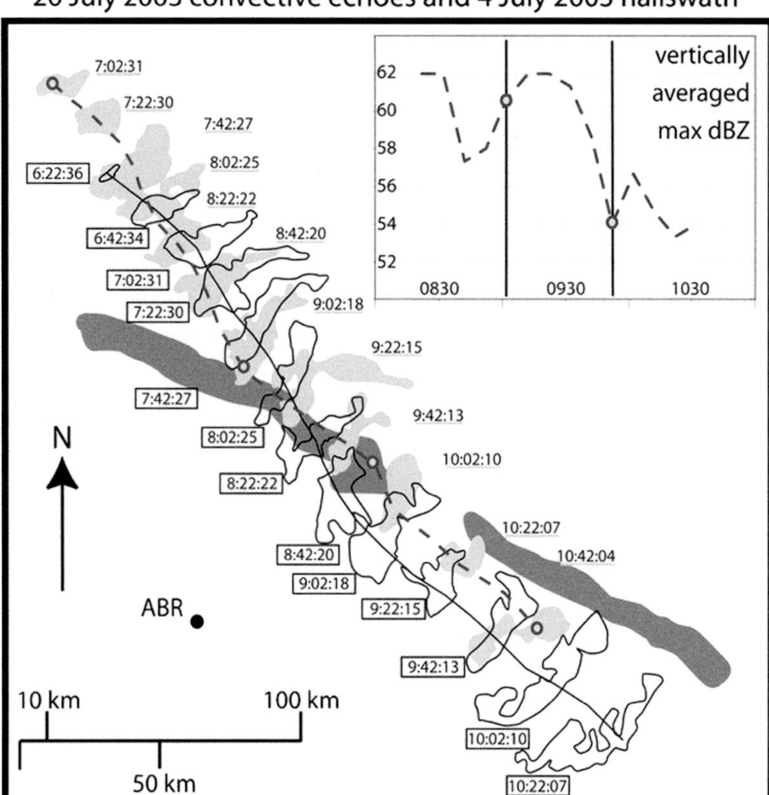

Figure 9.12 Radar echo tracks of two convective cells (depicted by contours and light shadings at 45 dBZ) on July 20, 2003, plotted relative to the devegetated hailswath (dark shading) created on July 4, 2003. Reference times of the cells are indicated, as are the tracks connecting their maximum reflectivity centroids. The time series of vertically averaged maximum reflectivity is given for the second (shaded) cell. From Parker et al. (2005).

branch can have speeds of a few m s^{-1}, and the circulation itself can have depths on the order of 1 km. Convective cloud formation in the rising air depends on the rising-branch characteristics as well as on the ambient static stability and moisture; the latter can be modified locally by the circulation itself, however.[36] The synoptic-scale forcing and winds play an additional role in the circulation development and organization. Reorientation and horizontal advection of the circulation are examples, and can explain why, despite the static forcing due to the landscape, circulation patterns may differ from day to day.[37]

The landscape is not always static, however: anthropogenic as well as natural modifications to the land surface can occur over a short time scale, with potentially longer-term impacts. Consider the case of a severe hailstorm that significantly damaged an extensive swath of cropland (Figure 9.12). The devegetated hailswath, of

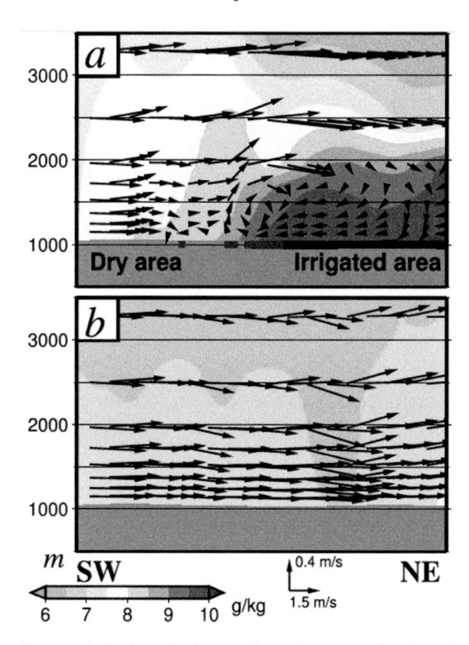

Figure 9.13 Mesoscale model simulations showing (a) the landscape-induced circulation that arises between an irrigated and nonirrigated land area, and (b) the lack of such a circulation when the land areas are not irrigated. Vectors indicate wind in a vertical cross section, and red and blue shadings are of water vapor mixing ratio. The light gray shows topography. From Kawase et al. (2008). (For a color version of this figure, please see the color plate section.)

230-km length and 12.5-km width, persisted for well over one month. It was associated with a measurable increase (decrease) in surface temperature (dewpoint), and consequently with an increase in Bowen ratio.[38] The hailswath, its gradient in F_{SH}, and an inferred mesoscale circulation appeared to play a role in the initiation of an intense convective storm roughly two weeks after the hailstorm, and perhaps contributed to the motion of yet another convective storm a few days later (Figure 9.12). Notably, this natural modification to the landscape triggered a feedback that transcended rainfall. Similar modifications and feedbacks associated with other convective weather hazards, such as tornadoes, are conceivable.

The mesoscale circulation at the interface between irrigated and nonirrigated land exemplifies an impact of human-induced landscape alternation. The nonirrigated land area has the relatively higher surface sensible heat flux and thus hosts the rising branch of this circulation (Figure 9.13). The irrigated land area hosts the sinking branch. Cloud formation and rainfall over the irrigated land should therefore be suppressed, causing a need to irrigate further, and in turn reinforcing the landscape-induced circulation.[39] Even though this negative feedback is associated with nonhazardous weather, it does raise the practical issue of water resource management.

9.6 Convective Processes and the Global Climate

The preceding section described how changes to the landscape, both by humans and naturally, lead to mesoscale circulations that affect convective cloud formation and precipitation. Over the long term, such *local* forcing modifies the local climate, in temperature, humidity, winds, and of course precipitation. *Global* forcing via modes of natural climate variability, as well as through the radiative effects of anthropogenically enhanced greenhouse gases (GHGs), also has a potential effect on the local frequency and intensity of convective clouds and precipitation. Chapter 10 will consider connections between modes of natural climate variability and deep moist convection, from the perspective of long-lead weather and climate prediction. In this section, we explore possible connections between anthropogenic climate change and convective processes.

There are many starting points to this extremely complex problem. We begin here with the influence of elevated GHG concentrations on globally averaged surface temperature. Recall from the surface energy balance equation (9.10) that increases in net surface radiative flux are compensated in part by increases in sensible heat flux; the radiative-flux increases come primarily from increased downward longwave radiation owing to the GHG increase. Further compensation comes from increases in latent heat flux: the associated evaporation from bodies of water, the soil, and indirectly from vegetation, yields an increase in near-surface water vapor. With vertical mixing, this increase in water vapor is realized throughout the ABL, but not necessarily in the free troposphere. The potential extent of the vapor increase is determined by the Clausius-Clapeyron equation,

$$\frac{d \ln e_s}{dt} = \frac{L_v}{R_d T^2}, \tag{9.12}$$

which shows the strong dependence of saturation vapor pressure (e_s) on temperature (often colloquially referred to as the atmosphere's "water holding capacity"). Indeed, observations indicate a strong coupling between temperature and water vapor, implying that much of the potential is realized.[40]

Our focus turns now to the hydrological cycle, because the increased water vapor must ultimately be processed through precipitating clouds to maintain a global hydrological balance; that is,

$$E - R = 0, \tag{9.13}$$

where E is an evaporation rate and R is a precipitation rate. Both are integrated over a sufficiently long time and large area to allow the neglect of runoff and storage from this balance.[41] Evaporation rates, and hence atmospheric water vapor, are nonuniformly distributed around the globe. The global distribution of precipitation, which also is highly nonuniform, does not correspond directly to that of evaporation, because convective and nonconvective precipitating systems utilize locally available moisture as well as that transported from some distance.[42] Nevertheless, the precipitating systems that contribute significantly to the hydrological balance have large rain rates and thus are often convective in nature. The projection of more atmospheric moisture in association with anthropogenic climate change implies even larger rain rates and extreme precipitation. Observational data are already showing the existence of such trends in precipitation character.[43]

It is natural to inquire about the convective-storm modes that will generate (and are generating) the projected extreme precipitation. This is particularly compelling because even supercells have been implicated in heavy precipitation and flash flooding.[44] Indeed, this extends our inquiry to questions about how anthropogenic climate change connects to convective-storm intensity, frequency, and severe-weather generation. The current generation of climate models provides no direct answers, because even large MCSs are effectively subgrid-scale processes. Consequently, we resort to *downscaling techniques* (see Chapter 10).

In one example of such a technique, environmental parameters evaluated at climate model gridpoints are used as a proxy for storm occurrence in the neighborhood of the grid points. Changes are then assessed by comparing environmental parameter values from climate-model integrations based on historical GHG concentrations to integrations based on anticipated future GHG concentrations (scenarios).

As alluded to previously, a consistent result in climate-change projections is an increase in surface temperature and, especially in regions proximate to water bodies and/or vegetation, an increase in boundary-layer moisture. CAPE increases are found particularly where moisture and temperature both increase (Figure 9.14).[45] Enhanced CAPE in projected future climates indicates stronger potential updrafts, and hence supports the previous arguments about larger rain rates and heavier precipitation. Vertical wind shear, on the other hand, tends to decrease in projected future climates. This is explained by the theorized and modeled decrease in the extratropical meridional temperature gradient, which relates to wind shear by virtue of the *thermal wind equation*,

$$\frac{\partial \vec{V}_g}{\partial \ln p} = -\frac{R_d}{f}\hat{k} \times \nabla_p T, \tag{9.14}$$

where subscript g denotes the geostrophic wind, and subscript p denotes evaluation on an isobaric surface.[46]

Depending on the region and season, climate model simulations indicate that future decreases in 0–6 km vertical wind shear (S06; see Chapter 7) tend to be more than compensated by future increases in CAPE, resulting in a net future increase in a CAPE $\times$ S06 product.[47] These results, as shown in Figure 9.14, are interpreted to mean that the *frequency of environmental conditions supportive of severe convective*

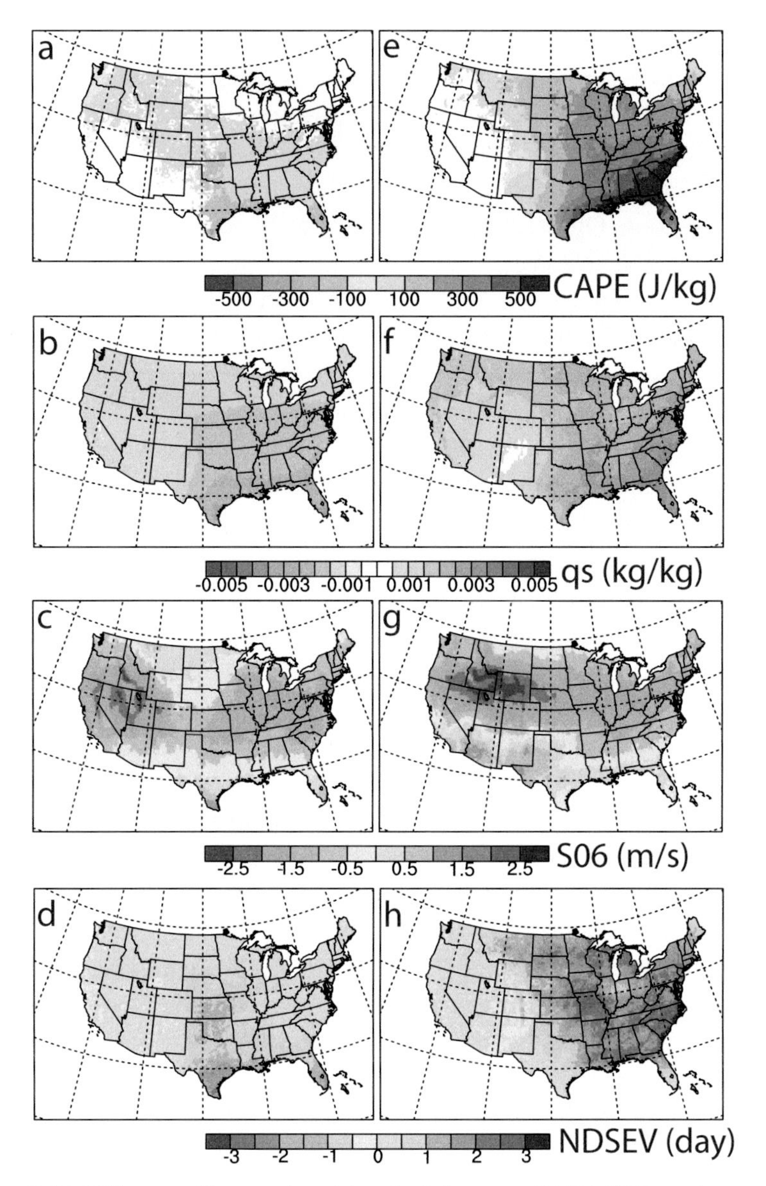

Figure 9.14 Difference (future minus historical) in mean CAPE, S06, surface specific humidity, and occurrence of the product CAPE $\times$ S06 $\geq$ 10,000. The latter is treated as the frequency of a severe convective storm environment. The future integration period is 2072–2099, the historical integration period is 1962–1989, and the analyses are valid for March-April-May (a–d), and June-July-August (e–h). From Trapp et al. (2007a). Copyright 2007 National Academy of Sciences, USA. (For a color version of this figure, please see the color plate section.)

storms has the potential to increase under anthropogenic climate change. The level of increase varies significantly with region and season (and also with climate model and GHG scenario). It should be cautioned that convection initiation is not explicitly treated by this proxy approach, and hence the *frequency of storm occurrence* would be some unknown percentage of the frequency of environmental conditions, and would have its own regional and seasonal variation. This limitation will be removed, and more satisfactory answers will be provided, when the environmental proxy approach is superceded by a more direct approach in which convective storms are explicitly represented in global climate models with (nested) convection-permitting domains (see also Chapter 10).[48]

The preceding text describes a one-way linkage between anthropogenic climate change and convective storms. However, we must also account for the feedback of clouds onto the climate system, which is a significant piece of the climate-change puzzle. Indeed, in addition to their associated latent heating (as discussed in previous sections), clouds interact with atmospheric radiation: all clouds absorb and reflect some percentage of incoming solar radiation, and all absorb and re-emit longwave radiation. Their net radiative effect, and thereby their net effect on the global energy balance, depends on cloud height, thickness, and microphysical composition, and hence varies across cloud type.

Satellite-based measurements of incoming and outgoing radiation at the top of the atmosphere (see Chapter 3) provide a means to estimate the cloud radiative forcing for the current climate. The (average) net radiative flux at the top of atmosphere is

$$F^{top} = F_{sun} \cos(Z) - F_{SW}^{top} - F_{LW}^{top}, \tag{9.15}$$

where F_{sun} is the solar constant adjusted for distance from the sun, Z is the solar zenith angle (see Chapter 4), F_{SW}^{top} is the shortwave radiant exitance (reflected solar radiation), and F_{LW}^{top} is longwave radiant exitance (emitted terrestrial radiation, and also known as *outgoing longwave radiation*).[49] The variables in this energy budget can be used to calculate the planetary albedo, which includes the albedo of clouds. Our primary goal, however, is to isolate the effect of clouds. This is done by calculating

$$\textit{Net cloud radiative forcing} = F^{top} - F_{clear-sky}^{top}, \tag{9.16}$$

where $F_{clear-sky}^{top}$ is net clear-sky radiation at the top of the atmosphere, determined from techniques that employ visible brightness and infrared temperature thresholds, for example.[50]

Figure 9.15 reveals that in the tropics, the net cloud radiative forcing is negative, although relatively small in magnitude.[51] Over the middle and high latitudes, and especially when solar insolation is large, the net cloud radiative forcing is also negative, but comparably larger in magnitude. Contributing to such net negative forcing are optically thick stratiform clouds that form in association with extratropical synoptic-scale weather systems.[52] Accordingly, much of the research to date on extratropical cloud radiative forcing has focused on nonconvective cloud feedbacks and their relationship to future projections of extratropical cyclone frequency and intensity.[53]

For clues on how convective cloud feedbacks might relate to anthropogenic climate change in the extratropics, we consult theories developed for tropical convection, bearing in mind the important differences between the tropical and extratropical

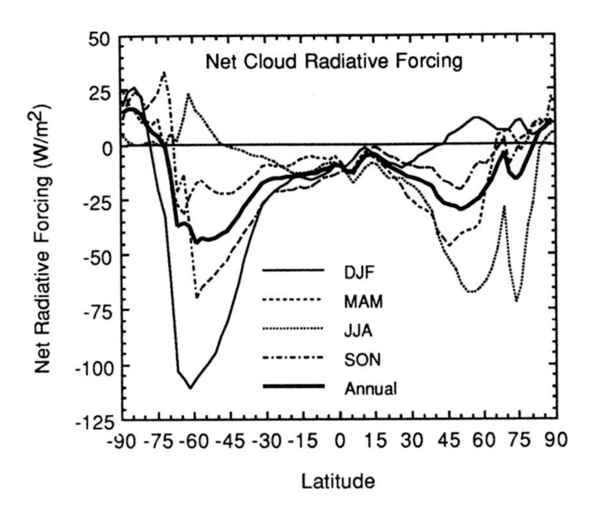

Figure 9.15 Zonally averaged net cloud radiative forcing, based on the Earth Radiation Budget Experiment (ERBE) data. From Hartmann (1993). Used with permission, Elsevier Press.

large-scale dynamics. The first is the *adaptive iris hypothesis*.[54] It draws an analogy to the iris of the human eye, which closes to reduce the amount of light passing through the pupil. Underlying this hypothesised negative feedback is an increase, with increasing (sea) surface temperature, in the precipitation efficiency of tropical cumulonimbus, and thus a decrease in the water substance available to be detrained from the cumulonimbi as cirrus.[55] Owing to this assumption, the atmospheric response to increasing surface temperature is effectively a decrease in the upper-level humidity, thereby promoting more outgoing longwave radiation to space, which then acts to lower the surface temperature. A different negative feedback to tropical warming is predicted by the *thermostat hypothesis*.[56] Here, increasing boundary-layer humidity leads to thicker, more extensive cirrus anvils that have dominant shortwave radiative forcing and hence contribute to surface cooling.

Both the "thermostat" and the "iris" serve to regulate maximum tropical surface temperatures. Regulation of extratropical surface temperatures by such negative feedbacks associated with enhanced cumulonimbi and associated anvils would seem plausible. We await further data analysis and modeling experiments for confirmation. For now, we are left to conclude that the magnitude and even the sign of the extratropical cloud radiative feedback is an unknown convolution of global-scale radiative forcing, synoptic-scale dynamics, and cloud and precipitation microphysics.

Supplementary Information

For exercises, problem sets, and suggested case studies, please see www.cambridge.org/trapp/chapter9.

Notes

1 Once the boundary-generating convection is no longer active, the boundary can be regarded as a storm remnant, and accordingly, as an external boundary.
2 Maddox et al. (1980).
3 Documentation of such a storm–boundary interaction is found in Wakimoto et al. (1998), with the caveat that the boundary in this case is apparently a synoptic-scale feature.
4 Markowski et al. (1998).
5 Atkins et al. (1999).
6 Knopfmeier (2007).
7 Przybylinski (1995).
8 Based on a survey of all MCS cases in the central United States during the warm season of 1998, such persistent MCVs occurred in 12% of the MCSs; Trier et al. (2000a).
9 Skamarock et al. (1994), Weisman and Davis (1998).
10 Brandes and Ziegler (1993).
11 Holton (2004).
12 Davis and Trier (2007).
13 Holton (2004), Bluestein (1993).
14 As proposed by Raymond and Jiang (1990), and subsequently explored by a number of other authors; e.g., Trier et al. (2000a).
15 Trier et al. (2000b).
16 Trier et al. (2000b), Trier and Davis (2007).
17 Hoskins et al. (1985).
18 This is based on the invertibility principle proposed by Hoskins et al. (1985).
19 Martin (2006).
20 Pauley and Smith (1988), Smith (2000).
21 Stensrud (1996b).
22 Ibid.
23 Thorncroft et al. (2008), Hsieh and Cook (2005), Berry and Thorncroft (2012).
24 Montgomery et al. (2006).
25 See Holton (2004) for more discussion of this and other cases.
26 Hoskins and Karoly (1981).
27 Stensrud and Anderson (2001).
28 Soil moisture is the total amount of water in an unsaturated soil.
29 Pal and Eltahir (2001).
30 Ibid.
31 Ibid.
32 Namias (1991).
33 Pal and Eltahir (2002).
34 Segal and Arritt (1992).
35 Weaver and Avissar (2001). These have also been referred to as *nonclassical mesoscale circulations* by Segal and Arritt (1992).
36 Garcia-Carreras et al. (2011).
37 Weaver and Avissar (2001), Carleton et al. (2008).
38 Parker et al. (2005). This impact on the surface energy budget is shown in analyses and also simulations of other hailswaths, such as by Segele et al. (2005).
39 Kawase et al. (2008).
40 Wentz and Schabel (2000).
41 Peixoto and Oort (1998).
42 Trenberth (1998).
43 Karl and Knight (1998).
44 Smith et al. (2001).
45 Trapp et al. (2007a).
46 Holton (2004).
47 This parameter set derives from the analysis of a large sample of observed events by Brooks et al. (2003). The statistical relationship represents the best discriminator between environments yielding significantly severe thunderstorms and those of generally nonsevere thunderstorms.

48 Trapp et al. (2010).
49 Kidder and Vonder Haar (1995).
50 Ibid.
51 Hartmann (1993).
52 Bony (2006).
53 Norris and Iacobellis (2005).
54 Lindzen et al. (2001).
55 See Hartmann and Michelsen (2002), and Del Genio and Kovari (2002), for alternative views on this hypothesis.
56 Ramanathan and Collins (1991).

10

Mesoscale Predictability and Prediction

Synopsis: The general focus of this final chapter is on numerical weather prediction (NWP) at the mesoscale. NWP models have theoretical limits imposed by the nonlinear governing equations and approximations, as well as by errors in the initial and boundary conditions. These theoretical limits are viewed in the context of actual applications of mesoscale forecast models. Consideration is given to deterministic forecasts, and then to the use of model ensembles to produce probabilistic forecasts. Measures needed to evaluate the accuracy and skill of these forecasts, especially on the scale of convective precipitating storms, are also considered. The chapter concludes with a section devoted to possible approaches to – and the feasibility of – longer-range prediction of mesoscale-convective processes.

10.1 Introduction

This chapter focuses on the limits and uses of numerical weather prediction (NWP) on the mesoscale. Because of an intrinsic link to digital computing technology, mesoscale predictability and prediction comprise rapidly advancing areas of basic and applied research. An attempt is made herein to introduce the reader to enough material to appreciate the directions of these advances.

Recognizing that "rapid advancement" demands a dynamic terminology, we will use *fine-* or *high-resolution* to refer to models with grid lengths of a few kilometers or less, and thus in which convective processes are represented without a parameterization scheme (see also Chapter 4). *Coarse-*, *low-*, or *medium-resolution* will refer to models that have parameterized convection, as necessitated by their grid lengths of tens of kilometers or more.

10.2 On Predictability and Its Theoretical Limits

In this section we are concerned with physical systems described by a finite number of differential equations. Let us begin by considering a simple periodic system, modeled exactly by

$$\frac{dA}{dt} = A_0 \sin(t), \tag{10.1}$$

where A is some relevant state variable, and A_0 is a constant amplitude. Equation (10.1) is *deterministic*, because the exact state at the present time completely describes the exact state at some future time.[1] Contingent on how well the present state is known, the modeled system also has *predictability*, defined herein as the extent to which some future state of the system can be determined, given knowledge of the physical laws that govern the system, of its current and past states.[2] A quantitative definition is given in the text that follows.

For complex physical systems like the atmosphere, the governing equations certainly are not as simple as (10.1), and conclusions about determinism and predictability and are not as easily reached. Indeed, the model equations introduced in Chapter 4 will ultimately have a nonlinear response even to linear forcing; exemplifying such behavior are the initiation and evolution of model-simulated convective clouds resulting from a constant surface heat flux (Chapter 5). Yet in practical applications of NWP models, even nonperiodic, nonlinear mesoscale-convective processes have been shown to possess some level of predictability. Thus we ask: How might this be possible, and under what constraints?

The possibility of limited deterministic prediction of nonperiodic flows is not a new inquiry. In his seminal study that helped popularize chaos theory, E. Lorenz pursued this using a low-order model of Rayleigh convection:

$$\frac{dX}{dt} = s(Y - X), \tag{10.2}$$

$$\frac{dY}{dt} = -XZ + rX - Y, \tag{10.3}$$

$$\text{and} \quad \frac{dZ}{dt} = XY - bZ, \tag{10.4}$$

where t is a nondimensional time, and dependent variables X, Y, and Z are proportional to the intensity of the convective motions, the temperature difference between ascending and descending motions, and the change of the temperature profile from linearity, respectively.[3] The remaining variables s, r, and b are constant coefficients.

A numerical integration of (10.2), (10.3), and (10.4) with initial conditions $(X_0, Y_0, Z_0) = (0, 1, 0)$ and coefficient values $(s, r, b) = (10, 28, 8/3)$ gives a solution that initially $(0 < t < 1650)$ appears quasiperiodic and generally is well behaved

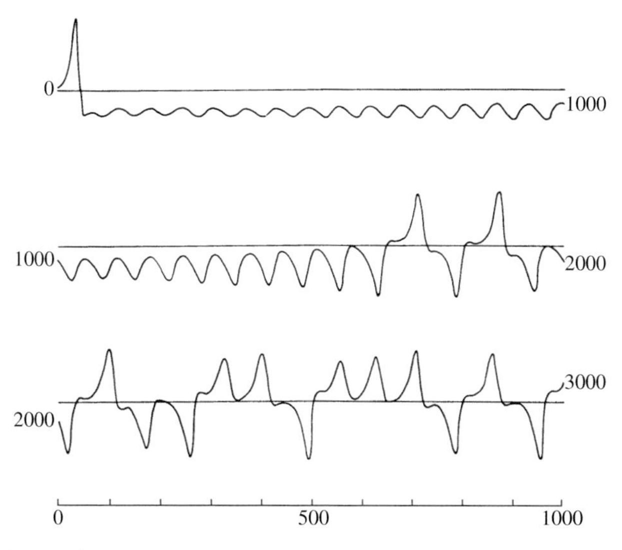

Figure 10.1 Numerical solution of a low-order model of Rayleigh convection. Shown is the time series of variable Y, which represents the temperature difference between ascending and descending convective motion. From Lorenz (1963).

(Figure 10.1). It is suggested that this initial period is one of limited predictability. Beyond $t = 1650$, however, the irregular fluctuations exhibited in the solution in Y imply a lack of predictability.

For more insight into the time evolution of the solution, let us consider the *solution trajectory* $P(t) = [X(t), Y(t), Z(t)]$ in the *phase space* described by the dependent variables (X, Y, Z). Each point on P(t) describes an instantaneous solution and, thus, the state of the system. A trajectory representing nonperiodic flow will not pass through a point through which it had passed previously, although it may approach its past behavior over some finite amount of time. This is illustrated well in the portion $(1400 \le t \le 1900)$ of the trajectory shown in Figure 10.2.[4] The trajectory that begins near reference point C′, which is a state of steady thermal convection, is consistent with the near-steady evolution portrayed in Figure 10.1. The subsequent fluctuating state seen in Figure 10.1 is manifest as a trajectory that spirals outward from C′, crosses the plane, encircles the state C, and then recrosses the plane.

The solution presented in Figures 10.1 and 10.2 is said to be bounded, yet unstable.[5] The instability is such that two integrations of the same set of governing equations, but with initial conditions differing by a small amount ε, will have respective solution trajectories that ultimately differ by an amount much larger than ε. As alluded to previously, this behavior is a consequence of the nonlinearity of the model and physical system.

In actual NWP, ε traditionally is the observational error in some state variable such as 500-hPa geopotential height; the error growth is the subsequent difference between two (or more) model solutions that have initial conditions that span the

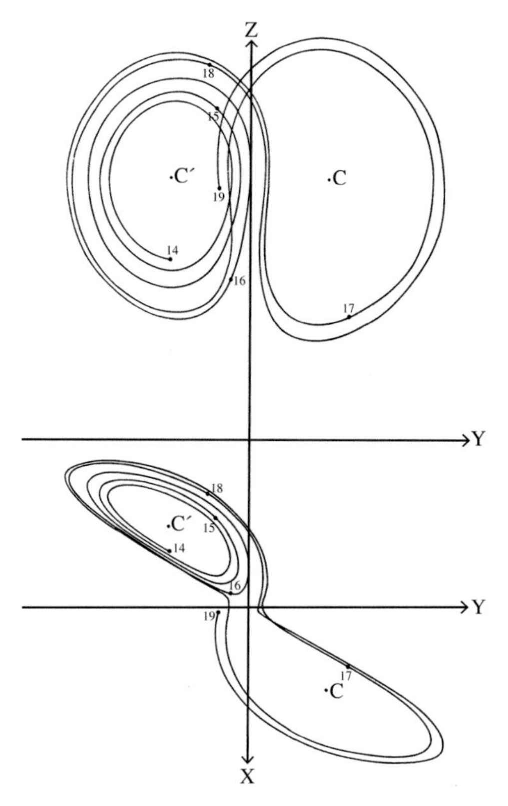

Figure 10.2 Phase-space trajectory of a solution of a low-order model of Rayleigh convection. Shown is the solution trajectory P(t) = [X(t), Y(t), Z(t)], as projected onto the XY and YZ planes, over the interval $1400 \leq t \leq 1900$. Included for reference are points C and C′, which are states of steady thermal convection. From Lorenz (1963).

observational error. The magnitude of the error growth imposes the temporal *limit of predictability* t_p, or, the time during which the error remains below some arbitrary threshold such as the climatological standard deviation of the state variable.[6] One might ask whether a reduction of ε through improved observations (e.g., an advance in observing technology) would, in turn, lead to a reduction in solution error growth, and thus to an increase in t_p. For weather prediction models, the answer depends primarily on the atmospheric scales of consideration.

A central point in discussions of atmospheric predictability is that the atmosphere possesses a range of scales that interact nonlinearly. It has been established that error growth at some wavenumber k is due in part to initial errors in that scale (and its forcings) but additionally due to errors originating in the smaller scales that then progress upscale. In other words, error or uncertainty in the initial state of small eddies contaminates slightly larger eddies, which in turn contaminates still larger eddies, and so on, until the error spreads throughout the spectrum (Figure 10.3).[7]

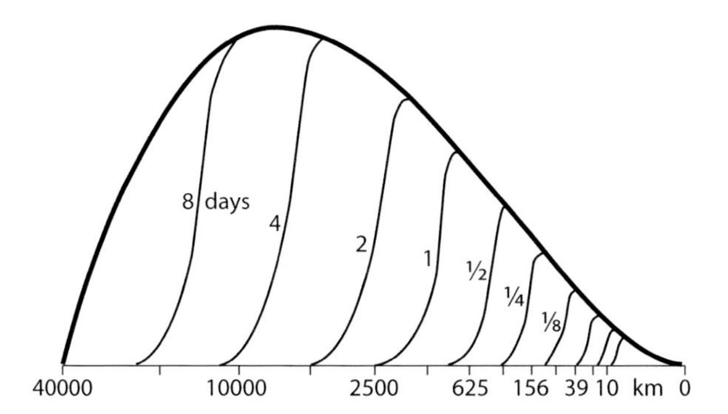

Figure 10.3 Time-evolution of errors in a theoretical model of atmospheric flow. Upper curve represents the full kinetic energy spectrum of atmospheric motion, and bottom line provides the horizontal scale. After Lorenz (1969), with adjustments from Lorenz (1984). Used with permission.

In models of 2D (and 3D) homogeneous, isotropic turbulence and other simple flows, the rate at which this *inverse cascade of errors* occurs is a function of the eddy turnover time t_e, given by[8]

$$t_e(k) = [kV(k)]^{-1} = [k^3 E(k)]^{-1/2}, \tag{10.5}$$

where $V(k)$ is a scaling velocity and $E(k)$ is the kinetic energy, represented through a power law of the form $E(k) \sim k^{-n}$. The predictability limit is the integrated effect of the eddy exchange through the spectrum:

$$t_p(k) = \int_k^{k_r} (t_e/k)\, dk = \int_k^{k_r} \left[k^5 E(k)\right]^{-1/2} dk, \tag{10.6}$$

where k_r is the wavenumber of the smallest observationally resolved eddy or scale; it is assumed that the spectrum of consideration is such that $k_r \gg k$ (or $\lambda_r \ll \lambda$). The eddy turnover time, and therefore the predictability limit, depend on the slope $-n$ of the energy spectrum. Over the range of the spectrum occupied by the synoptic scale, the slope has a characteristic value of –3.[9] Using $E(k) \sim k^{-3}$ in (10.5) we find that t_e is a scale-independent constant, meaning that the growth of error in all scales across this range is constant. We now ask: Can this error growth be slowed so that the predictability limit is increased? Integrating (10.6) with a constant t_e we find that

$$t_p(k) \sim \ln(k_r/k), \tag{10.7}$$

and thus conclude that it is indeed possible to gain a logarithmic increase in predictability t_p, for a given wavenumber k, if the wavenumber (wavelength) of the smallest resolved eddy k_r is increased (decreased).[10]

To apply this theoretical analysis to the predictability on the mesoscale, recall from Chapter 1 that the range of the energy spectrum occupied by the mesoscale has a spectral slope of –5/3.[11] In this spectral range:

(1) The kinetic energy is characterized by $E(k) \sim k^{-5/3}$.
(2) The corresponding eddy turnover time is not constant, but rather $t_e \sim k^{-4/6}$.
(3) The rate of the inverse cascade of errors is controlled by the smallest eddies (largest wavenumbers), because they have the smallest eddy turnover times.
(4) Predictability is limited by errors in the smallest scales (see (10.6)), which happen to be the scales observed with the least amount of accuracy.

We now seek whether the predictability limit in this spectral range can be increased.[12] Using the general form of $E(k) \sim k^{-n}$ in (10.5), we find that (10.6) integrates to

$$t_p(k) \sim k_r^{-(3-n)/2} - k^{-(3-n)/2}. \tag{10.8}$$

When $n < 3$, which includes the current case of $n = 5/3$, the first right-hand term in (10.8) is negligible relative to the second right-hand term, per our assumption that $k_r \gg k$. The predictability limit becomes

$$t_p(k) \sim k^{-(3-n)/2}, \tag{10.9}$$

and thus we conclude that, because (10.9) is independent of k_r, the predictability for each wavenumber in the –5/3 spectral range has a set upper limit and cannot be improved. Theoretical estimates for scales of several hundred kilometers yield a $t_p \approx$ 10 h (see Figure 10.3); for thunderstorm scales of tens of kilometers, t_p is an order of magnitude less.[13]

It is instructive at this point to examine the consistency between the results of this theoretical analysis and experiments using higher-order models of the atmospheric mesoscale. Consider the following comparison between a control integration of a mesoscale model and integrations that are identical, aside from slightly different initial conditions.[14] The model-solution differences (or "errors") depicted in Figure 10.4 arise from the addition of a sinusoidal disturbance – initial error– to the initial horizontal distribution of temperature. The disturbance wavelength is 85 km, equal to the hypotenuse of a 2×2 grid-cell square, and roughly twice the horizontal grid length of 30 km. Other experiments with the same disturbance wavelength, but different amplitudes, are used to simulate the effects of different initial-error magnitude.

The meteorological context of the experiments in Figure 10.4 is a baroclinic wave amplifying in a conditionally unstable atmosphere. The growth of the errors – in temperature, precipitation, and so on – first occurs at and near the scale of the imposed disturbance, and then progresses upscale as anticipated from theory. Let us first investigate the error-growth behavior in physical space. In this meteorological setting, the errors are manifest initially in the timing and location of convective clouds and

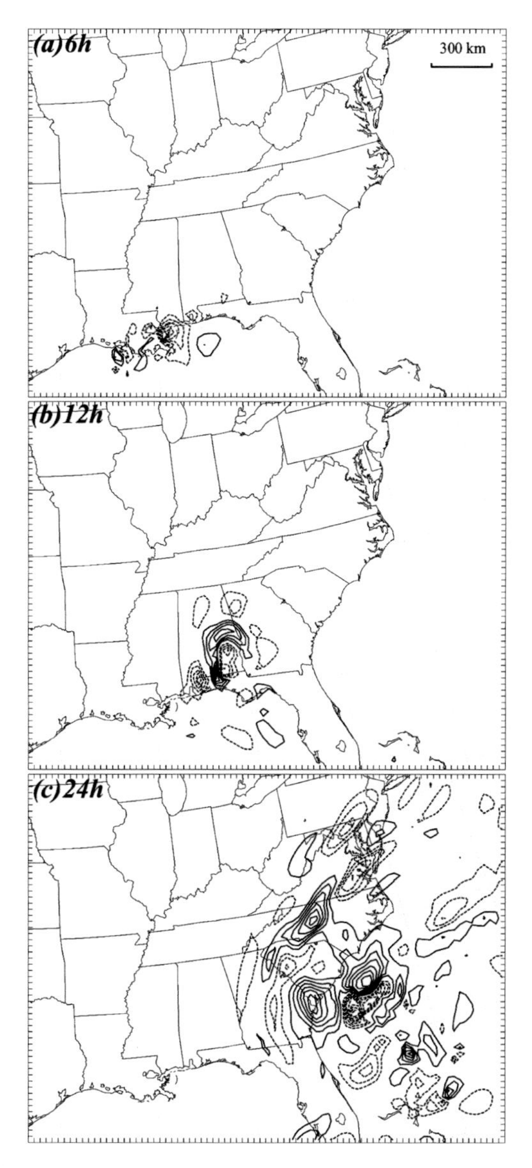

Figure 10.4 Difference in 300-hPa temperature between the control and initially perturbed simulations of the "surprise" snowstorm of January 24–25, 2000, at (a) 6 h, (b) 12 h, and (c) 24 h. The contour interval in (a)–(b) is 0.1 K, and in (c) is 0.2 K. From Zhang et al. (2003).

precipitation. This is in part because the disturbance causes thresholds in parameterizations such as that of ABL processes to be met in one location but not another, which ultimately influences where and when parameterized convection is triggered. Subsequent feedbacks between the subgrid- and resolved-grid-scale processes aid in the geographical spread of the convective-precipitation errors. The spread is promoted by gravity waves and convective outflow, but is limited to grid points where the model

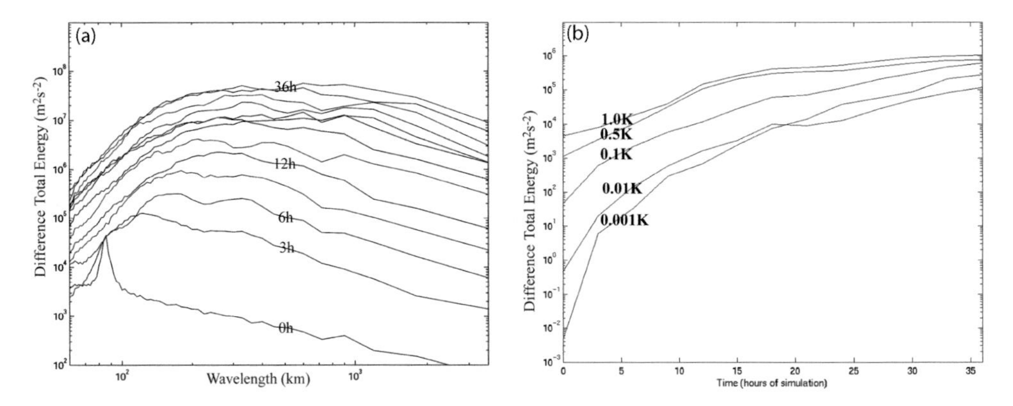

Figure 10.5 Analysis of domain total DTE (a) as a function of wavelength, and simulation time, and (b) as a function of simulation time and magnitude of initial temperature disturbance amplitude. The DTE is based on differences between the control and initially perturbed simulations of the "surprise" snowstorm of January 24–25, 2000. From Zhang et al. (2003).

atmosphere supports moist convection; once convection has been realized (triggered) at all possible grid points, error at the scale of the convective precipitation is said to be saturated. Meanwhile, the differences in the convectively generated diabatic heating lead to differences in potential vorticity, which ultimately influences the amplitude and track of the synoptic-scale baroclinic wave (see Chapter 9). Said another away, locational errors in convection beget errors in diabatic heating which are then transmitted to errors in the synoptic-scale system.

The error growth can be quantified by the difference total energy (DTE),

$$DTE = \frac{1}{2}\left[(\delta u)^2 + (\delta v)^2 + \frac{c_p}{R_d}(\delta T)^2\right], \tag{10.10}$$

where δu, δv, and δT, respectively, are the gridpoint differences in the x-component of velocity, y-component of velocity, and temperature, between two simulations.[15] The time-dependent DTE field is spectrally transformed and then summed over discrete wavelengths. The 0-h DTE has a peak at the wavelength of the initial disturbance (Figure 10.5a). The spectral peak at subsequent times in the simulation occurs at progressively larger wavelengths, as can be surmised from Figure 10.4. Over the entire spectrum, the fastest growth of DTE occurs during the first few simulation hours, in association with the parameterized convection (Figure 10.5b). In experiments with initially smaller- (larger-) amplitude temperature disturbances, this initial growth rate is actually faster (slower). Thus, consistent with theoretical arguments, reducing initial error by a factor of 2, for example, does *not* result in a doubling of the predictability time. Furthermore, consistent with theoretical arguments and (10.9), the mesoscale predictability limit does indeed appear to be bounded, with reductions in initial error doing little to extend this bound.

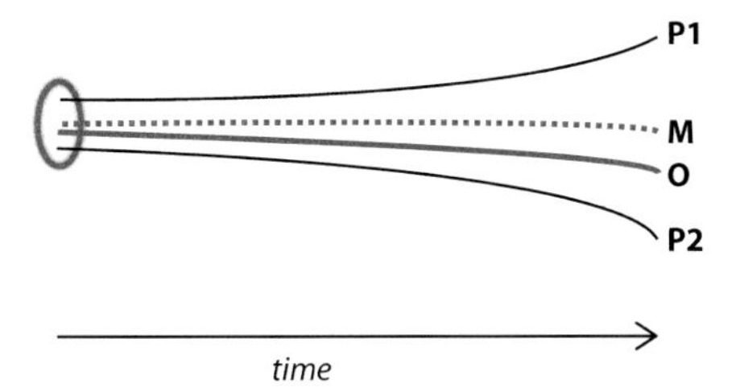

Figure 10.6 Concept of stochastic-dynamic weather prediction. The lines represent solution trajectories, plotted here as a function of time. The perturbed initial-condition ensemble members are P1 and P2, the ensemble mean is M, and the observed state is O. Based on Kalnay (2003).

Although the focus of the preceding experiments is the effects of errors in the initial conditions (ICs) on the model prediction, the prediction is also affected by errors admitted through the boundary conditions (BCs). In limited-area models, for example, errors may arise from the larger-scale model driver and associated imperfect conditions specified at the lateral boundaries.[16] BC (and IC) errors also may arise from an imperfect analysis of observational data, instrument-measurement errors, improper interpolation of the data to the model grid, or nonrepresentative data (see Chapter 3).[17]

The numerical model itself is imperfect and yet another source of uncertainty. The uncertainty is due in part to approximations of physical processes in the model equations, and in part to the discrete representation of these continuous equations (see Chapter 4). For example, physical-process parameterizations that are not completely appropriate for the particular model application might still be used because of the lack of a viable alternative. A finite-difference approximation or other numerical scheme might be used because of its computational efficiency, despite the fact that it introduces computational artifacts (e.g., noise). Finally, relatively coarse grid point spacings might be used because of computational resource limitations, leaving highly relevant processes unresolved and a numerical solution that does not converge.

Statistical techniques can be employed in an attempt to estimate error sources, and, more generally, to provide information about the bulk error growth and predictability of a particular forecast situation. As considered in the next section, probabilistic forecasts and ensemble prediction systems derive from such techniques.

10.3 Probabilistic Forecasts and Ensemble Prediction Systems

Here we discuss *stochastic-dynamic weather prediction*, in which a collection or *ensemble* of integrations of an NWP model is used in place of a single model integration. This approach, as illustrated conceptually in Figure 10.6, treats the atmosphere as a deterministic system, but recognizes that the atmospheric state is not observed perfectly and thus cannot be known exactly.[18] Recall from Section 10.2 that initial uncertainty – visualized here as a sphere or cloud of possible initial states – will be manifest in model solutions that diverge with time in phase space. With a well-posed model ensemble, the expectation is that the actual atmospheric evolution (or *nature*, in the NWP parlance) will lie within the range of evolutions given by the ensemble solutions, and in fact can be approximated by the stochastic properties of the solution distribution, such as the mean (Figure 10.6). Also intrinsic to this approach is that clustering of the solutions in sub-areas of phase space provides a means to quantify probabilistic forecasts. A lack of clustering, or least a large spread in the ensemble of solutions, suggests that the ensemble mean approximates the actual evolution with low probability. Indeed, high (low) ensemble spread is equated to low (high) predictability of the specific situation, or at least to low (high) confidence in the ensemble forecast.

A challenge in stochastic-dynamic weather prediction is how to create an ensemble appropriate for the specific prediction problem. The computational cost of integrating a model multiple times is always a practical concern. Another consideration of an *ensemble prediction system* (EPS), however, is how to sample the relevant range of possible atmospheric evolutions. One basic approach is to subject the prediction model to some uncertainty in the initial conditions. Ensemble members are generated by adding (and/or subtracting) perturbations to the initial conditions, with amplitudes comparable to the magnitude of observational errors. The perturbations can be placed randomly throughout the entire computational domain, as was done in the mesoscale predictability experiments just described in Section 10.2. Alternatively, the initial conditions can be perturbed in an optimal way to reveal the fastest-growing errors. Toward this end, *breeding* and *singular vector* methods[19] are commonly employed in operational NWP.

The perturbed members, as well as a nonperturbed or control member, are used to compute the sample statistics, such as the ensemble mean and some measure of the ensemble spread. A "spaghetti plot" is particularly useful in graphically highlighting the geographical regions where the ensemble spread is large or small.[20] A typical spaghetti plot is composed of a specific 500-hPa geopotential height contour line for each member (Figure 10.7). An animation of such a plot over the forecast period reveals, qualitatively, when (and where) the solutions have sufficiently diverged such that predictability limits have been reached.[21] For mesoscale-model EPSs, "postage-stamp plots" have become a popular means of conveying the ensemble spread. This type of graphical presentation shows well the intra-ensemble range of convective-storm organization, for example, as produced by a high-resolution EPS (Figure 10.8). The ensemble information can also be summarized quantitatively through statistics such as the ensemble-based probabilities of exceedances of forecast variables (and derived fields) (Figure 10.9).

An ensemble of hundreds, or even thousands, of perturbed members will fail to contain the truth if the error growth is dominated by the uncertainty of the model, rather

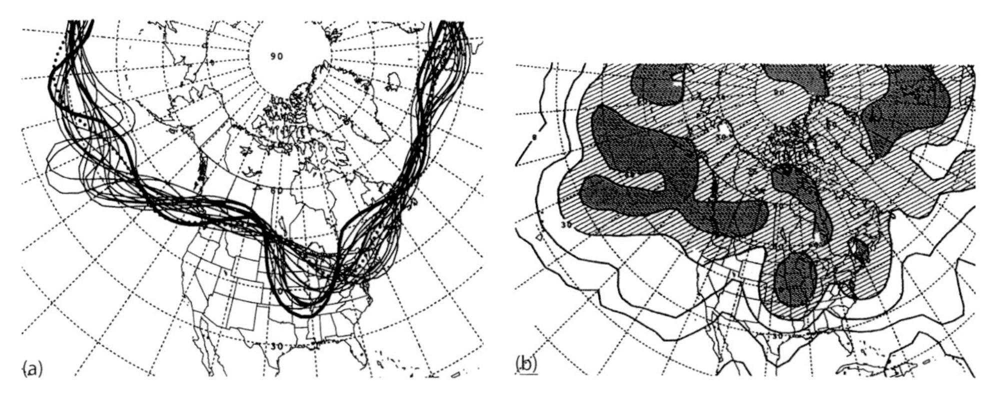

Figure 10.7 Example of a (a) spaghetti plot, and (b) the corresponding ensemble spread. In (a), lines are of 5640-m geopotential height at 500-hPa height, from the 108-h forecast of 17 ensemble members. The dotted line indicates a high-resolution control forecast, and the heavy solid line is from observational data valid at this time (the verification). In (b), contours and shading are the spread of ensemble members around the ensemble mean. From Toth et al. (1997).

than by the uncertainty in the initial conditions (Figure 10.10). Model uncertainty is often associated with parameterizations (see Chapter 4), and thus it is common to include "physics members" in an EPS.[22] Simply, the physics members: span available parameterization schemes, such as precipitation microphysics; span a range of settings of a particular scheme, such as the slope or intercept value in the microphysical dropsize distribution; and/or even stem from some introduction of randomness into a particular parameterization scheme.[23] Individual NWP models have biases that may not be related to the parameterizations and, therefore, ensembles are also formed from different NWP models. Exemplifying multisystem ensembles is the "grand global ensemble" associated with THORPEX (The Observing System Research and Predictability Experiment).[24]

10.4 Forecast Evaluation

Forecasts, and implicitly predictability limits, are assessed or *evaluated* by comparing the model predictands to an appropriate set of observations.[25] Forecast-evaluation methods that are objective and quantitative are especially desired. The method of choice depends in part on whether the predictands are continuous (atmospheric state variables such as temperature and pressure); dichotomous (yes/no, such as the occurrence of a tornado); expressed as a probability (as provided, for example, by the EPS); or given as a distribution function. The purpose of the evaluation also influences the method: commonly, evaluation serves to motivate improvements to an operational forecast system, but it also can provide support for decisions related to economic and public policy, as well as give information that can be used to understand the basic behavior of the atmosphere.[26]

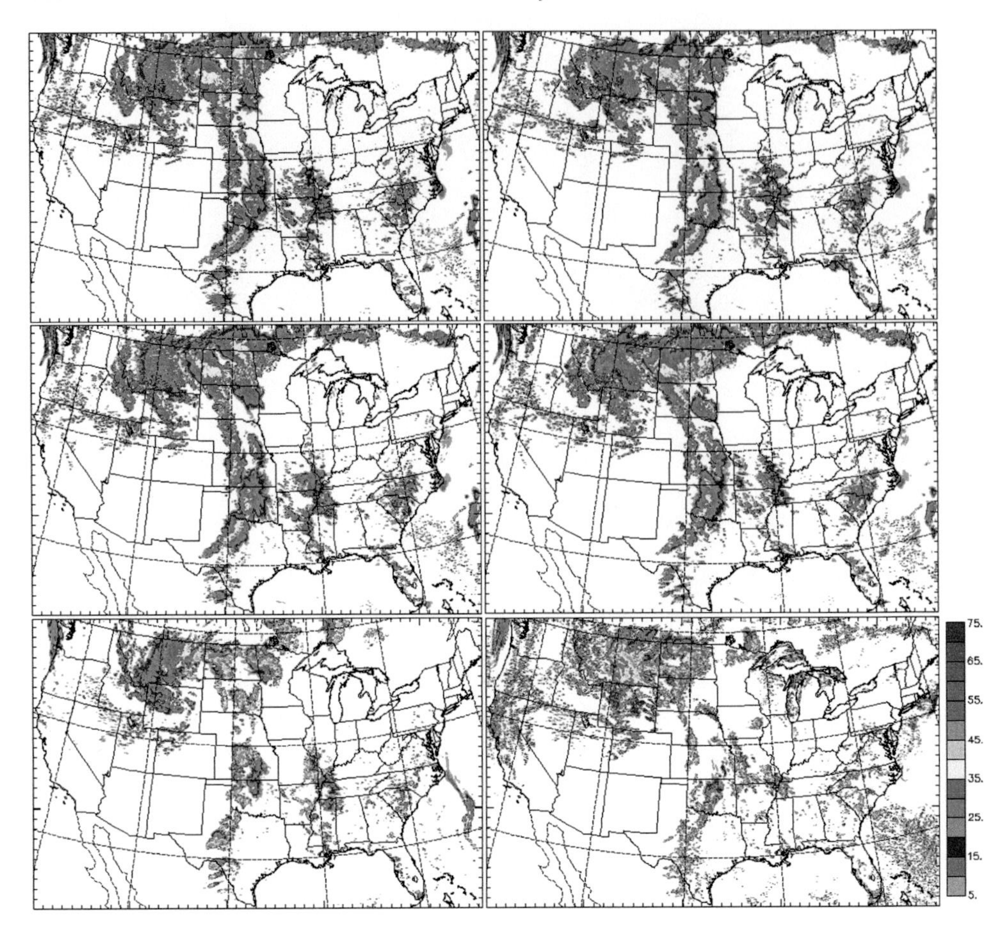

Figure 10.8 Postage-stamp plot of output from a high-resolution, multimodel ensemble system composed of 26 members. Shown is simulated radar reflectivity factor, from six members, at the forecast hour valid 0000 UTC on May 25, 2010. The system was initialized on May 24, 2010 at 0000 UTC. Courtesy of Dr. Fanyou Kong and the Center for Analysis and Prediction of Storms, University of Oklahoma. (For a color version of this figure, please see the color plate section.)

Let us focus here on evaluation of forecasts of continuous predictands at the mesoscale. A typical mesoscale NWP model predicts wind, temperature, pressure, and water (in the vapor, liquid, and ice phases). Each of these predictands has space and time dependency. Clearly this is a multidimensional problem, but fortunately is one that can reduced to one or more scalar attributes or *metrics*. A simple example of a metric is the magnitude of the error vector, also known as the Euclidean norm,

$$\mathrm{E} = |\vec{y} - \vec{o}|\,, \tag{10.11}$$

where $\vec{y} = (y_1, y_2, y_3, \ldots, y_n)$ represents the n values of predictand y within some domain (e.g., all values on a vertical model level at some time), and $\vec{o} = (o_1, o_2, o_3, \ldots, o_n)$ represents the corresponding observations.[27] For the purpose

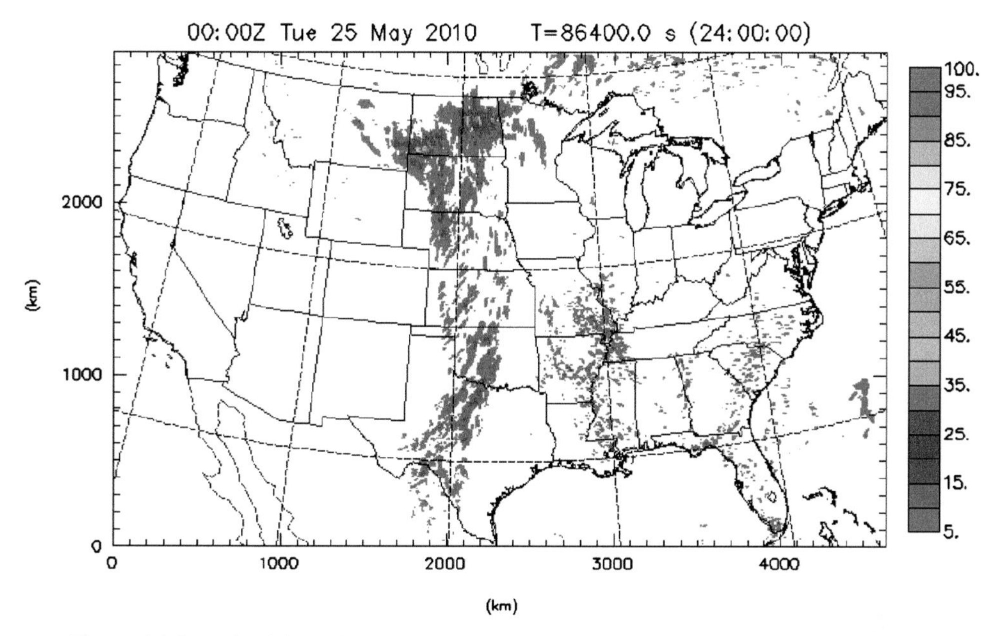

Figure 10.9 Probability of 1-h accumulated precipitation exceeding 0.5 in., at the forecast hour valid 0000 UTC on May 25, 2010. This is derived from output of a high-resolution, 26-member, multimodel ensemble system, initialized on May 24, 2010 at 0000 UTC (see Fig. 10.8). Courtesy of Dr. Fanyou Kong and the Center for Analysis and Prediction of Storms, University of Oklahoma. (For a color version of this figure, please see the color plate section.)

of the present discussion, we find it sufficient to consider the mean-squared error (MSE),

$$\mathrm{MSE} = \frac{1}{n}\sum_{i=1}^{n}(y_i - o_i)^2. \tag{10.12}$$

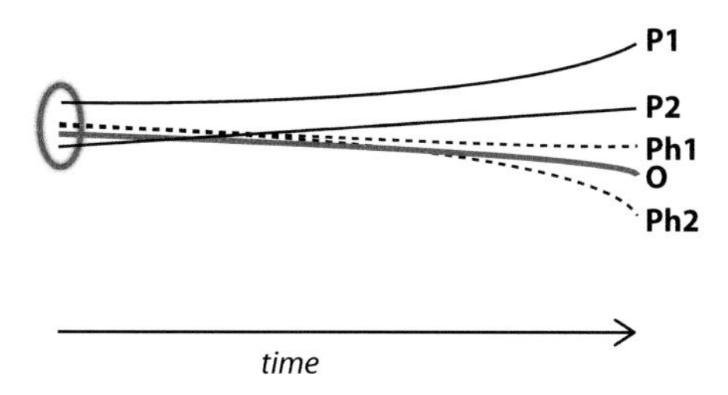

Figure 10.10 An ensemble prediction system comprised of "model physics" members (Ph1 and Ph2) in addition to members with perturbed initial conditions (P1 and P2). As in Figure 10.6, the lines represent solution trajectories plotted as a function of time. The perturbed initial-condition (physics) members are represented by thin solid (dashed) lines. The observed state (O) is the bold solid line.

MSE is a commonly used metric that addresses *forecast accuracy*, defined as the average correspondence between a prediction and the actual occurrence. Thus, the accuracy of an operational NWP system might involve the MSE of 500-hPa geopotential height at specific forecast hours. The calculation of (10.12) (and, in turn, the root-mean squared error RMSE $= \sqrt{\text{MSE}}$) uses observations at the equivalent (real) times; interpolation of model output to the n observation locations, or of observations interpolated to the n model grid points, is required. Clearly, MSE $= 0$ indicates a perfect forecast, but this and other conclusions drawn from MSE apply only to the times and spatial domains over which observations are available. Indeed, data availability is an especially problematic issue in evaluation of forecast models run at high resolution.

Forecast skill can be quantified when measures of forecast accuracy, such as MSE, are given with respect to some reference value, such as a long-term mean ("climatology"), or a persistence forecast. Consider the forecast skill score relative to climatology:

$$\text{SS} = 1 - \frac{\text{MSE}}{\text{MSE}_{\text{clm}}}, \tag{10.13}$$

where

$$\text{MSE}_{\text{clm}} = \frac{1}{n}\sum_{i=1}^{n}(o_i - \overline{o}_i)^2 \tag{10.14}$$

and where $\overline{o}_i$ is the climatological mean of the observed variable. A perfectly accurate forecast (MSE $= 0$) has the highest possible skill score, but otherwise, skill depends on how much the observations deviate from their corresponding climatological mean, and hence on the relative difficulty or ease of the forecast.

Accuracy and skill are not absolutes, however, even when evaluated objectively and quantitatively. High-resolution model predictions of precipitation, as idealized in Figure 10.11, provide support for this statement. Notice that the precipitation field in Figure 10.11 has a very large MSE, owing to a lack of local correspondence between forecast and observation. Subjectively, the forecast would still appear to have some value: the predicted and observed fields both are composed of mesoscale entities with identical areal coverage, magnitudes, and structural characteristics, but with slightly different locations. If we adopt the philosophy that slight displacement errors are acceptable provided that there is agreement in other characteristics, we can quantitatively assess the forecast using alternatives to the traditional measures.

One alternative is known as *fuzzy verification*, in which the forecasts at the m grid points in the small neighborhood of point i are compared to corresponding observations in that neighborhood (Figure 10.12).[28] Typically, the fuzzy verification metrics are based on dichotomous forecasts of events; we introduced the example of a yes/no forecast of a tornado, but dichotomous forecasts of precipitation (and other

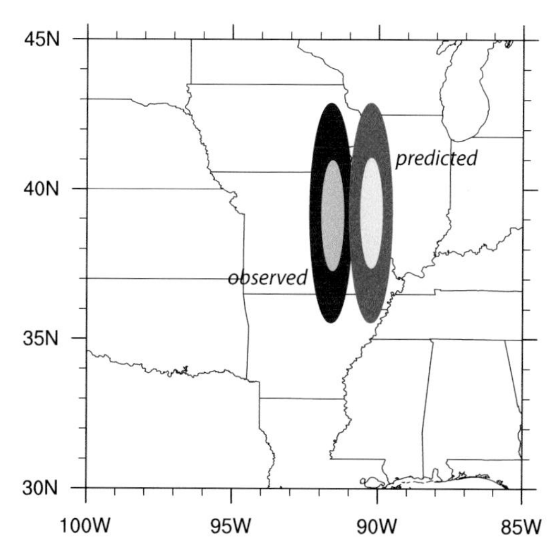

Figure 10.11 Hypothetical example of a forecast field of 1-h precipitation (red/yellow), with the verifying observations (blue/cyan). Assume that the red and blue (yellow and cyan) contours correspond to the same amounts. According to traditional measures (such as MSE), this forecast would have little to no skill, owing to a lack of local correspondence between the forecast and observation. Based on Gilleland et al. (2010). (For a color version of this figure, please see the color plate section.)

continuous predictands) can be constructed by defining an event as an exceedance of some threshold. Thus, let I_y (I_o) indicate a predicted event (observed event) at a grid point, and then assign $I = 1$ ($I = 0$) for an occurrence (nonoccurrence). The respective probabilities of a predicted and observed event within the neighborhood of the grid point are then

$$\langle P_y \rangle_s = \frac{1}{m} \sum_m I_y \tag{10.15}$$

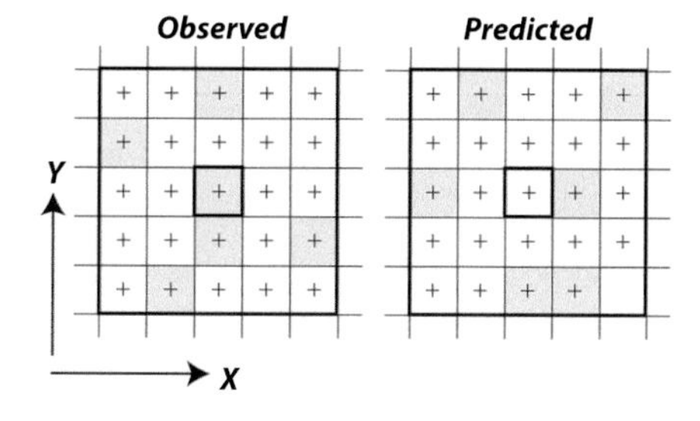

Figure 10.12 Example application of fuzzy verification. Shown is a subset of the computational domain. Shading in gr. cells indicates an occurrence, such as an exceedance of 1-h precipitation over some threshold value. The forecast in the central grid cell is incorrect, but in the 5 × 5 neighborhood of that grid cell, the forecast field has the exact same number of occurrences (6) than does the observational field. After Roberts and Lean (2008).

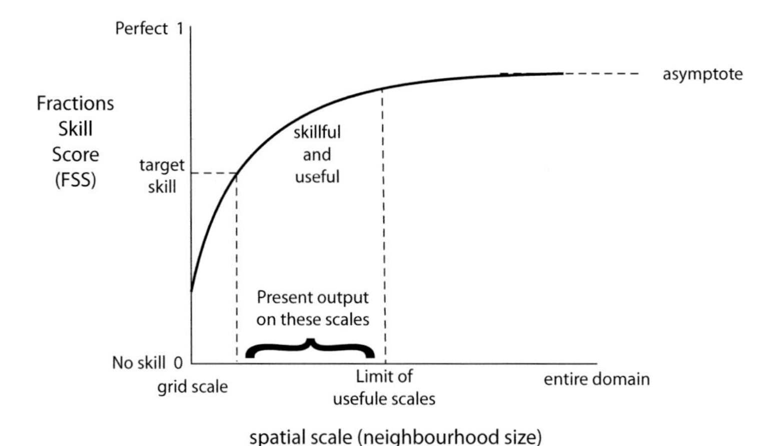

Figure 10.13 Hypothetical FSS evaluated over multiple scales or neighborhood sizes s. Notice that FSS asymptotes to a value ≤ 1 when neighborhood size approaches the domain size. The range of useful scales and the target skill are user defined. After Roberts and Lean (2008).

and

$$\langle P_o \rangle_s = \frac{1}{m} \sum_m I_o, \tag{10.16}$$

where $\langle\rangle_s$ denotes the neighborhood and its size or scale. Equations (10.15) and (10.16) also describe the respective *fractions* of events within the neighborhood, and provide, when summed over all n grid points of the evaluation domain, a measure of forecast skill,

$$\mathrm{FSS_s} = 1 - \frac{\frac{1}{n}\sum_n \left[\langle P_y \rangle_s - \langle P_o \rangle_s\right]^2}{\frac{1}{n}\sum_n \left[\langle P_y \rangle_s^2 + \langle P_o \rangle_s^2\right]}, \tag{10.17}$$

where FSS is the *fractions skill score*. FSS is evaluated over neighborhoods of increasing scale s, with s typically given as a multiple of grid length.[29] As demonstrated in Figure 10.13, the expectation is that FSS (and thus skill) will be a minimum at small s, and thereafter will increase to its asymptotic value FSS ≤ 1 as s is increased to a domain-length equivalent. One may use this information to evaluate whether a predetermined level of skill is reached at a scale commensurate with the grid point spacing. For example, if a model with 1-km grid point spacing is found to be skillful only at scales greater than 100 km, one would search for possible model pathologies, reassess the predictability of problem, and/or reconsider how the computational resources are allocated according to gridpoint spacing, parameterization complexity, and so forth (see Chapter 4).

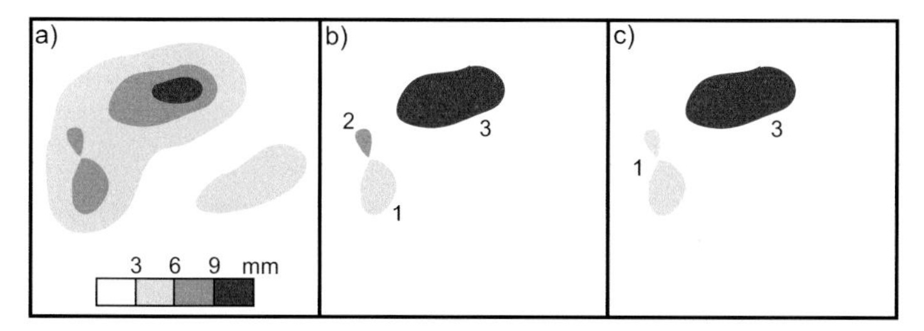

Figure 10.14 Illustration of an object-oriented approach to forecast evaluation, as applied to (a) the observational or forecast field. The initial objects in (b) are constructed from the field in (a) after a user-defined threshold is applied (here, 6 units). Notice that the initial objects 1 and 2 are separated by only 1 pixel. In this case, the separation is at the user-defined "search radius" threshold, and hence these objects are combined into a single object; the final objects are shown in (c). After Hitchens et al. (2012), and based on the algorithm of Baldwin et al. (2005).

For the precipitation field in Figure 10.11, fuzzy verification would have rewarded a forecast that had an areal coverage similar to that of the observations, but would not have accounted for how the precipitation was distributed within the neighborhood. If correspondence in spatial structure – shape, size, orientation – is valued, one might adopt a measure based on a decomposition of the field into "objects." The objects are groupings or regions of contiguous predictand values that meet user-defined criteria.[30] In the example in Figure 10.14, convective-precipitation objects are constructed from regions where grid point values exceed a specific 1-h precipitation accumulation, such as 6 mm; these regions may truly be contiguous, or have small permissible gaps. The evaluation metrics then involve differences between quantitative attributes of the predicted and observed objects. In the Euclidean norm (Eq. (10.11)), predictand vector $\vec{y}$ might now have components such as y_1 = object area, y_2 = maximum precipitation rate within the object, y_3 = location of object centroid, y_4 = object ellipticity (length of major axis/length of minor axis), y_5 = track of object, and so on.

A particular issue with this approach is that there will not necessarily be a one-to-one correspondence between every predicted and observed object. One might choose to compute E (or an equivalent metric) using predicted-observed objects that do correspond, but inevitably there will be some objects that are unpaired and thus unevaluated. Another implementation issue is how (and whether) to weight the individual attributes (e.g., object area, location, magnitude) in the metric. The ultimate use of the evaluation usually provides guidance on these decisions, with some applications valuing spatial proximity the greatest, others valuing predictand magnitude, and still others valuing predictand area. A final issue, which affects fuzzy verification as well, is the need for an observational dataset with properties (e.g., domain size, grid point spacing) that are consistent with those of the prediction model. In the case of high-resolution precipitation forecasts, data from the U.S. network of rain gages – and indeed from most gage networks worldwide – have unacceptably low spatial resolution. Evaluators of precipitation forecasts must therefore resort to alternatives, such as a blend of gage observations with composite estimates from radar reflectivity factor (see Chapter 3), which are available on a sufficiently fine grid.[31]

Object-based techniques have promising applications beyond forecast evaluation. Consider their use in the "mining" of large datasets. In what otherwise would have been an extremely labor-intensive project, the Baldwin Object Oriented Identification Algorithm (BOOIA) was employed to extract, from a precipitation dataset, the characteristics of 3,484 objects associated with short-duration extreme rainfall.[32] The quantitative information was then used to statistically describe the convective-storm morphology of a very large sample of events.

The BOOIA has also been adapted for use in the forecasts themselves. As will be described next, a "feature-specific" prediction system exemplifies the forecasting strategies that employ the new generation of high-resolution NWP models.

10.5 Forecast Strategies with Mesoscale Numerical Models

The scale and detail in the output generated by high-resolution, convective-permitting NWP models presents the model users with a number of challenges and opportunities. Besides the basic logistics (e.g., data post-processing, storage) associated with the copious amount of data generated by these models, one major challenge regards the appropriate level of interpretation of the fine structure in fields such as rainwater mixing ratio; another regards the appropriate use of this information in the human-produced forecasts. Indeed, the inaugural use of the high-resolution model output was largely qualitative, with forecasters interrogating the precipitation fields for locations of convection initiation and suggestions of convective-storm type.[33] It is now recognized that a combination of predicted and derived variables can be used quantitatively to identify "features" of forecast interest. This is the essence of *feature-specific prediction* (FSP).

Consider the example of an FSP system configured to identify and predict supercell storms.[34] The system commences with a search of fields of simulated radar reflectivity factor (SRF; see Chapter 4), using the functionality of an object-based forecast evaluation technique such as BOOIA. To constrain the search to convective features, the FSP system requires that objects have SRF $\geq$ 40 dBZ. Other model fields are then used to narrow the search further to only those convective features exhibiting updraft rotation. A diagnostic field developed for this purpose is *updraft helicity* (UH), defined as

$$\mathrm{UH} = \int_{z_B}^{z_T} w\zeta dz, \tag{10.18}$$

where the integration limits are often set as $z_1 = 2$ km, $z_2 = 5$ km. Equation (10.18) derives from (7.28) and helps to quantify midlevel updraft rotation.[35] Note that ζ is not a model variable, but it can be computed easily from predicted winds using

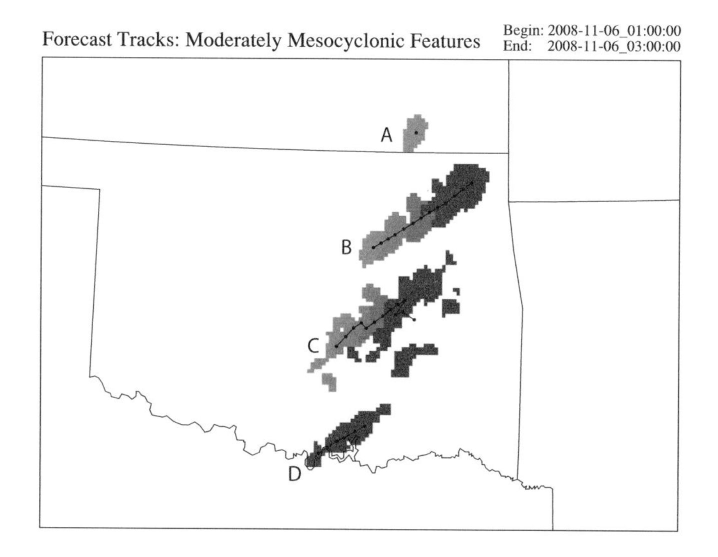

Time (UTC) 01:00 01:30 02:00 02:30 03:00

Feature ID	Start Time (UTC)	End Time (UTC)	Temporal Persistence (minutes)	Max dBZ (time UTC)
A	01:00:00	01:10:00	10	54 (0100)
B	01:00:00	03:00:00	120	59 (0140)
C	01:00:00	03:00:00	120	59 (0140)
D	02:10:00	03:00:00	50	57 (0210)

Figure 10.15 Example a feature-specific system applied to supercell prediction. Tracks and shadings indicate locations of supercells, as objectively determined from high-resolution model output. The table provides attributes of the individual supercells. From Carley et al. (2011). (For a color version of this figure, please see the color plate section.)

a finite difference approximation. Based on empirical evidence, a reasonable FSP requirement is that UH ≥ 50 m^2 s^{-2} at one or more grid points within the feature for it to be considered a supercell.[36]

Figure 10.15 displays output from this FSP application. The FSP system is used to diagnose tracks of individual supercells during the forecast interval, as well as to compute feature attributes that reveal predicted storm intensity. Figure 10.15 also demonstrates how FSP is amenable to the use of ensembles: member agreement (or lack thereof) in feature locations and tracks is readily displayed, as are quantitative attributes. Conversion of this information into probabilities is straightforward. Thus, an ensemble-aided FSP forms the basis for a decision support system.

These capabilities of high-resolution NWP models support the premise of the *warn-on-forecast* paradigm, which proposes the use of short-term model predictions as the

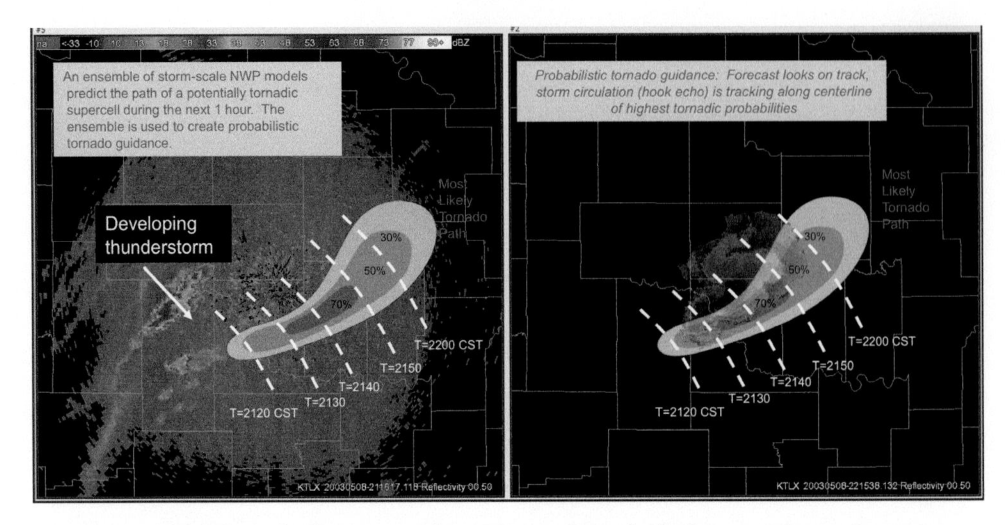

Figure 10.16 Tornado forecast guidance from a hypothetical convective-scale warn-on-forecast. Blue shadings show areal probabilities of tornado occurrence. White dashed lines indicate predicated storm locations. Color fill is of radar reflectivity factor. From Stensrud et al. (2009). (For a color version of this figure, please see the color plate section.)

basis for public warnings of ongoing or imminent severe convective storms.[37] It is prudent to recall here that the intended outcome of a warning is some immediate action, such as evacuation to an alternative shelter to mitigate personal injury. Warnings currently are activated by the detection of some phenomenon, such as a tornadic thunderstorm, by way of Doppler-radar scans and/or visual sighting. The warning timeliness therefore depends on how well the phenomenon is observed, and then on whether the phenomenon has evolved yet to its damage-generating state. The warn-on-forecast paradigm offers a potentially significant increase in warning lead time, because warning decisions are based on predictions rather than detections.

Owing to its focus on the 0–6-h forecast interval, warn-on-forecast requires the best possible estimate of the initial state of the atmosphere. This includes the state of ongoing precipitating storms, and implies a need to assimilate radar data. Inherent in assimilation techniques such as the ensemble Kalman filter (Chapter 4) is a measure of forecast uncertainty. Uncertainty measures – and information from ensembles in general – can be exploited to produce probabilities of the spatial location and occurrence of a specific hazard such as a tornado (Figure 10.16). In this regard, FSP is an example of implementation of the warn-on-forecast paradigm.

There are other strategies involving the use of mesoscale models for short-term forecasts. Consider that the initialization fields are themselves invaluable forecast tools. Indeed, it is possible to produce wind fields with high temporal ($\sim$5 min) and spatial ($\sim$1 km) resolution by coupling coarser model output with data from multiple Doppler radars, via a data assimilation system such as 3DVAR (Chapter 4).[38] The resultant gridded wind data are (2D and 3D) vector winds rather than single-Doppler radial winds, which reduces ambiguity in interpretation and in automated detection algorithms (Chapter 3). Furthermore, gridded thermodynamic variables are produced that are physically consistent with the winds, and thus can be incorporated into the diagnoses.

Certainly, relatively coarse NWP model data also play a variety of roles in forecasts of mesoscale-convective phenomena. A human forecaster may use model-predicted environmental information (e.g., CAPE and vertical wind shear) to subjectively assess the likelihood of the formation and subsequent behavior of specific phenomenon (e.g., tornadic supercells), taking into account the perceived accuracy of the current forecast as well as other factors. More formally, the forecaster may employ *statistical downscaling* of the coarse model data. In essence, a predictand (y) is quantitatively related through a statistical model to one or more predictors (x_i) from the coarse model (and perhaps from other data sources). Symbolically, this is expressed as

$$y = a_i x_i, \quad i = 1, \ldots, n, \tag{10.19}$$

where the a_i are empirical coefficients, as obtained through linear regression for example. The forecaster may even use coarse model data as the IC/BC on a dynamical model to numerically generate the otherwise unresolved predictand. This is one example of *dynamical downscaling*, which in fact pervades limited-area applications of high-resolution mesoscale models, particularly in the absence of data assimilation and other special initialization procedures. Hence, given predominantly lower-resolution IC/BC, integration of the mesoscale model results in mesoscale detail, as exemplified by the appearance of convective precipitating storms, and as manifest in the mesoscale range of the energy spectrum.

Dynamical downscaling approaches are common to studies of climate variability and change, and indeed have proven to be an effective means of linking large temporal- and spatial-scale processes down to the short time and length scales of deep moist convection. Figure 10.17 shows an example of the mean annual frequency of heavy precipitation (1-h rainfall exceeding 1 in.) based on dynamically downscaled global reanalysis data, from April through June, 1991–2000.[39] The downscaled precipitation compares favorably with existing observations, supporting conclusions drawn about the sufficiency of the information present in the IC/BC. Although this particular downscaling technique is still under refinement, there does appear to be the potential for its application to seasonal and longer-term climate forecasts of convective-scale weather. Because of the computational expensive of this technique, however, alternative approaches for such applications are still valuable.

10.6 Long-Range Prediction

For myriad reasons it has become desirable for operational meteorological centers and other agencies to offer seasonal (and longer-range) outlooks in addition to short-range

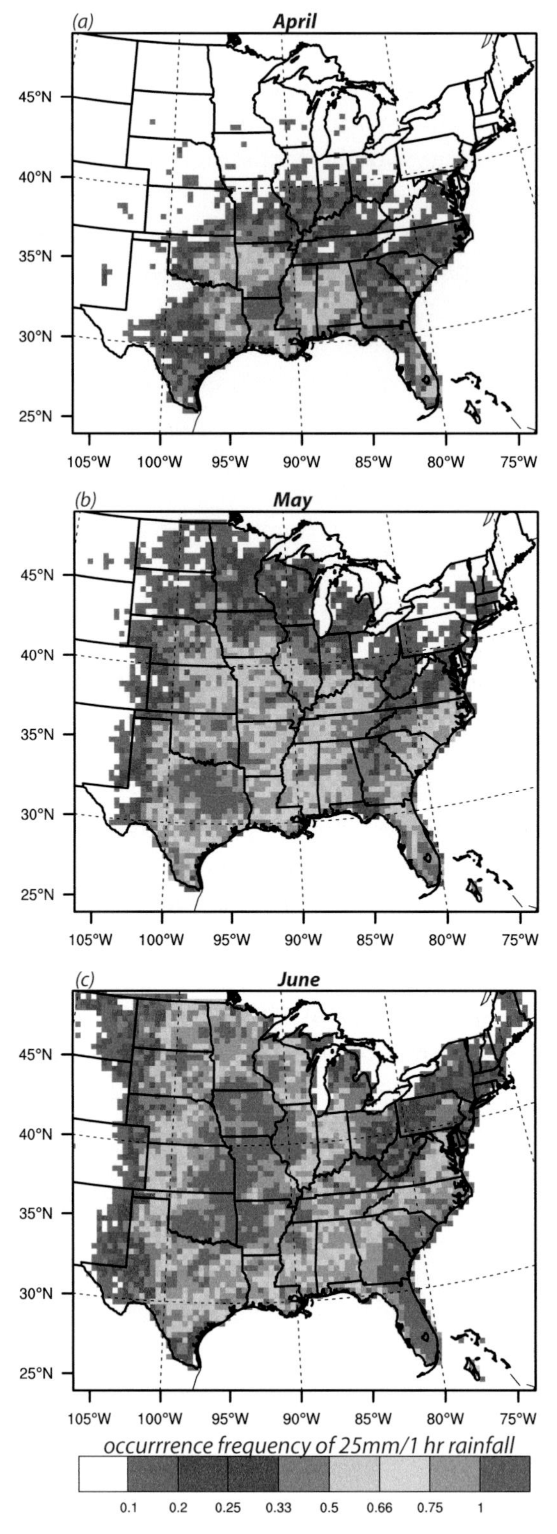

Figure 10.17 Mean frequency of 1-h rainfall exceeding 1 in., during the warm season months of April–June, 1991–2000, from a dynamical downscaling method. From Trapp et al. (2010). Used with kind permission from Springer Science and Business Media. (For a color version of this figure, please see the color plate section.)

forecasts. A typical outlook product is the probability of a seasonal departure from the long-term mean of temperature and precipitation.

The long lead time of such outlooks suggests the importance in accounting for "slow" or low-frequency processes on the planetary scale. Particularly relevant is the irregular, interannual variability of the atmosphere and ocean known as the El Niño-Southern Oscillation (ENSO). As can be surmised, long-range predictions must explicitly couple the global atmosphere to the oceans and to other components of the Earth system. The resultant complexity places practical limitations on the scales of motion that can be resolved in coupled prediction models. It is still possible, however, to gain information from these models and available data about mesoscale phenomena. The relative coarseness of the models and observations requires the use of statistical, dynamical, or mixed dynamical-statistical approaches.

To illustrate these approaches, let us consider predictions of seasonal hurricane activity in the Atlantic basin. Several-month-lead forecasts issued by Colorado State University and other groups are based on regression equations with predictors such as sea-surface temperature (SST), 200-hPa zonal wind, and sea-level pressure over specific geographical regions and time periods.[40] Each predictor has a physical link to hurricane formation, but also a statistical one, in the form of high linear correlation with indexed values of Atlantic hurricane activity. The predictors are based on observational data, and also derived from dynamical forecast models such as the Climate Forecast System (CFS) of the U.S. National Centers for Environmental Prediction (NCEP). The CFS is a global, fully coupled, ocean-atmosphere-land dynamical model, and has been shown to skillfully predict SSTs (and vertical wind shear) in the tropical Pacific and tropical North Atlantic.[41]

Similar forecast systems maintained by the European Centre for Medium-Range Weather Forecasts (ECMWF) and others worldwide belong to the family of coupled atmosphere-ocean general circulation models (GCMs). Although the effective resolution of these global dynamical models is still too coarse to provide details of tropical cyclones, many GCMs are able to resolve, even in their current versions, enough of the characteristic structure so that cyclones can be explicitly identified in the model output.[42] Indeed, the seasonal numbers of such structures predicted by ECMWF models have been shown to correlate well with the number of observed tropical cyclones, with an estimate of uncertainty provided by ensemble members.[43] Confidence in this application of GCMs has grown so that even multiyear predictions are now being made.[44]

Retrospective model forecasts, or *reforecasts*, are particularly useful in the development of the forecast methodologies.[45] A typical reforecast dataset consists of several tens of years of individual integrations using a "frozen" version of the model; a model ensemble is often constructed for each integration interval. In practice, the forecast methodology (e.g., regression model using forecast variables, or some feature detection scheme; Section 10.5) would be trained and tested with the reforecast dataset.

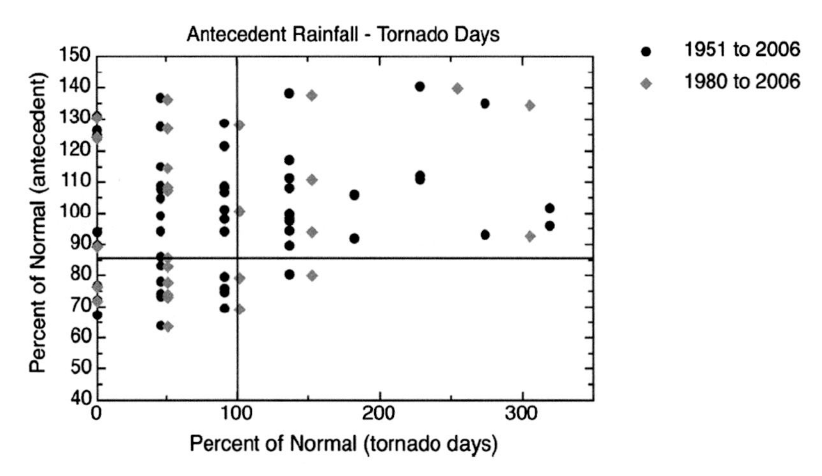

Figure 10.18 Relationship between 6-month antecedent rainfall and tornado occurrence over roughly the northern half of the U.S. state of Georgia. From Shepherd et al. (2009). Used with permission.

A sufficiently extensive dataset would also allow for a demonstration of skill and diagnoses of biases even before the methodology is applied in real time.

The focus in this section has been on tropical phenomena because, to date, the published literature on seasonal predictions of midlatitude mesoscale-convective phenomena is quite sparse.[46] A few attempts have been made to statistically relate tornado activity to rainfall, with the premise that drought conditions should equate to low tornado activity.[47] The somewhat limited data suggest only slight correlations between these two variables when examined contemporaneously over annual and seasonal time periods. Significant *time-lagged* correlations between drought and tornado activity, however, have been shown to exist within specific geographic regions. In northern Georgia, for example, autumn-to-winter drought conditions correlate well with below-normal tornado activity during the following spring (Figure 10.18). The assertion is that the drought is translated to the subsequent tornado season through anomalously dry soil. The open question is whether the reduced tornado activity is largely a result of the surface forcing associated with the soil moisture and its long memory, or of a feedback between the soil moisture and regional atmospheric circulation, or of an anomalous planetary-scale circulation that affects both (see Chapter 9).

It is in this light that the effect of ENSO on tornado occurrence within the contiguous United States continues to be explored. Although a consensus has yet to be reached on the overall statistical relation between ENSO phase and deep convective storms, recent work has shown that the total number of days of tornado occurrence in winter (January–March) is highest during the neutral phase of ENSO and lowest during the warm or El Niño phase.[48] The reduced activity appears to be influenced by a southward displacement of the mean winter jet stream and of the associated track of synoptic-scale cyclones, thus biasing tornadic-storm formation toward the Gulf Coast states (and away from the southern Great Plains, the climatologically favored region for tornado occurrence). This is consistent with separate observations relating cool, wet winters in the Gulf Coast states to the El Niño phase (and warm, dry winters to the La Niña phase).[49] Of course, a favorable jet-stream position and cyclone track do not alone guarantee that all meteorological conditions necessary for tornadic storm formation will be in place. However, these factors do at least promote a higher likelihood for the possibility of tornadic storms (than in a case of unfavorable jet-stream position), and in that regard could aid in their long-lead prediction.

The persistent, anomalously warm water in the eastern Pacific ocean during El Niño is an example of what can be considered *long-lasting surface forcing*.[50] The associated SST anomalies cause seasonal and interannual variability of the atmospheric state relative to some seasonal and annual mean. This variability can be realized, for example, as anomalous sea-level pressure over the same or remotely different region; similar comments can be made regarding the effect of the forcing associated with soil moisture and snow-cover anomalies. Intermittent, yet relatively higher-frequency weather systems simultaneously contribute to the atmospheric anomalies. Imagine the occurrence, within the central United States, of two or three strong synoptic-scale cyclones during some month. These would dominate the monthly, and perhaps seasonally averaged pressure field over that region, and thus add to the variability over these periods. The cyclones are due to the naturally varying atmosphere on the synoptic-scale, and thus their effect – and the cyclones themselves – are unpredictable beyond roughly two weeks. For this reason, the cyclones are considered *noise*. If the contribution of such noise is small relative to the variance attributed to the long-lasting surface forcing, the predictability could potentially exceed the limit of determininstic predictabililty. Tropical latitudes tend to be affected less by the weather noise, and thus have relatively larger potential predictability over long ranges, given sufficient long-lasting surface forcing. The proviso is that the surface forcing such as SST anomalies must also be predictable, returning us back to the need for coupled models.

The boundaries of long-range prediction using coupled models are quickly advancing. On the near horizon is a wider use of global, cloud-system-resolving models akin to the Nonhydrostatic Icosahedral Atmospheric Model (NICAM) developed at the Frontier Research Center for Global Change in Japan. Icosahedral grid structures and regional grid stretching techniques[51] allow for improved efficiency and accuracy in the global simulations. With continuous integrations over periods of at least a month, and global grid spacings of several kilometers, NICAM has successfully predicted the genesis, motion, and structures of tropical cyclones.[52]

The future of long-range (and perhaps even short-term) prediction of midlatitude convective weather will likely involve ensembles of models such as NICAM. The modeling systems will exploit data assimilation techniques, and will require clever and efficient analysis of the predicted model fields. History will be the judge of whether the modeling systems of the future are able to significantly advance the skill and value of predictions of mesoscale-convective processes.

Supplementary Information

For exercises, problem sets, and suggested case studies, please see www.cambridge.org/trapp/chapter10.

Notes

1 See Lorenz (1969); a general overview of important issues in atmospheric predictability can be found in Hacker et al. (2005).
2 This is based on Thompson (1957) and also the *Glossary of Meteorology* (Glickman 2000).
3 The model description is from Lorenz (1963), but the model itself derives from the work of Saltzman (1962).
4 One should bear in mind that Figure 10.2 gives 2D projections of the 3D trajectory behavior, and thus that apparent trajectory intersection are an artifact of this presentation.
5 Lorenz (1963).
6 Anthes (1986).
7 Lorenz (1969), Lorenz (1984).
8 The following is from an interpretation of Lorenz (1969) by Lilly (1990).
9 Nastron and Gage (1985).
10 For more on predictability at these scales, see Tribbia and Baumhaufner (2004).
11 Gage (1979), Lilly (1983).
12 Lilly (1990).
13 By Lorenz (1969), for example.
14 The following example and accompanying analysis is from Zhang et al. (2003).
15 Zhang et al. (2003).
16 Trapp et al. (2007b).
17 Tribbia and Baumhefner (1988).
18 The following draws from Kalnay (2003).
19 See Kalnay (2003) for a detailed treatment of these and other methods.
20 Toth et al. (1997).
21 http://www.emc.ncep.noaa.gov/gmb/ens/fcsts/ensframe.html.
22 Stensrud et al. (2000).
23 As summarized by Kalnay (2003); see also Houtekamer et al. (1996).
24 Bougeault (2010).
25 Based on arguments made by Oreskes et al. (1994), the term *evaluation* is used here instead of *verification*.
26 For more discussion on such purposes, see Hacker et al. (2005).
27 See Wilks (2006), which forms the basis of much of this discussion.
28 Ebert (2008).
29 Ebert (2008), Roberts and Lean (2008).
30 Baldwin et al. (2005), Gilleland et al. (2010).
31 An example is the U.S. National Centers for Environmental Prediction stage II and stage IV multisensor precipitation dataset (hereafter, the ST2/ST4 dataset); see Fulton et al. (1998).
32 Hitchens et al. (2012), Hitchens et al. (2010).
33 Kain et al. (2006).
34 Carley et al. (2011).
35 Kain (2008).
36 Threshold values of UH will depend on the horizontal gridpoint spacings in the model; see Carley et al. (2011), also Trapp et al. (2010).
37 See Stensrud (2009); this paradigm has long been in place for hurricane prediction, in which several days of lead time are needed for evacuation and other readiness preparations.

38 Gao et al. (2009).
39 Trapp et al. (2010).
40 Klotzbach and Gray (2003), Klotzbach (2007).
41 Wang (2009).
42 This statement excludes the experimental global models that are being run with high-enough resolution to explicitly represent tropical cloud systems; see Fudeyasu et al. (2010).
43 Vitart (2007).
44 Smith (2010).
45 Hamill et al. (2006).
46 At the time this chapter was being finalized, efforts involving the author and several others were being launched to develop methods for long-lead predictions of hazardous convective storm activity.
47 Galway (1979), Shepherd et al. (2009).
48 Cook and Schaefer (2008).
49 Ropelewski and Halpert (1987), Halpert and Ropelewski (1992).
50 The following discussion stems from Kalnay (2003).
51 Markovic et al. (2010).
52 Fudeyasu et al. (2010).

References

Adlerman, E. J., K. K. Droegemeier, and R. Davies-Jones, 1999: A numerical simulation of cyclic mesocyclogenesis. *J. Atmos. Sci.*, **56**, 2045–2069.

Anderson, D. A., J. C. Tannehill, and R. H. Pletcher, 1984: *Computational Fluid Mechanics and Heat Transfer*. Hemisphere Publishing, New York.

Anthes, R. A., 1986: The general question of predictability. *Mesoscale Meteorology and Forecasting*, American Meteorological Society, Boston, 636–656.

Armijo, L, 1969: A theory for the determination of wind and precipitation velocities with Doppler radars. *J. Atmos. Sci.*, **26**, 570–573.

Arnup, S. J., and M. J. Reeder, 2007: The diurnal and seasonal variation of the northern Australian dryline. *Mon. Wea. Rev.*, **135**, 2995–3008.

Arritt, R. W., 1993: Effects of the large-scale flow on characteristic features of the sea breeze. *J. Appl. Meteor.*, **32**, 116–125.

Atkins, N. T., C. S. Bouchard, R. W. Przybylinski, et al., 2005: Damaging surface wind mechanisms within the 10 June 2003 Saint Louis bow echo during BAMEX. *Mon. Wea. Rev.*, **113**, 2275–2296.

Atkins, N. T., and J. J. Cunningham, 2006: The influence of low-level stable layers on damaging surface winds within bow echoes. Preprints, *23rd Conf. on Severe Local Storms*, St. Louis, MO, Amer. Meteor. Soc., (6.4) CD-ROM.

Atkins, N. T., and M. St. Laurent, 2009: Bow echo mesovortices. Part II: Their genesis. *Mon. Wea. Rev.*, **137**, 1514–1532.

Atkins, N. T., R. M. Wakimoto, and T. M. Weckwerth, 1995: Observations of the sea-breeze front during CaPE. Part II: Dual-Doppler and aircraft analysis. *Mon. Wea. Rev.*, **123**, 944–968.

Atkins, N. T., R. M. Wakimoto, and C. L. Ziegler, 1998: Observations of the finescale structure of a dryline during VORTEX 95. *Mon. Wea. Rev.*, **126**, 525–555.

Atkins, N. T., M. L. Weisman, and L. J. Wicker, 1999: The influence of preexisting boundaries on supercell evolution. *Mon. Wea. Rev.*, **127**, 2910–2927.

Augustine, J. A., and F. Caracena, 1994: Lower-tropospheric precursors to nocturnal MCS development over the central United States. *Wea. Forecasting*, **9**, 116–135.

Balaji, V., and T. L. Clark, 1988: Scale selection in locally forced convective fields and the initiation of deep cumulus. *J. Atmos. Sci.*, **45**, 3188–3211.

Baldwin, M. E., J. S. Kain, and S. Lakshmivarahan, 2005: Development of an automated classification procedure for rainfall systems. *Mon. Wea. Rev.*, **133**, 844–862.

Banacos, P. C., and D. M. Schultz, 2005: The use of moisture flux convergence in forecasting convective initiation: Historical and operational perspectives. *Wea. Forecasting*, **20**, 351–366.

Bannon, P. R., 1996: On the anelastic approximation for a compressible atmosphere. *J. Atmos. Sci.*, **53**, 3618–3628.

Bannon, P. R., 2002: Theoretical foundations for models of moist convection. *J. Atmos. Sci.*, **59**, 1967–1982.

Banta, R. M., and C. Barker Schaaf, 1987: Thunderstorm genesis zones in the Colorado Rocky Mountains as determined by traceback of geosynchronous satellite images. *Mon. Wea. Rev.*, **115**, 463–476.

Barnes, S. L., 1964: A technique for maximizing details in numerical weather map analysis. *J. Appl. Meteor.*, **3**, 396–409.

Barnes, S. L., 1973: Mesoscale objective analysis using weighted time-series observations. NOAA Tech. Memo. ERL NSSL-62, National Severe Storms Laboratory, Norman, OK [NTIS COM-73-10781].

Barnes, G. M., K. Sieckman, 1984: The environment of fast- and slow-moving tropical mesoscale convective cloud lines. *Mon. Wea. Rev.*, **112**, 1782–1794.

Batchelor, G. K., 1967: *An Introduction to Fluid Dynamics*. Cambridge University Press.

Battan, L. J., 1973: *Radar Observation of the Atmosphere*. University of Chicago Press.

Bedka, K., J. Brunner, R. Dworak, et al., 2010: Objective Satellite-based detection of overshooting tops using infrared window channel brightness temperature gradients. *J. Appl. Meteor. Climatol.*, **49**, 181–202.

Beer, T., 1974: *Atmospheric Waves*. Wiley, New York.

Bell, G. D., and J. E. Janowiak, 1995: Atmospheric circulation associated with the Midwest floods of 1993. *Bull. Amer. Meteor. Soc.*, **76**, 681–695.

Benjamin, S. G., K. A. Brewster, R. L. Brummer, et al., 1991: An isentropic three-hourly data assimilation system using ACARS aircraft observations. *Mon. Wea. Rev.*, **119**, 888–906.

Benjamin, S. G., B. E. Schwartz, S. E. Koch, and E. J. Szoke, 2004: The value of wind profiler data in U.S. weather forecasting. *Bull. Amer. Meteor. Soc.*, **85**, 1871–1886.

Benjamin, T. B., 1968: Gravity currents and related phenomena. *J. Fluid Mech.*, **31**, 209–248.

Berry, G. J., and C. D. Thorncroft, 2012: African easterly wave dynamics in a mesoscale numerical model: The upscale role of convection. *J. Atmos. Sci.*, **69**, 1267–1283.

Biggerstaff, M. I., et al., 2005: The Shared Mobile Atmospheric Research and Teaching Radar: A collaboration to enhance research and teaching. *Bull. Amer. Meteor. Soc.*, **86**, 1263–1274.

Bluestein, H. B., 1993: *Observations and Theory of Weather Systems. Vol. 2, Synoptic–Dynamic Meteorology in Midlatitudes*, Oxford University Press.

Bluestein, H. B., and M. H. Jain, 1985: Formation of mesoscale lines of precipitation: Severe squall lines in Oklahoma during the spring. *J. Atmos. Sci.*, **42**, 1711–1732.

Bluestein, H. B., E. W. McCaul, Jr., G. P. Byrd, and G. R. Woodall, 1988: Mobile sounding observations of a tornadic storm near the dryline: The Canadian, Texas storm of 7 May 1986. *Mon. Wea. Rev.*, **116**, 1790–1804.

Bluestein, H. B., and M. L. Weisman, 2000: The interaction of numerically simulated supercells initiated along lines. *Mon. Wea. Rev.*, **128**, 3128–3148.

Bluestein, H. B., and G. R. Woodall, 1990: Doppler-radar analysis of a low-precipitation severe storm. *Mon. Wea. Rev.*, **118**, 1640–1664.

Blyth, A. M., W. A. Cooper, and J. B Jensen, 1988: A study of the source of entrained air in Montana cumuli. *J. Atmos. Sci.*, **45**, 3944–3964.

Blyth, A. M., S. G. Lasher-Trapp, and W. A. Cooper, 2005: A study of thermals in cumulus clouds. *Quart. J. Roy. Meteor. Soc.*, **131**, 1171–1190.

Bohme, T., T. P. Lane, W. D. Hall, and T. Hauf, 2007: Gravity waves above a convective boundary layer: A comparison between wind-profiler observations and numerical simulations. *Quart. J. Roy. Meteor. Soc.*, **133**, 1041–1055.

Bolton, D., 1980: The computation of equivalent potential temperature. *Mon. Wea. Rev.*, **108**, 1046–1053.

Bony, S., et al., 2006: How well do we understand and evaluate climate change feedback processes? *J. Climate*, **19**, 3445–3482.

Bougeault, P., et al., 2010: The THORPEX interactive grand global ensemble. *Bull. Amer. Met. Soc.*, **91**, 1059–1072.

Brady, R. H., and E. J. Szoke, 1989: A case study of nonmesocyclone tornado development in northeast Colorado: Similarities to waterspout formation. *Mon. Wea. Rev.*, **117**, 843–856.

Brandes, E. A., 1977: Flow in severe thunderstorms observed by dual-Doppler radar. *Mon. Wea. Rev.*, **105**, 113–120.

Brandes, E. A., 1978: Mesocyclone evolution and tornadogenesis: Some observations. *Mon. Wea. Rev.*, **106**, 995–1011.

Brandes, E. A., and C. L. Ziegler, 1993: Mesoscale downdraft influences on vertical vorticity in a mature mesoscale convective system. *Mon. Wea. Rev.*, **121**, 1337–1353.

Brock, F. V., and S. J. Richardson, 2001: *Meteorological Measurement Systems*. Oxford University Press.

Brock, F. V., K. C. Crawford, R. L. Elliott, et al., 1995: The Oklahoma Mesonet: A technical overview. *J. Atmos. Oceanic Technol.*, **12**, 5–19.

Brock, F. V., G. Lesins, and R. Walko, 1987: Measurement of pressure and air temperature near severe thunderstorms: An inexpensive and portable instrument. *Extended Abstracts, Sixth Symp. on Meteorological Observations and Instrumentation*, New Orleans, LA, American Meteorological Society, Boston, 320–323.

Brooks, H. E., C. A. Doswell III, and R. B. Wilhelmson, 1994: The role of midtropospheric winds in the evolution and maintenance of low-level mesocyclones. *Mon. Wea. Rev.*, **122**, 126–136.

Brooks, H. E., J. W. Lee, and J. P. Craven, 2003: The spatial distribution of severe thunderstorm and tornado environments from global reanalysis data. *Atmos. Res.*, 67–68, 73–94.

Brown, R. A., and V. T. Wood, 1991: On the interpretation of single-Doppler velocity patterns within severe thunderstorms. *Wea. Forecasting*, **6**, 32–48.

Brown, R. A., and V. T. Wood, 2007: A guide for interpreting Doppler velocity patterns: Northern Hemisphere Edition. NOAA National Severe Storms Laboratory document, 55 pp. (Available from http://publications.nssl.noaa.gov/.)

Browning, K. A., 1964: Airflow and precipitation trajectories within severe local storms which travel to the right of the winds. *J. Atmos. Sci.*, **21**, 634–639.

Browning, K. A., 1986: Conceptual models of precipitation systems. *Wea. Forecasting*, **1**, 23–41.

Browning, K.A. and F. H. Ludlam, 1962: Airflow in convective storms. *Quart. J. Roy. Meteor. Soc.*, **88**, 117–135.

Browning, K. A., and R. J. Donaldson, 1963: Airflow and structure of a tornadic storm. *J. Atmos. Sci.*, **20**, 533–545.

Bryan, G. H., 2008: On the computation of pseudoadiabatic entropy and equivalent potential temperature. *Mon. Wea. Rev.*, **136**, 5239–5245.

Bryan, G. H., and J. M. Fritsch, 2000: Moist absolute instability: The sixth static stability state. *Bull. Amer. Meteor. Soc.*, **81**, 1207–1230.

Bryan, G. H., and J. M. Fritsch, 2002: A benchmark simulation for moist nonhydrostatic numerical models. *Mon. Wea. Rev.*, **130**, 2917–2928.

Bryan, G. H., J. C. Knievel, and M. D. Parker, 2006: A multimodel assessment of RKW theory's relevance to squall-line characteristics. *Mon. Wea. Rev.*, **134**, 2772–2792.

Bryan, G. H., and R. Rotunno, 2008: Gravity currents in a deep anelastic atmosphere. *J. Atmos. Sci.*, **64**, 536–556.

Bunkers, M. J., B. A. Klimowski, J. W. Zeitler, et al., 2000: Predicting supercell motion using a new hodograph technique. *Wea. Forecasting*, **15**, 61–79.

Byers, H. R., and R. R. Braham, Jr., 1949: *The Thunderstorm*. U.S. Department of Commerce, Weather Bureau, Washington D.C.

Carbone, R. E., 1983: A severe frontal rainband. Part II: Tornado parent vortex circulation. *J. Atmos. Sci.*, **40**, 2639–2654.

Carleton, A. M., D. J. Travis, J. O. Adegoke, et al., 2008: Synoptic circulation and land surface influences on convection in the midwest U.S. "Corn Belt," summers 1999 and 2000. Part II: Role of vegetation boundaries. *J. Climate*, **21**, 3635–3659.

Carley, J. R., B. R. J. Schwedler, M. E. Baldwin, et al., 2011: A proposed model-based methodology for feature-specific prediction for high impact weather. *Wea. Forecasting*, **26**, 243–249.

Carlson, T. N., and F. H. Ludlam, 1968: Conditions for the formation of severe local storms. *Tellus*, **20**, 203–226.

Carlson, T. N., S. G. Benjamin, G. S. Forbes, and Y.-F. Li, 1983: Elevated mixed layers in the severe-storm environment: Conceptual model and case studies. *Mon. Wea. Rev.*, **111**, 1453–1473.

Chen, F., and J. Dudhia, 2001: Coupling an advanced land-surface/ hydrology model with the Penn State/NCAR MM5 modeling system. Part I: Model description and implementation. *Mon. Wea. Rev.*, **129**, 569–585.

Chisholm, A. J., and J. H. Renick, 1972: The kinematics of multicell and supercell Alberta hailstorms. Alberta hail studies, Research Council of Alberta Hail Studies, Rep. 72–2, 24–31.

Cohen, A. E., M. C. Coniglio, S. F. Corfidi, and S. J. Corfidi, 2007: Discrimination of mesoscale convective system environments using sounding observations. *Wea. Forecasting*, **22**, 1045–1062.

Coniglio, M. C., S. F. Corfidi, and J. S. Kain, 2011: Environment and early evolution of the 8 May 2009 derecho-producing convective system. *Mon. Wea. Rev.*, **139**, 1083–1102.

Coniglio, M. C., D. J. Stensrud, and M. B. Richman, 2004: An observational study of derecho-producing convective systems. *Wea. Forecasting*, **19**, 320–337.

Cook, A. R., and J. T. Schaefer, 2008: The relation of El Nino-Southern Oscillation (ENSO) to winter tornado activity. *Mon. Wea. Rev.*, **136**, 3121–3137.

Corfidi, S. F., 2003: Cold pools and MCS propagation: Forecasting the motion of downwind-developing MCSs. *Wea. Forecasting*, **18**, 997–1017.

Cotton, W. R., and R. A. Anthes, 1989: *Storm and Cloud Dynamics*. Academic Press, San Diego, CA.

Cressman, G. P., 1959: An operational objective analysis system. *Mon. Wea. Rev.*, **87**, 367–374.

Crook, N. A., 1988: Trapping of low-level internal gravity waves. *J. Atmos. Sci.*, **45**, 1533–1541.

Dailey, P. S., and R. G. Fovell, 1999: Numerical simulation of the interaction between the sea-breeze front and horizontal convective rolls. Part I: Offshore ambient flow. *Mon. Wea. Rev.*, **127**, 858–878.

Daley, R., 1991: *Atmospheric Data Analysis*. Cambridge University Press.

Damiani, R., G. Vali, and S. Haimov, 2006: The structure of thermals in cumulus from airborne dual-Doppler radar observations. *J. Atmos. Sci.*, **63**, 1432–1450.

Davies, H. C., 1994: Theories of frontogenesis. *The Life Cycles of Extratropical Cyclones*. S. Gronas and M. A. Shapiro (eds.), Vol. I, University of Bergen, 182–192.

Davies-Jones, R. P., 1974: Discussion of measurements inside high-speed thunderstorm updrafts. *J. Appl. Meteor.*, **13**, 710–717.

Davies-Jones, R. P., 1979: Dual-Doppler radar coverage area as a function of measurement accuracy and spatial resolution. *J. Appl. Meteor.*, **18**, 1229–1233.

Davies-Jones, R. P., 1984: Streamwise vorticity: The origin of updraft rotation in supercell storms. *J. Atmos. Sci.*, **41**, 2991–3006.

Davies-Jones, R. P., 1988: Tornado interception with mobile teams. Chapter 2 in *Measurements and Techniques for Thunderstorm Observations and Analysis*, Vol. 3, of *Thunderstorms: A Social, Scientific, and Technological Documentary*. E. Kessler (ed.), Univ. of Oklahoma Press, Norman, OK, 23–32.

Davies-Jones, R., 2002: Linear and nonlinear propagation of supercell storms. *J. Atmos. Sci.*, **59**, 3178–3205.

Davies-Jones, R., and H. E. Brooks, 1993: Mesocyclogenesis from a theoretical perspective. *The Tornado: Its Structure, Dynamics, Prediction, and Hazards, Geophys. Monogr.*, No. 79, American Geophysical Union, 105–114.

Davis, C. A., 1992: Piecewise potential vorticity inversion. *J. Atmos. Sci.*, **49**, 1397–1411.

Davis, C., et al., 2004: The bow echo and MCV experiment: Observations and opportunities. *Bull. Amer. Meteor. Soc.*, **85**, 1075–1093.

Davis, C. A., and S. B. Trier, 2007: Mesoscale convective vortices observed during BAMEX. Part I: Kinematic and thermodynamic structure. *Mon. Wea. Rev.*, **135**, 2029–2049.

Del Genio, A. D., and W. Kovari, 2002: Climatic properties of tropical precipitating convection under varying environmental conditions. *J. Climate*, **15**, 2597–2615.

Derber, J. C., and W.-S. Wu, 1998: The use of TOVS cloud-cleared radiances in the NCEP SSI analysis system. *Mon. Wea. Rev.*, **126**, 2287–2299.

Dial, G. L., J. P. Racy, and R. L. Thompson, 2010: Short-term convective mode evolution along synoptic boundaries. *Wea. Forecasting*, **25**, 1430–1446.

Diffenbaugh, N.S., R. J. Trapp, and H. E. Brooks, 2008: Challenges in identifying influences of global warming on tornado activity. *Eos Trans.*, **89**, 553–554.

Doswell, C. A., III, 1985: The operational meteorology of convective weather. Vol. II: Storm scale analysis. NOAA Technical Memorandum ERL ESG-15.

Doswell, C. A., III, 1987: The distinction between large-scale and mesoscale contribution to severe convection: A case study example. *Wea. Forecasting*, **2**, 3–16.

Doswell, C. A. III, 1991: A review for forecasters on the application of hodographs to forecasting severe thunderstorms. *Nat. Wea. Dig.*, **16** (1), 2–16.

Doswell, C. A., III, 2001: Severe convective storms – An overview. *Severe Convective Storms*, Meteor. Monogr., No. 50, American Meteorological Society, Boston, 1–26.

Doswell, C. A., III, and D. W. Burgess, 1993: Tornadoes and tornadic storms: A review of conceptual models. *The Tornado: Its Structure, Dynamics, Hazards, and Prediction* (Geophys. Monogr. 79), C. Church, D. Burgess, C. Doswell, and R. Davies-Jones (eds.), American Geophysical Union, 161–172.

Doswell, C. A., III, and E. N. Rasmussen 1994: The effect of neglecting the virtual temperature correction on CAPE calculations. *Wea. Forecasting*, **9**, 625–629.

Doswell, C. A. III, H. E. Brooks, and R. A. Maddox, 1996: Flash flood forecasting: An ingredients-based methodology. *Wea. Forecasting*, **11**, 560–580.

Doswell, C. A., III, and L. F. Bosart, 2001: Extratropical synoptic-scale processes and severe convection. *Severe Convective Storms*, Meteor. Monogr., No. 50, American Meteorological Society, Boston, 27–70.

Doswell, C. A., III, and P. M. Markowski, 2004: Is buoyancy a relative quantity? *Mon. Wea. Rev.*, **132**, 853–863.

Doswell, C. A., III, H. E. Brooks, and N. Dotzek, 2009: On the implementation of the enhanced Fujita scale in the USA. *Atmos. Res.*, **93**, 554–563.

Dotzek, N., P. Groenemeijer, B. Feuerstein, and A. M. Holzer, 2009: Overview of ESSL's severe convective storms research using the European Severe Weather Database ESWD. *Atmos. Res.*, **93**, 575–586.

Doviak, R. J., and D. S. Zrnic, 1993: *Doppler Radar and Weather Observations, Second Edition*. Academic Press, San Diego, 562 pp.

Dowell, D. C., and H. B. Bluestein, 1997: The Arcadia, Oklahoma, storm of 17 May 1981: Analysis of a supercell during tornadogenesis. *Mon. Wea. Rev.*, **125**, 2562–2582.

Dowell, D. C., H. B. Bluestein, and D. P. Jorgensen, 1997: Airborne Doppler radar analysis of supercells during COPS-91. *Mon. Wea. Rev.*, **125**, 365–383.

Dowell, D. C., L. J. Wicker, and C. Snyder, 2011: Ensemble Kalman Filter assimilation of radar observations of the 8 May 2003 Oklahoma City supercell: Influences of reflectivity observations on storm-scale analyses. *Mon. Wea. Rev.*, **139**, 272–294.

Drazin, P. G., 2002: *Introduction to Hydrodynamic Stability*. Cambridge University Press.

Droegemeier, K. K., and R. B. Wilhelmson, 1985a: Three-dimensional numerical modeling of convection produced by interacting thunderstorm outflows. Part I: Control simulation and low-level moisture variations. *J. Atmos. Sci.*, **42**, 2381–2403.

Droegemeier, K. K., and R. B. Wilhelmson, 1985b: Three-dimensional numerical modeling of convection produced by interacting thunderstorm outflows. Part II: Variations in vertical wind shear. *J. Atmos. Sci.*, **42**, 2404–2414.

Droegemeier, K. K., and R. B. Wilhelmson, 1987: Numerical simulation of thunderstorm outflow dynamics. Part I: Outflow sensitivity experiments and turbulence dynamics. *J. Atmos. Sci.*, **44**, 1180–1210.

Droegemeier, K. K., S. M. Lazarus, and R. Davies-Jones, 1993: The influence of helicity on numerically simulated convective storms. *Mon. Wea. Rev.*, **121**, 2005–2029.

Dudhia, J., 1989: Numerical study of convection observed during the winter monsoon experiment using a mesoscale two-dimensional model, *J. Atmos. Sci.*, **46**, 3077–3107.

Dworak, R., J. Brunner, W. Feltz, and K. Bedka, 2012: Comparison between GOES-12 overshooting top detections, WSR-88D radar reflectivity and severe storm reports. *Wea. Forecasting*, **27**, 684–699.

Ebert, E. E., 2008: Fuzzy verification of high-resolution gridded forecasts: a review and proposed framework. *Meteorol. Appl.*, **15**, 51–64.

Etling, D., and R. A. Brown, 1993: Roll vortices in the planetary boundary layer: A review. *Boundary-Layer Meteorology*, **65**, 215–248.

Emanuel, K. A., 1986: Overview and definition of mesoscale meteorology. In *Mesoscale Meteorology and Forecasting*, American Meteorological Society, Boston, 1–17.

Emanuel, K. A., 1994: *Atmospheric Convection*. Oxford University Press, Oxford.

Evans, J. S., and C. A. Doswell, III, 2001: Examination of derecho environments using proximity soundings. *Wea. Forecasting*, **16**, 329–342.

Ferrier, B. S., 1994: A double-moment multiple-phase four-class bulk ice scheme. Part I: Description. *J. Atmos. Sci.*, **51**, 249–280.

Fiedler, B. H., and R. J. Trapp, 1993: A fast dynamic grid adaption scheme for meteorological flows. *Mon. Wea. Rev.*, **121**, 2879–2888.

Fiedler, F., and H. A. Panofsky, 1970: Atmospheric scales and spectral gaps. *Bull. Amer. Meteor. Soc.*, **51**, 1114–1120.

Fovell, R. G., and P. S. Dailey, 1995: The temporal behavior of numerically simulated multicell-type storms. Part I: Modes of behavior. *J. Atmos. Sci.*, **52**, 2073–2095.

Fovell, R. G., and P.-H. Tan, 1998: The temporal behavior of numerically simulated multicell-type storms. Part II: The convective cell life cycle and cell regeneration. *Mon. Wea. Rev.*, **126**, 551–577.

French, A. J., and M. D. Parker, 2010: The response of simulated nocturnal convective systems to a developing low-level jet. *J. Atmos. Sci.*, **67**, 3384–3408.

Fritsch, J. M., and G. S. Forbes, 2001: Mesoscale convective systems. *Severe Convective Storms*, American Meteorological Society, Boston, 323–358.

Fudeyasu, H., Y. Wang, M. Satoh, et al., 2010: Multiscale interactions in the life cycle of a tropical cyclone simulated in a global cloud-system-resolving model. Part II: System-scale and mesoscale processes. *Mon. Wea. Rev.*, **138**, 4305–4327.

Fujita, T. T., 1979: Objective, operation, and results of Project NIMROD. Preprints, *11th Conf. on Severe Local Storms*, Kansas City, MO, American Meteorological Society, Boston, 259–266.

Fujita, T. T., 1981: Tornadoes and downbursts in the context of generalized planetary scales. *J. Atmos. Sci.*, **38**, 1512–1534.

Fujita, T. T., 1986: Mesoscale classifications: Their history and their application to forecasting. In *Mesoscale Meteorology and Forecasting*, American Meteorological Society, Boston, 18–35.

Fulton, R. A., J. P. Breidenbach, D.-J. Seo, et al., 1998: The WSR-88D rainfall algorithm. *Wea. Forecasting*, **13**, 377–395.

Gage, K. S., 1979: Evidence for a *k*-5/3 law inertial range in mesoscale two dimensional turbulence. *J. Atmos. Sci.*, **36**, 1950–1954.

Gage, K. S., and G. D. Nastrom, 1986: Theoretical interpretation of atmospheric wavenumber spectra of wind and temperature observed by commercial aircraft during GASP. *J. Atmos. Sci.*, **43**, 729–740.

Gal-Chen, T., 1978: A method for the initialization of the anelastic equations: implications for matching models with observations. *Mon. Wea. Rev.*, **106**, 587–606.

Galloway, J., A. Pazmany, J. Mead, et al., 1997: Detection of ice hydrometeor alignment using an airborne W-band polarimetric radar. *J. Atmos. Oceanic Technol.*, **14**, 3–12.

Gallus, W. A., Jr., and R. H. Johnson, 1991: Heat and moisture budgets of an intense midlatitude squall line. *J. Atmos. Sci.*, **48**, 122–146.

Galway, J. G., 1979: Relationship between precipitation and tornado activity. *Water Resources Research*, **15**, 961–964.

Galway, J. G., 1992: Early severe thunderstorm forecasting and research by the United States Weather Bureau. *Wea. Forecasting*, **7**, 564–587.

Gao, J., D. Stensrud, and M. Xue, 2009: A 3DVAR application to several thunderstorm cases observed during VORTEX2 field operations and potential for real-time warning. Preprints, *34th Conf. on Radar Meteorology*, Williamsburg, VA, Amer. Meteor. Soc., CD-ROM.

Garcia-Carreras, L., D. J. Parker, and J. H. Marsham, 2011: What is the mechanism for the modification of convective cloud distributions by land surface–induced flows? *J. Atmos. Sci.*, **68**, 619–634.

Geerts, B., Q. Miao, and J. C. Demko, 2008: Pressure perturbations and upslope flow over a heated, isolated mountain. *Mon. Wea. Rev.*, **136**, 4272–4288.

Gilleland, E., D. A. Ahijevych, B. G. Brown, and E. E. Ebert, 2010: Verifying forecasts spatially. *Bull. Amer. Meteor. Soc.*, **91**, 1365–1373.

Gilmore, M. S., and L. J. Wicker, 1998: The influence of midtropospheric dryness on supercell morphology and evolution. *Mon. Wea. Rev.*, **126**, 943–958.

Gilmore, M. S., J. M. Straka, and E. N. Rasmussen, 2004: Precipitation and evolution sensitivity in simulated deep convective storms: Comparisons between liquid-only and simple ice and liquid phase microphysics. *Mon. Wea. Rev.*, **132**, 1897–1916.

Glickman, T. S., Ed., 2000: *Glossary of Meteorology*. 2d ed. Amer. Meteor. Soc.

Goff, R. C., 1976: Vertical structure of thunderstorm outflows. *Mon. Wea. Rev.*, **104**, 1429–1440.

Goody, R. M., and Y. L. Yung, 1989: *Atmospheric Radiation, Theoretical Basis*. Oxford University Press.

Griffiths, M., A. J. Thorpe, and K. A. Browning KA, 2000: Convective destabilization by a tropopause fold diagnosed using potential-vorticity inversion. *Quart. J. Roy. Meteor. Soc.*, **126**, 125–144.

Guralnik, D. B., Ed., 1984: *Webster's New World Dictionary of the American Language*. Simon and Schuster, New York.

Hacker, J., et al., 2005: Predictability. *Bull. Amer. Meteor. Soc.*, **86**, 1733–1737.

Halpert, M. S., and C. F. Ropelewski, 1992: Surface temperature patterns associated with the Southern Oscillation. *J. Climate*, **5**, 577–593.

Haltiner, G. J., and R. T. Williams, 1980: *Numerical Prediction and Dynamic Meteorology*. Wiley, New York.

Hamill, T. M., J. S. Whitaker, and S. L. Mullen, 2006: Reforecasts, an important dataset for improving weather predictions. *Bull. Amer. Meteor. Soc.*, **87**, 33–46.

Hane, C. E., R. B. Wilhelmson, and T. Gal-Chen, 1981: Retrieval of thermodynamic variables within deep convective clouds: Experiments in three dimensions. *Mon. Wea. Rev.*, **109**, 564–576.

Hane, C. E., and P. S. Ray, 1985: Pressure and buoyancy fields derived from Doppler radar data in a tornadic thunderstorm. *J. Atmos. Sci.*, **42**, 18–35.

Härtel, C., F. Carlsson, and M. Thunblom, 2000: Analysis and direct numerical simulation of the flow at a gravity-current head. Part 2. The lobe-and-cleft instability. *J. Fluid Mech.*, **418**, 213–229.

Hartmann, D. L., 1993: Radiative effects of clouds on Earth's climate. In *Aerosol-Cloud-Climate Interactions*, P. V. Hobbs (ed.), Academic Press, San Diego, CA.

Hartmann, D. L., and M. L. Michelsen, 2002: No evidence for Iris. *Bull. Amer. Meteor. Soc.*, **83**, 249–254.

Hendricks, E. A., M. T. Montgomery, and C. A. Davis, 2004: On the role of "vortical" hot towers in formation of tropical cyclone Diana (1984). *J. Atmos. Sci.*, **61**, 1209–1232.

Hess, S. L., 1959: *Introduction to Theoretical Meteorology*. Robert E. Krieger Publishing, Huntington, NY.

Heymsfield, G. M., L. Tian, A. J. Heymsfield, et al., 2010: Characteristics of deep tropical and subtropical convection from nadir-viewing high-altitude airborne Doppler radar. *J. Atmos. Sci.*, **67**, 285–308.

Hildebrand, P. H., C. A. Walther, C. L. Frush, et al., 1994: The ELDORA/ASTRAIA airborne Doppler weather radar: Goals, design, and first field tests. *Proc. IEEE*, **82**, 1873–1890.

Hitchens, N. M., R. J. Trapp, M. E. Baldwin, and A. Gluhovsky, 2010: Characterizing subdiurnal extreme precipitation in the midwestern United States. *J. Hydrometeor.*, **11**, 211–218.

Hitchens, N. M., M. E. Baldwin, and R. J. Trapp, 2012: An object-oriented characterization of extreme precipitation-producing convective systems in the Midwestern United States. *Mon. Wea. Rev.*, **140**, 1356–1366.

Hjelmfelt, M. R., H. D. Orville, R. D. Roberts, et al., 1989: Observational and numerical study of a microburst line-producing storm. *J. Atmos. Sci.*, **46**, 2731–2743.

Hoch, J., and P. Markowski, 2004: A climatology of springtime dryline position in the U.S. Great Plains region. *J. Climate*, **18**, 2132–2137.

Hock, T. F., and J. L. Franklin, 1999: The NCAR GPS dropwindsonde. *Bull. Amer. Meteor. Soc.*, **80**, 407–420.

Holland, G. J., et al., 2001: The Aerosonde robotic aircraft: A new paradigm for environmental observations. *Bull. Amer. Meteor. Soc.*, **82**, 889–901.

Holton, J. R., 2004: *An Introduction to Dynamic Meteorology*. 4th ed. Elsevier, Burlington, MA.

Hong, S.-Y., and H.-L. Pan, 1996: Nonlocal boundary layer vertical diffusion in a medium-range forecast model. *Mon. Wea. Rev.*, **124**, 2322–2339.

Hooke, W. H., 1986: Gravity waves. In *Mesoscale Meteorology and Forecasting*, American Meteorological Society, Boston, 272–288.

Hoskins, B. J., and D. Karoly, 1981: The steady linear response of a spherical atmosphere to thermal and orographic forcing. *J. Atmos. Sci.*, **38**, 1179–1196.

Hoskins, B. J., M. E. McIntyre, and A. W. Robertson, 1985: On the use and significance of isentropic potential vorticity maps. *Quart. J. Roy. Meteor. Soc.*, **111**, 877–946.

Houtekamer, P. L., L. Lefaivre, J. Derome, et al., 1996: A system simulation approach to ensemble prediction. *Mon. Wea. Rev.*, **124**, 1225–1242.

Houze, R. A., Jr., 1993: *Cloud Dynamics*. Academic Press, San Diego, CA.

Houze, R. A., S. A. Rutledge, M. I. Biggerstaff, and B. F. Smull, 1989: Interpretation of Doppler weather radar displays of midlatitude mesoscale convective systems. *Bull. Amer. Meteor. Soc.*, **70**, 608–619.

Houze, R. A., Jr., B. F. Smull, and P. Dodge, 1990: Mesoscale organization of springtime rainstorms in Oklahoma. *Mon. Wea. Rev.*, **118**, 613–654.

Hsieh, J.-S., and K. H. Cook, 2005: Generation of African easterly wave disturbances: Relationship to the African easterly jet. *Mon. Wea. Rev.*, **133**, 1311–1327.

Jacobson, M. Z., 2005: *Fundamentals of Atmospheric Modeling*. Cambridge University Press.

James, R. P., J. M. Fritsch, and P. M. Markowski, 2005: Environmental distinctions between cellular and slabular convective lines. *Mon. Wea. Rev.*, **133**, 2669–2690.

James, E. P., and R. H. Johnson, 2010: Patterns of precipitation and mesolow evolution in midlatitude mesoscale convective vortices. *Mon. Wea. Rev.*, **138**, 909–931.

Johns, R. H., 1993: Meteorological conditions associated with bow echo development in convective storms. *Wea. Forecasting*, **8**, 294–299.

Johns, R. H. and W. D. Hirt, 1987: Derechos: Widespread convectively induced windstorms. *Wea. Forecasting*, **2**, 32–49.

Jorgensen, D. P., P. H. Hildebrand, and C. L. Frush, 1983: Feasibility test of airborne pulse Doppler meteorological radar. *J. Climate Appl. Meteor.*, **22**, 744–757.

Jorgensen, D. P., P. Zhaoxia, P. O. G. Persson, and W.-K. Tao, 2003: Variations associated with cores and gaps of a Pacific narrow cold frontal rainband. *Mon. Wea. Rev.*, **131**, 2705–2729.

Joss, J., and A. Waldvogel, 1970: Raindrop size distribution and Doppler velocities. Preprints, *14th Conf. Radar Meteorology*, American Meteorological Society, Boston, 153–156.

Kain, J. S., and J. M. Fritsch, 1990: A one-dimensional entraining/detraining plume model and its application in convective parameterization, *J. Atmos. Sci.*, **47**, 2784–2802.

Kain, J. S., S. J. Weiss, J. J. Levit, M. E. Baldwin, and D. R. Bright, 2006: Examination of convection-allowing configurations of the WRF model for the prediction of severe convective weather: The SPC/NSSL Spring Program 2004. *Wea. Forecasting*, **21**, 167–181.

Kain, J. S., et al., 2008: Some practical considerations regarding horizontal resolution in the first generation of operational convection-allowing NWP. *Wea. Forecasting*, **23**, 931–942.

Kalnay, E., 2003: *Atmospheric Modeling, Data Assimilation, and Predictability*. Cambridge University Press.

Karl, T. R., and R. W. Knight, 1998: Secular trends of precipitation amount, frequency, and intensity in the U.S.A. *Bull. Amer. Meteor. Soc.* **79**, 231–242.

Kawase, H., T. Yoshikane, M. Hara, et al., 2008: Impact of extensive irrigation on the formation of cumulus clouds, *Geophys. Res. Lett.*, **35**, L01806, doi:10.1029/2007GL032435.

Kelly, G. A., P. Bauer, A. J. Geer, P. Lopez, and J-N. Thépaut, 2008: Impact of SSM/I observations related to moisture, clouds, and precipitation on global NWP forecast skill. *Mon. Wea. Rev.*, **136**, 2713–2726.

Kessinger, C. J., P. S. Ray, and C. E. Hane, 1987: The Oklahoma squall line of 19 May 1977. Part I: A multiple Doppler analysis of convective and stratiform structure. *J. Atmos. Sci.*, **44**, 2840–2865.

Kessler, E., 1969: On the distribution and continuity of water substance in atmospheric circulation, *Meteor. Monogr.*, **32**, Amer. Meteor. Soc.

Khairoutdinov, M., and D. Randall, 2006: High-resolution simulation of shallow-to-deep convection transition over land. *J. Atmos. Sci.*, **63**, 3421–3436.

Kidder, S. Q., and T. H. Vonder Haar, 1995: *Satellite Meteorology: An Introduction*. Academic Press, San Diego, CA.

Kirshbaum, D. J., 2011: Cloud-resolving simulations of deep convection over a heated mountain. *J. Atmos. Sci.*, **68**, 361–378.

Klemp, J. B., 1987: Dynamics of tornadic thunderstorms. *Ann. Rev. Fluid Mech.*, **19**, 369–402.

Klemp, J. B., and R. Rotunno, 1983: A study of the tornadic region within a supercell thunderstorm. *J. Atmos. Sci.*, **40**, 359–377.

Klemp, J. B., and R. B. Wilhelmson, 1978: The simulation of three dimensional convective storm dynamics. *J. Atmos. Sci.*, **35**, 1070–1096.

Klemp, J. B., R. Rotunno, and W. C. Skamarock, 1994: On the dynamics of gravity currents in a channel. *J. Fluid Mech.*, **269**, 169–198.

Klimowski, B. A., M. R. Hjelmfelt, and M. J. Bunkers, 2004: Radar observations of the early evolution of bow echoes. *Wea. Forecasting*, **19**, 727–734.

Klotzbach, P. J., 2007: Recent developments in statistical prediction of seasonal Atlantic basin tropical cyclone activity. *Tellus A* **59**, 511–518.

Klotzbach, P. J., and W. M. Gray, 2003: Forecasting September Atlantic basic tropical cyclone activity. *Wea. Forecasting*, **18**, 1109–1128.

Knopfmeier, K. H., 2007: Real-data and idealized simulations of the 4 July 2004 bow echo event. M.S. Thesis, Purdue University.

Knupp, K. R., 2006: Observational analysis of a gust front to bore to solitary wave transition within an evolving nocturnal boundary layer. *J. Atmos. Sci.*, **63**, 2016–2035.

Koch, S. E., 1984: The role of an apparent mesoscale frontogenetic circulation in squall line initiation. *Mon. Wea. Rev.*, **112**, 2090–2111.

Koch, S. E., M. DesJardins, and P. J. Kocin, 1983: An interactive Barnes objective analysis scheme for use with satellite and conventional data. *J. Climate Appl. Meteor.*, **22**, 1487–1503.

Koch, S. E., B. Ferrier, M. Stolinga, et al., 2005: The use of simulated radar reflectivity fields in the diagnosis of mesoscale phenomena from high-resolution WRF model forecasts. Preprints, *12th Conf. on Mesoscale Processes*, Albuquerque, NM, Amer. Meteor. Soc., J4J.7. (Available online at http://ams.confex.com/ams/pdfpapers/ 97032.pdf.)

Kogan, Y. L., 1991: The simulation of a convective cloud in a 3-D model with explicit microphysics. Part I: Model description and sensitivity experiments. *J. Atmos. Sci.*, **48**, 1160–1189.

Kundu, P., 1990: *Fluid Mechanics*. Academic Press, San Diego, CA.

Laing, A. G., and J. M. Fritsch, 1997: The global population of mesoscale convective complexes. *Quart. J. Roy. Meteor. Soc.*, **123**, 389–405.

Lane, T. P., and M. J. Reeder, 2001: Convectively generated gravity waves and their effect on the cloud environment. *J. Atmos. Sci.*, **58**, 2427–2440.

Lee, B. D., and R. B. Wilhelmson, 1997: The numerical simulation of non-supercell tornadogenesis. Part I: Initiation and evolution of pretornadic miscocyclone circulations along a dry outflow boundary. *J. Atmos. Sci.*, **54**, 32–60.

Lemon, L. R., and C. A. Doswell III, 1979: Severe thunderstorm evolution and mesocyclone structure as related to tornadogenesis. *Mon. Wea. Rev.*, **107**, 1184–1197.

Leon, D., G. Vali, and M. Lothon, 2006: Dual-Doppler analysis in a single plane from an airborne platform. *J. Atmos. Oceanic Technol.*, **23**, 3–22.

Leslie, L. M., and R. K. Smith, 1978: The effect of vertical stability on tornadogenesis. *J. Atmos. Sci.*, **35**, 1281–1288.

Lewis, J. M., S. Lakshmivarahan, and S. K. Dhall, 2006: *Dynamic Data Assimilation: A Least Squares Approach*. Cambridge University Press.

Ligda, M. G. H., 1951: Radar storm observation. In *Compendium of Meteorology*, American Meteorological Society, Boston, 1265–1282.

Lilly, D. K., 1979: The dynamical structure and evolution of thunderstorms and squall lines. *Annu. Rev. Earth Planet. Sci.*, **7**, 117–161.

Lilly, D. K., 1982: The development and maintenance of rotation in convective storms. *Intense Atmospheric Vortices*, L. Bengtsson and J. Lighthill (eds.), Springer-Verlag, Berlin/Heidelberg/New York, 149–160.

Lilly, D. K., 1983: Stratified turbulence and the mesoscale variability of the atmosphere. *J. Atmos. Sci.*, **40**, 749–761.

Lilly, D. K., 1986a: The structure, energetics and propagation of rotating convective storms. Part I: Energy exchange with the mean flow. *J. Atmos. Sci.*, **43**, 113–125.

Lilly, D. K., 1986b: The structure, energetics and propagation of rotating convective storms. Part II: helicity and storm stabilization. *J. Atmos. Sci.*, **43**, 126–140.

Lilly, D. K., 1990: Numerical prediction of thunderstorms – has its time come? *Q. J. Roy. Meteor. Soc.*, **116**, 779–798.

Lima, M. A., and J. W. Wilson, 2008: Convective storm initiation in a moist tropical environment. *Mon. Wea. Rev.*, **136**, 1847–1864.

Lindborg, E., 1999: Can the atmospheric kinetic energy spectrum be explained by two-dimensional turbulence? *J. Fluid Mech.*, **388**, 259–288.

Lindzen, R. S., M.-D. Chou, and A. Y. Hou, 2001: Does the earth have an adaptive infrared iris? *Bull. Amer. Meteor. Soc.*, **82**, 417–432.

Liou, K. N., 2002: *An Introduction to Atmospheric Radiation*. Academic Press, San Diego, CA.

Loehrer, S. M., and R. H. Johnson, 1995: Surface pressure and precipitation life cycle characteristics of PRE-STORM mesoscale convective systems. *Mon. Wea. Rev.*, **123**, 600–621.

Loftus, A. M., D. B. Weber, and C. A. Doswell, III, 2008: Parameterized mesoscale forcing mechanisms for initiating numerically simulated isolated multicellular convection. *Mon. Wea. Rev.*, **136**, 2408–2421.

Long, A. B., R. J. Matson, and E. L. Crow, 1980: The hailpad: Materials, data reduction and calibration. *J. Appl. Meteor.*, **19**, 1300–1313.

Lorenz, E., 1963: Deterministic nonperiodic flow. *J. Atmos. Sci.*, **20**, 130–141.

Lorenz, E. N., 1969: The predictability of a flow which possesses many scales of motion. *Tellus*, **21**, 289–307.

Lorenz, E. N., 1984: Estimates of atmospheric predictability at medium range. *Predictability of Fluid Motions: A.I.P. Conference Proceedings*, No. 106, American Institute of Physics, La Jolla Institute, 133–140.

Lucas, C., E. J. Zipser, and M. A. LeMone, 1994: Vertical velocity in oceanic convection off tropical Australia. *J. Atmos. Sci.*, **51**, 3183–3193.

MacDonald, A. E., 2005: A Global profiling system for improved weather and climate prediction. *Bull. Amer. Meteor. Soc.*, **86**, 1747–1764.

Maddox, R. A., 1976: An evaluation of tornado proximity wind and stability data. *Mon. Wea. Rev.*, **104**, 133–142.

Maddox, R. A., 1980a: Mesoscale convective complexes. *Bull. Amer. Meteor. Soc.*, **61**, 1374–1387.

Maddox, R. A., 1980b: An objective technique for separating macroscale and mesoscale features in meteorological data. *Mon. Wea. Rev.*, **108**, 1108–1121.

Maddox, R. A., L. R. Hoxit, and C. F. Chappell, 1980: A study of tornadic thunderstorm interactions with thermal boundaries. *Mon. Wea. Rev.*, **108**, 322–336.

Mahoney, W. P., III, 1988: Gust front characteristics and the kinematics associated with interacting thunderstorm outflows. *Mon. Wea. Rev.*, **116**, 1474–1491.

Mahoney, K. M., G. M. Lackmann, and M. D. Parker, 2009: The role of momentum transport in the motion of a quasi-idealized mesoscale convective system. *Mon. Wea. Rev.*, **137**, 3316–3338.

Malkus, J. S., and R. S. Scorer, 1955: The erosion of cumulus towers. *J. Meteor.*, **12**, 43–57.

Mapes B. E., 1993: Gregarious tropical convection. *J. Atmos. Sci*, **50**, 2026–2037.

Markovic, M., H. Lin, and K. Winger, 2010: Simulating global and North American climate using the global environmental multiscale model with a variable-resolution modeling approach. *Mon. Wea. Rev.*, **138**, 3967–3987.

Markowski, P. M., E. N. Rasmussen, and J. M. Straka, 1998: The occurrence of tornadoes in supercells interacting with boundaries during VORTEX-95. *Wea. Forecasting*, **13**, 852–859.

Markowski, P. M., J. M. Straka, and E. N. Rasmussen, 2002: Direct surface thermodynamic observations within the rear-flank downdrafts of nontornadic and tornadic supercells. *Mon. Wea. Rev.*, **130**, 1692–1721.

Marquis, J. N., Y. P. Richardson, and J. M. Wurman, 2007: Kinematic observations of misocyclones along boundaries during IHOP. *Mon. Wea. Rev.*, **135**, 1749–1768.

Marshall, J. S., and W. McK. Palmer, 1948: The distribution of raindrops with size. *J. Meteor.*, **5**, 165–166.

Marsham, J. H., and D. J. Parker, 2006: Secondary initiation of multiple bands of cumulonimbus over southern Britain. II: Dynamics of secondary initiation. *Quart. J. Roy. Meteor. Soc.*, **132**, 1053–1072.

Marsham, J. H., S. B. Trier, T. M. Weckwerth, and J. W. Wilson, 2011: Observations of elevated convection initiation leading to a surface-based squall line during 13 June IHOP_2002. *Mon. Wea. Rev.*, **139**, 247–271.

Martin, J. E., 2006: *Mid-Latitude Atmospheric Dynamics: A First Course*. Wiley, New York.

Marwitz, J. D., 1972: The structure and motion of severe hailstorms. Part II: Multi-cell storms. *J. Appl. Meteor.*, **11**, 180–188.

May, P. T., and D. K. Rajopadhyaya, 1999: Vertical velocity characteristics of deep convection over Darwin, Australia. *Mon. Wea. Rev.*, **127**, 1056–1071.

McCaul, E. W., Jr., 1987: Observations of the Hurricane "Danny" tornado outbreak of 16 August 1985. *Mon. Wea. Rev.*, **115**, 1206–1223

McCaul, E. W. Jr., and M. L. Weisman, 1996: Simulations of shallow supercell storms in landfalling hurricane environments. *Mon. Wea. Rev.*, **124**, 408–429.

Miller, L. J., and S. M. Fredrick, 1998: CEDRIC: Custom Editing and Display of Reduced Information in Cartesian space. User's Manual, National Center for Atmospheric Research, Boulder, CO, 130 pp.

Miller, S. T. K., B. D. Keim, R. W. Talbot, and H. Mao, 2003: Sea breeze: structure, forecasting, and impacts. *Rev. Geophysics*, **41**, 1–31.

Mitchell, E. D., S. V. Vasiloff, G. J. Stumpf, et al., 1998: The National Severe Storms Laboratory Tornado Detection Algorithm. *Wea. Forecasting*, **13**, 352–366.

Mohr, C. G., and R. L. Vaughan, 1979: An economical procedure for Cartesian interpolation and display of reflectivity factor data in three-dimensional space. *J. Appl. Meteor.*, **18**, 661–670.

Moller, A. R., C. A. Doswell, III, M. P. Foster, and G. R. Woodall, 1994: The operational recognition of supercell thunderstorm environments and storm structures. *Wea. Forecasting*, **9**, 327–347.

Moncrieff, M. W., 1992: Organized convective systems: Archetypal dynamical models, mass and momentum flux theory, and parameterization. *Quart. J. Roy. Meteor. Soc.*, **118**, 819–850.

Moninger, W. R., R. D. Mamrosh, and P. M. Pauley, 2003: Automated meteorological reports from commercial aircraft. *Bull. Amer. Meteor. Soc.*, **84**, 203–216.

Montgomery, M. T., M. E. Nicholls, T. A. Cram, and A. B. Saunders, 2006: A vertical hot tower route to tropical cyclogenesis. *J. Atmos. Sci.*, **63**, 355–386.

Musil, D. J., A. J. Heymsfield, and P. L. Smith, 1986: Microphysical characteristics of a well-developed weak echo region in a High Plains supercell thunderstorm. *J. Clim. Appl. Meteor.*, **25**, 1037–1051.

Nachamkin, J. E., R. L. McAnelly, and W. R. Cotton, 1994: An observational analysis of a developing mesoscale convective complex. *Mon. Wea. Rev.*, **122**, 1168–1188.

Namias, J., 1991: Spring and summer 1998 drought over the contiguous United States – causes and prediction. *J. Climate*, **4**, 54–65.

Nastrom, G. D., and K. S. Gage, 1985: A climatology of atmospheric wavenumber spectra of wind and temperature observed by commercial aircraft. *J. Atmos. Sci.*, **42**, 950–960.

Neiman, P. J., and R. M. Wakimoto, 1999: The interaction of a Pacific cold front with shallow air masses east of the Rocky Mountains. *Mon. Wea. Rev.*, **127**, 2102–2127.

Newton, C. W., 1976: Severe convective storms. *Advances in Geophysics*, Vol. **12**, Academic Press, 257–303.

Newton, C. W., and H. R. Newton, 1959: Dynamical interactions between large convective clouds and environments with vertical shear. *J. Meteor.*, **16**, 483–496.

Nicholls, M. E., and R. A. Pielke, 2000: Thermally induced compression waves and gravity waves generated by convective storms. *J. Atmos. Sci.*, **57**, 3251–3271.

Nieman, S. J., W. P. Menzel, C. M. Hayden, et al., 1997: Fully automated cloud-drift winds in NESDIS operations. *Bull. Amer. Meteor. Soc.*, **78**, 1121–1133.

Norris, J. R., and S. F. Iacobellis, 2005: North Pacific cloud feedbacks inferred from synoptic-scale dynamic and thermodynamic relationships. *J. Climate*, **18**, 4862–4878.

Oreskes, N., K. Shrader-Frechette, and K. Belitz, 1994: Verification, validation, and confirmation of numerical models in the earth sciences. *Science*, **263**, 641–646.

Orlanski, I., 1975: A rational division of scales for atmospheric processes. *Bull. Amer. Meteor. Soc.*, **56**, 527–530.

Pal, J. S., and E. A. B. Eltahir, 2001: Pathways relating soil moisture conditions to future summer rainfall within a model of the land-atmosphere system. *J. Climate*, **14**, 1227–1242.

Pal, J. S., and E. A. B. Eltahir, 2002: Teleconnections of soil moisture and rainfall during the 1993 midwest summer flood. *Geophys. Res. Lett.*, **29**, doi:10.1029/2002GL014815.

Palencia, C., A. Castro, D. Giaiotti, et al., 2011: Dent overlap in hailpads: Error estimation and measurement correction. *J. Appl. Meteor. Climatol.*, **50**, 1073–1087.

Parker, M. D., 2008: Response of simulated squall lines to low-level cooling. *J. Atmos. Sci.*, **65**, 1323–1341.

Parker, M. D., 2010: Relationship between system slope and updraft intensity in squall lines. *Mon. Wea. Rev.*, **138**, 3572–3578.

Parker, M. D., and R. H. Johnson, 2000: Organizational modes of midlatitude mesoscale convective systems. *Mon. Wea. Rev.*, **128**, 3413–3436.

Parker, M. D., and R. H. Johnson, 2004: Simulated convective lines with leading precipitation. Part II: Evolution and maintenance. *J. Atmos. Sci.*, **61**, 1656–1673.

Parker, M. D., I. C. Ratcliffe, and G. M. Henebry, 2005: The July 2003 Dakota hailswaths: creation, characteristics, and possible impacts. *Mon. Wea. Rev.*, **133**, 1241–1260.

Parsons, D. B., M. A. Shapiro, R. M. Hardesty, et al., 1991: The finescale structure of a West Texas dryline. *Mon. Wea. Rev.*, **119**, 1242–1258.

Parsons, D. P., et al., 1994: The integrated sounding system: Description and preliminary observations from TOGA COARE. *Bull. Amer. Meteor. Soc.*, **75**, 553–567.

Pauley, P. M., and P. J. Smith, 1988: Direct and indirect effects of latent heat release on a synoptic-scale wave system. *Mon. Wea. Rev.*, **116**, 1209–1235.

Peckham, S. E., R. B. Wilhelmson, L. J. Wicker, and C. L. Ziegler, 2004: Numerical simulation of the interaction between the dryline and horizontal convective rolls. *Mon. Wea. Rev.*, **132**, 1792–1812.

Peixoto, J. P. and A. H. Oort, 1998: *Physics of Climate*, American Institute of Physics.

Pielke, R. A., 1974: A three-dimensional numerical model of the sea breezes over south Florida. *Mon. Wea. Rev.*, **102**, 115–139.

Pielke, R. A., Sr. 2002: *Mesoscale Meteorological Modeling*. Academic Press, San Diego, CA.

Pielke, R. A., T. J. Lee, J. H. Copeland, et al., 1997: Use of USGS-provided data to improve weather and climate simulations. *Ecological Applications*, **7**, 3–21.

Proctor, F. H., 1989: Numerical simulations of an isolated microburst. Part II: Sensitivity experiments. *J. Atmos. Sci.*, **46**, 2143–2165.

Pruppacher, H. R., and J. D. Klett, 1978: *Microphysics of Clouds and Precipitation*. D. Reidel, Dordrecht, the Netherlands.

Przybylinski, R. W., 1995: The bow echo: Observations, numerical simulations, and severe weather detection methods. *Wea. Forecasting*, **10**, 203–218.

Ramanathan, V., and W. Collins, 1991: Thermodynamic regulation of ocean warming by cirrus clouds deduced from observations of the 1987 El Niño. *Nature*, **351**, 27–32.

Randall, D. A., M. Khairoutdinov, A. Arakawa, and W. Grabowski, 2003: Breaking the cloud parameterization deadlock. *Bull. Amer. Meteor. Soc.*, **84**, 1547–1564.

Rasmussen, E. N., and D. O. Blanchard, 1998: A baseline climatology of sounding-derived supercell and tornado forecast parameters. *Wea. Forecasting*, **13**, 1148–1164.

Rasmussen, E. N., and J. M. Straka, 1998: Variations in supercell morphology. Part I: Observations of the role of upper-level storm-relative flow. *Mon. Wea. Rev.*, **126**, 2406–2421.

Raymond, D. J., and H. Jiang, 1990: A theory for long-lived mesoscale convective systems. *J. Atmos. Sci.*, **47**, 3067–3077.

Redelsperger, J. L., and T. L. Clark, 1990: The initiation and horizontal scale selection of convection over gently sloping terrain. *J. Atmos. Sci.*, **47**, 516–541.

Rinehart, R. E., 1997: *Radar for Meteorologists, Third Edition*. Rinehart Publications, Columbia, MO.

Roberts, N. M., and H. W. Lean, 2008: Scale-selective verification of rainfall accumulations from high resolution forecasts of convective events. *Mon. Wea. Rev.*, **136**, 78–96.

Roebber, P. J., D. M. Schultz, and R. Romero, 2002: Synoptic regulation of the 3 May 1999 tornado outbreak. *Wea. Forecasting*, **17**, 399–429.

Rogers, R. R., and M. K. Yau, 1989: *A Short Course in Cloud Physics*. Pergamon Press, Elmsford, NY.

Ropelewski, C. F., and M. S. Halpert, 1987: Global and regional scale precipitation patterns associated with the El Niño/Southern Oscillation. *Mon. Wea. Rev.*, **115**, 1606–1626.

Ross, A. N., A. M. Tompkins, and D. J. Parker, 2004: Simple models of the role of surface fluxes in convective cold pool evolution. *J. Atmos. Sci.*, **61**, 1582–1595.

Rotunno, R., and J. B. Klemp, 1985: On the rotation and propagation of numerically simulated supercell thunderstorms. *J. Atmos. Sci.*, **42**, 271–292.

Rotunno, R., J. B. Klemp, and M. L. Weisman, 1988: A theory for strong, long-lived squall lines. *J. Atmos. Sci.*, **45**, 463–485.

Russell, A., G. Vaughan, E. G. Norton, et al., 2008: Convective inhibition beneath an upper-level PV anomaly. *Quart. J. Roy. Meteor. Soc.*, **134**, 371–383.

Ryzhkov, A. V., S. E. Giangrande, and T. J. Schuur, 2005: Rainfall estimation with a polarimetric prototype of WSR-88D. *J. Appl. Meteor.*, **44**, 502–515.

Saltzman, B., 1962: Finite amplitude free convection as an initial value problem – I. *J. Atmos. Sci.*, **19**, 329–341.

Schiffer, R. A., and W. B. Rossowe, 1983: The International Satellite Cloud Climatology Project (ISCCP): The first project of the World Climate Research Programme. *Bull. Amer. Meteor. Soc.*, **64**, 779–748.

Schultz, D. M., P. N. Schumacher, and C. A. Doswell, III, 2000: The intricacies of instabilities. *Mon. Wea. Rev.*, **128**, 4143–4148.

Schultz, D. M, C. C. Weiss, and P. M. Hoffman, 2007: The synoptic regulation of the dryline. *Mon. Wea. Rev.*, **135**, 1699–1709.

Schumacher, R. S., and R. H. Johnson, 2005: Organization and environmental properties of extreme-rain-producing mesoscale convective systems. *Mon. Wea. Rev.*, **133**, 961–976.

Schroeder, J. L., and C. C. Weiss, 2008: Integrating research and education through measurement and analysis. *Bull. Amer. Meteor. Soc.*, **89**, 793–798.

Scorer, R. S., and F. H. Ludlam, 1953: Bubble theory of penetrative convection. *Quart. J. Roy. Meteor. Soc.*, **79**, 94–103.

Segal, M., and R. W. Arritt, 1992: Nonclassic mesoscale circulations caused by surface sensible heat-flux gradients. *Bull. Amer. Meteor. Soc.*, **73**, 1593–1604.

Segel, Z. T., D. S. Stensrud, I. C. Ratcliffe, and G. M. Henebry, 2005: Influence of a hailstreak on boundary layer evolution. *Mon. Wea. Rev.*, **133**, 942–960.

Shabbott, C. J., and P. M. Markowski, 2006: Surface in situ observations within the outflow of forward-flank downdrafts of supercell thunderstorms. *Mon. Wea. Rev.*, **134**, 1422–1441.

Shapiro, M. A., T. Hampel, D. Rotzoll, and F. Mosher, 1985: The frontal hydraulic head: A micro-α scale ($\sim$ 1 km) triggering mechanism for mesoconvective weather systems. *Mon. Wea. Rev.*, **113**, 1166–1183.

Shepherd, M., D. Niyogi, and T. L. Mote, 2009: A seasonal-scale climatological analysis correlating spring tornadic activity with antecedent fall-winter drought in the southeaster United States. *Environ. Res. Lett.*, **4**, 1–7.

Sherwood, S. C., 2000: On moist instability. *Mon. Wea. Rev.*, **128**, 4139–4142.

Simpson, J. E. 1969 A comparison between laboratory and atmospheric density currents. *Quart. J. Roy. Meteor. Soc.*, **95**, 758–765.

Simpson, J., R. F. Adler, and G. R. North, 1988: A proposed tropical rainfall measuring mission (TRMM) satellite. *Bull. Amer. Meteor. Soc.*, **69**, 278–295.

Skamarock, W. C., 2004: Evaluating mesoscale NWP models using kinetic energy spectra. *Mon. Wea. Rev.*, **132**, 3019–3032.

Skamarock, W. C., M. L. Weisman, and J. B. Klemp, 1994: Three-dimensional evolution of simulated long-lived squall lines. *J. Atmos. Sci.*, **51**, 2563–2584.

Skamarock, W. C., et al., 2008: A description of the Advanced Research WRF Version 3. NCAR Technical Note NCAR/TN-475-STR.

Smith, A. M., G. M. McFarquhar, R. M. Rauber, J. A. Grim, M. S. Timlin, and B. F. Jewett, 2009: Microphysical and thermodynamic structure and evolution of the trailing stratiform regions of mesoscale convective systems during BAMEX. Part I: Observations. *Mon. Wea. Rev.*, **137**, 1165–1185.

Smith, D. M., et al., 2010: Skillful multi-year predictions of Atlantic hurricane frequency. *Nature-Geos.*, **3**, 846–849.

Smith, J. A., M. L. Baeck, Y. Zhang, and C. A. Doswell, III, 2001: Extreme rainfall and flooding from supercell thunderstorms. *J. Hydrometeor.*, **2**, 469–489.

Smith, P. J., 1971: An analysis of kinematic vertical motions. *Mon. Wea. Rev.*, **99**, 715–724.

Smith, P. J., 2000: The importance of the horizontal distribution of heating during extratropical cyclone development. *Mon. Wea. Rev.*, **128**, 3692–3694.

Smith, R. K., and L. M. Leslie, 1978: Tornadogenesis. *Quart. J. Roy. Meteor. Soc.*, **104**, 189–199.

Srivastava, R. C., 1985: A simple model of evaporatively driven downdraft: Application to microburst downdraft. *J. Atmos. Sci.*, **42**, 1004–1023.

Srivastava, R. C., 1987: A model of intense downdrafts driven by the melting and evaporation of precipitation. *J. Atmos. Sci.*, **44**, 1752–1773.

Stensrud, D. J., 1993: Elevated residual layers and their influence on boundary-layer evolution. *J. Atmos. Sci.*, **50**, 2284–2293.

Stensrud, D. J., 1996a: Importance of low-level jets to climate: A review. *J. Climate*, **9**, 1698–1711.

Stensrud, D. J., 1996b: Effects of persistent, midlatitude mesoscale regions of convection on the large-scale environment during the warm season. *J. Atmos. Sci.*, **53**, 3503–3527.

Stensrud, D. J. and J. L. Anderson, 2001: Is midlatitude convection an active or a passive player in producing global circulation patterns? *J. Climate*, **14**, 2222–2237.

Stensrud, D. J., 2007: *Parameterization Schemes: Keys to Understanding Numerical Weather Prediction Models*. Cambridge University Press.

Stensrud, D. J., and J. M. Fritsch, 1994: Mesoscale convective systems in weakly forced large-scale environments. Part II: Generation of mesoscale initiation condition. *Mon. Wea. Rev.*, **122**, 2068–2083.

Stensrud, D. J., and R. A. Maddox, 1988: Opposing mesoscale circulations: A case study. *Wea. Forecasting*, **3**, 189–204.

Stensrud, D. J., J.-W. Bao, and T. T. Warner, 2000: Using initial condition and model physics perturbations in short-range ensemble simulations of mesoscale convective systems. *Mon. Wea. Rev.*, **128**, 2077–2107.

Stensrud, D. J., M. C. Coniglio, R. P. Davies-Jones, and J. S. Evans, 2005: Comments on "'A theory for strong long-lived squall lines' revisited." *J. Atmos. Sci.*, **62**, 2989–2996.

Stensrud, D. J., et al., 2009: Convective-scale warn-on-forecast system: A vision for 2020. *Bull. Amer. Met. Soc.*, **90**, 1487–1499.

Straka, J. M., 2009: *Cloud and Precipitation Microphysics: Principles and Parameterizations*. Cambridge University Press.

Straka, J. M., E. N. Rasmussen, and S. E. Fredrickson, 1996: A mobile mesonet for finescale meteorological observations. *J. Atmos. Oceanic Technol.*, **13**, 921–936.

Stull, R. B., 1988: *An Introduction to Boundary Layer Meteorology*. Kluwer Academic Publishers, Dordrecht/Boston/London.

Stumpf, G. J., A. Witt, E. D. Mitchell, et al., 1998: The National Severe Storms Laboratory Mesocyclone Detection Algorithm for the WSR-88D. *Wea. Forecasting*, **13**, 304–326.

Tennekes, H., and J. L. Lumley, 1972: *A First Course in Turbulence*. MIT Press, Cambridge, MA.

Tepper, M., 1950: On the origin of tornadoes. *Bull. Amer. Meteor. Soc.*, **31**, 311–314.

Thompson, P., 1957: Uncertainty in the initial state as a factor in the predictability of large scale atmospheric flow patterns. *Tellus*, **9**, 275–295.

Thompson, R. L., and R. Edwards, 2000: An overview of environmental conditions and forecast implications of the 3 May 1999 tornado outbreak. *Wea. Forecasting*, **15**, 682–699.

Thompson, R. L., R. Edwards, J. A. Hart, et al., 2003: Close proximity soundings within supercell environments obtained from the Rapid Update Cycle. *Wea. Forecasting*, **18**, 1243–1261.

Thompson, R. L., C. M. Mead, and R. Edwards, 2007: Effective storm-relative helicity and bulk shear in supercell thunderstorm environments. *Wea. Forecasting*, **22**, 102–115.

Thomson, D. W., 1986: Systems for measurements at the surface. *Mesoscale Meteorology and Forecasting*. P. Ray (ed.), Amer. Meteor. Soc.

Thorncroft, C. D., N. M. J. Hall, and G. N. Kiladis, 2008: Three-dimensional structure and dynamics of African easterly waves. Part III: Genesis. *J. Atmos. Sci.*, **65**, 3596–3607.

Thorpe, A. J., M. J. Miller, and M. W. Moncrieff, 1982: Two-dimensional convection in non-constant shear: A model of midlatitude squall lines. *Quart. J. Roy. Meteor. Soc.*, **108**, 739–762.

Tompkins, A. M., 2001: Organization of tropical convection in low vertical wind shears: The role of cold pools. *J. Atmos. Sci.*, **58**, 1650–1672.

Toth, Z., E. Kalnay, S. M. Tracton, R. Wobus, and J. Irwin, 1997: A synoptic evaluation of the NCEP ensemble. *Wea. Forecasting*, **12**, 140–153.

Toth, M., R. J. Trapp, J. Wurman, and K. A. Kosiba, 2013: Improving tornado estimates with Doppler radar. *Wea. Forecasting*. doi:10.1175/WAF-D-12-00019.1.

Trapp, R. J., 1999: Observations of nontornadic low-level mesocyclones and attendant tornadogenesis failure during VORTEX. *Mon. Wea. Rev.*, **127**, 1693–1705.

Trapp, R. J., and B. H. Fiedler, 1995: Tornado-like vortexgenesis in a simplified numerical model. *J. Atmos. Sci.*, **52**, 3757–3778.

Trapp, R. J., and R. Davies-Jones, 1997: Tornadogenesis with and without a dynamic pipe effect. *J. Atmos. Sci.*, **54**, 113–133.

Trapp, R. J., and C. A. Doswell, III, 2000: Radar data objective analysis. *J. Atmos. Oceanic Technol.*, **17**, 105–120.

Trapp, R. J., and M. L. Weisman, 2003: Low-level mesovortices within squall lines and bow echoes: Part II: Their genesis and implications. *Mon. Wea. Rev.*, **131**, 2804–2823.

Trapp, R. J., G. J. Stumpf, and K. L. Manross, 2005a: A reassessment of the percentage of tornadic mesocyclones. *Wea. Forecasting*, **20**, 680–687.

Trapp, R. J., S. A. Tessendorf, E. Savageau Godfrey, and H. E. Brooks, 2005b: Tornadoes from squall lines and bow echoes. Part I: Climatological distribution. *Wea. Forecasting*, **20**, 23–34.

Trapp, R. J., D. M. Wheatley, N. T. Atkins, et al., 2006: Buyer beware: Some words of caution on the use of severe wind reports in post-event assessment and research. *Wea. Forecasting*, **21**, 408–415.

Trapp, R. J., N. S. Diffenbaugh, H. E. Brooks, et al., 2007a: Changes in severe thunderstorm environment frequency during the 21st century caused by anthropogenically enhanced global radiative forcing. *Proc. Natl Acad. Sci.*, **104**, 19719–19723, doi:10.1073/pnas.0705494104.

Trapp, R. J., B. Halvorson, and N. S. Diffenbaugh, 2007b: Telescoping, multimodel approaches to evaluate extreme convective weather under future climates. *J. Geophys. Res.*, **112**, D20109, doi:10.1029/2006JD008345.

Trapp, R. J., E. D. Robinson, M. E. Baldwin, et al., 2010: Regional climate of hazardous convective weather through high-resolution dynamical downscaling. *Clim. Dyn.*, doi: 10.1007/s00382-010-0826-y.

Trenberth, K. E., 1998: Atmospheric moisture residence times and cycling: Implications for rainfall rates and climate change. *Climatic Change*, **39**, 667–694.

Trenberth, K. E., and C. J. Guillemot, 1996: Physical processes involved in the 1988 drought and 1993 floods in North America. *J. Climate*, **9**, 1288–1298.

Tribbia, J. J., and D. P. Baumhefner, 1988: The reliability of improvements in deterministic short-range forecasts in the presence of initial-state and modeling deficiencies. *Mon. Wea. Rev.*, **116**, 2276–2228.

Tribbia, J. J., and D. P. Baumhaufner, 2004: Scale interactions and atmospheric predictability: An updated perspective. *Mon. Wea. Rev.*, **132**, 703–713.

Trier, S. B., C. A. Davis, and J. D. Tuttle, 2000a: Long-lived mesoconvective vortices and their environment. Part I: Observations from the central United States during the 1998 warm season. *Mon. Wea. Rev.*, **128**, 3376–3395.

Trier, S. B., C. A. Davis, and W. C. Skamarock, 2000b: Long-lived mesoconvective vortices and their environment. Part II: Induced thermodynamic destabilization in idealized simulations. *Mon. Wea. Rev.*, **128**, 3396–3412.

Trier, S. B., and C. A. Davis, 2007: Mesoscale convective vortices observed during BAMEX. Part II: Influences on secondary deep convection. *Mon. Wea. Rev.*, **135**, 2051–2075.

Tripoli, G. J., and W. R. Cotton, 1981: The use of ice-liquid water potential temperature as a thermodynamic variable in deep atmospheric models. *Mon. Wea. Rev.*, **109**, 1094–1102.

Velden, C. S., C. M. Hayden, S. J. Nieman, et al., 1997: Upper-tropospheric winds derived from geostationary satellite water vapor observations. *Bull. Amer. Meteor. Soc.*, **78**, 173–195.

Velden, C., and coauthors, 2005: Recent innovations in deriving tropospheric winds from meteorological satellites. *Bull. Amer. Meteor. Soc.*, **86**, 205–223.

Vinnichenko, N. K., 1970: The kinetic energy spectrum in the free atmosphere–one second to five years. *Tellus*, **22**, 158–166.

Vitart, F., et al., 2007: Dynamically-based seasonal forecasts of Atlantic tropical storm activity issued in June by EUROSIP. *Geophys. Res. Lett.*, **34**, L16815, doi:10.1029/2007GL030740.

Wakimoto, R. M., 2001: Convectively driven high wind events. *Severe Convective Storms*, American Meteorological Society, Boston, 255–298.

Wakimoto, R. M., and H. V. Murphey, 2010: Analysis of convergence boundaries observed during IHOP_2002. *Mon. Wea. Rev.*, **138**, 2737–2760.

Wakimoto, R. M., and J. W. Wilson, 1989: Non-supercell tornadoes. *Mon. Wea. Rev.*, **117**, 1113–1140.

Wakimoto, R. M., H. Cai, and H. V. Murphey, 2004: The Superior, Nebraska, supercell during BAMEX. *Bull. Amer. Meteor. Soc.*, **85**, 1095–1106.

Wakimoto, R. M., W.-C. Lee, H. B. Bluestein, C.-H. Liu, P. H. Hildebrand, 1996: ELDORA observations during VORTEX 95. *Bull. Amer. Meteor. Soc.*, **77**, 1465–1481.

Wakimoto, R. M., C-H. Liu, and H-Q. Cai, 1998: The Garden City, Kansas, storm during VORTEX 95. Part I: Overview of the storm's life cycle and mesocyclogenesis. *Mon. Wea. Rev.*, **126**, 372–392.

Wakimoto, R. M., H. V. Murphey, A. Nester, et al., 2006: High winds generated by bow echoes. Part I: Overview of the Omaha bow echo 5 July 2003 storm during BAMEX. *Mon. Wea. Rev.*, **134**, 2793–2812.

Wallace, J. M., and P. V. Hobbs, 2006: *Atmospheric Science: An Introductory Survey, 2nd Edition*. Elsevier, London.

Wang, H., et al., 2009: A statistical forecast model for Atlantic seasonal hurricane activity based on the NCEP dynamical seasonal forecast. *J. Clim.*, **22**, 4481–4500.

Wang, J., and D. B. Wolff, 2010: Evaluation of TRMM ground-validation radar-rain errors using rain gauge measurements. *J. Appl. Meteor. Climatol.*, **49**, 310–324.

Warner, J., 1970: On steady-state one-dimensional models of cumulus convection. *J. Atmos. Sci.*, **27**, 1035–1040.

Warner, T. T., and H.-M. Hsu, 2000: Nested-model simulation of moist convection: The impact of coarse-grid parameterized convection on fine-grid resolved convection. *Mon. Wea. Rev.*, **128**, 2211–2231.

Weaver, C. P., and R. Avissar, 2001: Atmospheric disturbances caused by human modification of the landscape. *Bull. Amer. Meteor. Soc.*, **82**, 269–281.

Weaver, J. F., J. A. Knaff, D. Bikos, et al., 2002: Satellite observations of a severe supercell thunderstorm on 24 July 2000 made during the *GOES-11* Science Test. *Wea. Forecasting*, **17**, 124–138.

Weber, B. L., et al., 1990: Preliminary evaluation of the first NOAA demonstration network wind profiler. *J. Atmos. Oceanic Technol.*, **7**, 909–918.

Weckwerth, T. M., 2000: The effect of small-scale moisture variability on thunderstorm initiation. *Mon. Wea. Rev.*, **128**, 4017–4030.

Weckwerth, T. M., and D. B. Parsons, 2006: A review of convection initiation and motivation for IHOP_2002. *Mon. Wea. Rev.*, **134**, 5–22.

Weckwerth, T. M., H. V. Murphey, C. Flamant, et al., 2008: An observational study of convection initiation on 12 June 2002 during IHOP_2002. *Mon. Wea. Rev.*, **136**, 2283–2304.

Weckwerth, T. M., J. W. Wilson, and R. M. Wakimoto, 1996: Thermodynamic variability within the convective boundary layer due to horizontal convective rolls. *Mon. Wea. Rev.*, **124**, 769–784.

Weckwerth, T. M., J. W. Wilson, R. M. Wakimoto, and N. A. Crook, 1997: Horizontal convective rolls: Determining the environmental supporting their existence and characteristics. *Mon. Wea. Rev.*, **125**, 505–526.

Weckwerth, T. M., et al., 2004: An overview of the International H2O Project (IHOP_2002) and some preliminary highlights. *Bull. Amer. Meteor. Soc.*, **85**, 253–277.

Weisman, M. L., 1992: The role of convectively generated rear-inflow jets in the evolution of long-lived mesoconvective systems. *J. Atmos. Sci.*, **49**, 1826–1847.

Weisman, M. L., 1993: The genesis of severe, long-lived bow echoes. *J. Atmos. Sci.*, **50**, 645–670.

Weisman, M. L., 2001: Bow echoes: A tribute to T. T. Fujita. *Bull. Amer. Meteor. Soc.*, **82**, 97–116.

Weisman, M. L., and C. A. Davis, 1998: Mechanisms for the generation of mesoscale vortices within quasi-linear convective systems. *J. Atmos. Sci.*, **55**, 2603–2622.

Weisman, M. L., and J. B. Klemp, 1982: The dependence of numerically simulated convective storms on vertical wind shear and buoyancy. *Mon. Wea. Rev.*, **110**, 504–520.

Weisman, M. L., and J. B. Klemp, 1986: Characteristics of isolated convective storms. In *Mesoscale Meteorology and Forecasting*. American Meteorological Society, Boston, 331–358.

Weisman, M. L., J. B. Klemp, and R. Rotunno, 1988: Structure and evolution of numerically simulated squall lines. *J. Atmos. Sci.*, **45**, 1990–2013.

Weisman, M. L., W. C. Skamarock, and J. B. Klemp, 1997: The Resolution Dependence of Explicitly Modeled Convective Systems. *Mon. Wea. Rev.*, **125**, 527–548.

Weisman, M. L., and R. Rotunno, 2000: The use of vertical wind shear versus helicity in interpreting supercell dynamics. *J. Atmos. Sci.*, **57**, 1452–1472.

Weisman, M. L., and R. Rotunno, 2004: "A theory for long-lived squall lines" revisited. *J. Atmos. Sci.*, **61**, 361–382.

Weisman, M. L., and R. J. Trapp, 2003: Low-level mesovortices within squall lines and bow echoes: Part I: Overview and dependence on environmental shear. *Mon. Wea. Rev.*, **131**, 2779–2803.

Weiss, C. C., and H. B. Bluestein, 2002: Airborne pseudo-dual Doppler analysis of a dryline-out flow boundary intersection. *Mon. Wea. Rev.*, **130**, 1207–1226.

Wentz, F. J. and M. Schabel, 2000: Precise climate monitoring using complementary satellite data sets. *Nature*, **403**, 414–416.

Westrick, K. J., C. F. Mass, and B. A. Colle, 1999: The limitations of the WSR-88D radar network for quantitative precipitation measurement over the coastal western United States. *Bull. Amer. Meteor. Soc.*, **80**, 2289–2298.

Wheatley, D. M., and R. J. Trapp, 2008: The effect of mesoscale heterogeneity on the genesis and structure of mesovortices within quasi-linear convective systems. *Mon. Wea. Rev.*, **136**, 4220–4241.

Wilks, D. S., 2006: *Statistical Methods in the Atmospheric Sciences*. 2nd ed. Academic Press.

Wilson, J. W., and R. D. Roberts, 2006: Summary of convective storm initiation and evolution during IHOP: Observational and modeling perspective. *Mon Wea. Rev.*, **134**, 23–47.

Wilson, J. W., and W. E. Schreiber, 1986: Initiation of convective storms at radar-observed boundary-layer convergence lines. *Mon. Wea. Rev.*, **114**, 2516–2536.

Wilson, J. W., N. A. Crook, C. K. Mueller, et al., 1998: Nowcasting thunderstorms: A status report. *Bull. Amer. Meteor. Soc.*, **79**, 2079–2099.

Winn, W. P., S. J. Hunyady, and G. D. Aulich, 1999: Pressure at the ground in a large tornado. *J. Geophys. Res.*, **104**, 22 067–22 082.

Wurman, J., M. Randall, and A. Zahari, 1997: Design and deployment of a portable, pencil-beam, pulsed, 3-cm Doppler radar. *J. Atmos. Oceanic Technol.*, **14**, 1502–1512.

Wurman, J., et al., 2012: The Second Verification of the Origins of Rotation in Tornadoes Experiment: VORTEX2. *Bull. Amer. Meteor. Soc.*, doi: 10.1175/BAMS-D-11-00010.1

Xue, M., K. K. Drogemeier, and V. Wong, 2000: The Advanced Regional Prediction System (ARPS) – A multiscale nonhydrostatic atmospheric simulation and prediction tool. Part I: Model dynamics and verification. *Meteor. Atmos. Physics*, **75**, 161–193.

Yuter, S. E., and R. A. Houze, Jr., 1995: Three-dimensional kinematic and microphysical evolution of Florida cumulonimbus. Part I: Frequency distributions of vertical velocity, reflectivity, and differential reflectivity. *Mon. Wea. Rev.*, **123**, 1941–1963.

Zhang, F., C. Snyder, and R. Rotunno, 2003: Effects of moist convection on mesoscale predictability. *J. Atmos. Sci.*, **60**, 1173–1185.

Ziegler, C. L., and E. N. Rasmussen, 1998: The initiation of moist convection at the dryline: Forecasting issues from a case study perspective. *Wea. Forecasting*, **13**, 1106–1131.

Ziegler, C. L., W. J. Martin, R. A. Pielke, and R. L. Walko, 1995: A modeling study of the dryline. *J. Atmos. Sci.*, **52**, 263–285.

Ziegler, C. L., E. N. Rasmussen, M. S. Buban, et al., 2007: The "triple point" on 24 May 2002 during IHOP. Part II: Ground radar and in situ boundary layer analysis of cumulus development and convection initiation. *Mon. Wea. Rev.*, **135**, 2443–2472.

Zipser, E. J., 1977: Mesoscale and convective-scale downdrafts as distinct components of squall-line structure. *Mon. Wea. Rev.*, **105**, 1568–1589.

Zipser, E. J., 2003: Some views on "hot towers" after 50 years of tropical field programs and two years of TRMM data. *Cloud Systems, Hurricanes, and the TRMM. Meteor. Monogr.*, No. 51, Amer. Meteor. Soc., 49–58.

Zrnic, D. S., A. V. Ryzhkov, 1999: Polarimetry for weather surveillance radars. *Bull. Amer. Meteor. Soc.*, **80**, 389–406.

Index